SPRINGER SERIES IN SURFACE SCIENCES 23

Dear MyCopy Customer,

This Springer book is a monochrome print version of the eBook to which your library gives you access via SpringerLink. It is available to you at a subsidized price since your library subscribes to at least one Springer eBook subject collection.

Please note that MyCopy books are only offered to library patrons with access to at least one Springer eBook subject collection. MyCopy books are strictly for individual use only.

You may cite this book by referencing the bibliographic data and/or the DOI (Digital Object Identifier) found in the front matter. This book is an exact but monochrome copy of the print version of the eBook on SpringerLink.

Physics and Astronomy ONLINE LIBRARY

Springer-Verlag Berlin Heidelberg GmbH

http://www.springer.de/phys/

SPRINGER SERIES IN SURFACE SCIENCES

Series Editors: G. Ertl, H. Lüth and D.L. Mills

This series covers the whole spectrum of surface sciences, including structure and dynamics of clean and adsorbate-covered surfaces, thin films, basic surface effects, analytical methods and also the physics and chemistry of interfaces. Written by leading researchers in the field, the books are intended primarily for researchers in academia and industry and for graduate students.

38 **Progress in Transmission Electron Microscopy 1**
Concepts and Techniques
Editors: X.-F. Zhang, Z. Zhang

39 **Progress in Transmission Electron Microscopy 2**
Applications in Materials Science
Editors: X.-F. Zhang, Z. Zhang

40 **Giant Magneto-Resistance Devices**
By E. Hirota, H. Sakakima, and K. Inomata

41 **The Physics of Ultra-High-Density Magnetic Recording**
Editors: M.L. Plumer, J. van Ek, and D. Weller

Series homepage – http://www.springer.de/phys/books/ssss/

Volumes 1–37 are listed at the end of the book

D.J. O'Connor, B.A. Sexton,
R.St.C. Smart (Eds.)

Surface Analysis Methods in Materials Science

Second Edition
With 272 Figures

Springer

Associate Professor D. John O'Connor
School of Mathematical and Physical Sciences
University Newcastle
Callaghan, NSW 2308
Australia

Dr. Brett A. Sexton
CSIRO Manufacturing Science and Technology
Private Bag 33
Clayton South, VIC 3169
Australia

Professor Roger St. C. Smart
South Australian Surface Technology Centre
University of South Australia
The Levels, Adelaide, SA 25095
Australia

Series Editors:

Professor Dr. Gerhard Ertl
Fritz-Haber-Institute der Max-Planck-Gesellschaft, Faradayweg 4-6,
14195 Berlin, Germany

Professor Dr. Hans Lüth
Institut für Schicht- und Ionentechnik
Forschungszentrum Jülich GmbH,
52425 Jülich, Germany

Professor Douglas L. Mills, Ph.D.
Department of Physics, University of California,
Irvine, CA 92717, USA

Cataloging-in-Publication Data applied for

Bibliographic information published by Die Deutsche Bibliothek
Die Deutsche Bibliothek lists this publication in the Deutsche
Nationalbibliografie; detailed bibliographic data is available in the
Internet at <http://dnb.ddb.de>.

ISSN 0931-5195

springeronline.com

DOI 10.1007/978-3-662-05227-3

Originally published by Springer-Verlag Berlin Heidelberg New York in 2003.
MyCopy version of the original edition 2003

Typesetting: camera ready copy by the authors
Cover concept: eStudio Calamar Steinen
Cover production: *design & production* GmbH, Heidelberg

Printed on acid-free paper 57/3111/mf - 5 4 3 2 1 SPIN 10977446
www.springer.com/mycopy

Preface

The success of the first edition of this broad appeal book prompted the preparation of an updated and expanded second edition. The field of surface analysis is constantly changing as it answers the need to provide more specific and more detailed information about surface composition and structure in advanced materials science applications. The content of the second edition meets that need by including new techniques and expanded applications.

Newcastle
Clayton
Adelaide
January 2003

John O'Connor
Brett Sexton
Roger Smart

Preface to the First Edition

The idea for this book stemmed from a remark by Philip Jennings of Murdoch University in a discussion session following a regular meeting of the Australian Surface Science group. He observed that a text on surface analysis and applications to materials suitable for final year undergraduate and postgraduate science students was not currently available. Furthermore, the members of the Australian Surface Science group had the research experience and range of coverage of surface analytical techniques and applications to provide a text for this purpose. A list of techniques and applications to be included was agreed at that meeting. The intended readership of the book has been broadened since the early discussions, particularly to encompass industrial users, but there has been no significant alteration in content. The editors, in consultation with the contributors, have agreed that the book should be prepared for four major groups of readers:

- senior undergraduate students in chemistry, physics, metallurgy, materials science and materials engineering;
- postgraduate students undertaking research that involves the use of analytical techniques;
- groups of scientists and engineers attending training courses and workshops on the application of surface analytical techniques in materials science;
- industrial scientists and engineers in research and development seeking a description of available surface analytical techniques and guidance on the most appropriate techniques for particular applications.

The contributors mostly come from Australia, with the notable exception of Ray Browning from Stanford University. Australia is a very large country with a relatively small population, so it is inevitable that the Australian surface science community is spread rather thinly across the country. One aim in producing this book has therefore been to bring together the breadth of expertise within this Australian scientific community. All of the authors have made significant contributions to the techniques and their applications, in many cases over a period of more than 20 years. A second aim is to emphasise by example the very wide spectrum of information that can now be obtained from the use of a variety of surface analytical techniques applied to the same material. Here, the intention has been to encourage people involved in research, development and process control to become aware of the increasing usefulness of surface analysis in their own fields of materials science.

Finally, we believe that the approach adopted here – namely descriptions of the basic techniques, their limitations and their applications – will be accessible and beneficial to people in any of the four groups of potential readers in any country of the world. The authors are all closely involved with the international scientific community in their own areas of research and this is reflected in their selection of examples. We hope that this book will fulfil a need by extending the range of techniques from those covered by more specialised texts confined to one or two techniques, and by providing examples of real applications to research, development and problem solving in materials science.

The strategy and structure of the book is as follows: the book is designed for use by all those interested in the surface characterisation of materials. This group is expected to include scientists (e.g. chemists, physicists, biochemists, biologists, geologists, geochemists), technologists, metallurgists, engineers and workers in the various biomedical and microelectronics applications areas. It is also intended for those requiring an introductory overview of surfaces and techniques for their analysis, in particular final year undergraduate and graduate students in any of the specialties listed above. It is important to state clearly that it is not intended for experienced practitioners in particular techniques of surface analysis. It does not attempt to critically review all techniques or all variations and restrictions on the use of a particular technique. For instance, we are well aware that there are many difficulties in procedures for quantification, chemical mapping and depth validation for particular types of sample that are not explained in the short presentations on each technique given in this book. They will need to be learnt when undertaking a specific investigation of a material using one of the techniques described here and will probably require further reading in more advanced, single-technique books.

The book has been deliberately structured in three parts.

Part I provides a descriptive overview of different materials, the properties of their surfaces, the range of techniques available to study their structure and composition and some guidance to the choice of particular techniques based on information they can provide and major limitations to their use. It also includes a description of the essential elements of vacuum technology required for many of these techniques.

Part II presents short descriptions of major techniques for use in materials science and technology. The selection has been made on the basis of the accessibility of the technique and the universality of its application to different types of materials, and is certainly not unequivocal. Several techniques mentioned in Part I are not included in Part II. It is likely that the relevance of the various techniques will change with time and that, in later editions, it may be necessary to delete or add techniques. The descriptions concentrate on the types of information provided by the technique, samples most readily analysed and limitations to their use. The chapters of Part II can be read separately and, in many cases, it will not be necessary for the reader to complete all chapters in the first instance.

Part III describes some major applications. Again, some judgement has been exercised in the choice of these applications, with emphasis placed on major areas in materials science and technology. Options for addition and deletion also exist in this section.

The book is best approached by initially reading the whole of Part I and selected sections from Parts II and III. The reader may wish to focus on particular materials of direct interest to his or her research, development or technological application. This itself will suggest particular techniques as being most appropriate initially. It may also require more detailed reading of single-technique descriptions in other books: the references in Parts I and II give specific guidance to books most suitable for this purpose.

Newcastle
Clayton
Adelaide
December 1991

John O'Connor
Brett Sexton
Roger Smart

Contents

Part I Introduction

1 Solid Surfaces, Their Structure and Composition
C. Klauber, R. St. C. Smart . . . 3

1.1 Importance of the Surface . . . 3
1.2 Solid Surfaces of Different Materials . . . 7
1.2.1 A Material Under Attack: Aluminium . . . 11
1.3 Methods of Surface Analysis . . . 12
1.3.1 Variety of Surface Analytical Techniques . . . 12
1.4 Structural Imaging . . . 13
1.4.1 Direct Physical Imaging . . . 13
1.4.2 Indirect Structural Imaging – Relaxation and Reconstruction . . . 20
1.5 Composition of the Surface Selvedge . . . 23
1.5.1 Electron Inelastic Mean Free Paths . . . 24
1.5.2 Variation of Elemental Sensitivities . . . 28
1.5.3 Practical Detection Limits . . . 31
1.5.4 Practical Spatial Limits . . . 32
1.5.5 Chemical State Information . . . 37
1.5.6 Laboratory Standards . . . 41
1.5.7 Inter-laboratory Errors . . . 41
1.6 Defect and Reaction Sites at Surfaces . . . 42
1.7 Electronic Structure at Surfaces . . . 46
1.8 Structures of Adsorbed Layers . . . 49
1.9 Structure in Depth Profiles Through Surfaces . . . 51
1.10 Specific Structures . . . 55
1.10.1 Grain Structures, Phase Distributions and Inclusions . . . 55
1.10.2 Fracture Faces and Intergranular Regions . . . 55
1.10.3 Pore Structures . . . 57

1.10.4 Precipitates, Reaction Products and Recrystallised Particles on Surfaces ... 60
1.10.5 Magnetic Domains ... 60
1.11 Technique-Induced Artifacts ... 61
1.11.1 Radiation Damage ... 61
1.11.2 Electrostatic Charging ... 64
References ... 65

2 UHV Basics
C. Klauber ... 71

2.1 The Need for Ultrahigh Vacuum ... 71
2.2 Achieving UHV ... 74
2.3 Specimen Handling ... 76
2.4 Specimen Handling: ASTM Standards ... 78
References ... 81

Part II Techniques

3 Electron Microscope Techniques for Surface Characterization
P.S. Turner, C.E. Nockolds, S. Bulcock ... 85

3.1 What Do We Need to Know About Surface Structures? ... 86
3.2 Electron Optical Imaging Systems ... 87
3.2.1 Electron Sources ... 89
3.2.2 Electron Lenses ... 89
3.2.3 Detection Systems ... 90
3.3 Scanning Electron Microscopy of Surfaces ... 91
3.3.1 The SEM ... 91
3.3.2 The Signals and Detectors ... 92
3.3.3 Resolution and Contrast in SEM Images ... 93
3.3.4 Variable Pressure SEM and Environmental SEM ... 97
3.3.5 Energy Dispersive X-Ray Spectrometry ... 98
3.4 Transmission Electron Microscopy of Surfaces ... 99
3.4.1 The Transmission Electron Microscope ... 99
3.4.2 Electron Diffraction ... 101
3.4.3 Image Contrast and Resolution in the TEM ... 101
3.4.4 Imaging Surface Structures in the TEM ... 102
3.4.5 Reflection Electron Microscopy ... 103
3.5 Other Developments ... 103
References ... 104

4 Sputter Depth Profiling
B.V. King 107

4.1 Analysis of a Sputter Depth Profile 108
4.1.1 Calibration of the Depth Scale 108
4.1.2 Calibration of the Concentration Scale 112
4.2 The Depth Resolution of Sputter Profiling 115
4.2.1 Specification of the Depth Resolution 115
4.2.2 Instrumental Factors Determining the Depth Resolution . . 117
4.2.3 Surface Effects Determining the Depth Resolution 119
4.2.4 Bulk Effects Affecting the Depth Resolution 120
4.2.5 Minimisation of the Depth Resolution 121
4.3 Conclusion 123
References 123

5 SIMS – Secondary Ion Mass Spectrometry
R.J. MacDonald, B.V. King 127

5.1 The Practice of SIMS 128
5.1.1 Overview 128
5.1.2 Advantages and Disadvantages of SIMS 130
5.1.3 The Yield of Secondary Ions 131
5.2 Construction of a Secondary Ion Mass Spectrometer 138
5.3 Topics in SIMS Analysis 145
5.3.1 Signal Enhancement by Surface Adsorption 145
5.3.2 Using Secondary Ion Energies in SIMS Analysis 147
5.3.3 The Relative Sensitivity Factor 149
5.4 Static SIMS Analysis 150
References 153

6 Auger Electron Spectroscopy and Microscopy – Techniques and Applications
P.C. Dastoor 155

6.1 Introduction 155
6.2 Fundamentals 155
6.3 Instrumentation 158
6.4 Quantification 159
6.5 Techniques 160
6.5.1 Spot Analysis Mode 161
6.5.2 Line Scan Mode 162
6.5.3 Scanning Mode 163
6.5.4 Scanning Auger Microscopy 164
6.5.5 Depth Profiling Mode 167
6.5.6 Preferential Sputtering 167
6.5.7 Attenuation Length 168

6.5.8 Chemical Effects . . . 168
6.6 Applications . . . 169
6.6.1 Thin Film Analysis . . . 169
6.6.2 Surface Diffusion and Segregation . . . 170
6.7 Future . . . 171
References . . . 171

7 X-Ray Photoelectron Spectroscopy
M.H. Kibel . . . 175

7.1 Basic Principles . . . 175
7.1.1 Theory . . . 175
7.1.2 Typical Spectrum . . . 176
7.1.3 Surface Specificity . . . 178
7.2 Instrumentation . . . 179
7.2.1 Essential Components . . . 179
7.2.2 Optional Components . . . 182
7.2.3 Synchrotron Radiation . . . 183
7.2.4 Imaging XPS . . . 183
7.3 Spectral Information . . . 184
7.3.1 Spin-Orbit Splitting . . . 184
7.3.2 Chemical Shifts . . . 185
7.3.3 Auger Chemical Shifts in XPS . . . 186
7.3.4 X-Ray Line Satellites . . . 187
7.3.5 "Shake-up" Lines . . . 187
7.3.6 Ghost Lines . . . 188
7.3.7 Plasmon Loss Lines . . . 189
7.4 Quantitative Analysis . . . 189
7.5 Experimental Techniques . . . 191
7.5.1 Variation of X-Ray Sources . . . 191
7.5.2 Depth Profiles . . . 191
7.5.3 Angular Variations . . . 194
7.5.4 Sample Charging . . . 194
7.6 Comparison with Other Techniques . . . 195
7.7 Applications . . . 197
7.8 Conclusion . . . 197
References . . . 198

8 Vibrational Spectroscopy of Surfaces
R.L. Frost, N.K. Roberts . . . 203

8.1 Introduction . . . 203
8.2 Surface Techniques . . . 206
8.2.1 Diffuse Reflectance Infrared Fourier Transform (DRIFT) . . 207
8.2.2 Attenuated Total Reflectance Spectroscopy (ATR) . . . 209
8.2.3 Photoacoustic Spectroscopy (PAS) . . . 216

8.2.4 Infrared Emission Spectroscopy (IES) 219
8.3 Fourier transform Raman spectroscopy 221
8.4 Raman Microscopy 224
References 227

9 Rutherford Backscattering Spectrometry and Nuclear Reaction Analysis
S.H. Sie 229

9.1 Introduction 229
9.2 Principles 231
9.2.1 Stopping Power 232
9.2.2 Straggling 233
9.3 Rutherford Backscattering Spectrometry 234
9.3.1 Experimental Considerations 236
9.3.2 Examples 236
9.3.3 Special Cases 238
9.4 Nuclear Reaction Analysis 240
9.4.1 Formalism 241
9.4.2 Experimental Considerations 242
9.4.3 Examples 243
9.5 Summary 245
References 246

10 Materials Characterization by Scanned Probe Analysis
S. Myhra 247

10.1 Introduction 247
10.2 The Surface Analytical Context 251
10.3 Generic SPM Systems 251
10.4 Physical Principles 253
10.4.1 STM/STS 253
10.4.2 Scanning Force Microscopy (SFM) 256
10.4.3 Intermittent Contact Mode 257
10.4.4 F–d Analysis 258
10.4.5 Lateral Force Microscopy (LFM) 260
10.5 Procedures for 'Best Practice' 262
10.5.1 Spatial Characteristics of Scanners 263
10.5.2 Determination of c_N and c_T 264
10.5.3 Determination of Spring Constants 265
10.5.4 Determination of Actual Tip Parameters in the Mesoscopic Regime 266
10.6 Illustrative Case Studies 267
10.6.1 Surface and Defect Structures of WTe_2 Investigated by UHV-STM 267

10.6.2 Organic Thin Film and Surface Mechanical Characterization . . . 269
10.6.3 AFM Analysis of 'Soft' Biological Materials . . . 273
10.6.4 Nanotribology of Solid Lubricants . . . 278
References . . . 283

11 Low Energy Ion Scattering
D. J. O'Connor . . . 287

11.1 Qualitative Surface Analysis . . . 287
11.2 Advantage of Recoil Detection . . . 289
11.3 Quantitative Analysis . . . 291
11.3.1 Scattered Ion Yield . . . 291
11.3.2 Differential Scattering Cross Section . . . 291
11.3.3 Charge Exchange . . . 292
11.3.4 Relative Measurements . . . 295
11.3.5 Standards . . . 297
11.4 Surface Structural Analysis . . . 299
11.4.1 Multiple Scattering . . . 299
11.4.2 Impact Collision Ion Surface Scattering (ICISS) . . . 300
11.5 Experimental Apparatus . . . 302
References . . . 304

12 Reflection High Energy Electron Diffraction
G.L. Price . . . 307

12.1 Theory . . . 309
12.2 Applications . . . 312
References . . . 318

13 Low Energy Electron Diffraction
P. J. Jennings, C.Q. Sun . . . 319

13.1 The Development of LEED . . . 319
13.2 The LEED Experiment . . . 320
13.2.1 Sample Preparation . . . 323
13.2.2 Data Collection . . . 323
13.3 Diffraction from a Surface . . . 324
13.3.1 Bragg Peaks in LEED Spectra . . . 325
13.4 LEED Intensity Analysis . . . 326
13.5 LEED Fine Structure . . . 328
13.6 Applications of LEED . . . 329
13.6.1 Determination of the Symmetry and Size of the Unit Mesh . . . 329
13.6.2 Unit Meshes for Chemisorbed Systems . . . 330
13.6.3 LEED Intensity Analysis . . . 330
13.6.4 Surface Barrier Analysis . . . 332

13.7 Conclusion . . . 333
References . . . 334

14 Ultraviolet Photoelectron Spectroscopy of Solids
R. Leckey . . . 337

14.1 Experimental Considerations . . . 339
14.2 Angle Resolved UPS . . . 340
14.3 Fermi Surface Studies . . . 344
References . . . 345

15 EXAFS
R.F. Garrett, G.J. Foran . . . 347

15.1 Introduction . . . 347
15.2 Experimental Details . . . 348
15.2.1 Synchrotron Radiation . . . 350
15.2.2 Synchrotron Beamlines for EXAFS . . . 352
15.2.3 Detectors . . . 354
15.2.4 The Sample . . . 355
15.2.5 Acquiring EXAFS Data . . . 357
15.3 Theory of X-ray Absorption . . . 359
15.3.1 EXAFS . . . 359
15.3.2 XANES . . . 361
15.4 EXAFS Analysis . . . 362
15.4.1 Data Reduction . . . 362
15.4.2 Conversion to *k*-space . . . 363
15.4.3 Background Subtraction . . . 363
15.4.4 Fourier Transformation . . . 364
15.4.5 Fourier Filtering and Back Transformation . . . 364
15.4.6 Modelling and Least Squares Fitting to the EXAFS Equation . . . 365
15.5 Case Studies . . . 366
15.5.1 Surface EXAFS of Titanium Nanostructure Thin Films . . . 366
15.5.2 Ion-Implantation Induced Amorphisation of Germanium . . 370
References . . . 371

Part III Processes and Applications

16 Minerals, Ceramics and Glasses
R.St.C. Smart ... 377

16.1 Minerals ... 380
16.1.1 Iron Oxides in Mineral Mixtures ... 380
16.1.2 Surface Layers on Minerals ... 381
16.1.3 Mineral Processing of Sulfide Ores ... 381
16.1.4 Adsorption and Reaction of Oxide and Clay Minerals ... 386
16.1.5 Surface Modification of Minerals ... 387
16.2 Ceramics ... 389
16.2.1 Leaching and Dissolution ... 390
16.2.2 Ceramic Surface Layers: Bioceramics ... 392
16.3 Glasses ... 395
16.3.1 Leached and Recrystallised surfaces ... 395
16.3.2 Surface Modification of Glass Surfaces ... 398
References ... 400

17 Characterization of Catalysts by Surface Analysis
N.K. Singh, B.G. Baker ... 405

17.1 Examples of Catalytic Systems Studied by XPS ... 408
17.1.1 Alumina ... 408
17.1.2 Tungsten Oxide Catalysts ... 409
17.1.3 Palladium on Magnesia ... 412
17.1.4 Cobalt on Kieselguhr Catalysts ... 414
17.1.5 Iron Catalysts ... 416
17.2 Examples of Catalytic Systems Studied by FT-Infrared Spectroscopy ... 420
17.2.1 ZSM-5 Zeolites ... 420
17.2.2 Thin Alumina Films ... 426
17.2.3 Rhodium Supported Alumina Films ... 428
17.2.4 Copper Supported Silica Films ... 429
17.3 Conclusion ... 432
References ... 432

18 Application to Semiconductor Devices
P.W. Leech, P. Ressel ... 435

18.1 Micro and Nano-Analysis of Integrated Circuits ... 436
18.2 Analytical Techniques in the Characterisation of Ohmic Metal/Semiconductor Contacts ... 443
18.3 Summary ... 452
References ... 452

19 Characterisation of Oxidised Surfaces
J.L. Cocking, G.R. Johnston 455

19.1 The Oxidation Problem 456
19.2 Oxidation of Co-22Cr-11Al 457
19.2.1 Chemical Characterisation 458
19.2.2 Scanning Auger Microscopy 458
19.2.3 Rutherford Backscattering Analysis 459
19.3 Oxidation of Ni-18Cr-6Al-0.5Y 464
19.3.1 Extended X-Ray Absorption Fine Structure 466
References 471

20 Coated Steel
R. Payling 473

20.1 Applications 474
20.1.1 Grain Boundaries in Steel 474
20.1.2 Steel Surface 475
20.1.3 Alloy Region 478
20.1.4 Metallic Coatings 478
20.1.5 Treated Metallic Coating Surface 483
20.1.6 Metal–Polymer Interface 483
20.1.7 Polymer Surface 484
20.2 Conclusion 486
References 486

21 Thin Film Analysis
G.C. Morris 489

21.1 Thin Film Photovoltaics 490
21.1.1 Use for Solar Electricity 490
21.1.2 The Thin Film Solar Cell: Glass/ITO/nCdS/pCdTe/Au 491
21.2 Film Purity 492
21.2.1 Low Level Impurities – Qualitative 492
21.2.2 Low Level Impurities – Quantitative 494
21.2.3 Doping Profiles in Thin Films 495
21.3 Composition and Thickness of Layered Films 496
21.3.1 Composition Gradation in Films, e.g. $Cd_xHg_{1-x}Te$ Films on Platinum 496
21.3.2 Thin Overlayers on Films 497
21.4 Beam Effects in Thin Film Analysis 501
21.5 Conclusion 502
References 502

22 Identification of Adsorbed Species
B.G. Baker . . . 505

22.1 Examples of Adsorption Studies . . . 505
22.1.1 Nitric Oxide Adsorption on Metals . . . 505
22.1.2 Aurocyanide Adsorption on Carbon . . . 511
22.1.3 Adsorbed Methoxy on Copper and Platinum . . . 514
22.2 Conclusion . . . 518
References . . . 518

23 Surface Analysis of Polymers
H.A.W. StJohn, T.R. Gengenbach, P.G. Hartley, H.J. Griesser . . . 519

23.1 Specific Properties of Polymers . . . 520
23.2 Surface Contamination and Additives . . . 525
23.3 Contact Angle Measurements . . . 529
23.4 X-Ray Photoelectron Spectroscopy (XPS) . . . 529
23.4.1 General Aspects . . . 529
23.4.2 Angle-Resolved XPS . . . 531
23.4.3 Inelastic Mean Free Path in Polymers . . . 532
23.4.4 Cold Stage XPS . . . 533
23.4.5 Derivatization of Chemical Groups . . . 534
23.5 Secondary Ion Mass Spectrometry (SIMS) . . . 537
23.6 Scanning Probe Microscopy Methods: Scanning Tunneling Microscopy (STM) and Atomic Force Microscopy (AFM) . . . 540
23.7 Specimen Damage . . . 543
23.8 Charge Correction . . . 547
23.9 Grazing Angle Infrared Spectroscopy . . . 548
References . . . 549

24 Glow Discharge Optical Emission Spectrometry
T. Nelis, R. Payling . . . 553

24.1 Instrument . . . 553
24.2 Theory . . . 555
24.3 Applications . . . 556
24.3.1 Near Surface . . . 556
24.3.2 Coatings . . . 556
24.3.3 Semiconductor Processing . . . 557
References . . . 559

Part IV Appendix

Acronyms Used in Surface and Thin Film Analysis 563

Surface Science Bibliography 569

Index 577

List of Contributors

Baker, Bruce G. *retired*
Former address:
School of Physical Sciences
Flinders University
Sturt Road
Bedford Park, SA 5042
Australia
bruce.baker@flinders.edu.au

Bulcock, Shaun
Electron Microscope Unit
University of Sydney
Sydney, NSW 2006
Australia
shaun@emu.usyd.edu.au

Cocking, Janis L.
DSTO Maribyrnong
Maritime Platforms Division
PO Box 4431
Melbourne, Victoria 3001
Australia
janis.cocking@dsto.defence.gov.au

Dastoor, Paul C.
School of Mathematical
and Physical Sciences
University of Newcastle
Callaghan, NSW 2308
Australia
phpd@alinga.newcastle.edu.au

Frost, Ray L.
School of Physical
and Chemical Sciences
Queensland University of Technology
GPO Box 2434
Brisbane, Queensland 4001
Australia
r.frost@qut.edu.au

Foran, Garry J.
Australian Synchrotron
Research Program
Physics Division, Australian
Nuclear Science
and Technology Organisation
Private Mail Bag 1
Menai, NSW 2234
Australia
foran@anbf2.kek.jp

Garrett, Richard F.
Australian Synchrotron
Research Program
Physics Division, Australian
Nuclear Science
and Technology Organisation
Private Mail Bag 1
Menai, NSW 2234
Australia
garrett@ansto.gov.au

Gengenbach, Thomas R.
CSIRO Molecular Science
Bag 10
Clayton South VIC 3169
Australia
thomas.gengenbach@csiro.au

Griesser, Hans J.
Ian Wark Research Institute
University of South Australia
Mawson Lakes Blvd
Mawson Lakes, SA 5095
Australia
hans.griesser@unisa.edu.au

Hartley, Patrick G.
CSIRO Molecular Science
Bag 10
Clayton South VIC 3169
Australia
patrick.hartley@csiro.au

Jennings, Philip J.
School of Mathematical
and Physical Sciences
Murdoch University
Murdoch, Western Australia 6150
Australia
P.Jennings@murdoch.edu.au

Johnston, Graham R.
DSTO Maribyrnong
Maritime Platforms Division
PO Box 4431
Melbourne, Victoria 3001
Australia
graham.johnston@dsto.defence.gov.au

Kibel, Martyn H.
Telstra Research Laboratories
770 Blackburn Road,
Clayton, Victoria 3168
Australia
m.kibel@trl.oz.au

King, Bruce V.
School of Mathematical
and Physical Sciences
University of Newcastle
Callaghan, NSW 2308
Australia
phbvk@alinga.newcastle.edu.au

Klauber, Craig
CSIRO Division of Minerals
c/o Curtin University of Technology
GPO Box U1987
Perth, WA 6001
Australia

Leckey, Robert
Department of Physics
LaTrobe University
Bundoora, Victoria 3083
Australia
r.leckey@latrobe.edu.au

Leech, Patrick W.
CSIRO Manufacturing and Infrastructure Technology
Private Bag 33
Clayton South, Victoria 3169
patrick.leech@csiro.au

MacDonald, Ron J.
The Chancellory
University of Newcastle
Callaghan, NSW 2308
Australia
dvc-research@newcastle.edu.au

Morris, Graeme C. *retired*
Former address:
Department of Chemistry
University of Queensland
St. Lucia, Queensland 4072
Australia

Myhra, Sverre
School of Science
Griffith University
Kessels Rd, Queensland 4111
Australia
S.Myhra@sct.gu.edu.au

Nelis, Thomas
Jobin-Yvon Emission
Horiba Group
16-18 rue du Canal
91165 Longjumeau
France

Nockolds, Clive E.
Electron Microscope Unit
University of Sydney
Sydney, NSW 2006
Australia
clive@emu.usyd.edu.au

O'Connor, D. John
School of Mathematical
and Physical Sciences
University of Newcastle
Callaghan, NSW 2308
Australia
john.oconnor@newcastle.edu.au

Payling, Richard
11 Rymill Place
Bundeena, NSW 2230
Australia
rpayling@ozemail.com.au
website: TheSpectroscopyNet.com

Price, Garth L.
Telecom Australia
Research Laboratories
770 Blackburn Road
Clayton, Victoria 3168
Australia

Ressel, Peter
Ferdinand-Braun-Institut
für Hochstfrequenztechnik
Albert-Einstein-Str.11
12489 Berlin
Germany

Roberts, Noel K. *retired*
Former address:
Department of Chemistry
University of Tasmania
Hobart, Tasmania 7001
Australia

St John, Heather A.W.
CSIRO Molecular Science
Bag 10
Clayton South, Victoria 3169
Australia
heather.stjohn@csiro.au

Sie, Soey H.
CSIRO Exploration and Mining
PO Box 136
North Ryde, NSW 1670
Australia
soey.sie@csiro.au

Singh, Nagindar K.
School of Chemical Sciences
University of New South Wales
Sydney, NSW 2052
Australia
N.Singh@unsw.edu.au

Smart, Roger St. C.
Ian Wark Research Institute
University of South Australia
Mawson Lakes Blvd
Mawson Lakes, SA 5095
Australia
roger.smart@unisa.edu.au

Sun, Chang Q.
School of Electric
and Electronic Engineering
Nanyang Technological University
Singapore 639798
ecqsun@ntu.edu.sg

Turner, Peter S.
CSIRO Textile and Fibre Technology
PO Box 21
Belmont, Victoria 3216
Australia
peter.s.turner@csiro.au

Part I

Introduction

1 Solid Surfaces, Their Structure and Composition

C. Klauber and R. St. C. Smart

1.1 Importance of the Surface

"When a plate of gold shall be bonded with a plate of silver, or joined thereto, it is necessary to be beware of three things, of dust, of wind and of moisture: for if any come between the gold and silver they may not be joined together ..."

Translated extract from "De Proprietatibus Rerum" (The Properties of Things) 1323 A.D. by Bartholomew [1]

Awareness of the important role played by surfaces in technology has existed for some time, although it is only in the past three decades that we have been able to establish an improved understanding of their properties. In everyday life our perceptions of solid materials, and in particular their surfaces, are strongly distorted by the limitations of visible light. These wavelengths are a thousand times larger than dimensions of the surface region in which well understood bulk properties of materials break down, making way for the transitional interface with another phase, which may be gaseous, liquid or solid. Such are the alteration of bulk properties, structural and compositional, that it is not unreasonable to consider surfaces as an additional phase of matter [2]. Whilst this may serve as a useful general concept for surface scientists, in the various fields of technological endeavour what is thought of as a surface varies enormously, particularly in depth characterisation. Accepting the simplest definition of the surface, as the boundary defined by the outermost atomic layer separating the bulk solid from an adjacent phase, is thus inadequate in the area of practical surface technology. A more meaningful approach is to consider a selvedge layer of variable depth. In fact, the different depth regimes of the surface are defined by *that* depth which actually plays *the* definitive role in the technological application (see Table 1.1). Obviously the boundaries between these surface selvedge depths are not always clear and overlap will exist between adjacent categories. The point of particular importance is to view the "surface" with a degree of flexibility, depending not only upon the nature of the material and its environmental history, but the role that it plays. Wide variations in selvedge depth, from the Ångstrom to the millimetre, fortunately do not pose insurmountable problems in surface analysis. This is due to the complementary nature of the techniques that are employed. Indeed, from the investigator's viewpoint, the "surface" in a practical sense is often dependent on the analytical technique being used.

Table 1.1. The importance of the various depth regimes

Selvedge* depth	Examples of the definitive role
Outer monolayer ~ 0.1 nm	Heterogeneous catalysis, surface tension (contact angle) control, selective adsorption, electrochemical systems, biological systems, sensors
Thin film ~ 0.1–100 nm	Emulsions, tribological (friction) control, anti-reflection coatings, lipid membranes, Langmuir-Blodgett films, interference filters, release agents
Near surface 0.1–10 μm	Semiconductor devices, surface hardening, polymer biodegradation, controlled release membranes, osmosis devices, photographic film, optical recording media, aerosols
Thick layers > 10 μm	Anti-corrosion coatings (anodizing, electroplates, paints), phosphors, adhesives, electrowinning, magnetic recording media, surface cladding

*We use the term "selvedge" to emphasize that the surface is often a region (with depth) rather than a two-dimensional layer.

In attempting to understand the surfaces of materials and methods by which they may be structurally and compositionally characterized, it is useful to remind ourselves of the basic categories of bulk solid materials. Structurally a material will either be crystalline or amorphous depending upon the extent of internal order, a crystalline solid having the same unit cell structure repeating over macroscopic dimensions, whereas the amorphous material possesses order only on the nearest neighbour or molecular dimension scale. The disorder in amorphous materials, such as glasses and some polymers, ensures a continuum in their structural nature, quite different from crystalline solids, where, due to thermodynamic constraints, the individual crystallites are generally small. Crystalline materials are usually encountered in a polycrystalline form, each crystalline grain abutting another of differing crystal orientation. This intergranular region is a classic solid-solid interface with the grain boundary surface properties having a direct bearing on the gross mechanical and thermal properties displayed by the material. The solid may, of course, be multi-phase, each in either amorphous or crystalline form as in ceramics. The polycrystalline material may be an element, compound or alloy, the alloy in turn a continuous solid solution or a mixture of precipitated phases. The "modern" engineering practice of using composite materials (laminates, fibre reinforcing etc.) to create particular properties is merely an extension of phase precipitation.

Technologically the importance of materials' surfaces cannot be understated since they influence so many facets of modern industry. Surfaces are found to:

- control material stability via corrosion and friction/wear characteristics in everything from automobile components to medical prostheses
- determine material adhesion characteristics
- be crucial to systematic process control in materials fabrication such as electronic devices and thin films
- play a vital role in heterogeneous catalytic processes for compound synthesis as in petrochemicals and fertilizers
- control mineral beneficiation via selective flotation and adsorption
- be germane to membrane processes important in numerous diverse fields from environmental control to medicine

The application of surface analysis to such a variety of materials science and technology problems has several purposes. Sometimes the technology may be old and operationally understood only on a folklore basis. If we are to progress it is essential that the processes be understood in detail even if no direct economic benefit is obvious. As part of industrial process control, problems may arise that require immediate solutions. Without knowledge of the factors and mechanisms involved, solutions must still be based on empirical testing and guesswork. The area of system and materials design is increasingly important. We may know the surface properties that we want, but how can they be achieved and how do we know if we have achieved them? It is the purpose of this and following sections to explain and evaluate some of the more common methods and applications of surface analysis used to achieve these purposes.

The primary aim of materials science is to engineer materials with specific mechanical, electrical, magnetic, optical, thermal and chemical properties. These materials need to be stable under the environmental pressures that they normally encounter. Provided that the material is not intrinsically unstable or that extremes of mechanical force or radiation are not disrupting bonds within the bulk, all processes for breakdown of that material will be initiated at a surface. That surface will either be directly exposed or will lie within an accessible pore or within an intergranular region. Each class of material can exhibit a characteristic set of surface selvedge features, with particular physical, chemical and electronic properties. Examples of these are summarized in Table 1.2.

More general structural features, such as epitaxial layer growth, superstructures, nucleation, coatings (e.g. adhesives, passivation layers), and thin films apply to a variety of materials. Additionally, there is another kind of surface structure, which can be given the generic term *electronic* surface structure, that is of central importance to many applications in such fields as microelectronics, catalysis and hydrometallurgy.

These features all play a role in the overall engineering properties of those classes of materials. A common and obvious example of a surface-initiated phenomenon is the corrosion of iron and steels. This is particularly dramatic since the corrosion product typically has an iron content molar volume up

Table 1.2. Surface features of a variety of materials

Material	Surface Features
Metals, alloys	Oxide coating (thickness, type); faceting; relaxation; reconstruction, elemental segregation; defects; fracture faces; corrosion layers and regions; adsorbed layers; interdiffusion and reaction (joining technology)
Semiconductors (including organic semiconductors)	Space-charge region; reconstruction; relaxation; defects; low coordination sites; segregation (multi-element); adsorbed layers; reaction profiles
Ceramics (including high T superconductors)	Grain structure; phase separation; intergranular regions; fracture faces; pores; amorphous regions; triple points; reaction products; elemental distributions; leaching profiles
Minerals (including soils, oxides, salts)	Grain structure; phase separation; altered surface layers (weathered, deposited, oxidised, reduced); faceting; reconstruction; defect sites (low coordination); intergrowth structures; leaching profiles; adsorbed layers (minerals processing)
Glasses	Hydrolysed layer; crystalline regions; inhomogeneities; elemental segregation; reaction products (distribution); mould materials transfer
Polymers	Altered surface layers (oxidised, reacted); excluded layers (lubricants, catalysts, unreacted monomer); phase separation (elastomers, monomers); surface segregation of reactive groups
Composites and natural materials (including wood, paper, paints, cement, fibreglass, tissue, blood, bone)	Distribution of materials; modified surface layers (bonding compatibility); interphase regions; elemental distributions; reaction profiles; leaching profiles

to four times that of the original reduced metal. Such enormous expansion accentuates the material's failure. Whilst such massive corrosion is controlled by mass and charge transport in the solid state, its initiation is certainly a surface process. The corrosion need not be visibly dramatic in order to eventually affect the materials performance. Steels are typically protected from corrosion by painting, yet these coatings will also eventually oxidize, crack and fail to perform their intended task.

Not only chemical stability, but mechanical stability is of engineering importance. Wear can be minimized not only be lubrication, but also by surface hardening. This might be achieved by methods of nitridization or carburization to create a refractory surface phase of extreme hardness. The surface modification of materials by ion beams is one area of active current interest [3]. As an example it has been recently found by *Rabalais* and *Kasi* [4] that when mass selected C^+ ion beams (20–200 eV) impinge upon atomically clean surfaces (Si, Ni, Ta, W and Au), they initially create a bonded carbide structure. However, of particular interest is the observation that, with continued deposition, the carbon layers build up into a diamond-like structure. This is significant because of the particular hardness of diamond combined with its low electrical and high thermal conductivity.

An interesting use of active surface chemistry to maintain a material's stability can be found in the simple quartz-halogen incandescent lamp, first marketed in 1959. In a conventional incandescent lamp the brightness is governed by the filament temperature, which is necessarily limited so as to maintain a useful lamp lifetime. The quartz-halogen variety has the tungsten filament encased in a clear fused quartz envelope containing halogen (F_2, Br_2 or I_2) gas. Any tungsten which evaporates from the filament deposits on the cooler envelope wall where it reacts with incident halogen molecules to form a volatile halide. Halide production and desorption is enabled due to the quartz envelope wall temperatures, typically in the range 250–600°C. The tungsten halide then diffuses back to the filament where it dissociates to re-form tungsten metal and halogen gas. Filament integrity is thus maintained at higher temperatures enabling more efficient light production and longer life – all due to two recycling surface chemical reactions. The internal surface of the envelope can be further improved by the addition of etch barriers (aluminium fluoride, aluminosilicate) and infrared reflectants (titanium silicate, zirconia) [5].

1.2 Solid Surfaces of Different Materials

Having reminded ourselves of the variety of forms in which solid materials can exist, it is a useful exercise to conceptually "create" a surface from one of these materials. Taking the simplest case of a crystalline solid, we might imagine an ideal case of a sudden termination of the crystalline periodicity at the solid-vacuum interface as in Fig. 1.1a. In layer, chain or sheet structures, where relatively weak dispersion forces hold the constituents of the solid together, such an "ideal" surface case might be envisaged. In strongly bonded solids such as metallic, covalent or ionic systems, the termination of the periodicity means that valence electrons will spill out into a continuum with no positive cores, freed covalent bonds will be "dangling" into space and the Madelung constant, evaluated on the basis of 3-dimensional symmetry, will no longer apply. Invariably this breakdown of previously balanced bonding forces leads to a rearrangement of the outer layer or layers and to a periodicity, and possibly unit cell structure, different from that previously existing in the bulk.

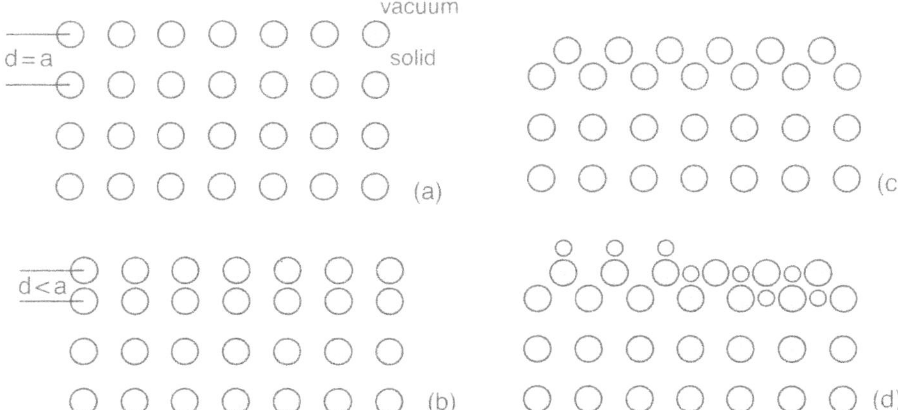

Fig. 1.1. Schematic of (**a**) a solid surface created by terminating the bulk of a crystalline solid. Bonding imbalances cause the outer layer to move (**b**) and even undergo reconstruction (**c**). When then exposed to a reactive medium, foreign atom adsorption can occur (**d**), possibly leading to surface compound formation

The surface in this sense becomes the ultimate defect that a solid material can have.

The simplest rearrangement is that of outer layer relaxation or contraction (Fig. 1.1b), the layer either moving away from or towards the inner layers. Dynamical low energy electron diffraction (LEED) calculations indicate that for metals this movement may be 0–25% of the normal interlayer spacing [6]. As no bond breaking is required this rearrangement is spontaneous. A more complex alteration occurs when the structure parallel to the surface undergoes what is known as reconstruction, shown schematically in Fig. 1.1c. This is particularly common in covalently bonded semiconductor materials. The surface atoms rearrange themselves so as to minimize the dangling bonds which arise from the original surface creation. Reconstruction can be spontaneous or may require heating to be induced. Surface reconstructions are dealt with further in Sect. 1.4 of this chapter and Part II of the book. A large amount of research work carried out in the discipline of surface science is concerned with the *in situ* creation of such virginal surfaces (atomically clean and ordered) and their subsequent controlled reaction with various gases. These can be created *in situ* by a number of methods:

- cleaving or fracturing in the vacuum system; useful for single crystal work, grain boundaries and fracture mechanics, but very limited for materials analysis
- evaporating of clean films (which can be substrate epitaxial); has application in nucleation, catalysis, adsorption and adhesion studies but is generally only useful for model systems

- high temperature flashing; generally limited to refractory materials and used to anneal out ion bombardment damage; heating usually causes surface segregation of bulk impurities
- field induced desorption; usefully limited to refractory metals in the form of a field tip
- ion beam cleaning (utilized in most surface analysis); has additional use of depth profiling, but surfaces and interfaces are damaged and smeared, often with preferential sputtering of a component
- gas-surface reaction to create an altered surface phase more easily removed by one of the above methods (surface titration).

The general techniques for the preparation of atomically clean surfaces have been reviewed by *Verhoeven* [7] and specific methods for selected elements by *Musket* et al. [8].

Our conceptual surface creation in Fig. 1.1 continues to a further stage if it is removed from vacuum and subjected to a reactive liquid or gas environment. The fluid molecules will interact with the surface to create an altered selvedge. This may begin with simple atomic adsorption, followed by incorporation into the near surface region and surface compound formation (Fig. 1.1d). Due to the number of possible surface-fluid couples the variety of reactions is immense, as evidenced just by the number of adsorbed monolayer systems that have been studied [9]. Of course for most "real" surfaces the situation is considerably more complex than depicted in Fig. 1.1d. Polycrystallinity leads to microfacets and domains on grains riddled with structural defects following a roughened topography. A schematic "snapshot" of part of a surface is shown in Fig. 1.2.

The structure of a surface is inextricably linked to its reactivity and, in many cases, its composition. At one extreme, we know that defects, such as dislocations, protusions, edge and corner sites on facets, on metal surfaces provide sites of enhanced reactivity in chemical attack [10]. For an ionic oxide salt, like MgO, it has been shown [11–13] that simple molecules (e.g. H_2, CO, H_2O) will not adsorb on the fully coordinated, five-fold ions in a $\{100\}$ face but will react readily with "kink" sites (i.e. edge and corner sites) of lower coordination as illustrated schematically in Fig. 1.2. At the other extreme, we find profound alterations in chemical composition between surface and bulk due to segregation (i.e. selective deposition of particular elements), surface rearrangement in the first few atomic layers, and consequent changes in the potential energy of surface sites e.g. [14,15]. This occurs in metal alloys, salts, minerals, doped elemental semiconductors and glasses. In polymers, we often see surface layers of material excluded during polymerisation. In ceramics, segregation of particular elements into narrow amorphous films between grains of different phase is exposed at the surface after preferential fracture. The structure of corroded material reveals depth profiles in which the progress of the reaction can be followed, particularly at structural features like grain boundaries, intergranular films, pores, precipitates, impurity

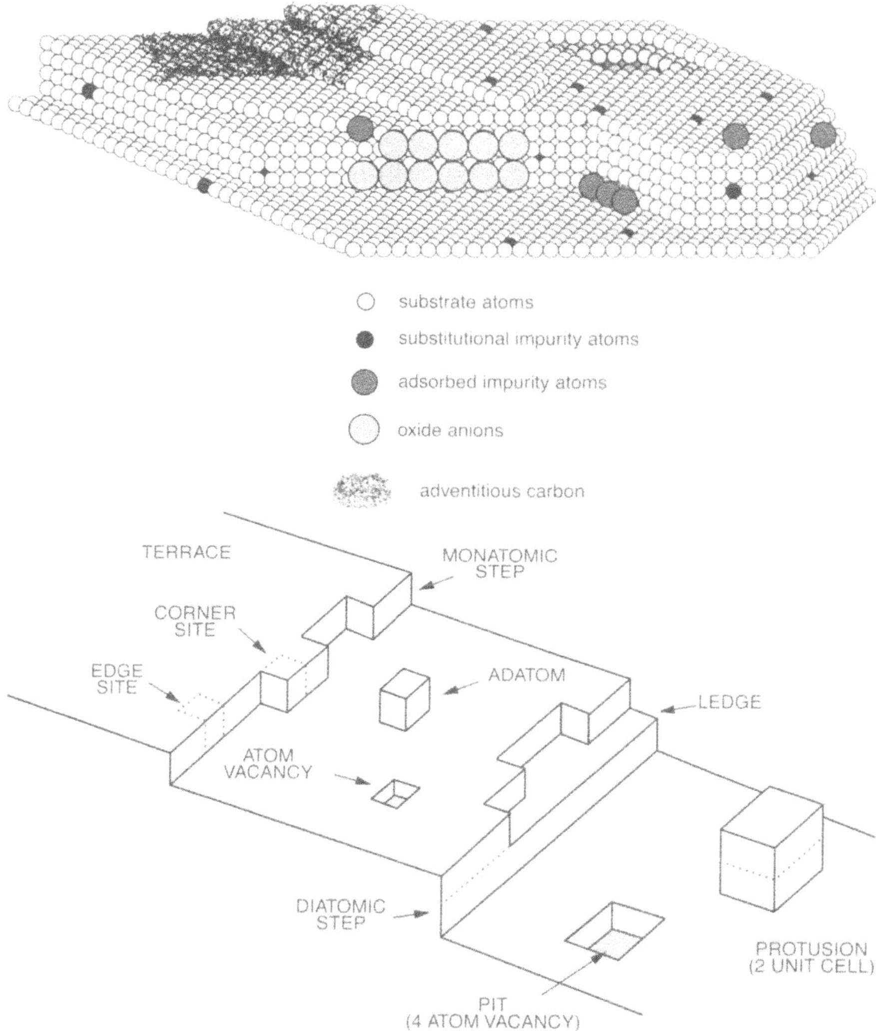

Fig. 1.2. Two representations of part of a solid surface depicting a variety of different surface sites. These sites are distinguishable by their number of nearest or coordinating neighbours. Low coordination sites (surface defects) are the preferred locations for adatom adsorption and can thus be the initiation point for a variety of surface processes. Also depicted are an oxide facet, a cubic etch pit and a region of extensive adventitious carbon contamination

regions and bonded interfaces. All of these examples relate particular surface structures to particular forms of reactivity. They are a few grains of sand on the seashore of surface structural features.

1.2.1 A Material Under Attack: Aluminium

The illustration of our conceptual surface and interface behaviour can be enhanced by reference to aluminium. Its importance as the most widely utilized non-ferrous metal is due to its many useful physical, chemical and metallurgical properties, not the least of which is its resistance to corrosion. Consider for the moment pure aluminium, in particular oxygen reaction at the Al(111) surface. As aluminium has a face centred cubic structure the (111) surface is a close packed plane. A wide variety of adsorption studies, e.g. [16–19] and references therein, conclude that, upon initial interaction with oxygen, a (1×1) overlayer forms, i.e. for every surface aluminium atom a dissociated ionised oxygen atom adsorbs in registry. The registry site is not directly atop the aluminium atoms, but rather in the shallow 3-fold hollow so as to maximize coordination. As oxygen coverage increases, another state forms that is thought to correspond to an oxygen underlayer, i.e. a layer beneath the first aluminium layer. With still heavier oxygen exposures, the resultant three-dimensional oxide layer that forms resembles amorphous Al_2O_3. This parallels the result observed at room temperature in dry air in which polycrystalline aluminium rapidly forms an oxide layer of Al_2O_3. Since the oxide is stable, tightly adherent and highly impervious, further corrosion is naturally resisted and the reaction is self-limiting, ceasing after producing a selvedge of about 5–10 nm in thickness, i.e. the surface passivates. If the film is disrupted, it begins to reform immediately in most environments. Of course aluminium's many properties do vary significantly both with its purity and with alloying. It forms a large variety of commercially useful binary, tertiary and quaternary alloys containing intermetallic phases with Mg, Si, Mn, Fe, Ni, Cu, Zn, Sn and other metals [20].

Not surprisingly the corrosion which can and does occur with aluminium and its alloys can be of many different types, such as intergranular, exfoliation, stress-corrosion cracking or even uniform attack in which the whole oxide film is destroyed. However, by far the most common form of corrosive attack on aluminium alloys is that of pitting corrosion [21]. Not unexpectedly, these pits will form at localized discontinuities in the oxide film. Frequently they possess a characteristic habit, indicating preferential attack on particular crystallographic planes e.g. dry hydrochloric acid gas produces cubic pits because it etches faster in the $\langle 100 \rangle$ direction whereas, in the presence of moisture, corrosion is more rapid in the $\langle 111 \rangle$ direction leading to octahedral pits [22]. An etch pit is schematically illustrated in Fig. 1.2. The vast metallurgical research effort into corrosion control of aluminium and its alloys is concerned essentially with improved coherence of the surface selvedge oxide. This can range from chemically producing a variety of conversion coatings, principally aluminium oxyhydroxides with specific substitutions of other ions such as chromate and phosphate, to engineering solutions such as cladding. In cladding, a thin layer (typically 5–10% of the nett sheet thickness) of pure aluminium or alloy, which is anodic to the base alloy, is attached to the structural sheet. The cladding provides electrochemical protection by corroding preferentially [19]. Cladding finds use in products such as aircraft skins.

It is the very aluminium alloy composition, conferring the desirable engineering characteristics, which is also the key to the corrosion, being strongly affected by the type, amount and properties of the intermetallic phases present in the matrix. A particularly interesting experimental approach taken by *Nisancioglu* et al. [23] for alloys with Fe-rich surface inclusions is their selective dissolution and removal by a controlled electrochemical etching process (Fig. 1.3). As this can be done with minimal dissolution of the surrounding matrix, i.e. without exposing additional intermetallics beneath the surface inclusions, an essentially pure aluminium surface with enhanced corrosion resistance is produced.

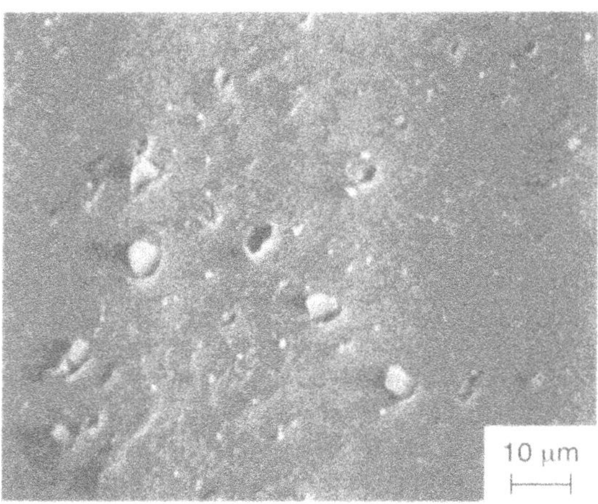

Fig. 1.3. Morphology of an electropolished aluminium alloy (6081) surface. The particles standing in relief are β-AlFeSi intermetallic phase particles with the Al and Fe components leached out [23]

1.3 Methods of Surface Analysis

1.3.1 Variety of Surface Analytical Techniques

To instrumentally probe a medium, in this case a solid surface, one of six basic probes may be applied to the surface: electrons, ions, neutrals, photons, heat or a field. The analysis consists of measuring the surface's response, also evident in one of these six ways. Combining the six probes and six responses gives 36 fundamental classes of experimental technique by which we may analyse a surface. By altering the energy and/or mass and/or character of these probes the variety of possible experimental methods increases even further. Obviously not all options will provide an experimentally viable pathway to gain useful surface information i.e. the surface structure (physical and electronic) and the surface composition (chemical). Whilst all possible variants have not yet been explored to the fullest extent, the number of surface analytical techniques currently in use is nonetheless quite impressive. Appendix A.1 lists over 130 technique acronyms; some of these are admittedly restricted to esoteric research endeavours, but many have gained widespread use in hundreds of laboratories throughout the world. As it is beyond the scope of this book to cover all of these, we will concentrate only on the more common and versatile, twenty techniques in all. These are outlined in Table 1.3 within a subset of 12 of the 36 fundamental classes. There is also a variety of more classical surface-specific techniques such as basic measurements of contact angle, zeta potential, particle size distributions, surface areas and porosities and also electrochemical methods such as cyclic voltammetry which are not included. They are generally covered in a number of good texts on colloid and surface chemistry (e.g. refer to the bibliography at the end of the book).

1.4 Structural Imaging

In approaching a problem of surface characterization the first need is usually physical imaging.

1.4.1 Direct Physical Imaging

By far the most important information on surface structure, at least in the first stage of examination of a sample, comes from the techniques that provide *images* of structural differentiation in the surface layers.

The simplest and most accessible of these techniques is scanning electron microscopy (SEM). Chapter 3 covers this technique in more detail. Secondary electron images can be obtained on all materials identifying surface features, on most instruments, to a practical limit of $\sim$ 100 nm. Despite the considerable depth of penetration of the incident primary electron beam (e.g. 0.5–5 μm), the re-emitted electrons (as secondary and backscattered electrons) come from mean depths of 50 nm – 0.5 μm depending on the density of the material. Hence, the technique is sensitive to the near-surface region. Scanning (or rastering) the beam over the surface minimises surface damage and surface charging. The surface features in this size range include extensive faceting, phase separation, morphology of crystals, the structure of fracture faces, precipitates, pores, distribution of materials in composites, bonding at interfaces and preferential reaction at different sites on the surface. Backscattered electron images can give contrast based on the average atomic number of the region or phase examined. Hence, it can be used to differentiate between grains in a multiphase ceramic, or mineral mixture, or between materials in

Table 1.3. Summary of the various techniques considered in this book for analyzing the structure and composition of the surface selvedge layer. The 12 basic categories, each of which may encompass a range of techniques, represent a subset of a possible 36

Summary of surface techniques * surface $<$ 10 nm $^\circ$ near surface $\sim$ 1 µm			
	Emitted analyzed response		
Incident excitation probe	Electrons	Ions	Photons
Electrons	**AES*** Auger electron spectroscopy **SAM*** Scanning Auger microscopy **SEM**$^\circ$ Scanning electron microscopy **TEM**$^{*\circ}$ Transmission electron microscopy **LEED*** Low energy electron diffraction **RHEED*** Reflection high energy electron diffraction **SPE**$^{*\circ}$ Spin polarised electron spectroscopy		**EDAX**$^\circ$ Energy dispersive analysis of X-rays
Ions		**SIMS*** Secondary ion mass spectrometry **LEIS*** Low energy ion scattering spectroscopy **RBS**$^\circ$ Rutherford backscattering spectroscopy	**NRA**$^\circ$ Nuclear reaction analysis
Photons	**XPS/ESCA*** X-ray photoelectron spectroscopy / Electron spectroscopy for chemical analysis **UPS*** Ultraviolet photoelectron spectroscopy **(AE)XAFS*** Auger emission extended X-ray absoption fine structure		**FTIR**$^\circ$ Fourier transform infrared spectroscopy **Raman**$^\circ$ Raman vibrational spectroscopy **XAFS**$^\circ$ X-ray absorption fine structure analysis
Electric / Magnetic Field	**STM*** Scanning tunnelling microscopy †**AFM*** Atomic force microscopy		**GDOES*** Glow discharge optical emission spectroscopy

†The AFM technique essentially analyses force due to electric and magnetic fields at surfaces

a composite. Topographic images, obtained by combining different backscattered electron images, can reveal detail of pits and protusions, precipitates and altered regions on the surface. Figure 1.4 shows examples of the three types of images – secondary electron, backscattered electron and topographic – from the same area of a ceramic surface. The SEM is relatively easy to use requiring straightforward specimen preparation and conventional vacuum. Its disadvantages are damage by the electron beam to polymer surfaces, some minerals and insulating materials (a minor effect in the scanning mode) and limited resolution.

Transmission electron microscopy (TEM) and scanning transmission electron microscopy (STEM), also covered in Chap. 3, both give much higher resolution of surface features down to unit cell level or < 2 Å in most cases. Hence, atomic steps, ledges, corners, defect sites, fine grains, intergranular films and interphase regions can be imaged. Diffraction contrast, phase contrast and defocus contrast can be used to enhance the images, particularly at edges and surfaces. SEM can also provide selected area electron diffraction (SAED) from crystalline regions. This is particularly important in distinguishing crystal structures of phases with closely similar (or the same) chemical composition. The major difficulty with these techniques is that the sample must transmit electrons before they are focussed to form an image. The sample thickness has to be usually < 300 nm for this to be achieved and the transmission (and image) is highly sensitive to this thickness and the material itself. Specimen thinning using accelerated ion beams introduces severe surface damage. TEM and STEM are most effectively used on very small particles and on the thinner (wedge-shaped) edges of larger particles where structural detail of major importance can often be obtained. An example of this kind of detail can be seen in Fig. 1.5.

Scanning Auger microscopy (SAM), covered in Chap. 6, combines physical imaging of the surface, as in SEM, with chemical analyses of spots, individual areas and chemical imaging (i.e. mapping). SAM uses a focussed electron beam, with energies in the 3–50 keV range, to cause ionisation of core levels in surface atoms. The ionised atoms relax via emission of Auger electrons in a two-electron process. Each element in these surface layers produces a unique set of Auger energies (from its set of energy levels) so that both chemical identification and composition (i.e. surface concentrations) can be determined. The electron mean free path in solids, discussed in detail in Sect. 1.5.1, reveals that Auger electrons with energies in the 20–1000 eV range escape from depths of only 5–20 Å. Additionally, of course, the primary electron beam produces secondary and backscattered electrons which can be used for imaging. Current instrumentation can achieve a beam size of 20 nm, rastered across the surface like a TV screen, to produce either a secondary (or backscattered) electron image or, using a selected Auger signal, an elemental distribution over a surface. Both types of image are illustrated in Fig. 1.6. SAM is an exceptionally useful tool also for spot (< 50 nm)

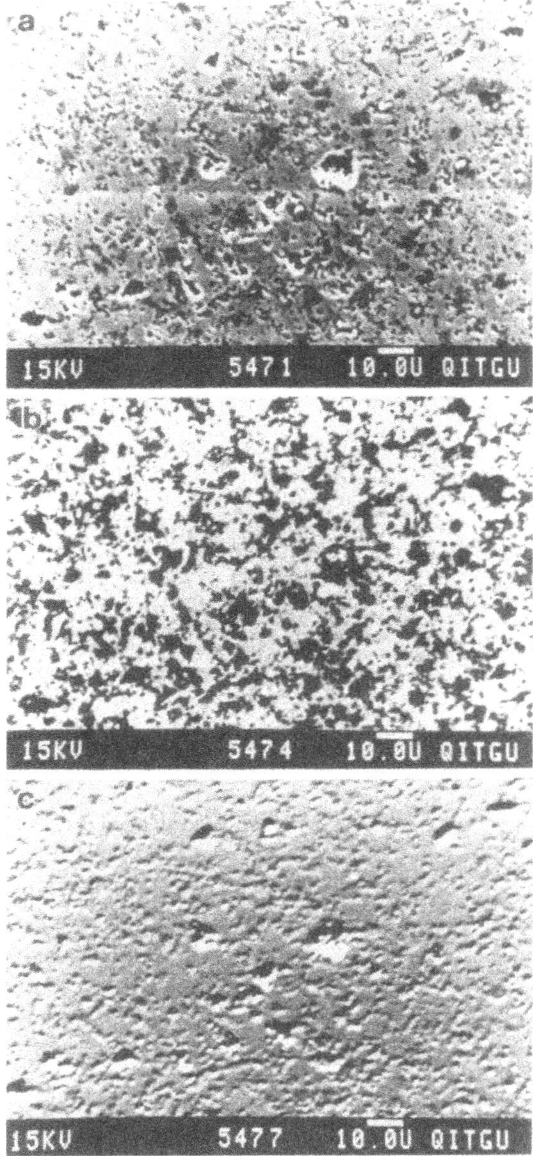

Fig. 1.4. Comparison of (**a**) secondary electron (**b**) backscattered electron (BSE) and (**c**) topographic images from scanning electron micrographs of the same area of a polished ceramic surface showing grain pull-out and porosity. The white bar indicates magnification of 10 μm. Note that the BSE image, in addition to porosity, indicates regions of low (dark) and high (white) average atomic number

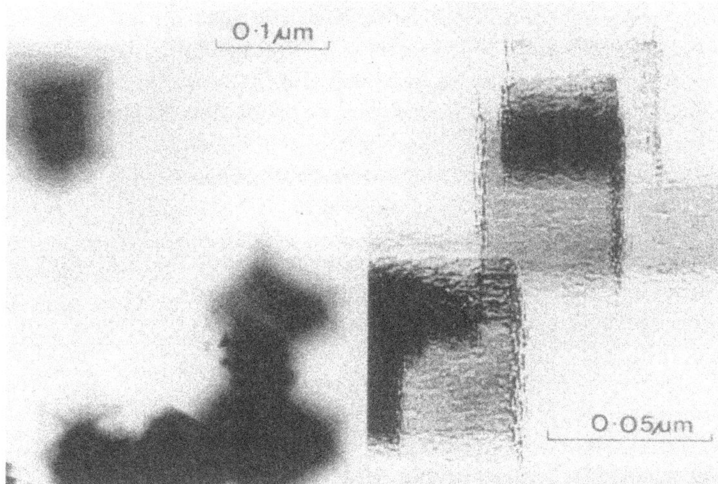

Fig. 1.5. Transmission electron micrographs of MgO "smoke" cubes formed by burning Mg in air. The left-hand image was recorded immediately after preparation; the right-hand image was recorded after < 5 sec immersion in pH 3 nitric acid. Phase contrast imaging reveals that the initially smooth cubes are roughened before dissolution commences. [Reprinted with permission from Fig. 13, C.F. Jones, R.L. Segall, R.St.C. Smart, and P.S. Turner, Proc. Roy. Soc. Lond. A **374**, 141 (1981)]

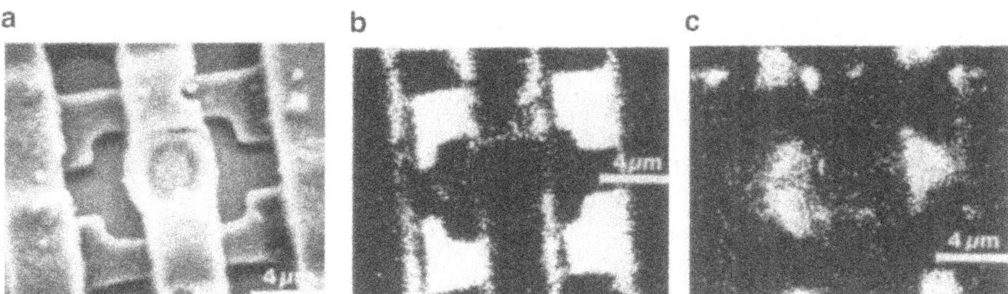

Fig. 1.6. Scanning Auger microscopy/spectroscopy from an area of a 64 K random access memory MOS device. (**a**) Secondary electron image, (**b**) elemental Si (LMM transition, 89 eV) distribution, (**c**) oxidised Si (LMM, 72 eV) distribution. [Reprinted with permission from Figs. 3, 4 K. Bomben, W. Stickle, J.J. Hammond, Proc. 1st Elec. Conf., Vol. 1, SADME (CA, USA 1987)]

analyses of the surfaces of individual particles, grains in ceramics, interiors of pores, precipitates, and other differentiated areas on a sample surface. The surface sensitivity of the technique, combined with structural images, gives us a powerful probe of the localised chemistry in surface layers. The limitations of the technique lie in three main areas:

- samples which are poor conductors or insulators cause major charging at high resolution (i.e. small beam size, high voltage and high current density) deflecting the beam onto adjacent areas
- chemical information on covalency, bonding or oxidation state of the element is limited or not available
- high current densities and voltages can cause decomposition of some samples as in electron microscopy.

Generally, however, SAM can be applied to all materials, with greater or less resolution, provided they are stable in vacuum.

The application of Auger spectroscopy to determination of chemical composition is discussed in Sect. 1.5 and Chap. 6. It is sufficient to note here that the combination of structural and chemical identification allows SAM to give direct information on most of the surface features previously listed with the exceptions only of reconstruction, relaxation, and space-charge regions. It is indeed a powerful surface technique in materials science.

A new and equally exciting technique for imaging surfaces has become available since 1981. The scanning tunnelling microscope (STM) uses the low field quantum mechanical tunnelling effect, where the tunnelling current is approximately related to the inverse exponential of the barrier thickness (gap width) and the square root of the work function, to produce images with resolution of individual atoms (see Chap. 10). The current dependence on gap width (d) is very sensitive (e.g. a ten-fold change for a change in d of 1 Å). An almost atomically sharp metal tip is used to scan areas as small as 10×10 Å to produce the most detailed images of atomic defects yet seen on material surfaces. The vertical resolution is even higher than the lateral resolution, in most cases being better than 0.1 Å. Figure 1.7 illustrates the level of detail seen in an STM image. The technique can also be used to study differences in local work function allowing, in principle, identification of chemically different species in the surface (e.g. impurities or defect sites) or adsorbed on the surface. The distribution of electronic states across surfaces can be defined in this way. The technical problems in achieving this resolution, like decoupling of the STM from external vibrations, controlling the tip scanning, measuring the current and graphically presenting the image, are non-trivial but now largely solved as explained in Chap. 10. The STM is rapidly becoming relatively accessible but, again, is limited in application to good semiconductors and metals. Contamination, normal on any real material surface, will obviously obscure information on the underlying material but may, of course, be directly studied with the STM for identification and distribution as in Fig. 1.7. The recently developed combination of STM with SEM, to allow an extreme range of imaging, promises to be of great benefit to the surface scientist.

There are other techniques for surface imaging deserving of mention but not included in our compilation. Field emission (FEM) and field ion (FIM) microscopes [24,25] both based on the tunnelling of electrons under high field

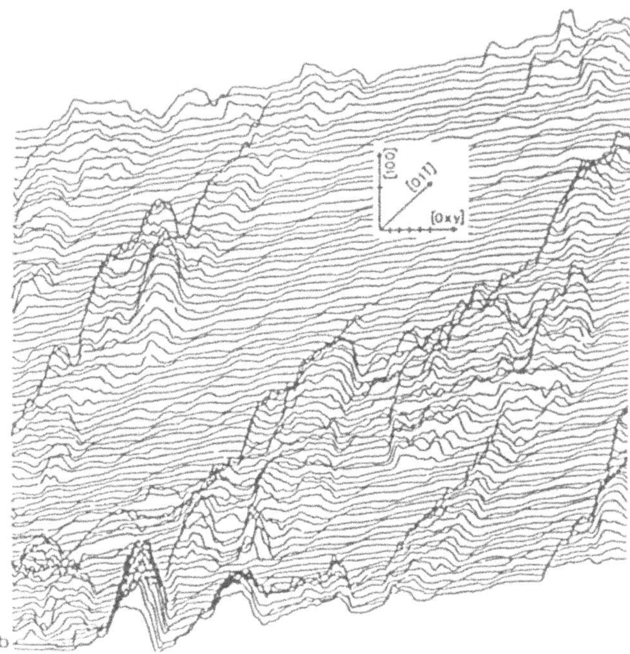

Fig. 1.7. Scanning tunnelling microscope image of carbon islands on a reconstructed Au (100) surface (smooth parts). Divisions on the (100) axis are 0.5 nm apart. [Reprinted with permission from Fig. 2, G. Binning and H. Rohrer, Proc. 9th Int. Vac. Cong.; 5th Int. Conf. Solid Surf., Madrid, Spain (Sept. 1983) (Ed. L. de Segovia) pp. 77–79 (Spanish Vac. Soc., ASEVA Publisher, 1984)]

conditions, and capable of atomic resolution, are now older techniques (with a large literature) to some extent superceded by STM. They apply only to refractory metals and require special sample preparation not normally applicable to materials science. They have greatly increased our understanding of surface structure but must be regarded as essentially fundamental research tools.

Low energy electron reflection microscopy (LEERM) is a relatively new technique [26], used in combination with LEED, giving images from elastically reflected electrons in the 0–300 eV range. Its practical resolution is about 50 nm but this can allow a combination of image and diffraction patterns from differentiated surface features, a result of considerable importance for surface phase formation in catalysis, reactivity, leaching, corrosion and surface segregation. It is not yet sufficiently accessible for most materials scientists to warrant a full discussion but may well be a technique for the future.

1.4.2 Indirect Structural Imaging – Relaxation and Reconstruction

Provided that low energies are used, the scattering of electrons, atoms and ions can be restricted to the top few atomic layers giving surface sensitive probes for atomic positions (interatomic spacings), periodic structures (via diffraction of the scattered beams), faceting, defect sites and disorder. Low energy electron diffraction (LEED) uses electrons in the 10 to 500 eV range with escape depths of 5–10 Å (Chap. 13). Atomic beam scattering (ABS) uses atomic beams of low-mass, unreactive gases (e.g. helium) monoenergetic in the range 30 to 300 meV, which scatter and diffract from literally the top atomic layer. Ions with energies less than 5 keV i.e. low energy ion scattering (LEIS) also scatter predominantly from the top layer due to the very large cross sections for ion-atom elastic scattering (Chap. 11). They can be used for studies of shadowing (of one atom by another) and atomic displacement (i.e. interplanar spacing and defect) effects. Scattering of ions with higher energies, i.e. medium energy MEIS, (20–200 keV) and high energy HEIS, or Rutherford backscattering RBS (200 keV–2 MeV), can give information on atomic positions and defects (e.g. interstitials) in the bulk of the solid. Penetration is deeper and blocking effects due to displaced atoms are more evident (Chap. 9). Similarly, medium (MEED, 500 eV to 5 keV) and RHEED reflection high energy electron diffraction (RHEED, 5–200 keV), using the same principles as LEED, give information particularly on surface topography and faceting but with less surface sensitivity (up to 10 nm depth) (Chap. 12). All of these techniques give primary data which can only be transformed to the surface lattice (LEED), surface atomic positions (ABS, LEIS) or interlayer spacings (MEIS, HEIS) by calculation using a model for the dynamical processes occurring at the surface. Programs for routine analysis in LEIS, MEIS and HEIS are readily available and reliable to a high level of accuracy. In the other cases, the final structural information is only as good as the assumptions the model allows and particularly for ABS and LEED, relatively sophisticated models are required for reliable results.

The electrons elastically scattered from surface atoms (i.e. $\sim$ 1–10% of the primary beam) in LEED studies of crystal surfaces will diffract sharply only from well ordered domains of 100 nm^2 or greater as determined by the coherence length over which the electrons will remain in phase. Disorder in the surface can thus give diffuse spots and loss of intensity, providing data for analysis of this disorder. Faceting appears as extra spots in the diffraction pattern and steps on the surface can produce multiple splitting of the primary spots in the pattern. The height of the steps can be inferred from the "appearance voltage" of the splitting. The accuracy of interatomic spacings determined from LEED calculations is about 0.1 Å if the surface is well-ordered and a dynamical model is used.

One of the major applications of LEED to surface structural elucidation has been in the determination of the nature and magnitude of surface relaxation and surface reconstruction with particular reference to verification of some now very sophisticated and highly accurate computer models of these processes [27]. For instance, it is now known that the neutral (100) planes of simple ionic oxides with NaCl structure, like MgO and NiO, have surface spacings almost unaltered from the bulk after relaxation and "rumpling". By contrast, surface planes which are charged [e.g. the (111) plane in a fluorite structure] or have net dipole moments [e.g. the (111) rocksalt plane or (001) fluorite plane] can show relaxation up to half the bond length and even reconstruction of (111) faces to stepped (100) facets a few unit cells long. Dynamical processes resulting from reaction or heating involving changes in surface topography via reconstruction can be followed from time-dependent LEED patterns using video recorders [28]. Segregation of impurities into (or out of) the surface can often be detected in LEED patterns as ordered overlayers producing streaks or new spots.

The combination of RHEED with LEED is particularly powerful for studying topography and reaction profiles propagating from the surface into the bulk of the material as in leaching, oxidation, corrosion, interdiffusion, joining technology, thin film reactions, elemental segregation, interphase regions and recrystallised (e.g. weathered) surfaces.

The LEED and RHEED techniques, however, are unfortunately limited to reasonably large ($>$ 2 mm) single crystals with relatively well-defined faces presented at the surface. Hence, metals, semiconductors, thin films and minerals can be studied – some multiphase ceramics are also possible – but powdered samples, glasses, polymers, composites and natural materials are not normally possible. For this reason, they tend to be available more for fundamental research and materials development rather than for general materials examination. The LEED technique is not normally destructive but RHEED can degrade some samples.

More information on relaxation, reconstruction and the positions of atoms in ordered adsorbed layers has been obtained using ABS. Helium atoms with energies below 300 meV have de Broglie wavelengths above 0.26 Å, they diffract from the top layer alone, and their angular intensities can be easily measured using a goniometer-mounted mass spectrometer. The measured intensities have to be matched with a calculated scattering contour of the surface. This contour is usually derived from a model of the surface structure and, if the model successfully predicts the diffraction pattern, this provides an indirect picture of the atomic arrangement in the surface. For instance, ABS has shown that substantial charge redistribution occurs in the NiO (100) (rocksalt structure) surface levelling the electron density contour with the oxygen 0.3 Å above the nickel atoms, i.e. "rumpled" [29]. An ab initio surface density calculation is usually necessary to obtain precise values of atomic positions and bond lengths. The distribution and concentrations of atomic defect sites can also be estimated from ABS patterns [30]. The technique has the major advantages of being non-destructive to samples and being applicable

Table 1.4. Comparison of the different surface analytical techniques (considered in further detail in Part II) with respect to the incident probe characteristics, their ability to obtain chemical information and experimental considerations [35]. For the latter categories the more dots the better. NA indicates not applicable

	Incident probe				
	Particle	Energy	Energy resolution	Current or flux	Beam diameter
AES	electron	0.5–10 keV	NA	0.1–500 μA	0.1–1 mm
SAM	electron	3–30 keV	NA	0.5 nA–2 μA	300–5000 Å
XPS	photon	1–15 keV	0.5–2 eV	10^{12}–10^{13} s^{-1}	0.2–6 mm
LEED/	electron	15–500 eV	0.5–1 eV	0.1–5 μA	0.2–1 mm
RHEED		2–30 keV		0.3–0.5 μA	50–100 μm
RBS	^1H, ^4He	1–3 MeV	keV		
NRA	^1H, ^4He, ^{19}F	0.3–6.4 MeV	keV		
FTIR	photon	0.05–0.5 eV	0.012–0.5 meV		∼ mm
SIMS	ion	0.5–30 keV	NA	1 pA–100 μA	500 Å–2 mm
static				1 pA–10 nA	
dynamic				> 1 μA	
ISS	ion	1–3 keV	5–10 eV	10 pA–1 μA	200 μm–2mm
STM	electron	0.1 eV	NA	0.5–1 nA	2 Å
EM	electron	10–50 keV	eV	0.1 μA–10 pA	2 nm–1 μm
UPS	photon	10–50 eV	3–20 meV	10^{11}–10^{12} s^{-1}	1–3 mm

to any sample, conducting or insulating, exhibiting an ordered surface layer, i.e. metals, semiconductors, ceramics, minerals and some crystalline polymers. ABS is not widely available mainly because the differentially-pumped high pressure nozzle sources to produce the monoenergetic atomic beam are not simple to make or operate.

Other techniques that can give indirect structural information include surface enhanced X-ray absorption fine structure (SEXAFS) [31], X-ray absorption near edge structure (XANES) [32] and surface extended energy loss fine structure (SEELFS) [33]. Both SEXAFS and XANES require synchrotron radiation (not accessible for most materials scientists) and very sophisticated calculations to obtain structural data – see Chap. 15. SEELFS can be used in a TEM, can be applied to almost all materials, and will certainly become more widely used as the instrumentation becomes available. At present, it is in a relatively early stage of development as described in Chap. 3.

Table 1.4. (cont.)

	Chemical information		Experimental considerations						
	Inner shells	Valence shells	Ease of interpretation	Ease of use	Ease of quantification	Lack of surface damage	Coping with charging	Speed and sensitivity of analysis	Elements of zero or low sensitivity
	..	..	..	....	...	..	..	.....	H, He
	..		..	...	..	..	.	..	H, He
	....	..	...	...	...	....	...	...	H, He
)/ ED	Structural techniques			..					
	–	–		....	...	...	.	..	Low z
	–	–		.	..	...	.	.	High z
	Vibrational states		..	....	..	....	....	...	Inorganics
c	Molecular			..		.	..	....	–
ımic	Elemental		..	...	..	NA	..	.....	–
	Elemental		..	...	...	..	...	....	H, He
	Structural technique		..	..				..	
	Elemental			...	..	..	.	..	Low z
	NA	....	...	...	..	.....	...	...	

1.5 Composition of the Surface Selvedge

Having established a picture of the simpler structural aspects of the outer atomic layers, the need for a more comprehensive understanding of the chemical composition of the surface becomes apparent. Defining the information that we require in order to determine a surface's composition is a simple exercise; actually obtaining that information is considerably more difficult. Indeed, very often only part of the picture will be fully understood. Beyond simply identifying the elements in the selvedge, their chemical states and atomic or molar proportions also need description. The lateral and depth distribution of each element is required i.e. a five dimensional problem per element. Up to atomic number 92 there are 90 elements which are naturally occurring. Even confining ourselves to the simple case of formal chemical states, those 90 elements represent about 250 possible variants [34]. Placing these variants in an atomically resolved matrix of macroscopic dimensions is not feasible. The pragmatic approach is thus to restrict our evaluation to the phases present and to the way they are distributed in the selvedge. This will be determined by the spatial resolution that the techniques applied can achieve. In order to establish phase identity a first approach might be to evaluate elemental stoichiometries. However, even the simple problem of ascertaining a mole percentage is subject to many errors. In the absence of due care these may be up to 20%. This would blur the distinction between, for example, a sesquioxide X_2O_3 and a dioxide XO_2. In such a case, the chemical state of X, i.e. whether it existed as X^{3+} or X^{4+}, would become a determining factor. Table 1.4 lists a comparison of the techniques examined in this book with respect to their ability to determine surface compositions [35].

The region represented by the selvedge is variable and is quite dependent upon the technique in question. For a low energy incident ion technique such as ion scattering spectroscopy (LEIS), projectiles entering below the outer surface are efficiently neutralised and are therefore, in effect, invisible to ion detection by electrostatic means. By contrast, a photon-excitation technique such as X-ray photoelectron spectroscopy (XPS), can reach thousands of nm's into the selvedge producing characteristic photoelectrons from a significant depth. However for a photoelectron from a specific element to be measured, characteristic electrons must reach the analyser without undergoing any energy loss. The inelastic mean free path (IMFP) of these electrons means that only electrons from the first few surface layers meet this requirement, ensuring the surface sensitivity of the technique.

1.5.1 Electron Inelastic Mean Free Paths

The inelastic mean free path (IMFP or λ) of an electron travelling within a solid can be defined as the mean distance it traverses before undergoing an inelastic event, i.e. some interaction whereby it loses energy. It is evident that, in any of the surface analytical techniques utilizing electrons as the excitation probe and/or the analyzed response, it is this path length which will govern the technique's surface sensitivity. The expectation of a complex dependence of electron IMFP upon the nature of the solid material is not found empirically and the surface electron spectroscopist can resort to the very useful so called "universal" IMFP curve. *Seah* and *Dench* [36] have compiled the most comprehensive electron IMFP data base to date. Shown in Fig. 1.8 is the resultant universal curve for the elements. Note that, for electrons in the 10–1000 eV range, the IMFP varies from about 2 nm through a minimum of 0.45 nm and back up to about 1.6 nm. In terms of monolayer equivalents (λ_m) the semi-empirical relationships found for electrons of energy E (in eV) above the Fermi level (between 1 and 10,000 eV) are:

$$\lambda_m = \frac{538}{E^2} + 0.41a^{1.5}E^{0.5} \quad \text{for elements} \tag{1.1}$$

$$\lambda_m = \frac{2170}{E^2} + 0.72a^{1.5}E^{0.5} \quad \text{for inorganic compounds} \tag{1.2}$$

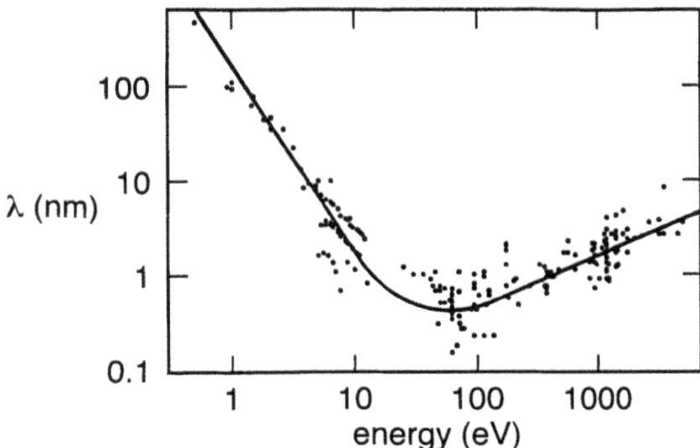

Fig. 1.8. Compilation of electron inelastic mean free path measurements (IMFP, λ) for elements from *Seah* and *Dench* [36]. Solid curve of best fit represents the "universal" IMFP applicable to all elements

where a, the solid material's particular monolayer thickness in nm, is given by:

$$a = \left(\frac{A}{\rho n N_{\mathrm{A}}}\right)^{1/3} \times 10^{8}\,\mathrm{nm} \quad , \tag{1.3}$$

where A is the molecular weight, ρ the bulk density in kg m^{-3}, n the number of atoms in the molecule and N_{A} is Avogadro's number.

Note that the Seah and Dench approach is essentially *structureless* as single crystal, polycrystalline and amorphous examples of the same substance are only differentiated on the basis of their bulk densities. More recent comments on the Seah and Dench formulae have been made by *Ballard* [37] and *Tanuma* et al. [38].

The electron IMFP is particularly useful since, by assuming a process of *homogeneous* attenuation, a Beer-Lambert type expression results for the description of electron flux reduction:

$$I = I_0 \exp\left(\frac{-z}{\lambda \cos\theta}\right) \quad , \tag{1.4}$$

where I_0 and I are the incident and emergent intensities and $z/\cos\theta$ is the path length for electrons travelling θ off-normal through a material of depth z along the normal. Although, as *Ballard* [37] points out, such an expression strictly only applies to electromagnetic radiation, in practical terms I/I_0 plots are found to be sufficiently linear to be useful for approximate estimates. The total electron intensity reaching the surface at an angle θ from *above* the depth z, as a fraction of intensity reaching the surface from *all* depths, can then be established from a simple integration to be:

$$F_{z,\theta} = 1 - \exp\left(\frac{-z}{\lambda \cos\theta}\right) \quad . \tag{1.5}$$

Table 1.5. Data set for the three-phase model of a passivated aluminium surface

Phase	A (g mol^{-1})	ρ (kg m^{-3})	n	a (nm)	Al $2p$ λ_m	O $1s$ λ_m	d_{Al}	d_O	z (nm)
AlO(OH)	59.99	3.01×10^3	4	0.202	2.25	1.76	0.501	1.002	0.202
α-Al_2O_3	101.96	3.5×10^3	5	0.213	2.43	1.90	0.686	1.029	5.325
Al	26.98	2.70×10^3	1	0.255	1.81		1.000		∞

Thus for $\theta = 0°$, 63% of all electron intensity reaching the surface comes from within one IMFP of the surface, 86% from within two IMFPs and 95% from within three. For the electron spectroscopies IMFP is not only relevant to surface sensitivity, but also to quantification of the analytical method.

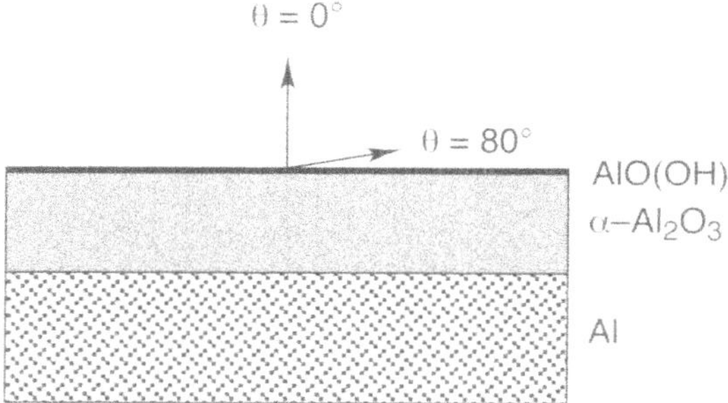

Fig. 1.9. Model of a passivated aluminium surface consisting of a 25-monolayer film of α-Al_2O_3 on bulk metal, with an outer monolayer of AlO(OH). Calculated IMFP-dependentrelative intensities of Al $2p$ and O $1s$ photoelectrons at the exit angles of $\theta = 0°$ and $80°$ are given in Table 1.6

Some of the implications of electron IMFPs in terms of the technique's analyzed response can be usefully illustrated by working through the case of a passivated aluminium surface. For simplicity let us consider a 3-phase planar film structure (Fig. 1.9) of bulk aluminium covered with a coherent but amorphous 25-monolayer film of Al_2O_3 capped in turn by a single monolayer of the monohydrate oxide AlO(OH). Only A, ρ and E need to be first established. All other parameters arise from the above equations, including λ in nm from λ_m and a. Approaching the analysis of this 3-phase selvedge with XPS, the characteristic photoelectrons of interest are the Al $2p$ and the O $1s$. From literature values for their kinetic energy (1178.9 and 1180.9 eV for Al $2p$ in the oxides and the metal respectively and 722.0 eV for O $1s$ in the oxides [39]) and estimates of a, values of λ follow Table 1.5.

As relative photoelectron intensities are a prime experimentally-measured quantity, it would be useful to estimate the effective relative photoelectron counts from each of the three phases. This consists of the initial photoelectron

Table 1.6. Relative Al 2*p* and O 1*s* photoelectron intensities for two escape angles from the passivated surface. No correction has been made for relative cross-sections and values are normalized for unit Al 2*p* emission at $\theta = 0°$

	Phase	$\theta = 0°$	$\theta = 80°$
Al 2*p* photoelectrons	AlO(OH)	0.059	0.048
from:	α-Al_2O_3	0.827	0.105
	Al	0.113	$< 10^{-6}$
O 1*s* photoelectrons	AlO(OH)	0.117	0.090
from:	α-Al_2O_3	1.002	0.107

production modified by its transport to the detector. Since the X-ray attenuation with depth can be considered negligible compared with the effects of λ, photoelectron production will be proportional to the relative atom density d_i and the layer thickness z, with subsequent escape dependent upon λ and z. Equation (1.4) is a statement of the probability of an electron on trajectory θ reaching $z = 0$ from a depth z. Hence, by integrating I/I_0 from $z = 0$ to ∞ we derive a maximum electron escape factor of $\lambda \cos\theta$. The fraction for a film of $z < \infty$ is given by (1.5). Combining both with d_i yields an expression for relative photoelectron production and escape from a selvedge of thickness z:

$$I_z = d_i \lambda \cos\theta \left[1 - \exp\left(\frac{-z}{\lambda \cos\theta}\right)\right] \quad . \tag{1.6}$$

It follows that electrons from layer 2 passsing through layer 1 will emerge with a relative intensity of:

$$\exp\left(\frac{-z_1}{\lambda_1 \cos\theta}\right) d_{i,2} \lambda_2 \cos\theta \left[1 - \exp\left(\frac{-z_2}{\lambda_2 \cos\theta}\right)\right] \tag{1.7}$$

with an additional product term for every subsequent attenuating layer.

Combining the data from Table 1.5 for the three-phase model and considering electron escape trajectories along $\theta = 0°$ and 80° off-normal, the relative fractions of Al 2*p* and O 1*s* photoelectrons can be evaluated from (1.7) and its extension. These are given in Table 1.6, normalized for unit total Al 2*p* emission at $\theta = 0°$. As would be expected the principal Al 2*p* and O 1*s* signals come from the α-Al_2O_3 layer. The underlying metallic substrate contributes only 11% to the total Al emission at 0° and effectively disappears at 80°. By contrast the relative contribution from the monohydrate outerlayer comparatively increases over six-fold for Al 2*p* and over seven-fold for O 1*s* (although the absolute signal strength decreases).

This simple quantitative model does not make allowance for complicating factors such as relative photoionization cross-sections or spectrometer characteristics. Also the simplicity overlooks aspects like film roughness, which can have a large bearing upon such a quantitative analysis [40]. However it does serve to illustrate the influence of λ upon surface analysis. Note that one consequence of λ is that isotropic electron production within a solid is transformed to an anisotropic $\cos\theta$ dependence with transport to the surface. Further general consideration to analyzing surface composition is given in the rest of Sect. 1.5, although formal treatment is left to the individual techniques in Part II.

Whether or not a component is detected thus depends initially upon its depth z from the outermost layer (assuming that it is within the lateral region being probed) and the probability of its responding with sufficient signal to be detected. If the concentration within the volume element falls below its sensitivity limit then that component will be invisible. The questions of varied elemental sensitivities, practical limits to detection, spatial resolution and the definition of chemical state information all affect how well the surface selvedge composition is determined. Before considering any examples of composition determination it is worth considering some of these basic constraints.

1.5.2 Variation of Elemental Sensitivities

For the unknown surface confronting the researcher, the first problem is that of detection of the elements present. Not all elements will be detected with equal ease and the variation of detectability changes between the various surface analytical techniques. Figures 1.10–1.12 illustrate the relative elemental sensitivities across most of the periodic table for three of the most common techniques i.e. AES, XPS and SIMS. The two electron spectroscopies, AES and XPS, both relying on ionization from particular energy levels and electron detection, have comparable relative sensitivity with variations of less than two orders of magnitude. However, things are quite different for ion spectrometry. In secondary ion mass spectrometry (SIMS), described in Chap. 5, an accelerated ion beam (e.g. Ar^+, Cs^+, O_2^+, 500 eV-5 keV) is focussed (down to < 40 nm) on the surface and the secondary ions, both positive and negative, sputtered from the surface are mass-analysed in a quadrupole or time-of-flight mass spectrometer. The beam can be rastered to produce chemical mapping, as in SAM, with very low (ppm) detection limits for most elements. So-called dynamic SIMS uses high current densities and sputters many monolayers per second whereas static SIMS uses low current densities and removes a single surface layer over periods up to an hour. The ease of ion detection and discrimination provides SIMS with an enormous dynamic range, easily encompassing five orders of magnitude in terms of relative sensitivity. Whilst that provides SIMS with its greatest strength, because atom *removal* is involved (rather than an intra-atomic excitation), these relative sensitivities are dramatically matrix dependent. For example a SIMS spectrum of GaAs based on positive ion detection would, on inspection, suggest a sample of pure Ga and a 0.1 atom % trace impurity of As. Conversely, the negative ion spectrum would suggest a sample of pure As [41]. Such widely varying matrix

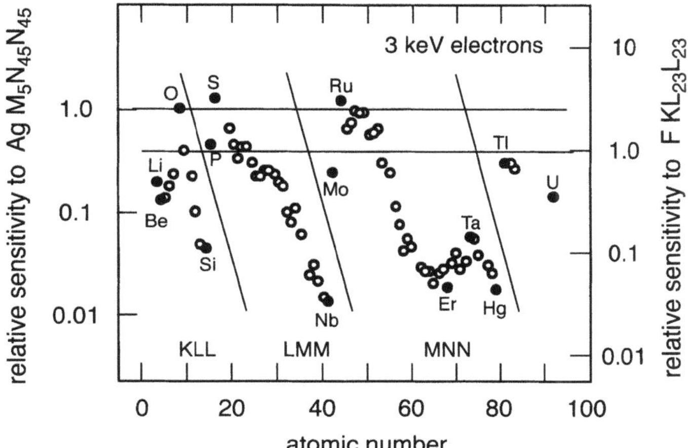

Fig. 1.10. Experimental elemental sensitivities relative to Ag $M_5N_{45}N_{45}$ and F $KL_{23}L_{23}$ across most of the periodic table for Auger electron spectroscopy (AES). Values are for an incident beam energy of 3 keV and include cylindrical mirror analyser transmission function. Not all transitions have been included. Those selected are the most intense, conveniently located transitions, e.g. intense, low-energy valence transitions are difficult to utilize quantitatively. [Compiled from L.E. Davis, N.C. McDonald, P.W. Palmberg, G.E. Riath, and R.E. Weber, Handbook of Auger Electron Spectroscopy, 2nd ed. (Perkin Elmer Corporation, Eden Prarie USA, 1976)]

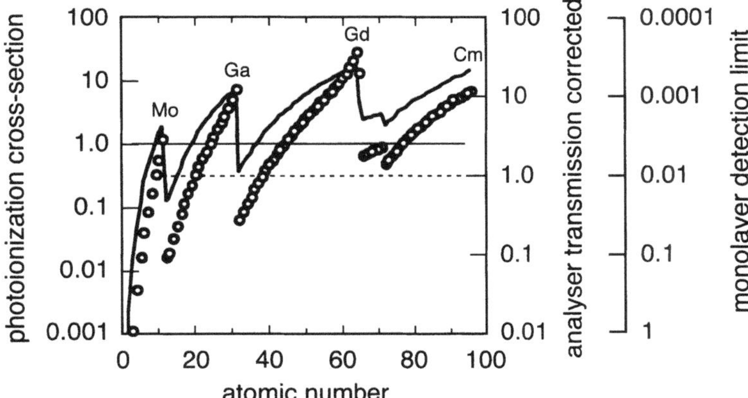

Fig. 1.11. Elemental sensitivities relative to F 1*s* across most of the periodic table for X-ray photoelectron spectroscopy (XPS). Values are based upon theoretical photoionization cross-sections with Mg K_α radiation for the most intense levels (–). The second data set (o) illustrates the influence of electron spectrometer transmission on the relative sensitivities. The transmission is for a concentric hemispherical analyser run at constant analyser energy (constant absolute resolution). [Compiled from J.H. Scofield, J. Electron Spec., Rel. Phen. **8**, 129 (1976); and A.E. Hughes and C.C. Phillips, Surf. Interface Anal. **4**, 220 (1982)]

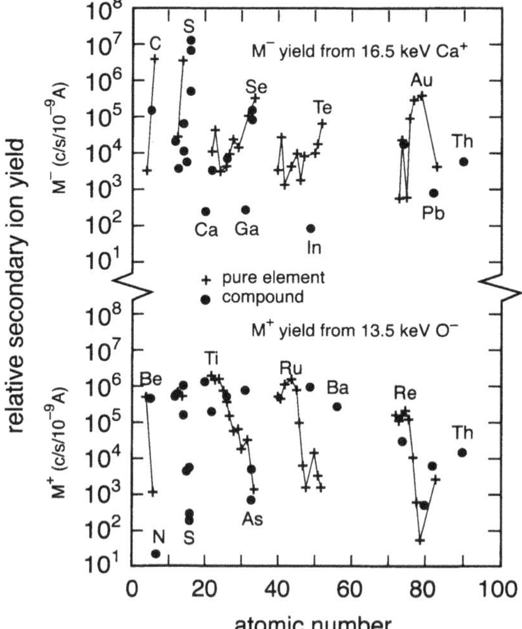

Fig. 1.12. Elemental sensitivities relative to F across most of the periodic table for secondary ion mass spectrometry (SIMS). Note that as these are for ion production they differ from the simple sputter yields shown in Fig. 1.13. [Compiled from H.A. Storms, K.F. Brown, and J.D. Stein, Anal. Chem. **49**, 2023 (1977)]

sensitivities have given SIMS a reputation of being nonquantitative. However, quantification can be achieved by the preparation of suitable standards, e.g. by moderate dose ion implantation into the correct matrix [42].

The problem of overlooking an elemental component remains even with a relatively easily-quantified technique such as XPS because elements with low photoionization cross-sections could easily be missed. For example, a 1 mole % content of Al or Si in a metal oxide matrix is detectable. However, the O 1s signal is $>$ 4 times more sensitive than either the Si 2p or Al 2p signal. With a signal-to-noise ratio S/N (i.e. ratio of O 1s peak height above background to peak-to-peak noise) of 100, a normal survey scan will indicate no Al or Si to be present. Indeed the peak heights will be roughly 1/4 the size of the noise. To be confident of confirming the presence of the Al or Si would mean improving S/N 12-fold. Statistically, for particle counting, the S/N improves as the square root of the number of counts. Hence the same survey scan would take 144 times longer to acquire, converting a 10 minute exercise into a 24 hour marathon (assuming that specimen contamination and damage did not occur). As this is impractical the best approach is to be aware of the elements of low sensitivity for the given technique and the likely level at which they will be detected. If it is thought that they might be present in the sample, then narrow width scans at the appropriate locations can always be acquired at the necessary S/N. Note that with XPS neither surface-associated H nor He can be detected in practice due to low cross-sections and likely overlap with substrate valence level features. For Auger spectroscopy neither H nor He can emit a characteristic electron due to an insufficient number of electrons.

1.5.3 Practical Detection Limits

Analytical detection limits are typically quoted as percent, ppm or ppb with no due care as to whether the units are in terms of moles or by mass. In surface analysis this is further complicated since it is often not stated whether only the outer layer is being referred to or indeed the detectable selvedge. A true inter-technique comparison is difficult to assess due to the wide variety of the probes and the wide variations in energy densities that can be applied. In Fig. 1.11 an approximate detection limit (under "typical" conditions) in terms of a monolayer fraction of the outermost surface layer is indicated for XPS. Detectability assumes a S/N of at least 2. Detection limits for the outermost surface layer and the selvedge probed are quite different as the quantity of material in the latter may be 10-fold that of the former. The stated limits are really only a guide and with due care a factor of 10 improvement ought to be possible in most circumstances. SIMS has a particularly low detection limit, comparable with that achieved in thermal desorption spectrometry (TDS) in which a mass spectrometer is also used for detection, but the reliance on thermal excitation greatly restricts the variety of molecules which will desorb.

As noted in the prior section, a comparison of SIMS "practical" detection limits has limited meaning at best. Some quoted limits can vary from 4×10^{12} atoms cm^{-3} (Mg in Si with O_2 source) to 3×10^{16} atoms cm^{-3} (C in Si with Cs source) [43]. Based on an atomic density in Si of 4.99×10^{22} atoms cm^{-3}, these limits translate to 0.1 ppb to 1 ppm (mole fractions). On average the density of atoms in a Si monolayer is 1.36×10^{15} atoms cm^{-2}. Since the optimum detection limits are based on multilayer analysis, converting from, say, a 50 Å selvedge to a single monolayer, then a feasible detection range would be $\sim 2.5 \times 10^{6}$ to $\sim 2.5 \times 10^{10}$ atoms cm^{-2} i.e. at the very least $< 10^{-4}$ of a monolayer. As pointed out by *Magee* [41], Rutherford backscattering spectroscopy (RBS) and nuclear reaction analysis (NRA – see Sect. 1.9 and Chap. 9) constitute with SIMS a complementary trio. Although detection sensitivities of RBS and NRA are inferior to SIMS e.g. RBS under favourable conditions can at best detect 10^{-4} of a monolayer of heavy atoms on silicon and NRA 3×10^{-2} of a monolayer of H atoms on silicon [41], they can provide superior quantitative capabilities. This is especially true of RBS, e.g. analysis of anodized films on aluminium [44] indicated the inner region to be composed of Al_2O_3 with an outer region of aluminium-deficient alumina incorporating electrolyte anions. The fraction of film thickness with this incorporation typically varied from 0.2 with molybdate at 0.61 ± 0.03

atom % to 0.7 with phosphate at 5.53 ± 0.24 atom %. Moreover RBS could determine the anion concentrations (as atomic % of the characteristic anions and Al) to an accuracy of a few percent.

The total quantity of material being probed can vary from a single atom as in STM STM to up to 10^{16} atoms in a broad area spectroscopy such as XPS. In order to detect a single element, XPS would need the presence of up to 10^{13}–10^{14} atoms so that STM would win in the adjusted sensitivity competition. The advantage in XPS is the wealth of chemical information provided. By comparison a destructive bulk analytical method such as atomic absorption spectroscopy may need as few as 10^{10} atoms for detection [45]. Despite their many advantages neither XPS nor AES display sufficiently high sensitivities for use in areas such as semiconductor doping. Here the highest impurity levels which affect performance can be below their detection limits. SIMS is generally used for this reason. In an alternative area, XPS has been applied to trace analysis by *Hercules* et al. [46]. Utilizing chelating glass surfaces it was found that heavy metals in solution could be easily analyzed down to 10 ppb, without any optimization of the method.

1.5.4 Practical Spatial Limits

The limit of spatial resolution for the selvedge volume element $\Delta x \Delta y \Delta z$ is invariably controlled by the lateral dimensions Δx and Δy. For many of the techniques considered in this book the analysed depth is a few atomic layers, which, with the exception of STM, is much less than the possible lateral dimensions, which can be up to mm. Improvements in lateral spatial resolution can be approached in two ways: either by controlling the dimensions of the incident probe beam or by adjusting the input optics of the analyser so that it views only a selected region of the irradiated surface. The first approach is the preferred method as it maintains high incident fluxes and consequently good S/N characteristics, although this can often be at the expense of considerable radiation damage. Generally this approach is only feasible for charged particle beams. The second method of selected area analysis is utilized when the incident probe cannot be easily focussed. In this case, generally poor S/N has to be accepted and, unless non-selected areas are protected, the whole surface is radiation exposed even if it is not contributing to the useful signal.

Using conventional thermionic emitters and simple electrostatic optics, electron beams can be relatively easily focussed down to 1000 Å spots with currents of up to 2 nA at 10 keV. Such electron guns are commerically available [42]. Importantly however, even at that resolution the current density involved is 6.4 A cm^{-2}. This equates to 4×10^4 electrons per surface site per second. This is a substantial radiation level considering the energy of the beam. Placing it into perspective the energy density is 6.4×10^4 W cm^{-2} or 4.7×10^5 times the solar radiation constant [47]. Finer electron beams down to 200 Å have been achieved using high brightness LaB_6 sources or field emission tips in combination with magnetic optics. The temptation to use beam energies higher than 30 keV for more precise focussing provides diminishing returns because of the reduced cross-sections for Auger electron production, especially for the lighter elements. Most electron excited Auger work at high spatial resolution is carried out in the 5–10 keV range. As noted in Chap. 6, the lateral resolution limit in SAM has not been reached and 30 Å, i.e. the IMFP limit, is not considered impossible. Currently, in terms of practical surface analysis in materials science, the "limit" is about 500 Å, see e.g. [48].

As with electrons, incident ion beams can also be focussed into fine beams. The popular technique of static SIMS has a scanning variation akin to SAM's relationship to conventional AES. A variety of means exist for the production of ion beams, but for microfocussing the most effective is the liquid metal field emission ion source. A liquid metal film, such as Ga, wets a solid needle and is drawn out into a fine tip at the needle's end. By the application of a high field, positive metal ions are extracted by field emission [49]. Such beams can be focussed down to 500 Å.

X-rays, in contrast to charged particles, are not easily focussed into small spots. A Fresnel phase plate lens has been constructed which can focus Al K_α radiation with 4000 Å spatial resolution [50]. However, this is yet to be applied in surface analysis. Less successful methods of localizing X-rays, such as the use of crystal optics to focus the beam down to 0.15 mm or the proximity method (generating the X-rays at a fine point, adjacent to a thin film to be analysed) down to 20 μm, have found some use to date [51]. Other than X-ray localization, several methods do exist to achieve selected area XPS (SAXPS), or the more general effect of photoelectron microscopy (PEM). These are considered by *Drummond* et al. [51]. The approach which is easiest to implement, because of its compatability with existing spectrometers, is the use of a transfer lens system to magnify the surface area projected onto the spectrometer entrance aperture [52]. Selection of a small entrance aperture then means the spectrometer is only viewing part of the surface. This can be further restricted by an additional aperture prior to the transfer lens. Selected areas down to 150 μm in diameter are available commercially [53]. Whilst relatively easy to implement, this approach does suffer from reduced sensitivity with the photoelectron count rate decreasing in proportion to the area being sampled. In practice this usually necessitates operating at reduced resolution and scan widths in order to obtain information in a viable time interval. By an extension of the detection electron optics it is feasible to produce an energy filtered photoelectron image of the surface with a resolution of 10 μm [54]. Note that, unlike SAM or imaging SIMS, the input beam is not rastered over the surface, and only a single selected area is analysed at a given time. A comparison of the possible different methods of selected area surface compositional analysis is given in Table 1.7.

Although in principle the depth, Δz, offers the highest spatial resolution (i.e. layer-by-layer) of the volume element dimensions, the achievable reso-

Table 1.7. Comparison of the possible different methods of selected area surface compositional analysis

Method	Information available	Spatial resolution	Advantages and disadvantages
Selected area XPS (SAXPS)	Surface elemental composition and chemical state analysis	20–150 µm depending upon method selected (see text)	Chemical shift information obtainable with a minimum of radiation damage. Physical movement of specimen may be required to obtain an image. Generally suffers from inferior signal to noise.
X-ray photoelectron microscope	Surface elemental composition and chemical state analysis	10 µm	Chemical shift information obtainable with a minimum of radiation damage. Image obtained by control of electron optics. Signal to noise superior to SAXPS.
Scanning Auger microscopy (SAM)	Surface elemental composition and some chemical state analysis via "fingerprinting"	30–500 nm	Chemical state information is difficult to obtain although phase identification is possible from a multielement correlation diagram analysis.
Secondary ion imaging mass spectrometry (SIIMS)	Surface elemental composition and isotope distributions	50 nm	Excellent elemental sensitivity. Based on fragments, some chemical state information is possible although the interpretation can be difficult.

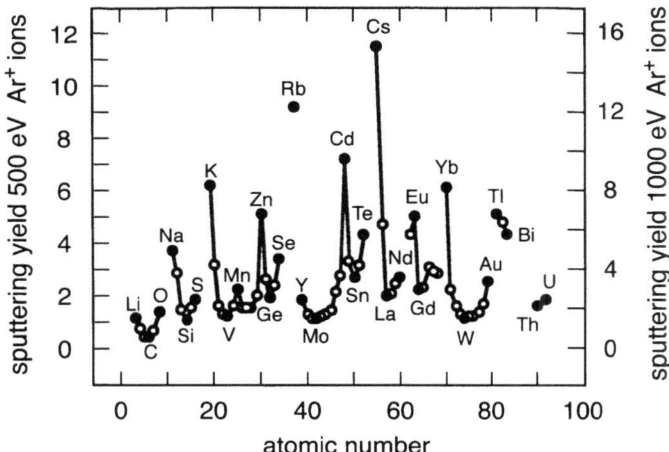

Fig. 1.13. Elemental sputtering yield across most of the periodic table calculated for Ar ions of energies 500 and 1000 eV [41]

lution with depth will depend upon the profiling approach. The commonly-employed method of sputter profiling can be fraught with pitfalls as outlined in Chap. 4. Of most significance is the variation of sputtering yields e.g. Fig. 1.13, leading to preferential sputtering which can dramatically alter relative selvedge compositions. Two alternative approaches exist to extract compositional information with depth. The first, principally applicable to the electron spectroscopies, simply relies upon altering the surface sensitivity by rotating the specimen such that the analyser (of narrow acceptance angle) receives electrons from increasing angles θ off-normal. As the IMFP does not alter, the analysed depth alters by cos θ and the technique thus becomes more surface sensitive at high θ. Angle-resolved XPS is explained in more detail in Chaps. 7 and 16. A typical example of this is illustrated in Fig. 1.14. It is the marked alteration in surface sensitivity which enables the detection of the variety of altered (oxidised) surface groups on plasma-treated polystyrene.

The second approach, which is especially useful for deeper interfaces, involves a mechanical lapping through the selvedge layer at some angle to the original surface i.e. bevelling. Analysis is then achieved by scanning across the lapped face, either mechanically or by beam raster. Depth resolution is controlled by the angle of the lap and the lateral resolution of the probe. *Tarng* and *Fischer* [56] point out that an ultimate depth resolution of $\sim$ 35 Å would be possible with AES using a 0.2° lap and a beam and manipulator of 0.5 µm resolution. A variation on straight lapping is ball cratering [57], in which a shallow spherical pit is ground into the surface and the spatially resolved probe is moved across the crater and thus to greater depths (Fig. 1.15). Craters are more rapidly produced than an angled lap, but both methods have a drawback in that the surfaces still have to be ion etched prior

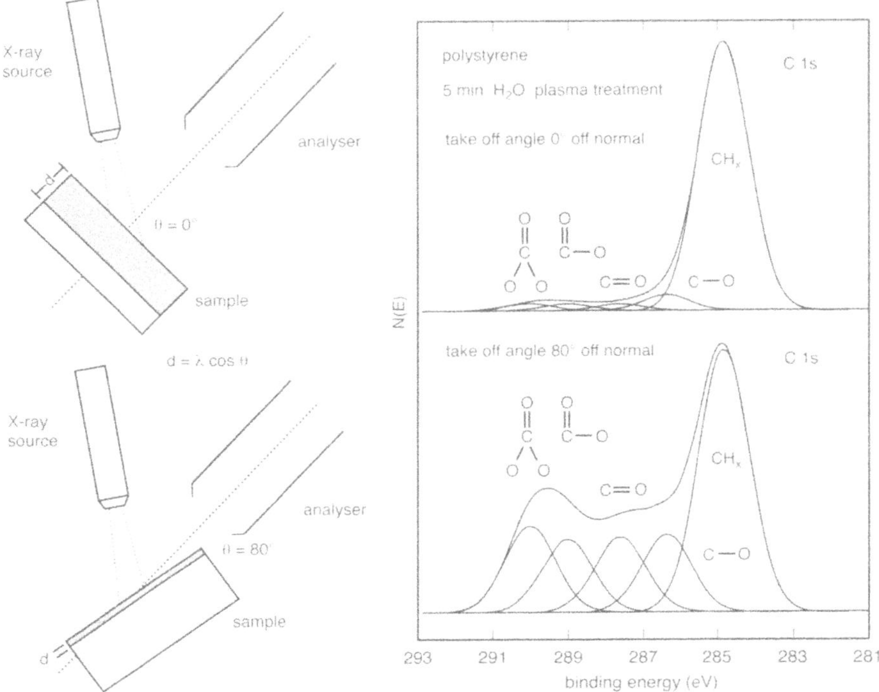

Fig. 1.14. Method of enhancing the surface sensitivity of the electron spectroscopies by increasing the take-off angle at which the analyser views the surface. Here the surface functionalities created by plasma treatment of a polystyrene surface are more readily detected. Adapted from *Evans* et al. [55]

to analysis due to contamination from the mechanical grinding. The latter can also smear out phase constituents if excessive pressures are applied.

One technique in which a relatively deep selvedge is probed is Fourier transform infrared spectroscopy (see Chap. 8). Absorption of the infrared radiation by the characteristic vibrations of the surface species usually provides direct evidence for their structure and bonding. Both FTIR and its electron analogue, electron energy loss spectroscopy (EELS), can provide very detailed structural information on atomic, molecular and multi-atomic ionic species. Surface sensitivity in FTIR is usually achieved using high specific surface area samples, diffuse, specular and attenuated total reflectance, or reflection-absorption methods, but the analysed depth is still roughly 1/10 to 1/4 of a wavelength (i.e. 0.1-5 μm) depending on the material, i.e. relatively deep compared with other surface techniques discussed here. Nevertheless, microprocessor-controlled signal averaging has allowed good spectra of submonolayer films to be recorded. FTIR has been extensively applied to studies of catalytic reactions, polymer and glass surfaces [58], mineral processing [59],

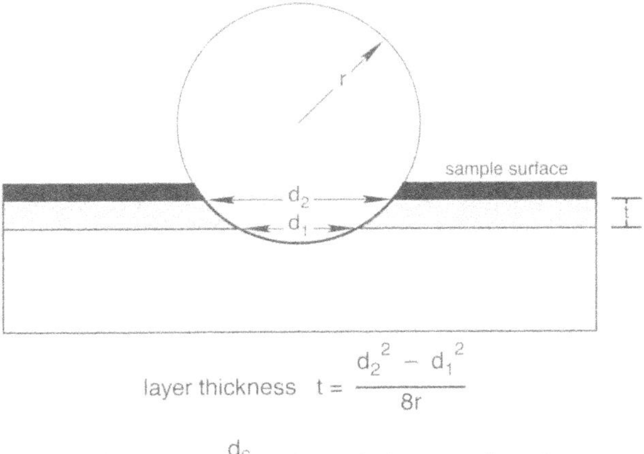

Fig. 1.15. Schematic cross-section of a ball cratered surface selvedge. Rastering the incident probe across the crater enables a profile analysis over a considerable depth

modification of oxide surfaces [60] and studies of most natural materials. It is an easy technique to use requiring little sample preparation and can be applied to any material in-situ i.e. in media other than gas or vacuum. It is also readily accessible in most laboratories.

Of course the extent to which a given technique probes the selvedge must be borne in mind when comparing results from the various techniques. For example the initial stages of 3-dimensional oxidation of Al(111) were examined by *Ocal* et al. [16] utilizing both ISS and XPS. With two slightly different oxidation schemes the outer layer of the oxide could be produced to be stoichiometric or reduced as determined by ISS. By contrast, XPS (probing well below the outer oxide layer, e.g. Tables 1.3 and 1.4) could not distinguish between the oxides, as the reduced outer layer only made a small contribution to the signal.

1.5.5 Chemical State Information

As mentioned at the beginning of this chapter, simply determining the elements present in the surface selvedge or even evaluating their mole proportions, does not guarantee phase identification. In any broad area probe the likelihood of multiple phases being present will confuse the result obtained. Stoichiometric errors aside, the "phase" will merely be an average of those actually present. The situation is improved at higher spatial resolutions when the physical dimensions of the phase become larger than the probe beam or the selected area. This is dramatically illustrated with the use of ratioed scatter diagrams in SAM ([49], see also Chap. 6) where accurate phase assessments and their distribution can be evaluated on the basis of stoichiometry.

Chemical state information greatly simplifies the identification of the phase or phases present, and is essential with multiple phases if a broad area is being analysed. Chemical state information in the technique response arises either from the indirect influence of chemical bonding on the core levels of the atoms concerned or from its more direct influence on the valence or molecular levels of those atoms or molecules. The photoelectron spectroscopies probe the levels directly, in the case of XPS the core or near-valence core levels, and in UPS UPS the occupied valence states. UPS (Chap. 14), uses excitation sources in the UV to produce photoionisation and is more sensitive to changes in valence band structure. For the core levels in XPS, chemical state interpretation is very easy to a first approximation. Core electrons ejected from an increasingly oxidised atom (i.e. a more positive atom) have a lower kinetic energy (and hence a higher "initial state" binding energy). For instance, the binding energy of the S $2p$ alters from sulphide (S^{2-}) through elemental S^0 to sulphate (S^{6+}) from 161 eV to 169 eV. Numerous examples of this chemical shift illustrate the original publication of *Siegbahn* et al. [61], which justly served to promote the value of electron spectroscopy for chemical analysis (ESCA). Although the emphasis throughout this book is on the current practice of surface analysis in materials science, Seigbahn's early examples eloquently argue the power of XPS. Reproduced in Fig. 1.16 are the C $1s$ lines from the sodium salts of the first four fatty acids. The carboxylic carbon attached to electronegative oxygen atoms appears at higher binding energy, well separated from the alkane carbons. Note the expected carbon atom ratios for each of the acids. Of course, it is immediately evident that a surface upon which fatty acids have adsorbed may not be amenable to a simple XPS analysis if several of the different acids are coexisting, thereby causing considerable C $1s$ peak overlap from different surface species.

The chemical information obtained from UPS can be less obvious. Although the occupied valence levels are probed directly, the valence region of the spectrum is narrow in energy terms, and the number of molecular levels increases with the complexity of the molecular unit. The final state of the UPS photoelectron is also perturbed by the unoccupied valence states. Except in special circumstances e.g. weakly adsorbed molecular oxygen on W(110) at 26 K [62], the vibrational level structure is completely smeared out for the selvedge constituents and a broadened multipeak spectrum is usually the result. In specific circumstances e.g. occupied valence band density of states measurements or simple molecular processes on well defined surfaces, UPS can be extremely valuable, but it is best utilized as a complementary technique. One unusual application of UPS has been to use adsorbed Xe atoms as an atomic size probe of the local surface work functions by following the $5p_{3/2,1/2}$ BEs. This in turn enables the evaluation of the microfacets that

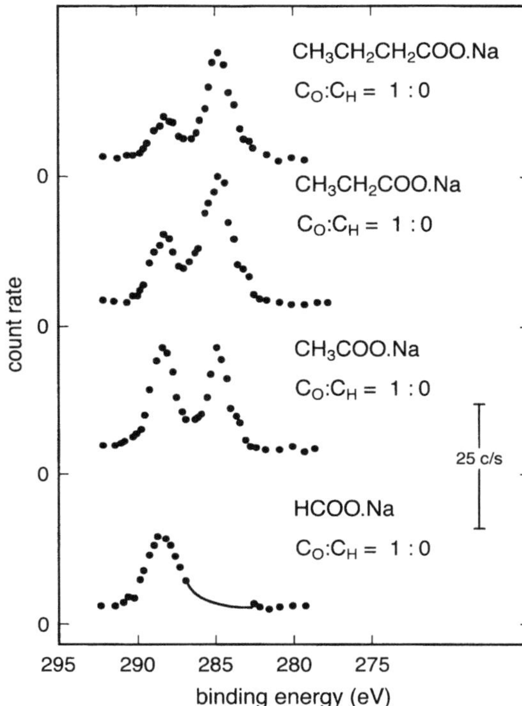

Fig. 1.16. C 1*s* photoelectron peaks for the sodium salts of the first four fatty acids. Adapted from the original work on ESCA by *Siegbahn* et al. [61]. The oxygen-attached carboxyl carbon appears at the higher binding energy

exist on polycrystalline metal surfaces [63]. UPS UPS is considered further in Chap. 14.

As might be expected the Auger electron also contains some chemical information. But since it results from a three-electron process (Chap. 6) that information is subtly convoluted into the spectral feature, which, in the differential mode of data acquisition (still commonly employed with AES), can nonetheless lead to striking differences in peak shapes and energies. Even if the reasons for the differences are unclear, they can be usefully employed on a "fingerprint" basis for the comparison of known and unknown phases. As illustrated by *Madden* [64], for even a "simple" process such as oxidation, the core-core-valence and core-valence-valence Auger peaks can display energy shifts from -15 to $+10$ eV between the metal and its oxidised surface. (By comparison, in XPS, the shift would be a kinetic energy decrease of a few eV). An example of Auger chemical information is given in Fig. 1.17 (from [64]). The Si $L_{23}VV$ peak is seen to shift to lower kinetic energies with increasing electronegativity difference in the series Si-X. Noteworthy is the influence of H, as it is itself invisible to Auger spectroscopy.

SIMS does not represent a simple case with respect to the extraction of chemical information (see Chap. 5). In order for useful chemical information to be obtained, molecular fragments need to be collected, with those

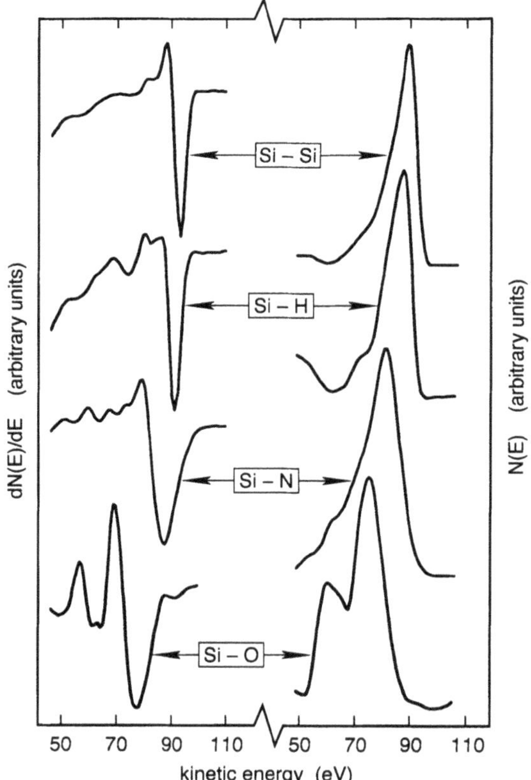

Fig. 1.17. Si $L_{23}VV$ Auger peaks [in differential $dN(E)/dE$ and integral $N(E)$ modes] from elemental silicon (Si-Si), chemisorbed hydrogen on Si(100) (Si-H), silicon nitride film (Si-N) and silicon dioxide film (Si-O) [64]

fragments representing the character of the selvedge i.e. without extensive molecular rearrangements being prompted by the act of sputtering. A variant of static SIMS which is capable of this is fast atom bombardment mass spectrometry (FABMS). This approach reduces the charging effects, ionic migration and radiation damage normally associated with ion bombardment, especially of insulators. FABMS was originally developed as a structural diagnostic for high molecular weight, nominally involatile, compounds [65].

Fourier transform infrared (FTIR) is quite different from the other analytical techniques in that it does not detect individual atoms, but rather bond vibrations due to the resonance absorption of infrared radiation at frequencies specific to the characteristic vibrations. In ascertaining chemical states at interfaces it can thus be a very powerful probe. Infrared techniques have been applied to surface studies for several decades and predate the electron and ion spectroscopies in extensive use. The generally poor signal strengths of the past have now been largely overcome with the improved sensitivity of FTIR over earlier dispersive instruments. In the attenuated total reflectance (ATR) mode it is particularly valuable for *in situ* solution studies of the solid-liquid interface. (This, in particular, will be considered in further detail in Chap. 8). This contrasts with most modern surface analytical methods where the studies are generally carried out *ex situ*.

1.5.6 Laboratory Standards

In spite of the present state of technological advancement of surface analysis, with system costs ranging up to millions of dollars, there is no guarantee of accuracy or precision. Problems still remain in the comparison of results between different laboratories and hence all data needs to be critically assessed. This is even more important for the non-surface specialist commissioning work performed elsewhere on a contract basis. Reliability of literature data is limited so that laboratory independence is still required and laboratory standards are indispensable. This is not to say that the wealth of literature data on photoelectron binding energies and cross-sections or modified Auger parameters etc., is inaccurate, but that, in practice, other factors can modify spectra. If only for reasons of different referencing methods, experimental set-ups and spectrometer transmission functions, there is no substitute for examining comparative standards on the same instrument and under the same conditions that the problem specimen is being examined. This is especially true for quantitative analyses. Whilst this can considerably lengthen the time taken to achieve a result, the correct interpretation is more likely to be made.

A fundamental problem in the choice of a standard is the selection of a material with known bulk composition which translates right through the surface selvedge. As pointed out already, very often the phase composition of the selvedge bears little resemblance to the bulk. The more surface sensitive the technique the more severe this problem. Those materials with greatest thermodynamic and kinetic stability, exhibiting the least tendency toward any environmental degradation are usually the most reliable standards and also the easiest to handle. If the material has a known susceptibility toward oxidation or hydrolysis, then in-situ fracture or glove box preparation is warranted. As a general rule standard materials are best obtained in the most pure state. However, lower qualities are quite acceptable if the impurities are evenly distributed throughout the bulk e.g. a 99% pure compound in this category would be far more acceptable than a 99.999% sample in which the remaining 0.001% was concentrated entirely in the surface.

1.5.7 Inter-laboratory Errors

The ASTM Committee E-42 on surface analysis has a variety of subcommittees covering AES, XPS, ISS, SIMS, ion beam sputtering, standard reference materials and standard reference data. Joint round-robin investigations sponsored by E-42 have been conducted for XPS [66] and AES [67] to compare interlaboratory performance. The XPS round robin compared results from Ni, Cu and Au foils on 38 different instruments manufactured by 8 companies and found a spread in reported binding energy (BE) values typically greater than 2 eV, whereas the individual imprecisions in measurements were believed to be less than or equal to 0.1 eV. Reported intensity ratios of photoelectron peaks for Cu and Au spectra (for clean surfaces) spread over a factor of ten, although reported imprecision was less than 10%. The subsequent AES round robin compared results from Cu and Au foils on 28 different instruments manufactured by 4 companies. Here the spread in reported kinetic energy (KE) as a function of kinetic energy varied from 7 eV at a KE of 60 eV, to 32 eV at a KE of 2 keV, with an imprecision range of $\sim 1-3$ eV. For an incident energy of 3 keV, the reported intensity ratios varied with a factor of $\sim$ 38 for Cu to one of $\sim$ 120 for Au, whilst again the individual imprecision was less than 10%.

At least part of the observed spreads could be ascribed to erratic instrumental performance, while in other cases it was believed that operator error was responsible. Both round-robins clearly demonstrated the need for improved calibration methods and operating procedures if these surface techniques are to produce accurate results on a routine basis. In the interim the only reliable approach is to utilize internal laboratory standards to aid the interpretation of the work performed in that laboratory.

1.6 Defect and Reaction Sites at Surfaces

The reactivity of a surface is nearly always increased if the concentration of surface defects is increased. In most cases, point defects, as in cation, anion (or atomic) vacancies and "kink" sites, provide sites with altered electronic structures permitting enhanced electron-transfer reactions not possible on the perfect surface. Extensive theoretical studies have successfully modelled the energies of vacancy and kink sites [11,14,15,68,69]. For instance, on ionic oxide surfaces with rock salt structure (e.g. MgO, NiO), when an oxide ion is missing from a (100) surface, the screening of the cations from each other by the large, polarizable oxide ions is much reduced. This produces changes in the electron energy levels of the surrounding, now 4-coordinate, cations and anions of the top layer such that molecular hydrogen can be dissociated at this site where it would not on a perfect 5-coordinate surface. For kink sites at steps and edges, coordination is reduced to four and, at corner sites, to three. The model also predicts substantial ionic displacements, tending to "smooth" the electron density contour across the step, as well as significantly altered energy levels for reaction.

Examples of changes in concentrations of such sites are myriad in materials science. Figure 1.18 presents TEM images using phase contrast from NiO

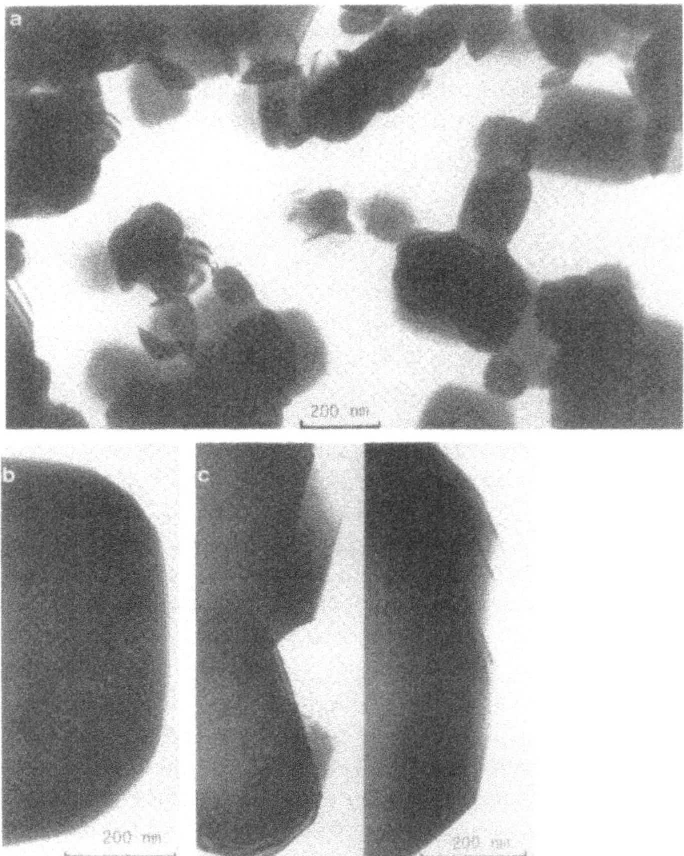

Fig. 1.18. Transmission electron micrographs of NiO crystals annealed in air for 4 *h* at (**a**) 700°C, (**b**) 1100°C, and (**c**) 1450°C (2 regions). The transition from defective, equiaxed crystals with rounded, relatively smooth surfaces to relatively perfect crystals with basically flat steps and ledges can be seen. [Reprinted with permission from Figs. 1b, c, C.F. Jones, R.L. Segall, R.St.C. Smart, and P.S. Turner,. J. Chem. Soc. Faraday Trans. I **73**, 1710 (1977)]

subjected to increasing annealing temperatures from 700°C to 1450°C. The initially rounded 700°C-formed particles exhibit relatively high concentrations of surface (and bulk) defects "frozen" into the structure during decomposition. The surface sites are disordered with variable coordination and high reactivity (e.g. in acidic dissolution kinetics). The 1450°C-formed oxide has grown quite extensive flat (100) faces with well-developed steps and ledges. The majority of the surface ions are in 5-fold coordination and the reactivity is accordingly reduced roughly ten-fold. These vacancy and kink sites in NiO surfaces can be directly correlated with the concentration of sites with Ni^{3+}

and O^- electronic structures derived from studies, a result in agreement with the theoretical model predictions [70].

Hence, three of the techniques most suited to studying defect sites to define their atomic positions and their electronic structure are high resolution, phase contrast TEM with XPS and/or UPS. TEM has been discussed in Sect. 1.4 above and XPS in Sect. 1.5. UPS has been used [71] to show that, although nearly perfect (110) TiO_2 surfaces do not dissociate H_2O on adsorption, oxygen vacancies interact strongly with H_2O dissociating the molecule to produce OH^- species [72]. Even more subtly, UPS shows that H_2O does not dissociate on the (100) NiO face and only dissociates to a very limited degree on the defective surface. Prior adsorption of O_2 however, giving O atoms at the defect sites, results in the immediate appearance of OH^- ions after H_2O admission. The role of oxygen in promoting H_2O dissociation through defect interaction is of major importance in catalysis [73] and energy conversions [68].

Two other techniques, already discussed in Sect. 1.4, are also of great value in defining defect sites on surfaces. The STM can image vacancies, low-coordinate kink sites at steps and ledges, and atomic displacements resulting from these defect structures. LEED has been used to study low-coordination defect sites through an approach based on stepped, high Miller index surfaces. The high index surfaces are formed on stable, low index surfaces, by atoms breaking away prior to desorption or migrating prior to condensation. In either case, kink sites result with different binding energies and reactivity [10,74]. In LEED patterns, multiple diffraction spots characterise the formation of steps with different orientations, and hence, high Miller indices as in Fig. 1.19. On the ZnO (0001) face, annealing at 900 K induced regular step arrays oriented in a single direction as evidenced by additional spots in the pattern which changed with changing electron energy [75]. The step height could be determined to be a single lattice unit (or two Zn-O layers). On the polar $(000\bar{1})$ face, step arrays oriented in three different directions were found and the streaky, diffuse form of the additional features in the pattern showed that the arrays were distinctly irregular [76]. The concentration of defect kink sites can be estimated from such observations.

FEM and FIM can, as with physical imaging, image defects and high Miller index planes on surfaces. However, the information is confined to the limited class of materials that can withstand the strong electric field (i.e. $\sim 10^{10}\,\mathrm{Vm}^{-1}$) at the sample tip without desorption, evaporation or destruction of the surface. The two techniques have been immensely useful in studying oxide structures and particularly stable atomic defect positions on refractory metals like W, Ta, Ir, Re (e.g. [77]) and some oxides like ZnO [78]. The work functions of individual planes can be determined as well as field migration of defects.

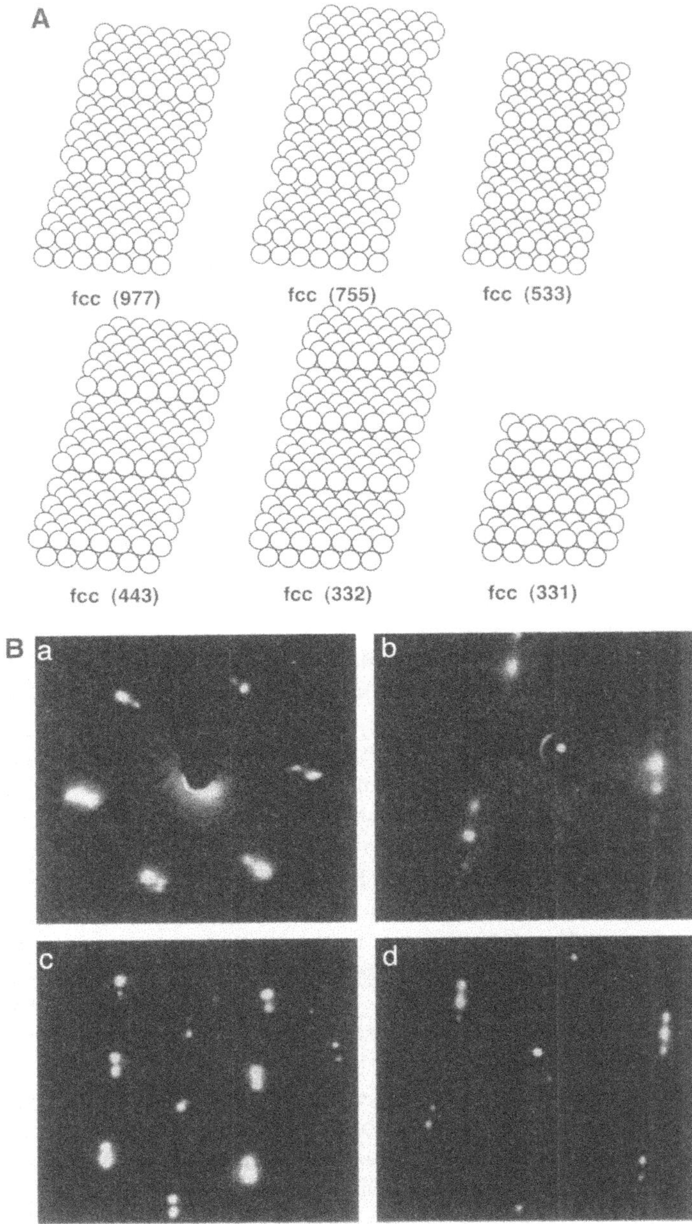

Fig. 1.19. (**A**) Structure of several high-index stepped surfaces with different terrace widths and step orientations. (**B**) LEED patterns of the (a) Pt(755), (b) Pt(679), (c) Pt(544), and (d) Pt(533) stepped surfaces. [Reprinted from Gabor A. Somorjai: Chemistry in Two Dimensions: Surfaces. Copyright (c) 1981 by Cornell University. Used by permission of the publisher, Cornell University Press]

1.7 Electronic Structure at Surfaces

The electronic structure of surfaces embraces a variety of concepts such as:

- the electronic barrier at the interface of a metal or semiconductor with another phase (vacuum, gas, liquid or solid);
- the space-charge (or diffuse) layer in semiconductor materials (including oxides, minerals and some ceramics);
- the electronic states (e.g. electron densities, valencies, bonding etc.) of surface atoms;
- defect sites (or states), active in adsorption and reaction, arising from the geometrical structure of the surface (e.g. faceting, vacancies, dislocations etc.);
- impurity segregation into or out of surfaces and consequent reactive surface sites;
- surface conduction and surface charging consequent upon any or all of the above considerations.

We have already described techniques to study the electronic structure of defect sites in the previous section. Information on the electronic (chemical) states of surface atoms, largely obtained from XPS, UPS, EELS and AES, has been discussed in Sect. 1.5.4 above. Impurity segregation and doped reactive sites will be covered in Sect. 1.8 below. Hence, we will limit ourselves here to techniques relevant to studies of the electronic surface barrier, space-charge layers and surface conduction or charging.

Schematically, the potential energy at the surface of a metal and a semiconductor in vacuum can be represented as in Fig. 1.20a and b, respectively showing the work function ϕ or potential that an electron at the Fermi level (average energy of the last electron added) must exceed in order to achieve complete ionisation with zero kinetic energy in the vacuum. The magnitude of this work function is related to interaction of the electron with the charge distribution at the surface (i.e. electrostatic, exchange and correlation energies) and changes as the surface charge is redistributed. Redistribution of this kind can be caused by changes in surface geometry (e.g. faceting, reconstruction, relaxation defects) surface composition and adsorption. In practice, measurements of $\Delta\phi$, the change in work function, are of most value. Dynamic or vibrating condenser methods, as in the Kelvin probe, are normally used to determine $\Delta\phi$ [79]. Uncertainties in value of the reference levels in the $\Delta\phi$ term, and relatively complicated charge transfer processes at surfaces have limited interpretation [10] and application [79]. Nevertheless, the method has been used to monitor adsorption in simple gas/solid systems where the solid is either a metal or semiconductor, oxidation of metals, oxygen interactions with oxides and ceramics, defect equilibria, phase transitions, surface segregation and catalysis.

LEED has been extensively applied to studies of the shape, extent and magnitude of the surface barrier, i.e. to the definition of the barrier itself,

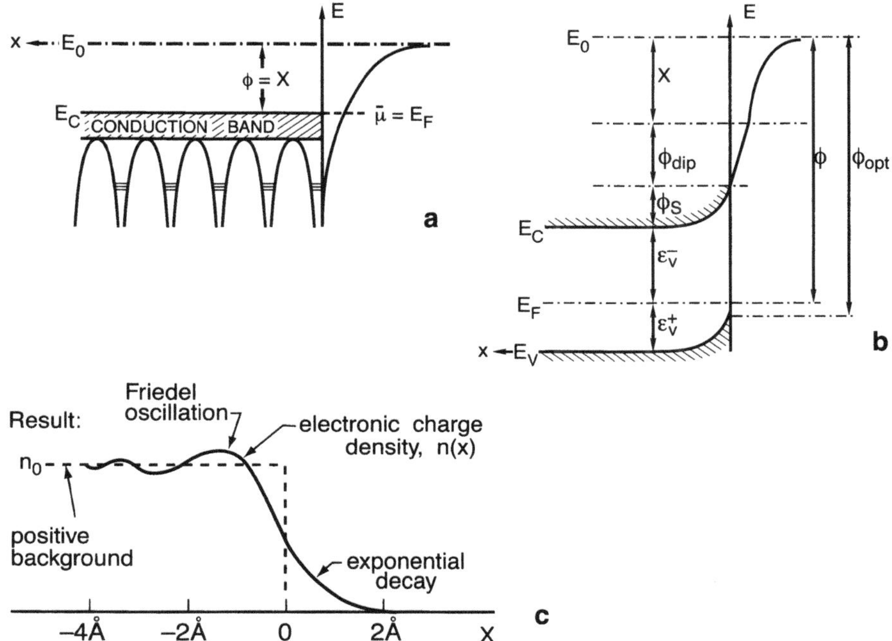

Fig. 1.20. (**a**) Schematic illustration of the band model of metals (without electric fields). The work function ϕ is defined relative to E_0. (**b**) The band model of a p-type semiconductor surface involving both the space charge (ϕ_s) and dipole moment (ϕ_{dip}) components. [Reprinted with permission from Figs. 7.1, 7.3, J. Nowotny and M. Sloma in "Surface and Near-Surface Chemistry of Oxide Materials", ed. by L.-C. Dufour and J. Nowotny, Ch. 7 (Elsevier, Amsterdam 1988) p. 281]. (**c**) Charge density oscillations and redistribution at a metal-vacuum interface. [Reprinted from Gabor A. Somorjai Chemistry in Two Dimensions: Surfaces. Copyright (c) 1981 by Cornell University. Used with permission of the publisher, Cornell University Press]

using the fine structure of intensities of the diffracted beams as a function of beam energy and direction (Chap. 13). It has elucidated barrier structures on metals and semiconductors in vacuum. The major limitation is that it can only be applied to extensive, single crystal surfaces in UHV conditions. It cannot be easily applied to granular (e.g. ceramics), multiphase or powdered material.

Another way of describing the surface barrier is associated with the concept of surface dipoles. The surface asymmetry (i.e. lattice termination) produces a redistribution of charge beyond the surface itself, into the vacuum (or other media) inducing a surface dipole which, in turn, modifies the barrier potential. This induced dipole varies in magnitude at different sites on the surface. On ionic surfaces, this variation is extreme, giving a net static charge separation. On the surfaces of semiconductors and insulators, separation of positive and negative charges results in a space-charge region and this partly controls the transport of charge (as electrons, holes, ions) and atomic species across the surface. These species, and adsorbed species, experience different dipoles (and hence, polarisation) as they move from site to site leading to different tendencies to bond formation. Techniques for monitoring these space-charge effects, conduction and charging, must contribute considerably to our understanding of the surface of any material. Descriptions of the older techniques, including work function measurements, can be found in references [80,81] but will not be explained in detail here.

The technique of STM allows an associated spectroscopy, scanning tunnelling spectroscopy (STS), in which the voltage dependence of the tunnelling current is used (Chap. 10). The densities of electronic states, surface band structures and band bending in the space charge region can be obtained from this dependence. Local work functions (or barrier heights ϕ), with atomic resolution, can be measured using the dependence of tunnelling current (I_t) on tip distance (d) from the surface through the (approximate) relationship $I_t \propto \exp(-\phi^{0.5}d)$. The STM/STS technique promises, with further development, images of the spatial distribution of both local work functions and electronic states including space-charge effects.

The high-field analogue of STS, field emission spectroscopy (FES), is of some value in defining electronic states of single-atom adsorbates (e.g. [82]) but, as with FEM, its application is limited to refractory metals and oxides. The other high-field technique, field ionisation spectroscopy (FIS), gives information, from the kinetic energy distribution of the ions, on the density of unoccupied states with similar limitations to those applying to FES, FEM and FIM.

Ion neutralisation spectroscopy (INS) is another electron tunnelling technique in which electrons neutralise He^+ ions directed at the surface inducing an Auger transition in which the second (Auger) electron is analysed for its kinetic energy [83,84]. A focussed ion beam gives lateral resolution to $< 100\,\mu m$ in a spectrometer set up for AES, XPS or UPS. INS has been used, for instance, to show that oxygen diffuses below the surface of Ni(100), a result which is supported by work function measurements of an inverted dipole and by UPS results. The electronic state of a surface space-charge layer is reflected in the tunnelling characteristics of the surface and, hence, in all of the tunnelling spectroscopies.

The other techniques that have been applied to studies of electronic states are not included in this compilation because the applications have so far been mainly in fundamental research. Surface photovoltage measures surface conductivity induced by transitions between electronic states (bands, polarons and discrete levels) under photoillumination. It can, particularly with elemental semiconductors, semiconducting compounds, oxides and some minerals, define these states on both single crystal and powder surfaces [85,86]. Photoluminescence, and related UV/visible optical spectroscopies, have produced new results correlating surface electronic states with changes in the coordination of oxide ions in metal oxide surfaces [87–89]. This work is paralleled by extensive studies using electron spin resonance (ESR) on the same topic [89]. The theoretical modelling of these surfaces referred to in Sect. 1.6, with these optical and ESR studies, appears to point the way to new methods of correlating defect structure and electronic states which may be applicable to a range of materials before long.

Surface conduction and surface charging are both considerations of singular importance in many of these measurements, and indeed, in materials science generally. Non-conducting or poorly conducting surfaces charge heavily under photoemission, electron emission or tunnelling ion beams and photovoltage measurements, in many cases severely limiting the data available. The level of charging can be used, however, to estimate surface conductivity and to infer information on surface electronic structure. In XPS, this approach has been used to show that an increased surface concentration of defects in NiO, as Ni^{3+} and O^-, leads to higher surface conductivity. For powdered samples the expression for surface charge:

$$V_{\mathrm{ch}} = \frac{K_1 - K_3}{B(\sigma_0)} \exp\big(eV_s/kT\big) \tag{1.8}$$

(where K_1 and K_3 are constants, V_s is the band bending and $B(\sigma_0)$ is a function of the sample conductivity) has been tested [90].

This is in accord with surface photovoltage results on the same surfaces correlating these defects with specific electronic states in the band structure [86]. The combination of XPS and photovoltage measurements is a useful approach to this problem. Information on surface conduction can also be found using the STM, SAM, SEELFS, UPS, INS and EELS techniques.

1.8 Structures of Adsorbed Layers

In addition to the structure of the surface of the solid material, it is obvious that we would like to know as much as possible about the structure of adsorbed or reacted layers in order to understand properties such as:

- coverage of the surface (continuous or discontinuous);
- degree of ordering in the layer(s) and effects of reaction conditions (particularly temperature) on this ordering;
- effects of the solid surface (e.g. defects, facets) on the adsorbed layer,
- local structure and geometry of adsorbed atoms, molecules or ions;
- multilayer formation and reaction profiles into the surface

The importance of this information is obvious in relation to contamination of surfaces, particularly by carbon, before reaction or coating. It is also central to materials technology in areas like corrosion and passivation, adhesion, minerals beneficiation, joining and plating technologies, catalysis, thin film technology, surface modification of materials for composite compatibility, slip (frictional) properties, wear resistance and reaction resistance. A good deal of materials research and development is concerned with the verification (or failure) of surface treatments of this kind.

Classical adsorption isotherms (e.g. Langmuir, BET) can be used to estimate the surface coverage and give some indirect information, from isotherm shapes, on monolayer or multilayer formation [91]. More directly, STM can image discontinuous carbon layers on surfaces, as in carbon islands and thread-like structures seen on the Au (100) surface (Fig. 1.7), as well as local structure and geometry of atoms, molecules and ions [92]. SAM is also directly applicable to studies of surface coverage, localised adsorption and reaction and changes in the structure of these layers with altered reaction conditions. It lacks resolution for structure of individual species as does SEM. Despite the high resolution of TEM, it is not much used to study adsorbed layers because the less stringent vacuum conditions in most instruments leads to surface carbon contamination and contrast differentiation from single, discontinuous layers of adsorbed species is not usually sufficient. TEM, however, gives direct imaging of surface structural alterations caused by adsorption and reaction as in the formation of extensive facetting, recrystallised overlayers and porosity induced by etching, oxidation or reduction on ZnO surfaces [93]. FEM and FIM, on the limited materials applicable to these techniques, give detailed images of adsorbed layer coverage, ordering, surface facets, effects of surface defects, and specific atom sites on the surface (e.g. [94]) but not on the structure of individual adsorbed species. ABS, using He beams, has been extensively used to study coverage, order (via diffraction of the beam) and disorder, atomic defects, and surface site configurations [30]. Ion scattering (LEIS, MEIS and HEIS) techniques are also directly applicable to these studies. LEIS has been used [95] to identify specific sites (and depths below the surface) for oxygen on mineral surfaces. HEIS (or RBS) is particularly suited to identification of the positions of adsorbates dissolving or reacting into the surface (Chap. 9).

Ordering beyond the atomic scale, i.e. surface crystallography in adsorbed or reacted layers, is generally studied using LEED. These applications of LEED are similar to those described in Sect. 1.4.2. Order and disorder are represented in the diffraction patterns by new spots, multiple splitting of spots, changes of intensity and alterations in the pattern as a function of beam energy and entrance (or exit) angles. As mentioned above, ABS and ion beam scattering give data indicative of surface ordering. SEXAFS, although less accessible since it requires synchroton radiation, and SEELFS should be added to this list because both can identify atomic positions and ordering of surface layers.

SIMS is also capable of producing useful information on the structure and composition of adsorbed layers. Static SIMS (Chap. 5), with its very slow sputter rate, can be used to determine: whether the surface is fully covered by an adsorbed or reacted layer, whether multilayer formation has occurred; the structure of molecular and ionic species in each layer; and the structure of reaction profiles in the surface [96,97].

For instance, a static SIMS study of an oxidised copper-zinc alloy [98] showed that the top layer only consisted of oxidised copper yet zinc was preferentially segregated into the first few surface layers below the top layer (as seen by XPS). The ability to see the structure layer-by-layer is a major feature of static SIMS applications. It is a relatively straightforward technique which can be applied to most materials although many polymers are decomposed by the beam and insulators are difficult to handle, requiring special discharging techniques. Its main disadvantage is that it is not reliably quantitative for most materials because sputter yields are radically altered by changes in the bonding and composition of the matrix. Multilayer formation and reaction profiles will be further considered in the next section where the use of (RBS), nuclear reaction analysis (NRA) and ellipsometry will be discussed.

Finally, EELS, like SEELFS, measures inelastic losses in the scattering of electrons from surfaces. These losses encompass: core level ionisation, valence electron excitation and ionisation, plasmon (collective) excitation of valence electrons and vibrational excitations at the surface. The last of these, like FTIR, can give molecular structure but very high resolution energy measurement of the electron loss is required. This technique is instrumentally sophisticated and not readily available. It is sometimes coupled with LEED apparatus.

1.9 Structure in Depth Profiles Through Surfaces

Very often in materials science, it is not the top atomic or molecular surface layer that we are concerned with but a near-surface region from 2–100 nm thick. Some examples of surface phenomena in this category are:

- segregation of particular elements of the solid or impurities (e.g. dopants) preferentially into or out of the surface region;
- thin, deposited films as in optical and microelectronic applications;
- reprecipitated (from solution) or recrystallised (in-situ) layers (for nucleated crystallites, see Sect. 1.10.4 below);
- reaction layers of different composition (e.g. hydrolysed, oxidised, reduced);
- leaching or dissolution profiles after aqueous attack;
- applied passivation layers (often multilayers).

The range of surface concentrations in these examples may vary from major matrix elements to ppm (as in n-type or p-type semiconductor dopants).

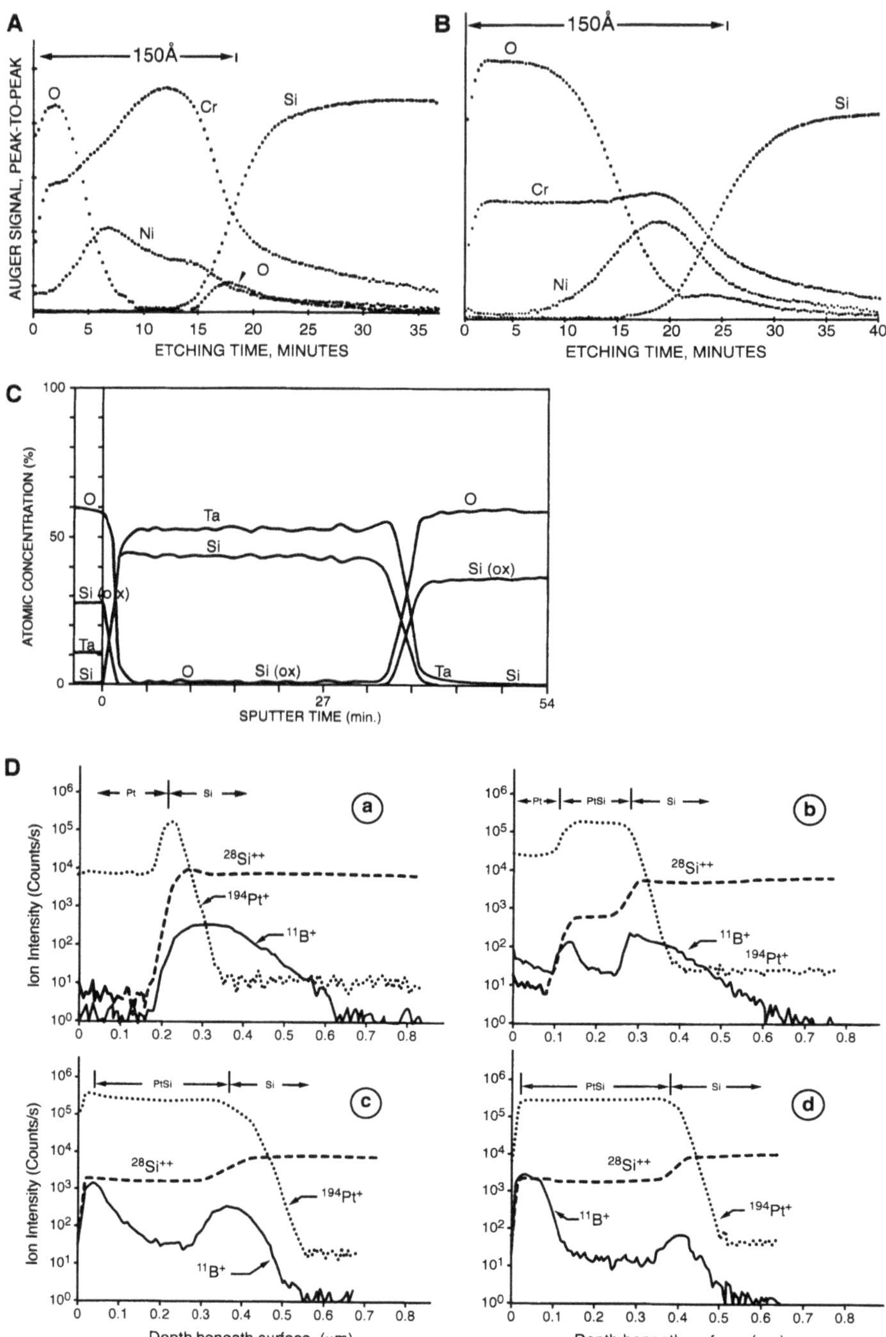
A
150Å
O
Cr
Si
Ni
O
AUGER SIGNAL, PEAK-TO-PEAK
0 5 10 15 20 25 30 35
ETCHING TIME, MINUTES
B
150Å
O
Si
Cr
Ni
0 5 10 15 20 25 30 35 40
ETCHING TIME, MINUTES
C
100
50
0
ATOMIC CONCENTRATION (%)
O
Ta
Si
Si (ox)
Ta
Si
O
Si (ox)
O
Si (ox)
Ta
Si
0 27 54
SPUTTER TIME (min.)
D
Pt
Si
a
$^{28}Si^{++}$
$^{194}Pt^{+}$
$^{11}B^{+}$
Ion Intensity (Counts/s)
Pt
PtSi
Si
b
$^{28}Si^{++}$
$^{11}B^{+}$
$^{194}Pt^{+}$
PtSi
Si
c
$^{28}Si^{++}$
$^{194}Pt^{+}$
$^{11}B^{+}$
Ion Intensity (Counts/s)
PtSi
Si
d
$^{28}Si^{++}$
$^{11}B^{+}$
$^{194}Pt^{+}$
Depth beneath surface (μm)
Depth beneath surface (μm)

Fig. 1.21. (previous page) (**A**) Auger (AES) depth profiles of 15 nm thick nichrome film deposited on a silicon substrate recorded immediately after deposition and (**B**) after heating in air at 450°C for 30 s. The heat treatment has produced both oxidation and migration of the Cr and Ni. (**C**) XPS depth profiles of a tantalum silicate film on SiO_2 substrate. The Si(ox) signal used was the Si(2*p*), 103 eV binding energy emission; the elemental Si is the Si(2*p*) intensity at 99 eV binding energy. (**D**) SIMS depth profiles showing redistribution of B implanted in Si during platinum silicate growth; (a) as-deposited; (b) after annealing at 400°C for 5 min; (c) after annealing at 400°C for 30 min; (d) after 600°C annealing for 30 min. [Reprinted with permission from Figs. 7, 8, 12, K. Bomben and W.F. Stickle, Surface Analysis Characterisation of Thin Films in "Microelectronic Manufacturing and Testing" (Lake Publishing, Libertyville, IL, USA 1987)]

In some materials (e.g. oxides, polymers, glasses), the immediate surface can be dominated by one particular functional group of the matrix (e.g. hydroxyls, alkyl groups or alkali metal cations) while excluding others (e.g. metal ions, amine groups or other cations respectively). This segregation in the first few layers can be studied non-destructively (i.e. without ion beam etching) using angle resolved XPS (ARXPS) as discussed above. Photoelectrons emitted from the surface at angles close to grazing exit angle (i.e. 80–85°) come almost entirely from the top monolayer. At 0° exit angle, the depth penetration is at a maximum corresponding to the IMFP in the solid. Comparison of spectra from 0°, 45° and 80° reveals structure of this type.

Segregation of impurities, positive or negative as in the Gibbs surface excess, is a characteristic of most materials and they are known to play crucial roles in surface reactivity and other properties. In ceramics, minerals and electronic materials this can be the single most important factor in their behaviour in application. Theoretical calculations similar to those used to model defects in oxide and ceramic surfaces [27,68], have been applied to impurity segregation [15]. Energetically preferred sites emerge from these studies. For instance, Ca^{2+} is found to preferentially incorporate into an MgO surface while Li^{+} segregates out of it. Experimental verification of this kind of structure has come from ARXPS, AES and LEIS. Static SIMS is also useful particularly for trace concentrations of impurities but, like LEIS, is more damaging to the surface. If impurity segregation results in ordered overlayer structures, as in K diffusing to ZnO surfaces [99,100], LEED patterns reveal this as new spots.

Depth profiles are most commonly generated using XPS, AES or SIMS with an ion beam to erode the surface under controlled conditions. The analysis may be recorded either intermittently by interrupting the sputtering or simultaneously. The rate of removal of surface material by the ion beam can vary from the equivalent of < 0.1 to > 35 nm min^{-1} so that quite different surface effects are found at the extremes of this range. At high beam currents the following difficulties apply: preferential sputtering giving compositional variation; enhanced diffusion, reconstruction and segregation; decomposition and loss of specific elements (e.g. F, S, Cl, O); spatial mixing distorting layer separation; and surface roughness (e.g. formation of cones, pyramidal pits and other specific structures). Removal is very often non-uniform. Despite all of these problems, under controlled conditions very good results can be obtained as illustrated in Fig. 1.21. A full discussion of ion beam effects and profiling is in Chap. 4.

These films and deposited layers can be very usefully analysed with HEIS (RBS), RHEED and ellipsometry. As explained in Sect. 1.7 and Chap. 12, RHEED is particularly suited to the study of near-surface layers in thin films and reaction profiles. The use of grazing angles of incidence and emergence, in order to minimise the depth of analysis in RHEED, places requirements on surface planarity of the samples that are often difficult to meet so that it is primarily a near-surface analysis technique. Changing these angles, however, generates data on segregation, corrosion, wear hardening, passivation and other systems where a concentration gradient through the surface applies. Alternatively, HEIS or RBS has some significant advantages for structure in depth profiles such as: cross sections and scattered ion yield proportional to Z^2; no sputter damage; and correlation of the energy of the backscattered ions with penetration depth. Thus, non-destructive profiles can be generated giving both thickness and composition of each layer. It is therefore applied not only to segregation of impurities but also to thin film determination.

Ellipsometry has not yet been introduced here but it is also valuable for studies of thin films, reprecipitated, recrystallised or reacted layers, passivation layers and leached layers. The amplitude and phase of polarised light reflected from a surface change when one or other of these layers is present. There is a linear relationship between these changes and optical thickness which, in turn, is related to the average thickness of the layer. Changes in the polarisation (i.e. parallel and perpendicular electric vectors) of the reflected light can be measured to give the refractive index and the absorption index of the film, values which can ideally be matched to those predicted for the film structure and composition. The technique is relatively simple to set up and use but requires surfaces with good reflectivity, i.e. fairly flat, and gives averaged values in multiphase systems. It is mostly applied to deposited (e.g. from vacuum, solution or electrochemical) films on extensive substrates before and after reaction.

In relation to depth profiling, we should also introduce another ion beam technique mentioned in Sect. 1.5.2 but not previously discussed, namely nuclear reaction analysis (NRA) (see Chap. 9). This, like HEIS, uses very energetic (i.e. > 1 MeV) ion beams from an accelerator and is not widely available. In NRA, it is the products of nuclear disintegration, i.e. γ-rays, α- or β-particles and nuclear fragments, that are analysed not the incident ions. These nuclear reactions occur only when the energy of the beam, which progressively attenuates as it penetrates the solid, resonates with the intranuclear process. Hence, by changing the beam energy, resonance occurs at different depths in the sample generating a profile of the material. It has been applied particularly to hydrogen impurity profiles in amorphous semiconductors.

1.10 Specific Structures

There are still a few surface structural features as yet not fully explored by the preceding sections. They tend to be specific or localised features on or in the surface and are largely associated with ceramic materials although, in all cases, they can be found in other materials as well. We will look at techniques for their examination.

1.10.1 Grain Structures, Phase Distributions and Inclusions

Ceramics, rocks, ores, and composites exhibit these features. They are of major importance in determining the surface properties of the material.

Large grain (i.e. > 1 μm) multiphase materials can be imaged and chemically mapped using SEM/EDS and backscattered electron images (see Section 1.4.1). The analysis depth is relatively large (i.e. 0.1–1 μm) in these modes. For more surface-sensitive structural differentiation (i.e. 1–10 nm) SAM can map features down to 50 nm. An example is shown in Fig. 1.22. If the structural differentiation is even finer than this, TEM or STEM, with EDS or EELS for analyses, must be used. These techniques can distinguish grain structures, phase distributions and inclusions down to unit cell level (i.e. 1–2 Å) but do require electron-transparent samples. STM, with atomic resolution, would rarely be justified for this application unless there is a need to image and analyse (via STS) sub-structure within one of these regions.

All of these techniques can be used on the majority of materials as listed in the beginning of this book. Extensive use of these techniques can be found in the literature of metals, alloys, catalysts, minerals (i.e. in rock-forming) and ceramics. A particularly valuable application is in the definition of regions of different structure in blended polymers. These can be: elastomeric inclusions; phase-separated regions; segregated layers or regions (e.g. silicones in polyester matrices); unreacted monomer inclusions; and regions containing oxidised products or unsaturated bonding (e.g. Fig. 1.23). Similar applications can be found for composites and many natural materials.

1.10.2 Fracture Faces and Intergranular Regions

Materials generally fracture along regions of least resistance, i.e. those across which bonding at the boundary is weakest. In single-phase materials, this may occur predominantly along grain boundaries (although trans-granular fracture also occurs under high-energy deposition). Surface analysis of the

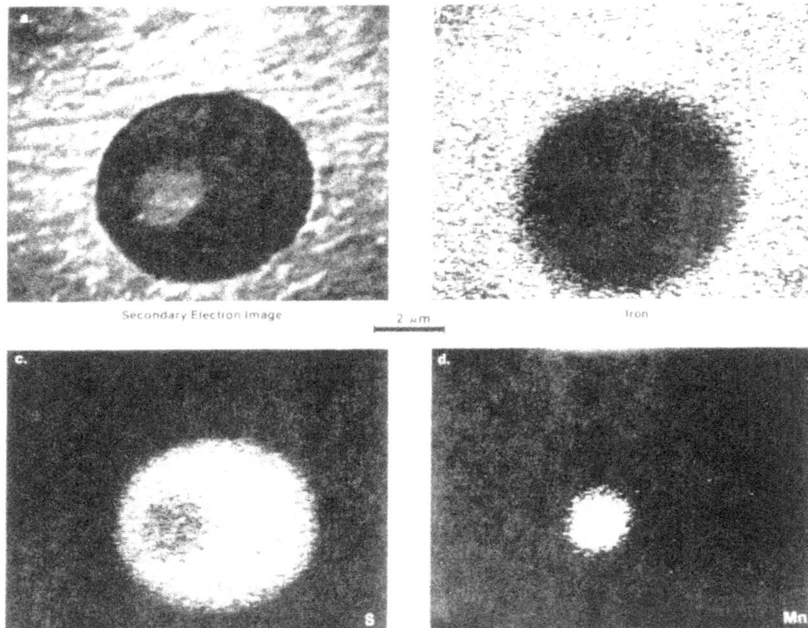

Fig. 1.22. SAM images from a precipitate in an Fe/Mn metal alloy (**a**) secondary electron image showing substructure in the precipitate; (**b**) Fe distribution showing absence of iron in the precipitate region; (**c**) S distribution with high concentration except in the substructure and; (**d**) Mn distribution showing separation of Mn into substructure of the precipitates. [Reprinted with permission from Applications Note No. 7903 (Perkin Elmer Physical Electronics Division, MN, USA)]

presented grain boundaries, using SAM, XPS and static SIMS, has revealed segregation of large impurity atoms into the atomic-width boundaries. The detailed structure of displaced atoms can be seen using TEM, STM, LEIS and ABS.

In multiphase materials, particularly multiphase ceramics, rocks and minerals, the region between grains of different phase is usually wider than a single atomic layer and has a composition different from that of either of the two adjacent phases. This integranular region can vary in width from 2 nm, as in the ceramic Synroc [101], to several micrometres, as in alumina-based ceramics. The structure of these intergranular films is often amorphous. Fracture in these materials is predominantly intergranular with the layer tending to "peel" off with one grain or the other leaving only a monolayer or two on the other surface. The material in these regions, the bonding and structure, can dominate such properties as mechanical strength, heat transfer, physical and chemical durability and mineral separation (i.e. processing).

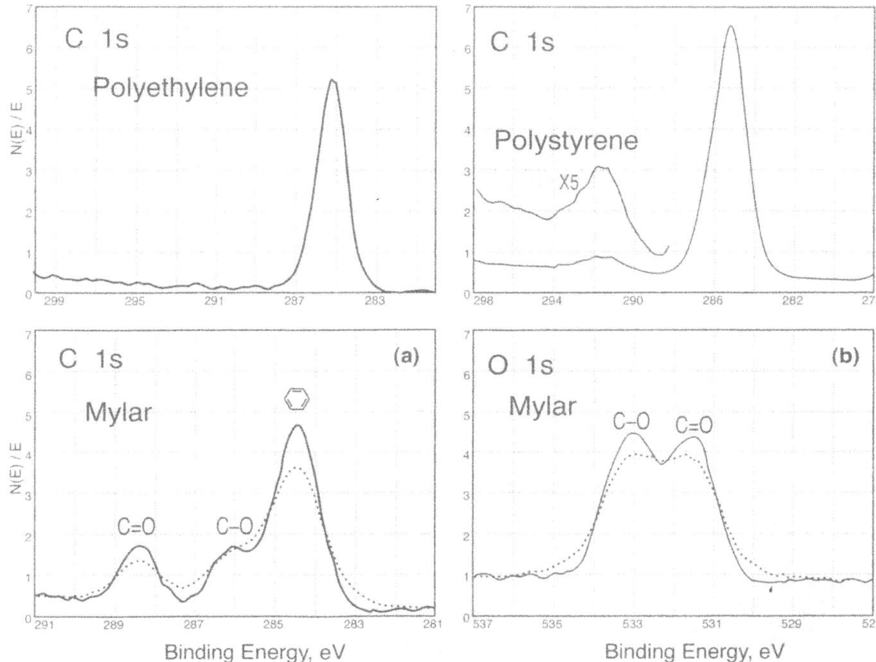

Fig. 1.23. C 1*s* XPS spectra from polyethylene and polystyrene. Note the $\pi \rightarrow \pi^*$ shake-up satellite in the polystyrene spectrum, at ca. 6.7 eV higher binding energy than the new signal, indicative of the phenyl groups. C 1*s* and O 1*s* XPS spectra from Mylar (polyethylene terephthalate) surface before (*dotted*) and after (*continuous line*) removal of the X-ray line width, i.e. deconvolution. Note the contributions from C–O and C = O groups in the structure. [Reprinted with permission from Figs. 1, 2, 4, Applications Note No. 7905 (Perkin Elmer Physical Electronics Division, MN, USA]

Figure 1.24a illustrates the use of SAM to study fracture faces and intergranular films from a ceramic material. A surface showing enhancement of alkali metal cations [e.g. (Na, Cs), Al and O is found, removable by etching with an ion beam to a depth of > 2 nm. These surface enhancements are confirmed by XPS and static SIMS as in Fig. 1.24b and c. TEM can be used to image the width of the film (Fig. 1.24d)]. A considerable body of work now exists on the study of intergranular films in ceramics using these techniques (e.g. [102]).

1.10.3 Pore Structures

In ceramics, rocks, polymers, composites and some natural materials, porous structures can intersect the surface in fracture faces or cut sections. The concentration, size and structure of these pores is important in determining

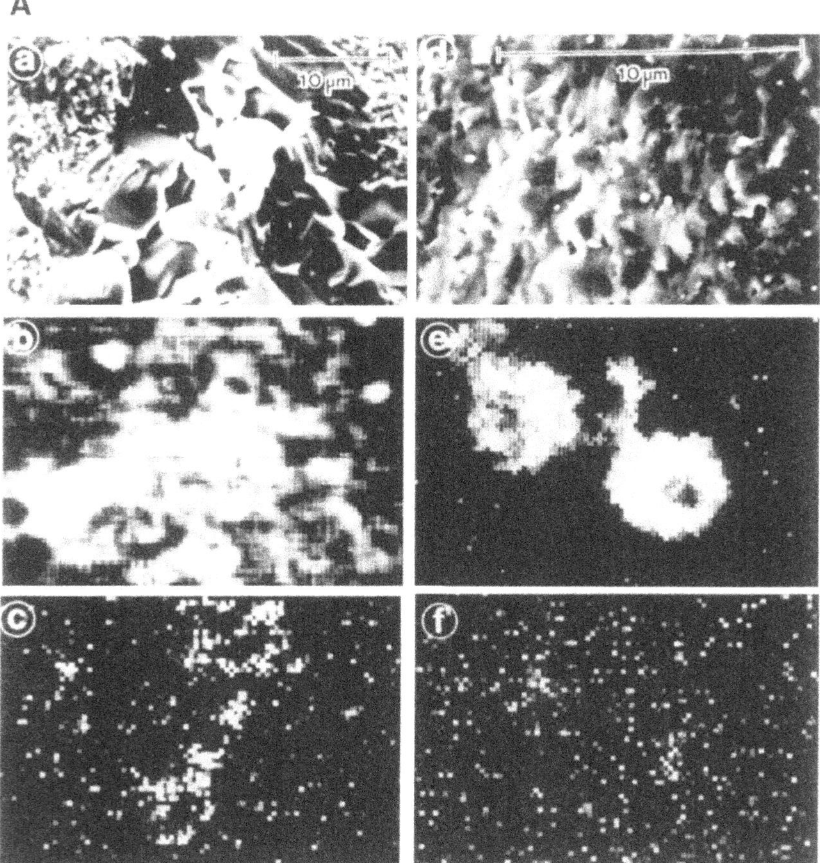

Fig. 1.24. (**A**) Secondary electron image (*a*) of a pore in a fracture face of the ceramic Synroc C with Cs maps of the same area (*b*) before and (*c*) after removal of ca. 15 nm by ion etching. Similar images (*d, e, f*) of a Cs-rich area (*e*) before and after (*f*) removal of ca. 375 nm. The fracture face exposes closed porosity and thin (< 2 nm) remnants of intergranular films in the ceramic. (**B**) XPS depth profiles from a Synroc C fracture face before (*full line*) and after (*dashed line*) immersion in doubly distilled water for ca. 30 s. An enhancement of Cs, Al and, possibly, Mo in the fresh fracture face is reduced by the water. The numbers on the right indicate averaged atomic % of each element in the bulk of the material. (**C**) Static SIMS depth profile of a fresh fracture face of Synroc C demonstrating enchancement of Cs and Na, but not Ca and Ti, in the intergranular region. (**D**) High resolution TEM of an integranular film at the interface between grains of perovskite and magnetoplumbite. Cleavage has been initiated, with the crack (arrowed) running between grains, the intergranular film adhering to the pervoskite grain. [Reprinted with permission from Figs. 1, 4, 5, 6; J.A. Cooper, D.R. Cousens, J.A. Hanna, R.A. Lewis, S. Myhra, R.L. Segall, R.St.C. Smart, P.S. Turner, and T.J. White, J. Amer. Ceram. Soc. **69**(4), 347–352 (1986)]

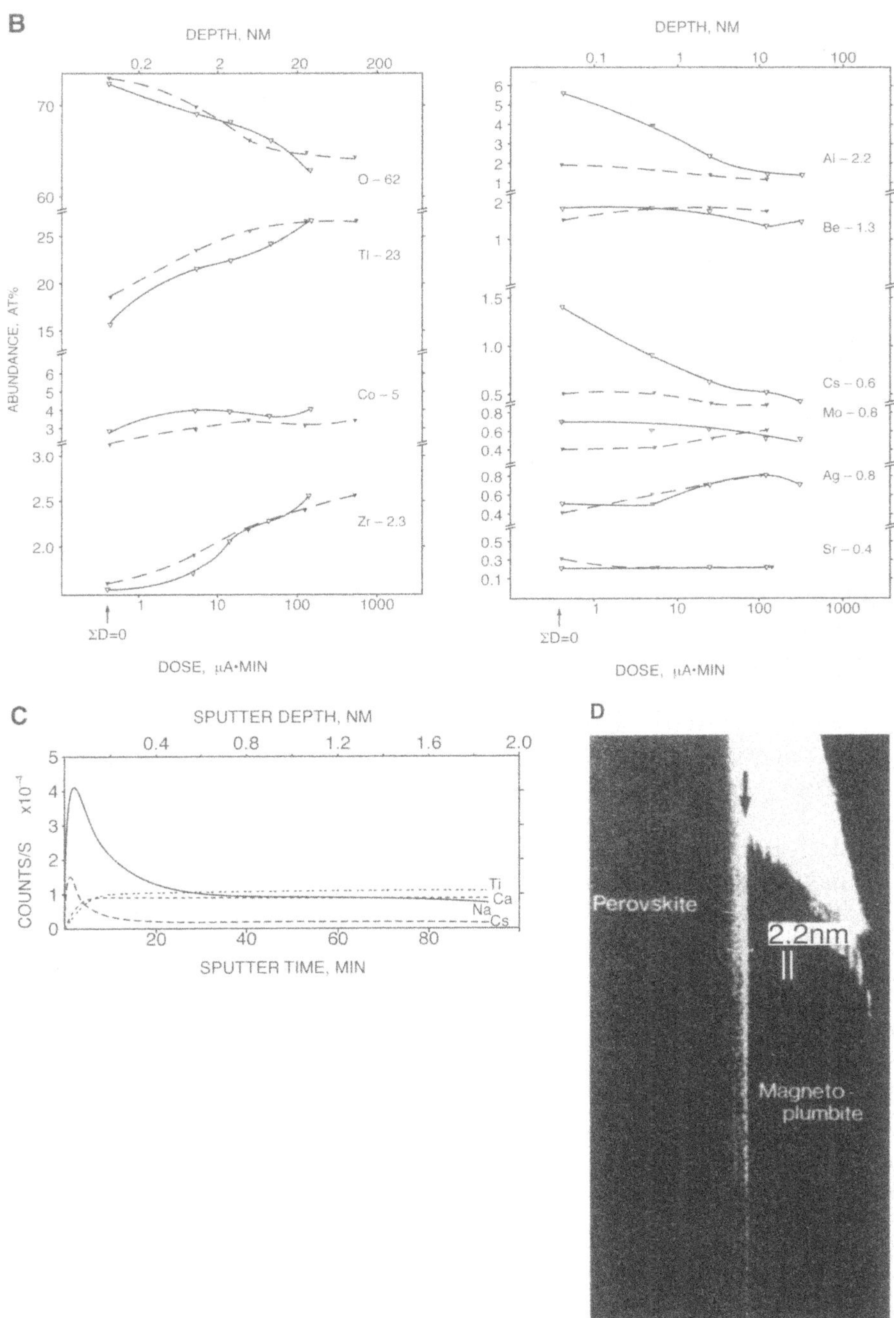
B
DEPTH, NM
0.2
2
20
200
70
60
25
20
15
6
5
4
3
3.0
2.5
2.0
ABUNDANCE, AT%
O – 62
Ti – 23
Co – 5
Zr – 2.3
1
10
100
1000
ΣD=0
DOSE, μA•MIN
DEPTH, NM
0.1
1
10
100
Al – 2.2
Be – 1.3
Cs – 0.6
Mo – 0.8
Ag – 0.8
Sr – 0.4
DOSE, μA•MIN
C
SPUTTER DEPTH, NM
0.4
0.8
1.2
1.6
2.0
COUNTS/S x10⁻⁴
Ti
Ca
Na
Cs
20
40
60
80
SPUTTER TIME, MIN
D
Perovskite
2.2nm
Magneto-
plumbite

the mechanical properties of these materials. Additionally, they may contain specific elements or compounds trapped during their formation. These species are often soluble or reactive leading to undesirable leaching and reactivity properties. Trapped (i.e. closed) pores below the surface can also affect these properties by diffusion down intergranular films to the surface.

SEM and SAM can image these structures since they are usually larger than 500 nm in major dimensions. SAM can also analyse condensed layers on their surfaces. An illustration of this application can be seen in Fig. 1.24a where a pore intersects the surface of a ceramic with a monolayer of volatile Cs condensed during pore formation and subsequent cooling.

1.10.4 Precipitates, Reaction Products and Recrystallised Particles on Surfaces

Separate particles on surfaces can be formed by: precipitation from saturated solution; reaction to form new phases; and leaching, dissolution or hydrolysis of a surface layer (i.e. loss of specific elements) followed by recrystallisation in situ to form new phases. In all three cases, nucleation occurs at specific sites on the surface rather than as a relatively uniform surface layer. Precipitation of sequentially supersaturated phases can sometimes be found in successive layers of the same crystallite at the same nucleation site (i.e. changing composition with depth through the crystallite). SAM can reveal these layers by imaging individual crystallites and using ion beam sputtering to develop depth profiles as the precipitate is removed. The spatial resolution and surface specificity of the technique are both essential to the structural analysis. Reaction at a surface can lead to extensive surface diffusion, facilitated by local exothermic processes, and structural rearrangement of its products. SAM, or in many cases SEM when the crystallites are $> 1\,\mu m$, can provide the required information on these new phases. The newest generation of XPS analysers can image regions $< 10\,\mu m$ in diameter. (Recent instrumental advances have also produced scanning XPS with resolution down to 500 nm but these instruments are as yet only available for fundamental research). Hence, larger crystals or collections of crystallites can be analysed in this field.

Sometimes, compositional information is not unequivocal and it is essential for structural identification to have diffraction patterns. TEM of thin sections with crystallite formation can provide this using selected area electron diffraction. An example is illustrated in Fig. 1.25 where TiO_2 crystallites in two different phases, as brookite and anatase, are found recrystallised in situ after hydrolysis of the surface of a $CaTiO_3$ perovskite.

1.10.5 Magnetic Domains

Chapter 15 explains techniques for measuring and mapping different domains of magnetic fields in the surface of materials. The techniques are relatively specific to this use and examples will be left to that section of the book.

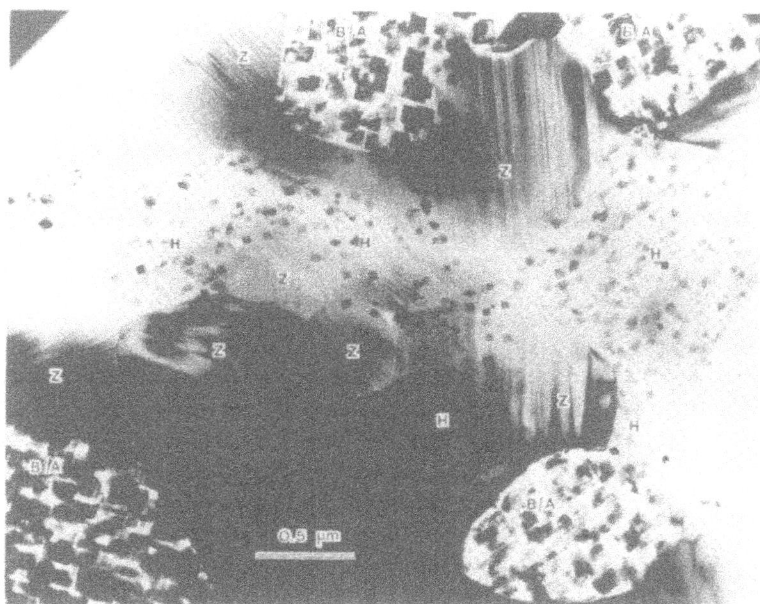

Fig. 1.25. TEM of an ion beam – thinned section of Synroc B (Synroc C without addition of simulated nuclear waste) after hydrothermal attack for 1 day. The perovskite grains have completely dissolved, recrystallising in-situ as brookite (B) and anatase (A). Hollandite (H) grains are slightly attacked with a few small TiO_2 crystallites on their surfaces. Zirconolite (Z) appears unaltered by this treatment. [Reprinted with permission from T. Kastrissios, M. Stephenson, P.S. Turner, and T.J. White, J. Amer. Ceram. Soc. **70**, 144–146 (1987)]

1.11 Technique-Induced Artifacts

Each of the techniques utilized in surface analysis in materials science will perturb the system being analysed to some extent. With methods such as SIMS, where the act of detection involves sputtering off layers of surface material, the initial perturbation can be obvious. However, more subtle secondary consequences may also arise. For example, in any ion beam technique, the preferential loss of one element and atomic (knock-on) mixing may induce a result difficult to relate to the original selvedge material. It is useful to consider two artifacts in particular, both commonly encountered, which can obscure useful information.

1.11.1 Radiation Damage

A chemical bond strength of 100 kJ mol^{-1} translates to a mere 1.036 eV. As can be seen from Table 1.4 incident probes can have energies up to MeV.

Hence it is not surprising that chemical bonds will be disrupted. As a general rule of thumb the higher the energy of the incident probe the greater the total damage, with photons being more benign than charged particles. In fact, the probability of local damage at a particular site is a function of the product of energy and cross section. For medium and high energy particles, this function is related to $E \times E^{-2}$ or E^{-1} so that, while the total damage increases, the local damage decreases with increasing energy. Logically, if a sample were to be analysed with a sophisticated multi-technique instrument, offering say, scanning Auger microscopy (SAM), XPS, SIMS and ultraviolet photoelectron spectroscopy (UPS), the rule of thumb would dictate performing the sequence of analyses: UPS, XPS, SAM and finally SIMS, i.e. in the order of increasing radiation damage.

The destructive technique of SIMS has historically developed along two pathways [103]. Originally intended as a non-surface technique for in-depth and microanalysis in materials science, the method of *dynamic* SIMS utilizes high incident ion fluxes ($> 1\,\mu\mathrm{A\,cm^{-2}}$) with consequently large sputtering rates (> 50 Å min^{-1}). The surface oriented method of *static* SIMS with considerably lower incident fluxes (< 1 nA cm^{-2}) has sub-monolayer sensitivity and results in much less damage than dynamic SIMS. To increase the signal level in static SIMS the area of analysis is usually increased, either by a broad beam or via rastering. The rastering can be synchronized with an analyser gate for a particular signal to provide the variant known as *imaging* SIMS. Although the atomic mixing in static SIMS can be reduced to sub-monolayer levels, it remains larger than the damage induced with incident electrons at the same current density. Dynamic SIMS has the disadvantages of rapid loss of surface layers, lateral and vertical (recoil) mixing of species, preferential sputtering of particular elemental species and sputter-induced cratering/protusions. By contrast, static SIMS removes a single surface layer over periods of hours with much less damage or mixing effects.

Considering electrons as the incident probe and an incident current density of 1 mA cm^{-2} or $10\,\mu\mathrm{A\,mm^{-2}}$ (this equates to about 6 electrons per surface site per second into the outermost layer), Table 1.8 shows, for a number of materials and electron energies, the accumulated threshold charge densities for detectable damage to occur [104]. Also indicated are the times that the current density can be maintained without detectable damage and the lateral spatial resolution that may be afforded in an Auger analysis (based on a beam of 100 nA residing on a spot for 5 minutes [105]). Certainly for a number of materials it is not viable to analyse them via electron excited Auger as damage will occur well before any meaningful data can be acquired. The radiation damage can occur as electron stimulated diffusion, reaction or desorption of adsorbed species, all leading to an altered selvedge. The action of electron stimulated desorption (ESD) can also be deliberately employed as an analytical method, and, in its more sophisticated form of electron stimulated desorption ion angular distribution (ESDIAD), yield information on

Table 1.8. Accumulated threshold charge densities D_c for detectable electron radiation damage to occur for a variety of materials and electron energies. The threshold levels relate to maximum exposure times and optimal spatial resolutions possible without damage occurring

Material	Energy (keV)	D_c ($C\,cm^{-2}$)	Time at $1\,mA\,cm^{-2}$	Spatial resolution (mm)
Si_3N_4	2	Stable	–	Small
Al_2O_3	5	10	3 h	0.02
Cu, Fe phthalocyanines	1	> 1	> 15 min	0.06
SiO_2	2	0.6	10 min	0.08
Li_2WO_4	1	0.5	8 min	0.09
NaF, LiF	0.1	0.06	60 s	0.25
$LiNO_3$, Li_2SO_4	1	0.05	50 s	0.28
KCl	1.5	0.03	30 s	0.36
TeO_2	2	0.02	20 s	0.44
$H_2O_{(s)}$	1.5	0.01	10 s	0.62
Native oxides	5	2×10^{-3}	2 s	1.4
$C_6H_{12(s)}$	0.1	3×10^{-4}	0.3 s	3.6
Na_3AlF_6	3	$10^{-4} - 10^{-3}$	0.1 s	2–6
$CH_3OH_{(s)}$	1.5	2.5×10^{-4}	0.3 s	3.9
Formvar	75	2×10^{-3}	0.2 s	1.4

bond geometries of adsorbed species. Complementary to ESD is the subtler form of damage known as electron stimulated adsorption [104]. The sticking probability of a vacuum residual is enhanced by radiation placing it in an excited state or causing its fragmentation. Note that the concept of electron beam radiation damage is equally applicable to the diffraction techniques of low energy electron diffraction (LEED) and reflection high energy electron diffraction (RHEED) and the electron microscopies. Scanning tunnelling microscopy (STM) is somewhat of a different case. Although the total tunnelling current in STM is very small (see Chap. 10), say 1 nA, this tunnels into about a 5 Å diameter area and represents a massive current density of about 5×10^5 A cm^{-2} or 3×10^9 electrons per surface site per second (compare this with 6!). The fact that apparently little damage occurs even after lengthy exposure can only be a consequence of the small tunnelling bias, typically 0.1 V, well below the threshold energy involved in a bond. Indeed the most severe damage from STM results from physical tip crashing.

As an incident photon method, XPS has the general reputation of being fairly benign. However, even beyond the photoionization of core levels and subsequent relaxation processes which may cause chemical alterations, the exiting photoelectrons and associated secondaries represent a definite, though small, electron flux. Due to small subtended solid angles, band pass electron spectrometers generally have small collection efficiencies. Based on detected count rates the actual total electron flux passing through the irradiated surface area is typically up to 10^{10} electrons s^{-1}. Over a 5 mm diameter circle that converts to about 5×10^{-5} electrons per surface site per second. Purely on the grounds of electron flux that makes XPS a factor of 10^5 less likely to produce radiation damage than conventional electron excited Auger. Nevertheless radiation damage can be observed with XPS e.g. alkali halides readily form colour centres, the damage being macroscopically visible after just a few minutes X-ray exposure. The Au(I) state in $KAu(CN)_2$ readily disproportionates under X-ray irradiation to Au(0) and Au(III), thereby forming a mixture of three states of gold. With extended exposure the quantity of metallic gold formed is also macroscopically visible [106].

In systems prone to radiation damage the effect cannot be entirely eliminated but the extent of damage can be minimized [107]. The occurrence of damage is readily established by sequential spectral acquisition during continuous irradiation, thereby enabling the threshold dose at a given flux to be estimated. It may then be possible not to exceed the dose by minimizing the incident flux (even at the expense of spatial resolution), turning off the radiation source except when data is being acquired and scanning over narrow energy ranges. In some instances it can also help to cool the specimen to liquid nitrogen temperatures since some radiation damage can be mass diffusion dependent. Cooling can also reduce the secondary effects of possible specimen heating, especially in the case of poor thermal conductors.

1.11.2 Electrostatic Charging

After radiation damage probably the most disruptive artifact is that of induced electrostatic charging of the specimen surface. Except for techniques utilizing neutral incident and exit probes, charging will always be experienced to some extent with specimens of low electrical conductivity, or with insulating selvedge layers e.g. oxide films on semiconductors (or even conductors). In the case of incident electrons, when the secondary electron yield is less than unity i.e. less than the nett incident current, the surface will accumulate negative charge. This accumulation and the resultant negative surface potential increases with time, often leading to values which induce dielectric breakdown. Influences may range from deflection of the incident beam to gross distortions in the energy of the characteristic electrons produced. Under these unstable conditions meaningful spectra cannot always be obtained. It is possible to reduce this type of charging by judicious sample mounting (e.g. in In foil) and by careful choice of both electron beam energy and angle of incidence, thereby altering the value of the secondary electron yield. These high surface potentials may also be capable of causing field-induced migration of mobile ions, thereby becoming another source of radiation damage [105].

Under conditions where the secondary electron yield is greater than unity the surface charge becomes positive rather than negative. For example, this will always be the case for XPS applied to insulators since photons arrive and electrons leave. The induced positive charge prevents the lowest kinetic energy secondary electrons from leaving, thereby rapidly establishing a stable equilibrium situation. In the case of Auger or XPS spectra the positive surface potential will shift the entire spectrum uniformly to lower kinetic energies, typically 2–15 eV. Since for a given specimen and given incident flux the static charge shift is constant, careful energy referencing, to a known spectral feature will enable a correction to be applied. Such energy referencing is particularly important in evaluating binding energies in XPS and considerable effort has gone into the problem. This will be considered in more detail in Chap. 7. In an alternative to referencing, the surface may be flooded with an adjustable flux of additional low energy electrons in order to balance out the positive charge.

References

1. D. Tabor: Surf. Sci. **89**, 1 (1979)
2. P.H. Abelson: Science **234**, 257 (1986)
3. W.A. Grant, R.P.M. Procter, J.L. Whitton (eds.): *Surface Modification of Metals by Ion Beams* (Elsevier Sequoia, Lausanne 1987)
4. J.W. Rabalais, S. Kasi: Science **239**, 623 (1988)
5. P. Danielson: In *Kirk-Othmer Encyclopedia of Chemical Technology*, Vol. 20 (Wiley, New York 1982) pp. 811, 812
6. F. Jona: J. Phys. C **11**, 4271 (1978)
7. J. Verhoeven: J. Environ. Sci. **22**, 24 (1979)
8. R.G. Musket, W. McLean, C.A. Colmenares, D.M. Makowiecki, W.J. Siekhaus: Appl. Surf. Sci. **10**, 143 (1982)
9. G.A. Somorjai, M.A. Van Hove: Structure and Bonding **38**, 1 (1979)
10. G.A. Somorjai: *Chemistry in Two Dimensions: Surfaces* (Cornell University Press, Ithaca 1981)
11. V.M. Bermudez: Electronic structure of point defects on insulator surfaces. Prog. Surf. Sci. **11**, 1 (1981)
12. E.A. Colbourn, W.C. Mackrodt: Theoretical aspects of H_2 and CO chemicsorption on MgO surfaces. Surf. Sci. **117**, 571 (1982)
13. C.F. Jones, R.A. Reeve, R. Rigg, R.L. Segall, R.St.C. Smart, P.S. Turner: Surface area and the mechanism of hydroxylation of ionic oxide surface. J. Chem. Soc. Faraday I, **80**, 2609 (1984)
14. E.A. Colbourn, W.C. Mackrodt: Irregularities at the (001) surface of MgO: topography and other aspects. Solid State Ionics **8**, 221 (1983)
15. P.W. Tasker, E.A. Colbourn, W.C. Mackrodt: The segregation of isovalent impurity cations at the surfaces of MgO and CaO. J. Am. Ceram. Soc. **68**, 74 (1985)
16. C. Ocal, B. Basurco, S. Ferrer: Surf. Sci. **157**, 233 (1985)
17. J.E. Crowell, J.G. Chen, J.T. Yates Jr.: Surf. Sci. **165**, 37 (1986)
18. C.F. McConville, D.L. Seymour, D.F. Woodruff, S. Bao: Surf. Sci. **188**, 1 (1987)
19. I.P. Batra: J. Electron. Spectrosc. Relat. Phen. **33**, 175 (1984)
20. W.A. Anderson, W.E. Haupin: In *Kirk-Othmer Encyclopedia of Chemical Technology*, 3rd ed., ed. by M. Grayson, D. Eckroth (Wiley, New York 1978) Vol. 2, pp. 181–183

21. W.W. Binger, E.H. Hollingsworth, D.O. Sprowls: In *Aluminum, Properties, Physical Metallurgy and Phase Diagrams*, Vol. I, ed. by K.R. Van Horn (American Society for Metals, Metals Park, Ohio 1967) pp. 226–235
22. L.F. Mondolfo: *Aluminum Alloys: Structure and Properties* (Butterworths, London 1976) pp. 123, 148
23. K. Nisancioglu, O. Lunder, H. Holtan: Corrosion **41**, 247 (1985)
24. E.W. Müller, T.T. Tsong: *Field Ion Microscopy* (Elsevier, New York 1969)
25. E.W. Müller: Ann. Rev. Phys. Chem. **18**, 35 (1967)
26. E. Bauer: Surf. Sci. **162**, 163 (1986)
27. A.M. Stoneham, P.W. Tasker: The Theory of Ceramic Surfaces, in *Surface and Near-Surface Chemistry of Oxide Materials*, ed. by J. Nowotny, L.-C. Dufour (Elsevier, Amsterdam 1988) pp. 1–22
28. K. Heinz, K. Müller: LEED-intensities – experimental progress, and new possibilities of surface structure determination, In *Structural Studies of Surfaces*, Springer Tracts Mod. Phys. 91 (Springer, Berlin, Heidelberg 1982) pp. 1–54
29. I.P. Batra, T. Engel, K.H. Rieder: In *The Structure of Surfaces*, Springer Ser. Surf. Sci., Vol. 2, ed. by M.A. van Hove, S.Y. Tong (Springer, Berlin, Heidelberg 1985) p. 251
30. T. Engel, K.H. Rieder: Structural studies of surfaces with atomic and molecular beam diffraction, In *Structural Studies of Surfaces*, Tracts Mod. Phys. 91 (Springer, Berlin, Heidelberg 1982) pp. 55–180
31. J. Stöhr: Surface crystallography by SEXAFS and NEXAFS, In *Chemistry and Physics of Solid Surfaces V*, ed. by R. Vanselow, R. Howe, Springer Ser. Chem. Phys. Vol. 35 (Springer, Berlin, Heidelberg 1984) pp. 231–256
32. W.M. Gibson: Determination by ion scattering of atomic positions at surfaces and interfaces, In *Chemistry and Physics of Solid Surfaces V*, ed. by R. Vanselow, R. Howe, Springer Ser. Chem. Phys. Vol. 35 (Springer, Berlin, Heidelberg 1984) pp. 427–454
33. B.K. Teo, D.C. Joy (Eds.): *EXAFS Spectroscopy – Techniques and Applications* (Plenum, New York 1981)
34. Periodic Table of the Elements, Sargent-Welch Scientific Company, Skokie, Illinois, Catalogue Number S-18806. Based upon National Standard Reference Data System material
35. Table 1.4 is drawn from a variety of sources including tables in the references listed below, manufacturers specifications and the contributions of other authors in this book. Due to the limited scope in presenting information in such tables they should be utilized only as a guide.
(a) H. Fellner-Feldegg, U. Gelius, B. Wannberg, A.G. Nilsson, E. Basilier, K. Seigbahn: J. Electron Spectrosc. Relat. Phen. **5**, 643 (1974)
(b) D. Roy, J.D. Carette: In *Electron Spectroscopy for Surface Analysis*, ed. by H. Ibach (Springer, Berlin, Heidelberg 1977) pp. 14, 15
(c) M.W. Roberts, C.S. McKee: *Chemistry of the Metal-Gas Interface* (Clarendon, Oxford 1978) p. 207
(d) M.P. Seah, D. Briggs: In *Practical Surface Analysis by Auger and X-ray Photoelectron Spectroscopy*, ed. by D. Briggs, M.P. Seah (Wiley, Chichester 1983) p. 12
(e) C.W. Magee: Nucl. Instrum. Meth. **191**, 297 (1981)
36. M.P. Seah, W.A. Dench: Surf. Interface Anal. **1**, 2 (1979)
37. R.E. Ballard: J. Electron Spectrosc. Relat. Phen. **25**, 75 (1982)

38. S. Tanuma, C.J. Powell, D.R. Penn: Surf. Sci. **192**, L849 (1987)
39. C.D. Wagner, W.M. Riggs, L.E. Davis, J.F. Moulder, G.E. Muilenberg (eds.): *Handbook of X-ray Photoelectron Spectroscopy* (Perkin-Elmer, Eden Prairie, Minnesota 1978) pp. 42, 50
40. M.F. Ebel: J. Electron Spectrosc. Relat. Rhenom. **22**, 157 (1981)
41. C.W. Magee: Nucl. Instrum. Meth. **191**, 297 (1981)
42. D.P. Leta, G.H. Morrison: Anal. Chem. **52**, 514 (1980)
43. Riber MIQ 256 SIMS/Ion microprobe
44. P. Skeldon, K. Shimizu, G.E. Thompson, G.C. Wood: Thin Solid Films **123**, 127 (1985)
45. Based on an absorption sensitivity for iron of 0.15 μg/ml/1% Abs (p. 825) and a sample volume of 5 μl (p. 355), In *Instrumental Methods of Analysis*, H.H. Willard, L.L. Merritt Jr., J.A. Dean (eds.) (Van Nostrand, New York 1974)
46. D.M. Hercules, L.E. Cox, S. Osnisick, G.D. Nichols, J.C. Carver: Anal. Chem. **45**, 1973 (1973)
47. R.C. Weast, M.J. Astle (eds.): *CRC Handbook of Chemistry and Physics*, 63rd edn. (CRC, Boca Raton 1982) F-161
48. R. Browning: J. Vac. Sci. Technol. A **2**, 1453 (1984)
49. P.D. Prewett, D.K. Jeffries: J. Phys. D **13**, 1747 (1980)
50. N.M. Ceglio, A.M. Hawryluk, M. Schattenburg: J. Vac. Sci. Technol. B **1**, 1285 (1983)
51. I.W. Drummond, T.A. Cooper, F.J. Street: Spectrochimica Acta **40B**, 801 (1985)
52. K. Yates, R.H. West: Surf. Interface Anal. **5**, 217 (1983)
53. For example, VG Scientific ESCALAB 20-X
54. For example, VG Scientific ESCASCOPE
55. J.F. Evans, J.H. Gibson, J.F. Moulder, J.S. Hammond. The PHI Interface **7**, 1 (1984) (Perkin Elmer, Eden Prairie, USA)
56. M.L. Tarng, D.G. Fischer: J. Vac. Sci. Technol. **15**, 50 (1978)
57. V. Thompson, H.E. Hintermann, L. Chollet: Surf. Technol. **8**, 421 (1979)
58. D.R. Clark, L.L. Hench: An overview of the physical characterisation of leached surfaces, Nucl. Chem. Waste Manag. **2**, 93 (1981)
59. J.W. Strojek, J. Mielczarski: Spectroscopic investigations of the solid-liquid interface by the ATR technique, Adv. Colloid Interf Sci. **19**, 309 (1983)
60. R.F. Willis (ed.): *Vibrational Spectroscopy of Adsorbates* (Springer, Berlin, Heidelberg 1980)
61. K. Siegbahn, C. Nordling, A. Fahlman, R. Nordberg, K. Hamrin, J. Hedman, G. Johansson, T. Bergmark, S.-E. Karlsson, I. Lindgren, B. Lindberg: *ESCA, Atomic, Molecular and Solid State Structure Studied by Means of Electron Spectroscopy* (Almqvist and Wiksells, Uppsala 1967) p. 79
62. R. Opila, R. Gomer: Surf. Sci. **105**, 41 (1981)
63. J. Hulse, J. Küppers, K. Wandelt, G. Ertl: Appl. Surf. Sci. **6**, 453 (1980)
64. H.H. Madden: J. Vac. Sci. Technol. **18**, 677 (1981)
65. M. Barber, R.S. Bordoli, G.J. Elliot, R.D. Sedgwick, A.N. Tyler: Anal. Chem. **54**, 645A (1982)
66. C.J. Powell, N.E. Erickson, T.E. Madey: J. Electron Spectrosc. Relat. Phen. **17**, 361 (1979)
67. C.J. Powell, N.E. Erickson, T.E. Madey: J. Electron Spectrosc. Relat. Phen. **25**, 87 (1982)

68. V.E. Henrich: Electronic and geometric structure of defects on oxides and their role in chemisorption, In *Surface and Near Surface Chemistry of Oxide Materials*, ed. by J. Nowotny, L.-C. Dufour (Elsevier, Amsterdam 1988) pp. 23–60
69. A.B. Kunz: Theoretical study of defects and chemisorption by oxide surfaces, In *External and Internal Surfaces in Metal Oxides*, ed. by L.-C. Dufour, J. Nowotny, Materials Science Forum (Trans. Tech. Publications, Claustal-Zellerfeld 1988) pp. 1–30
70. R.L. Segall, R.St.C. Smart, P.S. Turner: Oxide surfaces in solution, In *Surface and Near-Surface Chemistry of Oxide Materials*, ed. by J. Nowotny, L.-C. Dufour (Elsevier, Amsterdam 1988) pp. 527–576
71. V.E. Henrich: Ultraviolet photoemission studies of molecular adsorption on oxide surfaces, Prog. Surf. Sci. **9**, 143 (1979)
72. V.E. Henrich, G. Dresselhaus, H.J. Zelger: Chemisorbed phases of H_2O on TiO_2 and $SrTiO_3$, Solid State Commun. **24**, 623 (1977)
73. M.W. Roberts: Metal oxide overlayers and oxygen-induced chemical reactivity studied by photoelectron spectroscopy, In *Surface and Near-Surface Chemistry of Oxide Materials*, ed. by J. Nowotny, L.-C. Dufour (Elsevier, Amsterdam 1988) pp. 219–246
74. G.A. Somorjai: Adv. Catal. **26**, 1 (1979)
75. G. Heiland, H. Lüth: In *The Chemical Physics of Solid Surfaces and Heterogeneous Catalysis*, Vol. 3 , ed. by D.A. King, D.P. Woodruff (Elsevier, Amsterdam 1984) p. 147
76. S.C. Chang, P. Mark: Surf. Sci. **45**, 721 (1974); ibid. **46**, 293 (1974)
77. C.C. Schubert, C.L. Page, B. Ralph: Electrochim. Acta **18**, 33 (1973)
78. J. Marien: Phys. Status Solidi A **38**, 339, 513 (1976)
79. J. Nowotny, M. Sloma: Work function of oxide ceramic materials, In *Surface and Near-Surface Chemistry of Oxide Materials*, ed. by J. Nowotny, L.-C. Dufour (Elsevier, Amsterdam 1988) pp. 281–344
80. S.R. Morrison: *The Chemical Physics of Surfaces* (Plenum, New York 1977)
81. H. Wagner: *Physical and Chemical Properties of Stepped Surfaces*, Springer Tracts Mod. Phys. (Springer, Berlin, Heidelberg 1978)
82. H.E. Clark, R.D. Young: Surf. Sci. **12**, 385 (1968)
83. H.D. Hagstrum: Science **178**, 275 (1972)
84. H.D. Hagstrum: Phys. Rev. **150**, 495 (1966)
85. G. Heiland, H. Lüth: Adsorption on oxides, in *The Chemical Physics of Solid Surfaces and Heterogeneous Catalysis*, ed. by D.A. King, D.P. Woodruff (Elsevier, Amsterdam 1984) p. 156
86. N.S. Huck, R.St.C. Smart, S.M. Thurgate: Surface photovoltage and XPS studies of electronic structure in defective nickel oxide powders, Surf. Sci. **169**, L245 (1986)
87. E. Garrone, A. Zecchina, F.S. Stone: Philos. Mag. B **42**, 683 (1980)
88. W. Göpel: Prog. Surf. Sci. **20**, 9 (1985)
89. J. Cunningham: Photoeffects on metal oxide powders, in *Surface and Near-Surface Chemistry of Oxide materials*, ed. by J. Nowotny, L.-C. Dufour (Elsevier, Amsterdam 1988) pp. 345–412
90. M.W. Roberts, R.St.C. Smart: XPS determination of band bending in defective semiconducting oxide surfaces, Surf. Sci. **151**, 1 (1985)
91. A.W. Adamson: *Physical Chemistry of Surfaces* (Wiley, New York 1986)

92. R.J. Bohm, W. Höseler: Scanning tunneling microscopy – a review, in *Chemistry and Physics of Solid Surfaces VI*, ed. by R. Vanselow, R. Howe, Springer Ser. Surf. Sci. Vol. 5 (Springer, Berlin, Heidelberg 1986) pp. 361
93. W. Hirschwald: In *Current Topics in Materials Science*, ed. by E. Kaldis, Vol. 7 (1981) p. 143
94. C.C. Schubert, C.L. Page, B. Ralph: Electrochim. Acta **18**, 33 (1973)
95. W. Heiland, E. Taglauer: Surf. Sci. **68**, 96 (1977)
96. A. Benninghoven: Developments in secondary ion mass spectrometry and applications to surface studies, Surf. Sci. **53**, 596 (1975)
97. A. Brown, J.C. Vickerman: Static SIMS for applied surface analysis, Surf. Interface Anal. **6**, 1 (1984)
98. W. Hirschwald: Selected experimental methods on the characterization of oxide surfaces, In *Surface and Near-Surface Chemistry of Oxide Materials* ed. by J. Nowotny, L.-C. Dufour (Elsevier, Amsterdam 1988) pp. 140–141
99. M. Grunze, W. Hirschwald, D. Hoffman: J. Cryst. Growth **52**, 241 (1981)
100. W. Hirschwald, P. Bonasewicz, L. Ernst, M. Grade, D. Hopmann, S. Krebs, R. Littbarski, G. Neumann, M. Grunze, D. Kobb, H.J. Schultz: Zinc oxide, Current Topics Mater. Sci. **7**, 143 (1981)
101. J.A. Cooper, D.R. Cousens, J.A. Hanna, R.A. Lewis, S. Myhra, R.L. Segall, R.St.C. Smart, P.S. Turner, T.J. White: Intergranular films and pore surfaces in Synroc C: structure, composition and dissolution characteristics, J. Amer. Ceram. Soc. **69**, 347 (1986)
102. D.R. Clarke: Observation of microcracks and thin intergranular films in ceramics by transmission electron microscopy, J. Am. Ceram. Sco. **63**, 104 (1980)
103. N.H. Turner, B.I. Dunlap, R.J. Colton: Anal. Chem. **56**, 373R (1984)
104. C.G. Pantano, T.E. Madey: Appl. Surf. Sci. **7**, 115 (1981). Note error in Table 1 re. D_c value of Li_2WO_4
105. M.P. Seah: In *Surface Analysis of High Temperature Materials: Chemistry and Topography*, ed. by G. Kemeny (Elsevier, London 1984) p. 124
106. C. Klauber: Unpublished data
107. ASTM Standards on Surface Analysis (1986) ISBN 0-8031-0948-2

2 UHV Basics

C. Klauber

2.1 The Need for Ultrahigh Vacuum

Modern surface analytical methods over the last three decades have been dominated by those requiring ultrahigh vacuum (UHV) chambers in which to carry out the analyses. This is not a universal requirement for surface analysis and several of the techniques such as Fourier transform infrared (FTIR), scanning tunnelling microscopy (STM) and ellipsometry do not have a mandatory vacuum requirement. Vacuum is of course required by those techniques utilizing beams of particles and higher energy radiation so that the beams may be generated and travel undisturbed until intercepting the surface. The requirement for UHV or vacua of $\leq 10^{-10}$ mbar (10^{-8} Pa) is fundamental to surface analysis when those beams are employed. This arises due to the flux of residual gas molecules striking the surface i.e. the number of molecules per unit area per unit time, which is responsible for the pressure that those gas molecules exert upon the surface. By knowing the pressure the flux can be evaluated. From the kinetic theory of gases [1] the molecular flux Z is given by the Herz-Knudsen equation:

$$Z = \frac{Nc}{4V} , \tag{2.1}$$

where N/V is the number of molecules per unit volume and c is the average speed of the molecules. For a gas of molecular weight M at temperature T [1]:

$$c = \sqrt{\frac{8RT}{\pi M}} \tag{2.2}$$

where R is the gas constant. Combining the above with the ideal gas equation $PV = nRT$ and $N = nN_{\mathrm{A}}$ where N_{A} is Avogadro's number, gives:

$$Z = \frac{nN_{\mathrm{A}}P\sqrt{(8RT/\pi M)}}{4nRT} ,$$

and therefore

$$Z = \frac{N_{\mathrm{A}}P}{\sqrt{2\pi MRT}} . \tag{2.3}$$

For pressure P given in Pa (units $\mathrm{N\,m^{-2}}$) and M (units $\mathrm{g\,mol^{-1}}$), substitution of the relevant values for N_{A}, π and R gives:

$$Z = \frac{2.635 \times 10^{24}\, P}{\sqrt{MT}} \text{ collisions m}^{-2}\,\mathrm{s}^{-1} . \tag{2.4}$$

Substitution of P by a pressure-time integral gives the integrated gas flux or fluence (dose and exposure are used synonomously, though exposure is not technically equivalent).

The rate of molecular impingement can thus be seen to be dependent upon the molecular weight, temperature and pressure of the gas involved. At atmospheric pressure (1.01325×10^5 Pa) the flux is extremely high, e.g. for nitrogen at room temperature $Z = 2.91 \times 10^{27}$ collisions $\mathrm{m^{-2}\,s^{-1}}$. Since a typical surface might have an exposed atomic density $\sim 2 \times 10^{19}$ atoms $\mathrm{m^{-2}}$, each surface atom would be struck by $\sim 1.5 \times 10^8$ molecules every second. Surfaces in an atmospheric environment thus reach an almost instantaneous pseudo-equilibrium with the reactive gases present such as oxygen and water. If we created a virginal surface of some material by fracturing the bulk, then to maintain the virginal state we must reduce the gas pressure. At a pressure of 10^{-8} Pa the average time interval between a surface atom being struck by a gas molecule would be about 7×10^4 s or 19 hours. This is a useful time interval during which a variety of surface characterizations can be carried out, hence the need for UHV. In practical surface analysis often the surface of interest has been removed from a non-vacuum environment and placed in the instrumental system. The necessity for UHV remains since whatever reacted selvedge has formed it is essential that the residual gases in the vacuum system do not form an additional overlayer. In poorer high vacuum systems such as typically found in electron microscopy, reacted carbonaceous overlayers rapidly form due to the interaction of the impinging electron beam with adsorbed residual gases. Figure 2.1 illustrates the relationship between gas pressure, surface contamination times and mean free path lengths [2].

- Contamination time is the time taken for a perfectly clean surface to acquire a monolayer of contaminant. This time is dependent not only on the molecular weight, temperature and pressure, but also on the reactivity of the system. The latter can be simply expressed as a sticking probability s i.e. the probability that an incident molecule will remain on the surface (in some form) after collision, $0 \leq s \leq 1$.
- Mean free path λ is the average distance travelled by a particle before it undergoes a collision with another particle. This collision may or may not be inelastic i.e. involve the loss of energy. The latter case of inelastic mean free path (IMFP) is particularly important in surface analysis, not only for particles through gas, but particularly particles through solids as it is the crux of a technique's surface sensitivity (see Sect. 1.5.1).

The comparison with the variation of atmospheric pressure with altitude is interesting since it places the environment of earth orbit space flight in

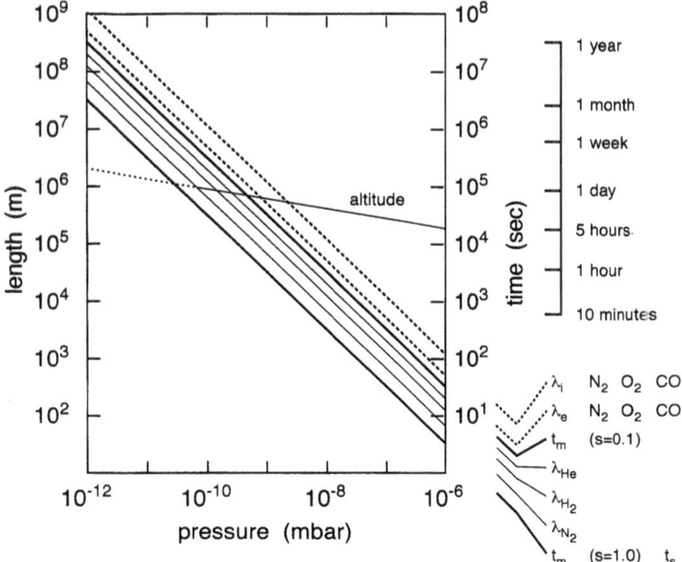

Fig. 2.1. Relationship between gas pressure, surface contamination times and mean free path lengths. The earth's atmospheric altitude variation is included for comparison. t_m is the time to form a monolayer of nitrogen with sticking probabilities $s = 0.1$ and 1.0, t_s is the average time before a surface atom is struck by a gas phase molecule. λ_e is the mean free path of a 100 eV electron in N_2, O_2 or CO, λ_i is the mean free path for that collision to be ionizing, λ_x is the mean free path of a molecule of X in X [2]

perspective. UHV of 10^{-10} mbar is attained at an altitude of about 900 km [3]. The American space shuttle has a maximum operational ceiling of about 1000 km.

Current practice is to measure pressure in millibar (mbar) units, with mbar × 100 yielding the SI derived unit of Pascals (Pa). This is convenient since mbar closely relates to the older metric, but non-SI unit, of the torr (1 torr = 1.333 mbar). The torr is still widely used, especially in the gas exposure unit of the Langmuir (L), where 1 L represents a pressure-time integral of 10^{-6} torr s (approximately the time for the formation of a monolayer). In view of the inaccuracies associated with pressure-time integrals the substitution of 10^{-6} mbar s is equally useful, though Pa s is more often used. *Menzel* and *Fuggle* [4] suggested replacing the concept of gas exposure with measurements of true gas fluence, the unit of fluence to be the Ex, such that 1 Ex is equal to 10^{18} collisions m^{-2}. Though quite logical, this suggestion has not been widely adopted.

2.2 Achieving UHV

In order to practically achieve the UHV regime the vacuum system must not only be pumped adequately but it must be free of true and virtual leaks. The latter arise from gases, especially water, adsorbed on the internal chamber walls and instruments. At room temperature these will slowly desorb constituting a large virtual leak which can continue for years. These are eliminated in a matter of hours by heating or baking the whole vacuum system, typically to 150–200°C. UHV chambers can be constructed from materials such as glass and aluminium, but most commonly the material employed is non-magnetic stainless steel. Where electrical insulation is required, such as in feedthroughs and lens supports, ceramics are usually utilized. Great care must be taken to ensure that no oil or high vapour pressure materials are used in the system. Some polymers and elastomers such as teflon and viton can be used provided that they are not overheated especially whilst under compression or tension. A diagrammatic elevation and cross section of a typical multi-technique surface analytical vacuum rig is shown in Fig. 2.2. The various flanges are attached by a plethora of bolts, making such systems look somewhat like pressure vessels. These are necessary to compress steel knife edges within the flanges into softer gaskets (oxygen free high conductivity copper) to achieve an all-metal leak proof seal. The flange design along with that of all mechanical, electrical and fluid feedthroughs and internal components are designed to remain leak-proof with repeated thermal cycling up to 250°C. Any true gas leaks that do occur are easily detected via the use of a quadrupole mass spectrometer (usually a mandatory attachment) tuned to He in residual gas analysis mode. A jet of He is directed externally at the suspected area until an influx is observed. The technology required to fabricate such components is the principal cause of the high cost of UHV systems. Several methods for baking exist, the most common being heater bands around various parts of the system, heating elements contained within an oven-like shroud or the more recent practice of internal quartz encased radiation bars. Whatever approach is taken it is essential that even temperatures are reached over all internal surfaces, so as to avoid condensation in cooler regions. Once a system has been up to atmosphere, pumpdown to high vacuum normally takes several hours, baking a further 10–12 hours with a similar time for all internal components to cool. After UHV has been achieved an analysis of the residual gases making up that low pressure invariably indicates a composition that is not related to atmospheric constituents. The dominant residuals are usually H_2, H_2O, CO, CO_2 and a variety of hydrocarbons. These promote a reducing environment, which given time, will lead to carbonaceous deposits on all internal surfaces, including introduced specimens. Depending upon the preferential nature of the pumping, what pump oils and lubricants are used and the system's past history, the precise makeup of the residuals will change from system to system.

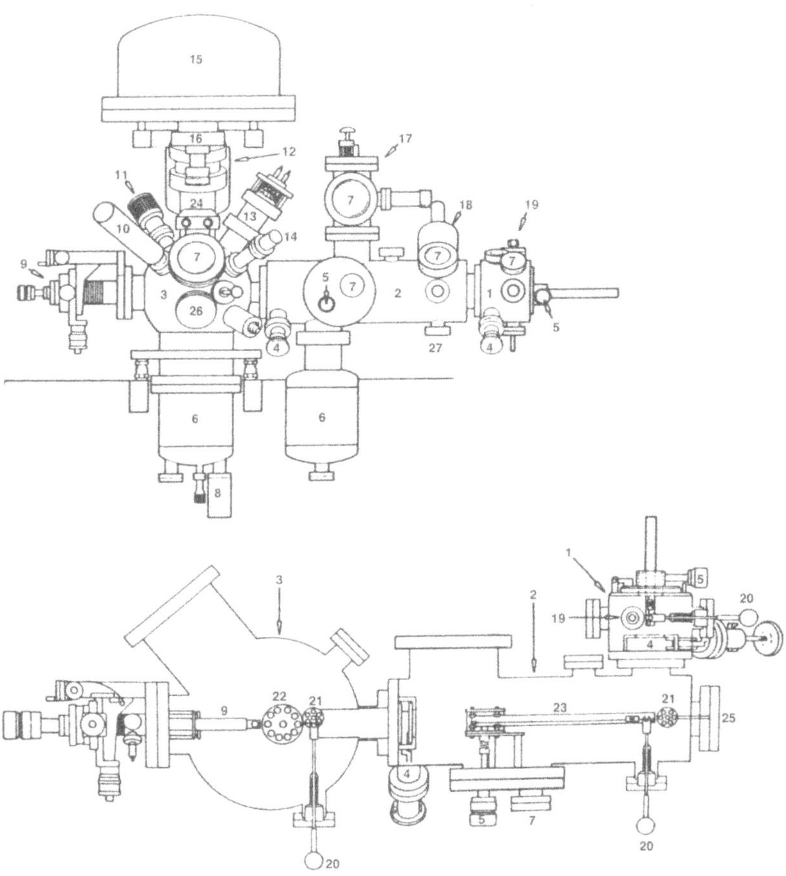

Fig. 2.2. Front elevation of a typical multitechnique surface analysis system (*top*) and a cross sectional view from above of specimen transporter mechanisms for such a system (*bottom*). Key to the diagrams: (*1*) Fast entry specimen insertion lock (stainless steel). (*2*) UHV specimen preparation chamber (stainless steel). (*3*) UHV experimental/analysis chamber (mu-metal). (*4*) Viton sealed gate valve. (*5*) Rotary drive to specimen transfer mechanism. (*6*) Titanium sublimation pump vessel (stainless steel). (*7*) Viewport. (*8*) Autocarousel motor drive. (*9*) High precision specimen translator (X, Y, Z translation and θ tilt). (*10*) Twin anode X-ray source (Al/Mg). (*11*) UV discharge source (UPS). (*12*) Monochromated X-ray source (Al/Mg). (*13*) 2000 Å electron source (AES, SEM, SAM). (*14*) Scanning ion source. (*15*) Electron energy analyzer vessel. (*16*) Detector (single or multichannel). (*17*) Specimen fracture stage. (*18*) Static, broad beam ion source. (*19*) High pressure gas reaction/catalysis cell. (*20*) Specimen transfer fork (wobble stick). (*21*) 6-specimen mobile carousel. (*22*) 10-specimen autocarousel. (*23*) Preparation vessel specimen transfer "railway". (*24*) Binocular microscope. (*25*) Alternative fast entry lock position or optional extension chamber port. (*26*) Port for monochromated electron source. (*27*) Port for specimen heating/cooling stage. Reproduced courtesy of VG Scientific Ltd, UK

A variety of vacuum pumps exist which can pump down to the UHV regime, each type has its own virtues. The most popular main chamber pumps are diffusion, turbomolecular or ion pumps. Diffusion pumps using modern fluids such as polyphenyl ethers are excellent performers but liquid nitrogen cooled traps are mandatory. Turbomolecular pumps can be used without traps, although trapping improves their ultimate performance. Titanium sublimation pumps, utilizing chemical gettering, are useful for additional pumping. Cryogenic based sorption pumps are particularly useful for rapidly evacuating large chambers. A comprehensive description of modern vacuum technology can be found in *Weissler* and *Carlson* [5].

Aspects of the various probes such as electron and ion guns, X-ray sources, mass analysers and electron spectrometers are dealt with in the subsequent chapters pertaining to the individual techniques. In commercially produced instruments these are normally produced by the one manufacturer. One-off research instruments often consist of a custom central chamber with a variety of accessories, often in-house constructed.

2.3 Specimen Handling

Means by which specimens of interest are introduced into vacuum can vary. Dedicated research instruments may need to be brought entirely up to atmosphere in order to change samples via unbolting flanges. The instrumental cost saving by eliminating a separate introduction chamber is impractical for analytical purposes as specimens need to be changed frequently. In any case, the subsequent baking required exposes the specimen to high pressures of residuals whilst at temperature, thereby forming a considerable contamination selvedge. The alternative fast entry systems are of two general types, one involves a specimen probe or rod onto which the sample is mounted, the rod then being pushed into the analytical region via a series of differentially pumped seals. The other, illustrated in Fig. 2.3, involves moving the specimen, mounted on a small stub, through one or two isolatable locks via rack and pinion or pulley specimen transporters. Stubs are a more convenient approach, but they can be less versatile with respect to heating and cooling of the specimen or making specific electrical connections e.g. as in attachment of a thermocouple. It is feasible to introduce up to a half-dozen stubs at once via a carousel system. With the specimen probe system electrical connections are made ex situ and the wiring travels with the specimen. The specimen translator for dedicated research instruments are usually superior at heating, cooling and can have additional degrees of mechanical freedom, such as azimuthal adjustments for angle-resolved studies. Figure 2.3 shows an internal view inside a vacuum system utilizing the stub method of sample transport.

The method of specimen mounting is largely governed by the nature of the material being examined. For most spectroscopies, good earthing of con-

Fig. 2.3. View inside a surface analysis vacuum system incorporating an auto-carousel and utilizing the stub method of sample transport. Reproduced courtesy of VG Scientific Ltd, UK

ducting specimens is essential to avoid static charging of the surface. Powders are often pressed into soft indium foil [6], whilst metallic specimens can be directly spot welded with suitable wire (typically nickel or stainless steel). Non-metallic conducting samples are usually silver dagged, and although quite conductive, the organic solvent in the dag can cause contamination problems. Bulk insulating samples can usually be readily glued with cyanoacrylate (super glue). Screws and clips are also employed. Indium foil is also useful for insulating powders simply as an adhesive system. Double-sided tapes have been profitably used for years, with X-ray photoelectron spectroscopy (XPS) in particular using the tape as an energy calibrant. The tape, however, is quite gaseous under X-ray irradiation and many workers avoid its use.

Viable specimen dimensions and masses are dependent on the vacuum system and technique(s) being employed. For instance, spot electron-excited Auger analyses can be performed quite easily on particles microns across, whereas signal-to-noise requirements for XPS mean that dimensions of 5–10 mm are routine. Large specimens that cannot be cut usually pose the greatest difficulty. Some systems will accept wafers up to 50 mm in diameter, but beyond a few mm in thickness they soon become unwieldy to handle. Of particular concern is a specimen's prehistory, especially the nature of fluids with which it has been in contact, as not all materials are UHV compatible. With, say, high vapour pressure lubricants, any attempts at solvent cleaning may disturb the surface that is really of interest and the information required may be lost.

2.4 Specimen Handling: ASTM Standards

In order to avoid erroneous results it is of critical importance to handle specimens correctly. Below is a condensed/adapted version of American Society for Testing and Materials (ASTM) standards designation E 1078–85 [7] for specimen handling in AES and XPS, although it is equally applicable to the other surface analytical techniques. As pointed out in the standard, researchers from more traditional analytical disciplines often need to be educated in the more stringent requirements in surface analysis. The technique sensitivities are such that levels of contamination can easily occur which will lead to severe perturbations. The key concept in specimen handling is *cleanliness*. All handling of the surface in question should be minimized or preferably eliminated whenever possible. If followed carefully, the procedures outlined below will reduce the chance of undetected errors, but no procedures are foolproof and, in some instances, innovation may be required depending on the nature of the specimen.

Visual inspection. Visual microscopic inspection prior to analysis is generally very useful. As contrast in secondary electron images may appear quite different from visual images it may be necessary to place an identifying scratch near to the region of interest in order to locate it. A visible light micrograph can also be helpful in correlating locations. Following analysis, reinspection may reveal signs of instrumentally induced artifacts, which will temper the assessment of results.

Specimen history. The environments the specimen has been subjected to will influence its handling e.g. an oil-coated mechanical component that has failed requires care more from the contamination it might impart to the spectrometer and other specimens. If the oil is removed prior to submission to the surface analyst, then the details of cleaning must be communicated. Often specimens are subjected to more conventional analytical procedures prior to surface analysis e.g. EDAX. In this case simply exposing the surface to radiation in a poor vacuum will ensure a contamination layer too thick for the surface techniques to probe. It is best to apply surface analysis prior to other techniques.

Sources of contamination. Specimens should only be handled with *clean* tools, avoiding the surface if possible. Regular cleaning of the tools with high purity solvents prior to use is essential. The necessity for increased dexterity often means that gloves need to be employed in lieu of tools. These need to be carefully selected as they can be a serious source of contamination. Although uncomfortable and awkward the disposable polyethylene gloves are generally the cleanest. Particulate debris can be imparted to the surface by compressed gas blasts intended to remove such items. Photographic quality canned gases are convenient and generally of a good standard, but any gas stream can charge a surface making it more susceptible to picking up particulates. An ionizing nozzle on the gas stream is recommended. Contamination can also arise from within the vacuum system. Hence the need for ultrahigh vacuum as already discussed. Of course total pressure is only a guide and a partial pressure analysis is very useful due to the vast differences in sticking probability that various molecules can have. Note that nearby hot filaments can alter the gas phase chemistry and/or enhance adsorption. Radiation damage has already been considered in Sect. 1.11.1. In situ sputtering can lead to direct contamination of other specimens which may be in the chamber e.g. as in carousel systems.

Of particular concern is not only specimen contamination but also analytical chamber contamination. A contaminated specimen represents a small loss compared to a chamber which, by comparison, may not be easily cleaned. High vapour pressure elements e.g. Hg, Te, Cs, K, Na, As, I, Zn, Se, P, S etc., or alloys thereof, should be analysed with caution. Silicone compounds can also pose a problem due to contamination by surface diffusion, despite their low vapour pressures. Apparently suitable specimens may also become unsuitable as a result of radiation damage.

Specimen storage and transfer. Storage is a problem even in very clean laboratories. Hence the best approach is to analyse specimens as soon as possible. Glove boxes, vacuum chambers and desiccators are generally the best means of storage as they can minimize oxidation/hydrolysis effects. These can also be adapted to directly connect to entry locks to avoid any atmospheric exposure of the specimen surface. The container itself should not be a source of contamination and materials with volatile components should be kept separate to avoid cross contamination. During shipping most factors can be easily controlled except temperature, which can be quite detrimental.

Overlayer removal. If the surface of interest lies below a coating or a layer of contamination then this outer layer will need to be removed. The most commonly employed technique is sputter profiling (see Chap. 4) through other approaches may be more suitable. Mechanical separation such as peeling is feasible if the interlayer bonding is sufficiently weak. Thick layers can also be removed by sectioning methods such as abrasive wheels, sawing, shearing, mechanical fracture, chemical or electropolishing. It needs to be remembered that the lubricants, abrasives and electrolytes all add another dimension to surface contamination, hence some in situ ion etching is invariably required. Most mounting materials used in sectioning e.g. thermoset plastics, are unsuitable for insertion into vacuum and need to be removed. Angle lapping and ball cratering have already been mentioned in the previous chapter.

In certain cases solvents and surfactant solutions can be employed to remove soluble contaminants and overlayers. Ultrasonic agitation vastly improves the effectiveness in these cases but care needs to be employed with liquid purities. Even high purity solvents can evaporate and leave behind tell-tale interference films. Although solvent soaked tissues are often used to clean tools etc., it is best to avoid tissue contact with specimen surfaces.

With refractory materials, heating can be successfully employed to remove contaminants. However, this finds most application in basic surface studies requiring initially clean and reproducible surfaces rather than in general materials analysis. For non-refractory systems the high temperatures would undesirably alter the specimen. Some overlayers may be sufficiently volatile that they can be pumped away at room temperature given sufficient time. Of course, even in UHV, if too long is required vacuum residuals end up replacing the original contamination. Degassing specimens is an extension of this problem. Normally this degassing is best performed in an auxillary chamber to prevent degrading the vacuum in the main analytical chamber.

In situ exposure techniques. Such exposure can be achieved by fracture, cleaving or scribing. Impact or tensile fracture can be achieved in specially designed chambers provided specimen geometry and materials characteristics criteria are met. Non-ideal geometries can often be accommodated by incorporation into or attachment to an ideal geometry. Ease of fracturing of metals can often be enhanced by lowering temperature or by charging intergranular regions with hydrogen. As electrical insulators may pose serious charging problems, exterior conductive coatings ought to be applied prior to in situ fracture. Scribing is useful provided the scribe mark is larger than the input probe and none of the constituents are smeared.

Specimen mounting. This has been briefly touched upon in Sect. 2.3. The most serious difficulty which might arise concerns specimens which are poor electrical conductors. In addition to the approaches pointed out in Sect. 1.11.2 to minimize charging it may also be feasible to place a conductive mask, wrap or coating about the specimen. This mask e.g. metal foil or colloidal graphite, needs to be effectively earthed and make contact as close as possible to the area to be analyzed. When sputtering is also used care must be taken not to sputter mask material onto the surface to be analyzed.

Given specimen geometries can require some innovation depending upon the nature of the analytical system. For instance, wire or fibre specimens are best clamped so as to suspend them free in space away from the main part of the sample stub. When the overhang is in the focal area of the spectrometer background, artifacts from the stub can then be avoided. Specific particles which cannot be grouped as a powder or compacted as a pellet can sometimes be floated on suitable liquids and picked up on conducting fibres.

References

1. W.J. Moore: *Physical Chemistry* (Longman, London 1972) p. 133
2. Adapted from chart devised by L. de Chernatony, GEC-AEI publication 2015–14
3. R.C. Weast, M.J. Astle (eds.): *CRC Handbook of Chemistry and Physics*, 63rd edn. (CRC Press, Boca Raton 1982) F-164–168
4. D. Menzel, J.C. Fuggle: Surf. Sci. **74**, 321 (1978)
5. G.L. Weissler, R.W. Carlson (eds.): *Vacuum Physics and Technology* (Academic, New York 1979)
6. G.E. Theriault, T.L. Barry, M.J.B. Thomas: Anal. Chem. **47**, 1492 (1975)
7. ASTM Standards on Surface Analysis (1986) ISBN 0-8031-0948-2

Part II

Techniques

3 Electron Microscope Techniques for Surface Characterization

P.S. Turner, C.E. Nockolds, and S. Bulcock

Electron microscopy in its various forms has developed over the past fifty years into one of the major techniques of materials science. Surface analytical techniques are more recent additions to the materials scientists' range of experimental methods for learning about materials properties. Microscopy and spectroscopy are complementary, and the use of one alone can result in an inadequate characterization of a material. In this chapter we will consider the various modes of electron microscopy and attempt to demonstrate the critical importance of applying electron optical imaging together with surface analytical techniques in studies of surfaces.

The textbook picture of a specimen to be examined using surface analytical techniques such as XPS or SIMS is typically a perfectly flat surface of a laterally homogeneous material (Fig. 3.1a). Of course, as the chapters of this book make clear, no real sample of any significant interest is like that, and techniques such as scanning Auger microscopy can provide information about variations in composition across a surface, with resolution to $< 100\,\mathrm{nm}$. Nevertheless, there is a tendency to interpret surface analytical data in terms of a postulated model of the surface. Only by looking at the surface, with whatever resolution is required in order to see the relevant detail, can one be confident of the interpretation of the surface analytical data. Thus microscopy – optical, electron, scanned probe, etc. – must be employed in conjunction with the surface spectroscopies. For ex-

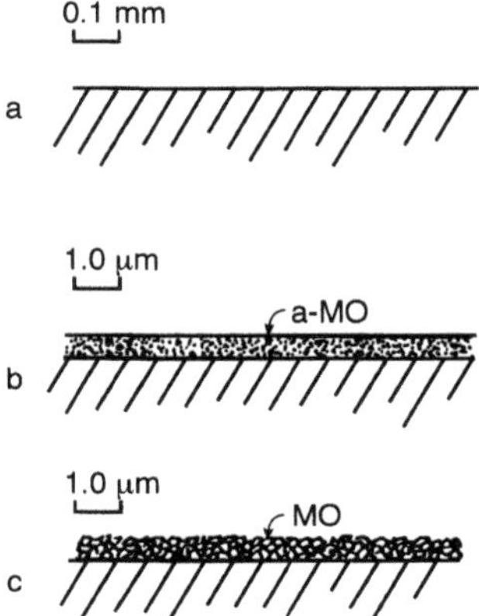

Fig. 3.1. Schematic illustration of some typical types of surface (see text for details)

ample, the two samples illustrated in Fig. 3.1b,c could give identical XPS spectra and depth profiles, whereas one consists of a continuous amorphous oxide film and the other of a fine dispersion of oxide crystallites across the surface.

The aim of this chapter is to introduce the basic concepts of electron microscopy of surfaces to readers who are not familiar with electron microscopy. We will consider the ways in which scanning and transmission electron microscopes can provide information about surfaces, and the nature of that information. Two other surface imaging systems – scanned probe microscopy, and scanning Auger microscopy – are covered elsewhere in this book (in Chaps. 10 and 6 respectively) and will not be considered here. An excellent introduction to more advanced aspects of the subject has been presented by *Venables* et al. [1]. The subject was covered in the proceedings of two NATO Advanced Study Institutes [2,3], and at the 1989 Wickenburg Workshop [4].

In this short article, we cannot cover all relevant aspects of the instruments and techniques in detail. There are many excellent texts on electron microscopy, some of which are listed at the end of this chapter as sources for further reading [5–11]. In particular, more detailed descriptions of the major components of electron microscopes, and more comprehensive explanations of their operation and performance, should be studied by anyone proposing to learn to use electron microscopy for the characterization of surfaces. A good starting point is the short book by *Goodhew* [5].

In the following sections, we will consider first the nature of the information we may require from electron microscopy, then the major characteristics of electron optical systems, and the factors which determine the nature and quality of electron images. The roles of the scanning electron microscope and the transmission electron microscope in surface studies will be described and illustrated. In the final section we will look briefly at other developments in the imaging of surfaces.

3.1 What Do We Need to Know About Surface Structures?

The variety of samples which are studied using surface analysis is immense, ranging from atomically flat semiconductors grown in situ using molecular beam epitaxy, to pieces of corroded brass taps and compressed powders. For the purposes of this discussion, we will consider the problem of investigating the reactivity of a multiphase ceramic exposed to chemical attack via either vapour or liquid, and of associated separate studies of the reactivities of the major component phases in this ceramic.

We could prepare surfaces of a single phase, polycrystalline ceramic by polishing, and also by fracturing it. In order to interpret our surface analytical spectra, we will need to know some or all of:

- the grain size;

- whether there are minor phases present in the nominally single phase material;
- the orientation of grains;
- whether chemical attack results in the growth of precipitates;
- whether attack occurs preferentially at boundaries, or on certain grains;
- the topography of fractured surfaces, and whether they reveal intergranular or transgranular fracture;
- whether the sample is porous.

These questions can probably be answered using scanning electron microscopy on the samples used for surface analysis. Scanned Probe Microscopy should also be considered (see Chap. 10). A higher image resolution would be required to determine

- the width and structure of intergranular films;
- the crystal structure of minor phase or precipitates;
- whether amorphous surface layers are formed during the reactions; and so on.

If transmission microscopy is required, the preparation of samples representative of those studied by surface analysis may provide difficulties.

Similar questions will arise for the multiphase ceramic, with additional points such as

- the degree of homogeneity of the multiphase material;
- different rates of attack on different phases;
- variations in the width of intergranular films, and preferential attack at grain boundaries.

Whatever the sample – ceramic, metallic, polymeric – similar questions can be identified to guide the selection of imaging techniques needed to complement the surface analyses.

3.2 Electron Optical Imaging Systems

There are two ways in which images can be obtained using electron beams. The traditional transmission electron microscope (TEM) is a close analogue of the light optical microscope, and involves the illumination of a transparent object by a beam of electrons, and the formation of a magnified image of the electron waves emerging from the object using an objective lens and two or more projector lenses. The scanning electron microscope (SEM) uses lenses to form a demagnified image of the electron source. This fine probe of electrons is scanned across the sample, and one of a number of possible signals arising from the electron-specimen interaction is detected, amplified, and used to modulate the intensity of a TV-like image tube, which is scanned at the same rate as the electron probe. Clearly there are fundamental differences between these methods, but both use electron sources, electron lenses and electron detectors, and therefore both are limited by the performance of such devices.

The TEM and SEM techniques converge in the STEM (scanning TEM) in which a sub-nanometre diameter probe is scanned across a thin sample and a transmitted electron signal – often also filtered to remove inelastically scattered electrons – used to form the image. Both TEM and STEM can be operated in reflection mode, in which the electron beam is incident at a glancing angle and reflected and/or diffracted from a flat sample (REM and SREM).

In forming an image, two factors govern the information which may be obtained about the object – the resolving power of the imaging instrument (the smallest resolvable separation of two distinct points in the object) and the contrast in the image (the difference in intensity which allows us to distinguish resolved detail from a background intensity level).

The best resolution attainable with a given instrument is determined by the quality of the objective lens in the TEM or the probe-forming lens in the SEM. The wave nature of the electrons leads to a fundamental diffraction-limited resolution, but the high aberrations of magnetic electron lenses determine the actual resolution. For a given lens, the TEM resolution and the smallest probe diameter which that lens could produce are essentially identical (e.g., $\sim 0.3\,\mathrm{nm}$ at $100\,\mathrm{keV}$), but in practice a typical commercial SEM has a best resolution significantly worse than a standard TEM, due to lower electron energies and larger focal lengths (e.g., $3\,\mathrm{nm}$ at $15\,\mathrm{keV}$). With the recent widespread availability of field emission sources for SEM, and probe sizes down to $1\,\mathrm{nm}$, greatly improved resolution of surface is possible, but is often not achieved due to limitations in contrast.

The contrast of an image is limited by the random nature of the emission of electrons from the source. If we consider a particular element in the image (a picture element or pixel) with an intensity corresponding to the detection of N electrons in that pixel, then the standard deviation in the intensity will be proportional to $\sqrt{N}$. The difference ΔN between signals from nearby pixels should be at least $3\sqrt{N}$ if contrast between these pixels is to be detected in the image [8,9]. In the TEM, all image points are detected simultaneously (parallel detection), and in practice it is usually easy to ensure that the noise level ($\sqrt{N}/N$) is very low. But in the SEM each pixel is detected in sequence as the probe scans across the sample (serial detection); the noise level in scanned images is a critical factor in determining the image contrast and detectable resolution.

3.2.1 Electron Sources

A critical factor in electron imaging is the current density of the electron beam at the object. The electron optical brightness, β of the system, is defined by

$$\beta = \frac{J}{\pi\alpha^2} \tag{3.1}$$

where J is the current density at a point in the illuminating beam, and α is the half-angle of the converging cone of illumination ($\pi\alpha^2$ steradian is the solid angle subtended by the illumination at that point). The significance of β is that it has a constant value below the electron gun, and it is proportional to the electron energy E. In a focussed probe of electrons we can express J in terms of the probe current I_{p} and the diameter d_{p} of the probe crossover, so

$$d_{\mathrm{p}}^2\,\alpha^2 \simeq \frac{4I_{\mathrm{p}}}{\pi^2\beta} \tag{3.2}$$

Thus for scanned imaging systems, with a given beam current, small probe sizes require high convergence angles, and vice versa.

There are three types of electron source used in contemporary EMs, differing in maximum brightness and in maximum beam current. The traditional electron gun uses a tungsten wire, heated to about 2700 K, from which electrons are emitted and accelerated to the required energy. Under optimal adjustment, a brightness $\beta = 5 \times 10^5\,\mathrm{A\,cm^{-2}\,sr^{-1}}$ is obtained at 100 keV, with a total current of up to 100 µA from an effective source diameter of about 30 µm. A more recent development is the lanthanum hexaboride (LaB_6) source, which, operating at lower temperatures, gives about ten times higher brightness, but requires better vacuum conditions. The field emission source, a pointed tungsten tip operated in UHV at room temperature, can provide brightness values as high as $10^9\,\mathrm{A\,cm^{-2}\,sr^{-1}}$ at 100 keV, but at much lower total beam currents. For details of the construction and operation of electron guns using these sources, the reader is referred to texts on electron microscopy, e.g. [6,9,10]. The standard guns for TEMs and SEMs, using W or LaB_6, are very flexible, and easily adjusted to provide for a range of beam currents at or close to the corresponding maximum brightness values.

These electron guns are almost monoenergetic, with a narrow range of electron energies around the mean, due to the thermal spread in energy of the emitted electrons. In the thermionic guns, this is about 1.5 eV, whereas for the field emission gun it is about 0.25 eV.

3.2.2 Electron Lenses

Electric and magnetic fields having cylindrical symmetry act as lenses for charged particles. Electrons emerging from a point in an object are brought to a focus in the corresponding image plane. In comparison with light optical lenses, electron lenses suffer severe, uncorrectable aberrations, particularly spherical and chromatic aberration, as a result of which a point image is blurred into a disc of diameter d_{ab} given by [9]

$$d_{ab}^2 = d_s^2 + d_c^2 + d_d^2 ,$$

with

$$d_s = 0.5C_s\alpha^3 , \; d_c = C_c\alpha \,\Delta E/E \quad \text{and} \quad d_d = 1.22\lambda/\alpha^2 , \tag{3.3}$$

where C_s, C_c are the spherical and chromatic aberration coefficients, α is the angle between the lens axis and the electron trajectory, ΔE is the effective energy spread of electrons having mean energy E, and λ is the electron wavelength. In addition to these inherent aberrations, all electron lenses suffer from quite severe astigmatism. Manufacturers provide compensating coils to counter this; in practical electron microscopy the correction of astigmatism is a critical factor in obtaining an optimum image.

With the exception of the electrostatic lens action of the guns, all electron microscopes use electro-magnetic lenses. These consist of closed coaxial soft iron cylinders enclosing the windings, with a specially shaped gap – the pole piece – at which a strong magnetic field is constrained within the bore of the lens. The shape of the pole piece depends on the purpose of the lens. A probe-forming lens for SEM may have a conical shape, and provide a range of focal lengths from 5–50 mm. The highest resolution TEM instruments have objective lens focal lengths down to 1 mm or less, with the object immersed in the magnetic field. The values of C_s and C_c depend upon the pole piece geometry, but are of the same magnitude as the focal length. The strength of a particular lens can be varied over a considerable range by varying the current in the magnet windings.

The trajectories of electrons in magnetic fields rotate around the axis of the lens, and hence electron images rotate as the strength of a lens in varied. In today's microprocessor-controlled instruments, groups of lenses are varied in such a way as to give almost zero total rotation while image magnification is changed.

3.2.3 Detection Systems

In transmission microscopy the image is conventionally viewed using a flat screen coated with a powder which fluoresces under the electron beam. Images may be recorded on photographic film placed below this screen. The photographic emulsions used have a fine grain size and are highly efficient and convenient. Recent developments in electronic detection systems, especially CCD arrays, have resulted in a change towards digital recording. These systems have improved to the point where their resolution is close to that of photographic film, with greater dynamic range and linearity. Digital storage systems allow integration of image intensity to reduce noise and enhance contrast. Subsequent digital image processing and direct importing into reports are further advantages.

A variety of detectors are used in the scanned image systems, in order to select and amplify the required signal. We will consider the individual detectors in Sect. 3.3, but may note some general common factors. The detectors are characterized by an efficiency factor, which can be $\ll 1$. The serial nature of scanned image formation, with electronic signal detection, allows for amplification and for background subtraction before the signal is applied to the image tube. All such detection systems can add to the image noise, and the quality and performance of amplifiers can effect the speed at which scanned images may be viewed and recorded. Thus TV rate display of images, while convenient, necessarily results in lower contrast images than do slow scan rates, since there are fewer electrons per pixel in TV images. Again, the use of digital frame stores which average successive frames has become common in SEM.

3.3 Scanning Electron Microscopy of Surfaces

The standard SEM is the most useful electron optical system for surface imaging, and should be applied routinely to characterize all surfaces which will be, or have been analyzed by XPS, SIMS, etc. (It will usually be preferable to image after surface analysis, to avoid the effects of contamination during SEM examination.)

3.3.1 The SEM

The essential elements of an SEM are shown schematically in Fig. 3.2. The electron gun, fitted with a W, LaB_6 or Field Emission (FE) gun operates typically over the range 0.1–30 kV accelerating voltage. A condenser lens produces a demagnified image of the source, which in turn is imaged by the

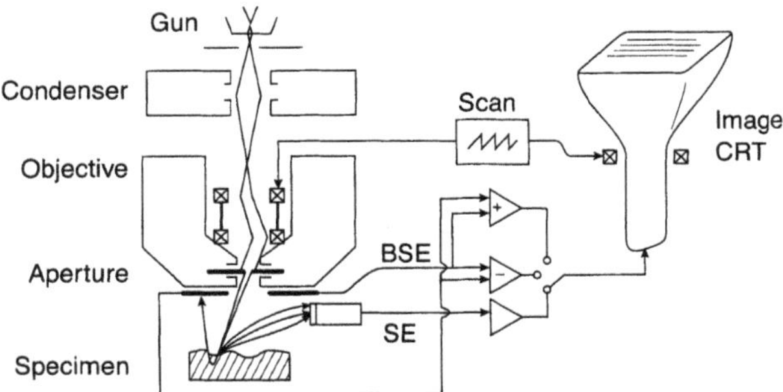

Fig. 3.2. Basic features of the SEM, showing electron gun, condenser and objective (probe-forming) lenses, scanning system, detectors and image display CRT. The diameter of the aperture and the working distance from lens to specimen determine the convergence angle of the probe. The probe at the specimen is the demagnified image of the source, broadened by lens aberrations

probe forming lens (often called the objective lens) onto the specimen. The electron path and sample chamber are evacuated. Scanning coils deflect the probe over a rectangular raster, the size of which, relative to the display screen, determines the magnification. Detectors collect the emitted electron signals, which after suitable amplification can be used to modulate the intensity of the beam of the display video screen, which is rastered in synchronism with the probe.

3.3.2 The Signals and Detectors

When an electron is incident on the surface of a thick specimen, several signals can be detected. The primary electron results in ionization of atoms along its path in the solid, which in turn can result in the ejection from the surface of secondary electrons, very close to the incident beam position. These have energies from 0–20 eV, and can be attracted to a positively charged detector with high efficiency. The secondary electron yield per primary electron is high and increases as the angle ϕ between the electron beam and the surface normal increases in proportion to $1/\cos\phi$. For example, for 10 keV electrons incident normally ($\phi = 0°$), the secondary emission coefficient is ~ 0.2 for Al, increasing to 1.0 at $\phi = 75°$ incident. The coefficient varies with incident electron energy E in proportion to $E^{-0.8}$ [8,9].

The primary electrons can be deflected through large angles, and hence emerge as backscattered electrons with high energy, from a region surrounding the incident probe. Where these backscattered electrons emerge from the surface, secondary electrons are generated, so a component of the secondary signal is proportional to the backscattered yield. At sharp edges, scattered primary electrons can give rise to very high secondary signals.

By placing a large annular detector below the final lens, facing towards the sample, a proportion of the backscattered electron (BSE) signal can be collected. The scattering of primary electrons within an elemental sample of atomic number Z increases with Z, but is relatively insensitive to E over the range 5–30 keV. Thus the BSE signal from a flat, polished sample provides contrast which is composition dependent. The BSE signal is also dependent on the orientation between the beam and surface. The typical BSE annular detector is split into two semicircles; the sum of the two signals is predominantly sensitive to composition, whereas the difference signal is sensitive to topography.

The resolution of the backscattered electron image is typically of the order of 0.1–1 μm, being determined by the volume within the sample from which most of the detected BSE signal comes. However, using high brightness guns and efficient backscattered electron detectors it is possible to image very finely spaced regions regions showing compositional contrast at resolutions down to 2–3 nm. It is clear from this work that there is a small high resolution component in the BSE signal which is generated by incident beam electrons being scattered out of the sample very early in their path. In principle the resolution of the whole BSE signal can be improved by reducing the beam energy but this is offset by the fact that the detectors become less efficient for the lower energy electrons. The current generation of SEMs have the capability of operating at low kV with very small beam diameters and high beam currents and this opens up new possibilities for the study of surface composition using BSE.

3.3.3 Resolution and Contrast in SEM Images

The best resolution attainable in principle is determined by the minimum probe size, which is obtained when the image of the electron source is demagnified to the greatest extent possible with the available lenses, using a final aperture angle α chosen to minimize the combined effects of spherical aberration ($\sim C_s\alpha^3$) and diffraction ($\sim \lambda/\alpha$). However, from the brightness relationship, (3.2), the resulting current I_p in the probe will be too low for useful imaging unless a field emission source is fitted. Thus in standard SEMs, using W or LaB_6 sources, it is usual to select the probe current required for the particular imaging condition of interest (i.e., required contrast level), so that the geometrical probe size d_0 and aperture angle α are related, from (3.3), by

$$d_0 = \sqrt{4I_p/\pi^2\beta}/\alpha \,. \tag{3.4}$$

The probe diameter d_p is then estimated by adding to d_0 the spherical, chromatic and diffraction aberration disk diameters d_s, d_c, and d_d [see (3.3)] in quadrature [9]:

$$d_p^2 = d_0^2 + d_d^2 + d_s^2 + d_c^2 \tag{3.5}$$

The dependence of probe diameter d_p on aperture α for a given set of operating parameters (gun voltage and brightness, objective lens focal length and aberrations) is illustrated in Fig. 3.3, for two values of the probe current. The resolution is poorer for higher beam currents, because d_0 is increased. For the higher current, the optimum resolution is achieved at a higher value for the angular aperture α. In practice α is set to one of several fixed values, determined by the diameters of the actual apertures and the lens-to-sample working distance, so the optimum operating conditions for a given probe current can only be approximated.

The resolution actually achieved in the secondary electron (SE) images can be close to the optimum, with contrast determined by the topography of the surface. For example, images of compressed pellets of a fine single phase powder (a common form of sample for routine XPS analysis) will show topographical contrast, revealing the particle size and shape, and the extent of packing of the powders (Fig. 3.4), and, at high magnification, fine details of surface topology. This information is clearly relevant to the interpretation of any depth profiles which might be recorded in XPS/SIMS/AES instruments.

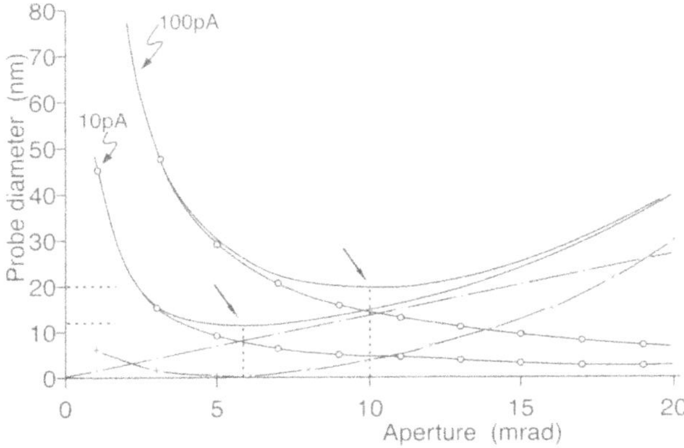

Fig. 3.3. Variation of SEM probe diameter with aperture angle, in accordance with (3.4). The contributions due to diffraction (d_d: —+—); spherical aberration (d_s: —×—); chromatic aberration (d_c: —·—) are shown for the case of 15 kV, 2 eV energy spread, C_s = 7.5 mm and C_c = 10 mm. Curves for the geometrical probe diameter d_0(—o—) and total probe diameter d_p (——) are shown for two probe current values and gun brightness 3×10^4 A cm^{-2} sr^{-1}. The optimum probe diameters of 12 and 20 nm are obtained for apertures of 6 and 10 mrad respectively (*arrows*)

Fig. 3.4. SEI micrograph of a crushed single phase mineral sample, pressed onto a conducting adhesive surface, as used for some XPS characterizations

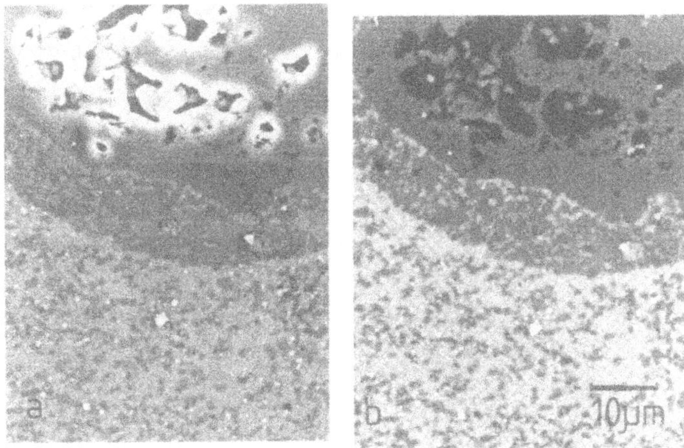

Fig. 3.5. (**a**) SE and (**b**) BSE SEM images of a polished sample of Synroc, showing the strong edge contrast in the SE image at pores; in both images, the contrast differences between grains is due to backscattered electrons

If there is little or no topographical contrast (as for a polished sample), then the SE image may have contrast determined by those SE's generated by the backscattered electrons; the image will also show correspondingly poorer resolution. This effect is shown in Fig. 3.5a, for a polished surface of Synroc, the multiphase titanate ceramic for immobilization of radioactive waste [12]. Here most of the contrast is compositional, arising from the atomic number dependence of the SE signal derived from BS electrons. The presence of, and topography of, pores and cracks is highlighted by the very strong topographical contrast at the edges of such features. The BSE image is similar but does not show up the structure of the pores (Fig. 3.5b).

Although the secondary electron image usually has better resolution and higher contrast (relative to noise levels), the backscattered images are often more valuable. This is illustrated in Fig. 3.6, showing SE and both topological and compositional BSE images of a fracture face from the sample of Synroc. The "Compo" BSE image shows the distribution of phases; only at large cracks is there some topographical contrast. The broad features of the topology of the fracture face are revealed in the difference between the two BE signals – the compositionally sensitive signal being suppressed by subtraction. This gain in information is at the cost of increased noise level, as shown in the line traces for all three signals. The SE image shows fine details of the fracture face topography, but this can be confused by the underlying compositional signal.

With the advent of high brightness field emission guns, low voltage SEM has become highly practical and is now widely used. The nominal resolution at 1 kV for a Field Emission gun SEM (FESEM) is typically 2–3 nm and with improved design of the lens systems high beam currents are available

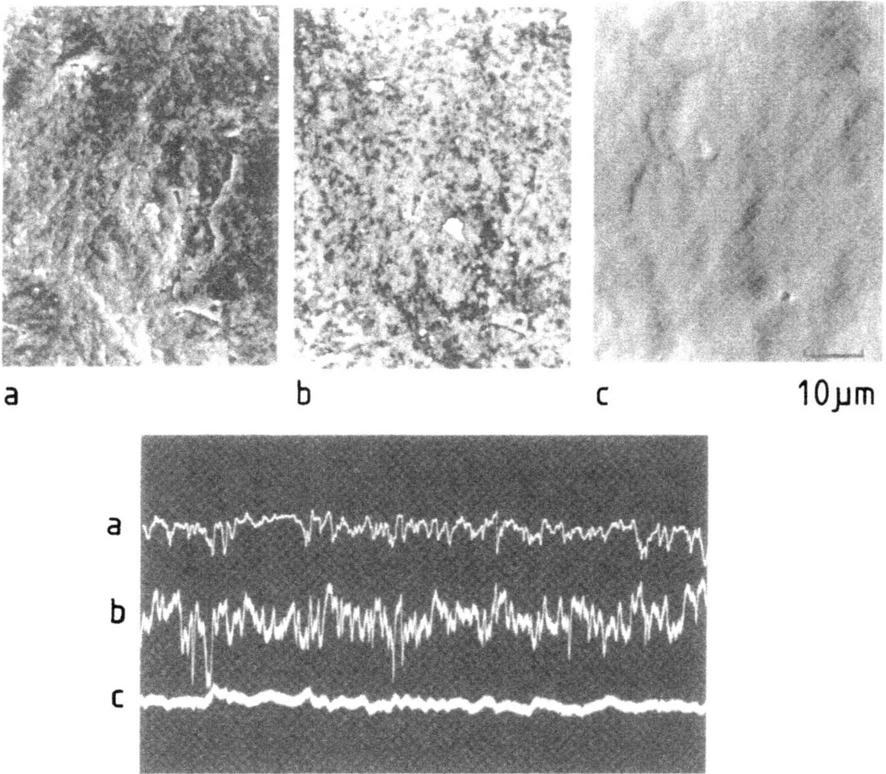

Fig. 3.6. SEM images and line traces from a fracture face of Synroc, for (**a**) SE (**b**) BSE compositional and (**c**) BSE topographical signals. The much greater noise levels in the BSE signals, are evident in the line traces; the topographical BSE image is formed by subtracting signals from the two halves of the BSE detector, thus removing the compositional contrast, but losing detail to enhanced noise levels

at low kV. One of the major advantages of operating at low beam energies is the reduction in the interaction volume. Monte Carlo calculations [26] show that at 10 keV the maximum depth of penetration of the beam in Si is approximately 1 μm, while at 1 keV it is close to 0.03 μm. This means that the secondary electron image at 1 kV gives a better representation of the surface detail. Another advantage of low kV operation is that the secondary electron yield increases very strongly, approximately as a function of the inverse of the beam energy.

Low voltage SEM has other advantages, especially for non-conducting materials for which a conducting coating (usually about 10 nm of gold) is necessary for conventional SEM imaging. There is an electron energy, usually around 1 kV, at which no charging occurs. Thus no metal coating of such samples is required, and fine surface detail is not obscured. In addition,

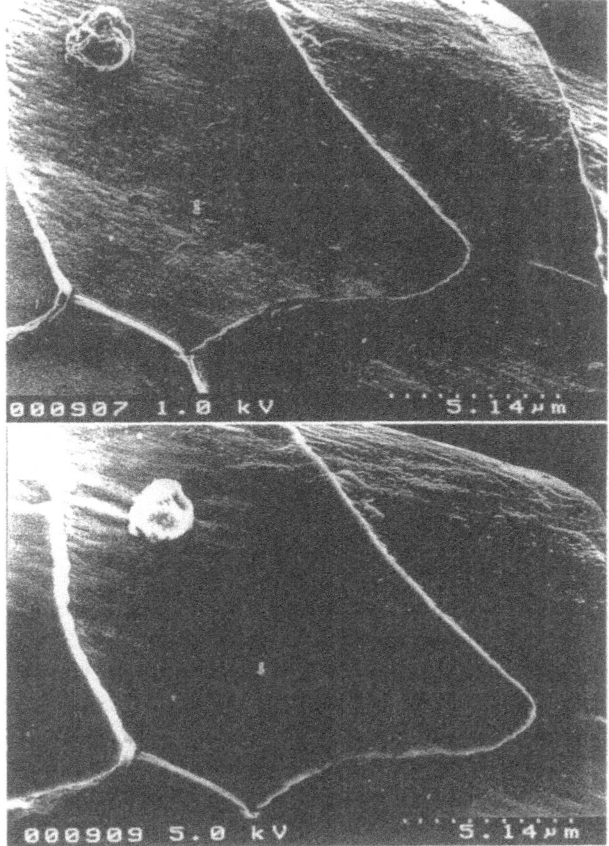

Fig. 3.7. FESEM images of scale structure of a Merino sheep wool fibre. At 1 kV on wool, no significant charging of the fibre occurs, allowing imaging without the need for a conductive coating. Fine details of the surface are revealed. At 5 kV, electrons penetrate further into the fibre, and the contrast of surface detail is greatly reduced. (This sample was coated with a thin film of evaporated carbon, to avoid charging at 5 kV.) Images courtesy of Mr David Watson, CSIRO Textile and Fibre Technology.

electron radiation damage to sensitive samples is reduced. An example of the improvements in surface image detail of wool fibres obtainable with the FESEM is shown in Fig. 3.7.

3.3.4 Variable Pressure SEM and Environmental SEM

In 1975 Robinson showed that by placing a small pressure limiting aperture between the specimen chamber and the gun chamber and pumping the two chambers separately it was possible to operate the gun at high vacuum while maintaining a much poorer vacuum around the specimen [27]. Under these conditions the secondary electron detector is disabled because of the high voltage bias on the detector surface, however backscattered electron detectors can still be operated. With this system Robinson was able to image a "wet" sample held at 5°C using backscattered electrons and a specimen chamber pressure of 5 mBar. At pressures in the region of 0.2 to 0.5 mBar, charge build up on the surface of insulating materials is neutralised by interaction between the gas molecules and the electrons, and in many cases it is possible to study these materials without applying a conductive coating. Most current SEMs now offer the option of operating at chamber pressures up to about 2 mBar.

Danilatos [28] realised that the high pressure specimen chamber was a potential ionisation chamber and by placing an electrode in the specimen chamber and applying a voltage he was able to obtain a strong signal on the electrode. This signal was found to show similar contrast to the conventional secondary electron signal. Using extra levels of differential pumping it is now possible to operate at pressures up to 20 mBar and form images using this "environmental secondary electron detector" (ESD) signal. Danilatos coined the name "environmental SEM" (ESEM) for this type of microscope. The ESEM has made it possible to study a wide range of specimens and to carry out experiments in the microscope, by varying, for example, the gas pressure, the gas mixture and the temperature of the specimen (melting and cooling).

Work is continuing on understanding the image formation processes in the ESEM. Images with different information can be obtained with the ESD by varying the operating conditions. Griffin has demonstrated a charge contrast image (CCI), that show defects in crystalline material, similar to cathodoluminescence imaging but with a much better signal to noise [29]. CCI also has the interesting characteristic that the image appears to show up very fine contamination on the surface, similar to that one would expect from a conventional secondary electron image obtained at < 0.5 kV. Another benefit of the ESEM in the study of surfaces relates to the problem of "carbon contamination". The build up of carbon on the surface is caused by an interaction between the electron beam and any hydrocarbon contamination on the specimen surface and this can have a severe effect on the image at low beam energies. This effect appears to be greatly reduced in the elevated pressure environment of the ESEM.

3.3.5 Energy Dispersive X-Ray Spectrometry

X-rays are generated by the incident probe within a volume similar to, but rather larger than, that for the backscattered electrons. Peaks at energies characteristic of the elements within that volume can be identified and the concentrations of the elements can be calculated. Thus the bulk composition of the sample, and of the individual grains in a polycrystalline sample (provided they are larger than the X-ray excitation volume) can be determined for comparison with surface analytical data.

The dimension of the interaction volume depends on the mean atomic number and the density of the material, and on the beam energy and the emitted X-ray energy. Diameters range from 0.03 μm to several μm [8]. Thus conventional EDX spectrometry will not reveal compositional changes due to surface segregation. It is possible, however, by analysing a point at several different beam energies, to determine the thicknesses and compositions of layers on substrates, and hence determine the composition of this surface layers [30]. X-ray maps of surfaces, with spatial resolutions of 0.1–1.0 μm are often valuable in revealing compositional variation across a sample, especially if scanning Auger microscopy is not available.

3.4 Transmission Electron Microscopy of Surfaces

The TEM is usually used to study the internal microstructure and crystal structure of samples which are thin enough to transmit electrons with relatively little loss of energy. This requires thicknesses in the range 20–300 nm, depending on the average atomic number of the material, using the typical 200 keV TEM. Clearly the bulk samples used in surface analysis cannot be examined directly this way. The transmission microscope is nevertheless an important adjunct to the SEM in studies of surfaces, because of its greater resolving power (down to crystal lattice dimensions), its ability to provide surface sensitive diffraction data and images, and associated analytical capabilities through X-ray and electron energy loss spectroscopies. In addition, images and diffraction patterns may be formed in the reflection mode, off smooth surfaces or facets. The challenge is to ensure that the surfaces of samples examined in the TEM are equivalent to those studied by surface analysis. The most satisfactory approach would be to combine the surface analytical and microscopy techniques in the one instrument; some progress has been made in this direction (Sect. 3.5). However, we will consider here the more typical situation of separate instrumentation, and the consequent requirements on sample preparation to achieve our aim.

3.4.1 The Transmission Electron Microscope

The TEM is very similar to the conventional light optical microscope in terms of optical principles. The electron source is followed by two condenser lenses to provide a uniform illumination of the specimen over the area of interest, adjustable as the magnification is changed. The sample is mounted on a stage to provide suitable movement. The primary image is formed by the objective lens, which determines the resolution obtainable. The final image is projected onto a viewing screen through two or more projection lenses, and can be recorded on film placed below the screen, or using CCD array detectors.

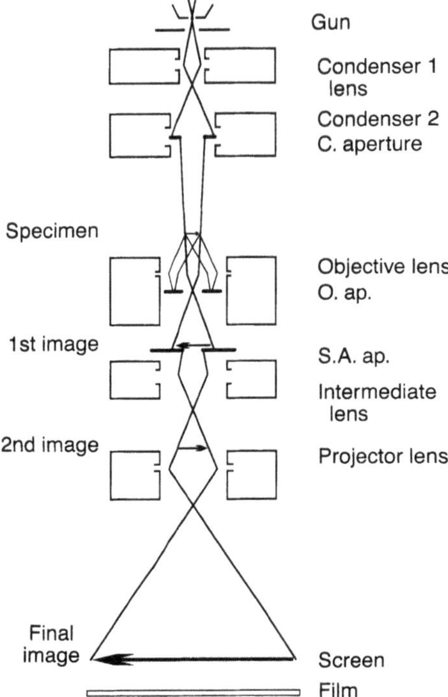

Fig. 3.8. The TEM column. The double condenser lens system forms a focussed beam at the specimen, which is immersed within the magnetic field of the objective lens. The objective aperture, placed at the back focal plane, controls the image contrast. The projector lenses magnify the image onto the screen

In practice, these basic similarities are obscured by the very different structure of the TEM. The electron path from gun to camera must be evacuated to 10^{-5} torr or better; the magnetic electron lenses are large iron cylinders, with substantial aberrations, which restrict the angular range of the electron beam and hence result in a very tall instrument, with resolution far worse than the inherent diffraction limit; the strength of the lenses is changed by changing the currents in the electromagnetic coils; the very thin samples need to be tilted and translated, with very high mechanical stabilities; the image is viewed by converting the electron image into light at a fluorescent screen; and so on.

The essential features of the TEM are shown in Fig. 3.8. In addition to providing images over a magnification range from about $100\times$ to $500,000\times$, the electron diffraction pattern, formed at the back focal plane of the objective lens, can be observed on the screen and recorded, to provide information about the crystalline structure of the sample. Adjustable apertures are used to control the angular aperture of the illumination; to provide image contrast by selecting some of the diffracted electrons and preventing the rest from contributing to the image; and to select specific areas of images from which to obtain diffraction information. Magnetic deflection coils are used not only to align the electron beam, but also to control it to provide specific image modes, and to correct for astigmatism in the lenses.

3.4.2 Electron Diffraction

The wavelength of 100 keV electrons is 37 pm, which is about 50 times smaller than the typical nearest-neighbour separation in many crystals. Electrons incident on a thin sample are diffracted though angles of order 10^{-2} radian; the resulting electron diffraction pattern can be observed by imaging the back focal plane of the objective lens. Using an aperture at the level of the first image (the "selected area aperture"), diffraction data from areas down to about 300 nm can be obtained; smaller areas may be selected for diffraction analysis if the TEM has the facility to form fine illumination probes (down to 2 nm).

An amorphous film shows broad rings in its diffraction pattern, with peaks corresponding to the most probable interatomic spacings. Crystalline materials show the sharp spots corresponding to Bragg diffraction by the periodic lattice. Thus information about the degree of crystallinity, the orientation and crystal structure of individual grains, etc., can be obtained from the electron diffraction patterns. Although the dominant features of the patterns will arise from the bulk crystal, weak diffraction spots arising from surface structures can often be detected, and may be used to form dark field images of the corresponding surface detail.

3.4.3 Image Contrast and Resolution in the TEM

If all the diffracted waves were to be recombined by a perfect objective lens into a perfectly focussed image, the image would show no contrast, because the electrons are not absorbed in the thin specimen and the differences in phase are lost in the observed intensities. Defocussing the lens gives phase contrast through interference between adjacent parts of the wave (Fresnel diffraction), but in order to obtain reasonable contrast levels in TEM images, it is usually necessary to insert an aperture at the level of the back focal plane of the objective lens, to remove some fraction of the diffracted electrons from the image. If only the unscattered electrons are permitted to pass, through a small aperture centred about the optical axis, a bright field image is formed. If the incident illumination is tilted, so that one of the diffracted beams is aligned down the optic axis and passes the objective aperture, a dark-field image may be recorded. Larger apertures may be used to pass several diffracted beams; the resulting image contrast arises from interference between these beams, and shows fringes with spacings equal to those of the lattice planes of the Bragg diffracted beams included in the image. In general such images are very complex, but under certain conditions (very thin crystals, electron beam orientated down a crystal symmetry axis, a specific defocus of the objective lens) "structure images" can be recorded which relate directly to the crystal structure with a resolution corresponding to the spacing of the planes giving rise to the Bragg reflections passing the aperture.

The resolution achieved in the images is determined by a balance between the diffraction limit ($\sim \lambda/\alpha$) and the effects of spherical aberration ($\sim C_s\alpha^3$) and chromatic aberration ($\sim C_c\alpha\,\Delta E/E$). The simplest estimates of resolution may be made using (3.3). More accurate estimates of resolution involve consideration of the effects on the phases of the electron waves of defocus, spherical and chromatic aberration, and the objective aperture which limits the angular range of diffracted waves contributing to the image (see [10,11] for detailed discussions of these topics).

3.4.4 Imaging Surface Structures in the TEM

The degree of roughness of the surfaces of perfect crystals can be revealed qualitatively using phase contrast imaging, with an objective aperture which just cuts out all the diffracted waves, so the resolution is typically around 0.3 nm. The classic examples of this mode are images of MgO cubes [13]; note that the crystals must be oriented to avoid any strong Bragg diffraction effects, so that there is only weak dependence of contrast on thickness (Fig. 3.9). More quantitative information can be obtained by setting the crystals to specific diffraction conditions, and recording either bright or dark field images; the resulting contrast can reveal monatomic surface steps through the thickness sensitivity of the diffracted intensity [14–16].

Reconstructed surfaces, having a different crystal structure from the bulk, can be imaged in dark field by selecting a surface specific diffracted beam with the objective aperture. The 7×7 superlattice structure on Si was first determined in this way [17].

Lattice images of crystals are not usually sensitive to minor (monatomic) changes in surface structures, since the diffracted waves will be largely determined by diffraction within the bulk of the crystals. The technique of "profile imaging" [18] shows directly the surface structures of facets imaged at the thin edges of crystals, in atomic detail. The accurate interpretation of such images requires the matching of the experimental images with theoretical ones computed from the full theory of dynamical scattering of electrons in crystals [11,19]. This method is a powerful complement to the use of scanning tunneling microscopy to determine surface structures at atomic resolution [2].

Fig. 3.9. Phase contrast image of an MgO crystal, after exposure to water vapour and electron beam irradiation; recorded at 100 keV in a TEM. Phase contrast is achieved by defocussing the image, revealing a pattern of small (1–2 nm) rectangular depressions and protrusions across the initially smooth {100} surfaces

3.4.5 Reflection Electron Microscopy

REM images and reflection electron diffraction patterns are obtainable in the TEM or STEM by mounting the sample so that the surface of interest is almost vertical. REM images of surfaces have played a very significant role in improving knowledge of the structure of extensive "flat" surfaces [20]. Although severely foreshortened, these images are easily interpreted in most cases, revealing surface steps, regions with different structure, dislocations, etc. [3]. The classic study of the growth of the 7×7 structure on the (111) surface of silicon, using dark field REM images, established the power of this technique [21]. Recently, using a high resolution TEM with UHV specimen stage, the mechanisms of growth of Au on the Si-7×7 superlattice have been elucidated, through video recordings of the growth process in which both Si and Au surface superlattices were resolved [22]. In this type of research, REM is a powerful complement to LEED and STM.

3.5 Other Developments

A rather different approach to surface imaging has been achieved with the development of the Low-energy and the PhotoEmission EM (LEEM/PEEM) [24]. The sample is held at a high negative potential, so that high energy incident electrons are decelerated to a few eV before hitting the surface. The reflected/diffracted electrons are re-accelerated to form an image with resolution as high as 2 nm, and high sensitivity to surface structures and composition. LEED patterns can be displayed, and dark field images formed using selected diffraction spots. In PEEM, the sample is illuminated with an intense light source, and the emitted photoelectrons are imaged.

Important advances in surface science are likely to come from new instruments in which one or more of the surface spectroscopies are combined with imaging techniques such as REM and high resolution SEM. Such instruments must have ultra high vacuum levels in the sample chamber and facilities for treatment of the sample (e.g. heating, evaporation onto substrates without withdrawal from UHV), while retaining high resolution surface spectroscopy (AEM, SIMS), imaging and diffraction. The development of a STEM-based instrument, with the facility to obtain high resolution secondary images from both sides of a thin sample, has been reported [25]; this system is designed to include an Auger spectrometer.

There is an exciting future for studies involving the combination of surface analytical techniques and high resolution electron microscopy of surfaces. However, at a more routine level, the interpretation of XPS and SIMS data from a multitude of surfaces of many material will always be assisted by – and frequently requires – the additional information provided by the scanning electron microscope. Imaging and spectroscopy should be regarded as essential partners in surface science.

References

1. J.A. Venables, D.J. Smith, J.M. Cowley: HREM, STEM, REM SEM and STM, Surf. Sci. **181**, 235 (1987), see also J.A. Venables: Ultramicroscopy **7**, 81 (1981)
2. A. Howie, U. Valdre (eds.): *Surface and Interface Characterization by Electron Optical Methods*; Proc. NATO ASI, Erice 1987 (Plenum, New York 1988)
3. P.K. Larson, P.J. Dobson (eds.): *Reflection High-Energy Electron Diffraction and Reflection Imaging of Surfaces*, Proc. NATO ASI, Veldhoven 1987 (Plenum, New York 1988)
4. J.A. Venables, D.J. Smith (eds.): Proceedings of Workshop on Surfaces and Surface Reactions, Wickenberg Inn, Arizona, 1989: Ultramicroscopy **31** (1989)
5. P.J. Goodhew: *Electron Microscopy and Analysis* (Wykeham, London 1975); P.J. Goodhew, F.J. Humphreys: *Electron Microscopy and Analysis*, 2nd edn. (Taylor and Francis, London 1988)
6. P.W. Hawkes: *Electron Optics and Electron Microscopy* (Taylor and Francis, London 1972)
7. I.M. Watt: *The Principles and Practice of Electron Microscopy* (Cambridge University Press, Cambridge 1985; Second edition 1997)
8. J.I. Goldstein, D.E. Newbury, P. Echlin, D.C. Joy, C. Fiori, E. Lifshin: *Scanning Electron Microscopy and X-ray Microanalysis* (Plenum, New York 1981)
9. L. Reimer: *Scanning Electron Microscopy*, Springer Ser. Opt. Sci. Vol. 45 (Springer Berlin, Heidelberg 1985; Second completely revised and updated edition 1998)
10. L. Reimer: *Transmission Electron Microscopy*, Springer Ser. Opt. Sci., Vol. 36 (Springer Berlin, Heidelberg 1984; fourth edition 1997)

11. J.M. Cowley: *Diffraction Physics*, 2nd edn. (North Holland, Amsterdam 1981)
12. A.E. Ringwood: *Safe Disposal of High Level Nuclear Reactor Wastes* (ANU, Canberra 1978); J.A. Cooper, D.R. Cousens, R.A. Lewis, S. Myhra, R.L. Segall, R.StC. Smart, P.S. Turner: J. Am. Ceram. Soc. **68**, 64 (1985)
13. A.F. Moodie, C.E. Warble: J. Cryst. Growth **10**, 26 (1971)
14. D. Cherns: Phil. Mag. **30**, 549 (1974)
15. K. Kambe, G. Lehmpfuhl: Optik **42**, 187 (1975); G. Lehmpfuhl, Y. Uchida: Ultramicroscopy **4**, 275 (1979)
16. M. Klaua, H. Bethge: Ultramicroscopy **11**, 125 (1983)
17. K. Takayanagi, Y. Tanashiro, M. Takahashi, S. Takahashi: J. Vac. Sci. Technol. **A3**, 1502 (1985)
18. L.D. Marks, D.J. Smith: Nature **303**, 316 (1983); D.J. Smith: Surf. Sci. **178**, 462 (1986)
19. L.D. Marks: Surf. Sci. **139**, 281 (1984)
20. For reviews of REM, see K. Yagi: J. Appl. Cryst. **20**, 147 (1987); K. Yagi: Electron microscopy of surface structure, Adv. Opt. Elec. Microsc. **11**, 57 (1989)
21. N. Osakabe, Y. Tanashiro, K. Yagi, G. Honjo: Japan J. Appl. Phys. **19**, L309 (1980); Surf. Sci. **109**, 353 (1981)
22. Y. Tanishiro, K. Takayanagi: Ultramicroscopy **31**, 20 (1989)
23. T. Nagatani, S. Saito: Proc. 11th Int. Conf. Elec. Microsc., Kyoto, Japan (1986) 2101; K. Ogura, M. Kersker: Proc. 46th EMSA (1988) p. 204
24. E. Bauer: Ultramicroscopy **17**, 51 (1985); E. Bauer, M. Mundschau, W. Swiech, W. Telieps: Ultramicroscopy **31**, 49 (1989)
25. G.G. Hembree, P.A. Crozier, J.S. Drucker, M. Krishnamurthy, J.A. Venables, J.M. Cowley: Ultramicroscopy **31**, 111 (1989)
26. D.C. Joy: *Monte Carlo Modelling for E.M. and Microanalysis* (Oxford University Press, Oxford, 1995)
27. V.N.E. Robinson: J. Microscopy, **103**, 71 (1975)
28. G.D. Danilatos: Micron and Microscopica Acta, **14**, 307 (1983)
29. B.J. Griffin: Scanning, **22**, 234 (2000)
30. J.L. Pouchou, F Pichoir: Scanning, **12**, 212 (1990)

4 Sputter Depth Profiling

B.V. King

The understanding and modification of surface properties of materials often requires a detailed knowledge of the spatial distribution of specific elements in both the surface plane and as a function of depth normal to the surface. This chapter will concentrate on ways of analyzing depth distributions of elements, up to a few microns deep, using ion beam sputter profiling. Sputter profiling uses the combination of a surface sensitive analytical technique, such as SIMS, LEIS, AES, XPS or SNMS [1], together with the continuous exposure of a new surface by ion beam sputtering. The sputtering ion beam typically is Ar^+, O_2^+ or Cs^+ with energy in the range 0.25–20 keV, although other inert gas (Kr^+ or Xe^+), liquid metal (Ga^+ or In^+) or, more recently, cluster (SF_6^+) ion sources are also used. Ar^+ is most commonly used with AES and XPS since it does not form compounds with target constituents and so does not significantly alter bulk atomic concentrations. O_2^+ and Cs^+ are favoured for SIMS since they increase secondary ion yields and reduce matrix effects. However, this is at the price of reduced sputter rates and profile distortion though compound formation and segregation. Liquid metal sources are normally used for imaging. Profiling rates of up to 2 µm/h can be achieved although 0.1 µm/h is more typical.

Sputter profiling has three advantages over other methods of elemental depth profiling. Firstly, sputter profiling uses the strengths of the particular analysis technique employed (e.g. the high sensitivity of SIMS, lateral resolution of AES or the chemical identification of XPS). Secondly, sputter profiling has excellent depth resolution. Other techniques, such as RBS or NRA, which sample the whole near surface region simultaneously, have comparable depth resolutions at the surface to sputter profiling (a few nm), but the depth resolutions of RBS and NRA degrade rapidly with depth and the elemental distribution within the sample. Thirdly, sputter profiles are not subject to unavoidable interferences between signals from different elements at varying depths.

Balanced against the above advantages are the problems of establishing depth and concentration scales. The characteristics of the various surface analytical techniques have been discussed in previous chapters. Their treatment in this chapter will be restricted to their influence on the practice of sputter profiling and the optimisation of technique.

4.1 Analysis of a Sputter Depth Profile

The task for the experimentalist is to deduce the structure of the target from the measured depth profile. For example, Fig. 4.1a shows a SIMS depth profile of a 16 period SiGe superlattice structure deposited onto a Si substate (Fig. 4.1b). Analysis of this depth profile requires the establishment of accurate concentration and depth scales as well as an appreciation of whether the features in the depth profile correspond to the structures in the target or to artifacts of the sputter profiling process. This chapter will provide a summary of possible artifacts and their effect on depth resolution.

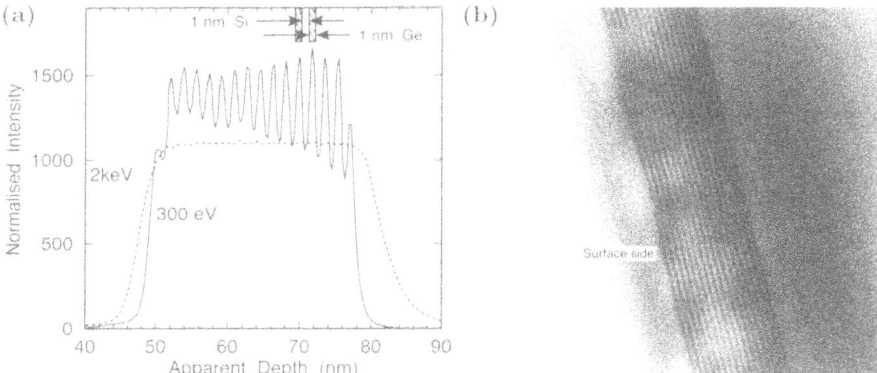

Fig. 4.1. (a) SIMS depth profiles of a MBE grown superlattice comprising 1 nm thick Ge layers separated by 1nm Si layers, together with **(b)** a TEM image of the same structure. The profile obtained with the 300 eV O_2^+ primary beam has far better depth resolution than the profile obtained with the 2 keV O_2^+ primary beam [60].

4.1.1 Calibration of the Depth Scale

In Fig. 4.1 the elemental concentrations are presented as a function of the sputtering depth. This depth scale has been obtained by either (i) measuring the depth of the eroded crater using a surface profilometer or interferometry or (ii) calculating the eroded depth, z, from the sputter time. Generally both measurements are done and compared.

In the second method, the depth sputtered during a depth profile is related to the product of the ion beam current density J ($\mathrm{Am^{-2}}$), the sputtering time t (s) and the sputter yield Y (atoms ion^{-1}) of the sample by

$$z = YJMt/1000e\rho N_{\mathrm{A}}n \tag{4.1}$$

where e is the electronic charge, M is the target molecular weight, ρ is the target atomic density (kg/m^3), N_A is Avogadro's number and n is the number of atoms in the molecule.

This method is accurate when Y does not change significantly during the analysis and so is most suitable for dilute ($< 10\%$) impurity concentrations in a uniform matrix. The multilayer shown in Fig. 4.1 comprises layers of GaAs and AlGaAs and so the sputter yield should change during analysis. Depth quantitation using (4.1) is then difficult. Care should also be taken in using this method near the surface of even elemental targets since the sputtering rate may change within a depth of the order of the ion range because of the implantation of the sputtering ions [2] or surface contamination. The two main areas limiting the accuracy of (4.1) are the measurement of J and unreliable data for Y even for suitable samples. The current density, J, cannot be found from simple target current integration because the emission of a considerable fraction of charged secondaries gives rise to large errors, typically a factor of 2 for keV Ar^+ on metals. For clean targets the contribution of secondary and scattered ions is small so the dominant contribution comes from secondary electrons with energies in the range 30–50 eV. These may be repelled back onto the target by biassing the target positively, +100 V, or they may be collected in a Faraday cup, either mounted on the target holder or retractable for separate insertion into the ion beam. In these ways accuracies of about 10%, limited by incomplete suppression or unknown Faraday cup aperture sizes, are achievable [3].

The sputter yield, Y, the ratio of the number of emitted particles to the number of incident ions, is estimated either theoretically, from available compilations [4–6] or by computer simulation . The theoretical predictions are based on linear cascade theory [7,8]. which gives a sputter yield Y_i for an ion i with energy E_i of the form [7,8]

$$Y(E_i, \theta_i) = K_{it}\, S_n(\xi) f(\theta_i) / U_0 \tag{4.2}$$

where $\xi = E_i/E_{it}$ and E_{it} (in keV) and K_{it} [9] are scaling constants depending on the atomic numbers and masses of the ion and target atoms

$$E_{it} = \left((1 + M_i/M_t) Z_i Z_t (Z_i^{0.66} + Z_t^{0.66})^{0.5}\right) / 32.5 \tag{4.3}$$

$$K_{it} = \left((Z_i Z_t)^{0.833}\right) / 3 \quad \text{for} \quad 1/16 < Z_t/Z_i < 5 \tag{4.4}$$

and U_0 is the surface binding energy in eV. U_0 is usually taken as the sublimation energy [10]. The reduced nuclear stopping cross-section, $S_n(\xi)$, has been estimated [11] by

$$S_n(\xi) = 0.5 \ln(1 + \xi) \;/\; (\xi + (\xi/383)^{0.375}) \tag{4.5}$$

whilst the angle of incidence dependence $f(\theta_i)$ is given by

$$f(\theta_i) = \cos^{-n}(\theta_i) \tag{4.6}$$

for $\theta_i < 80°$ with $n \geqslant 1$. For angles of incidence less than 45°, $n \approx 1$ and $J \propto \cos\theta$ so that the erosion rate does not vary significantly with angle. The results of these formulae are accurate to within a factor of two for typical sputtering energies and species but break down for very low ion energies (as used in SNMS), for light ions, when the energy transferred is insufficient to initiate a collision cascade or for high energy heavy ions when interactions occur between moving atoms in the cascade. They also underestimate the sputter yield of insulators when energy deposited into electronic excitation and ionization may lead to atom displacement and sputtering [12]. For keV energies, the yields for Ar^+ bombardment of most materials range from 0.5 to about 2 and are approximately linearly dependent on energy.

Sputter yields, energy deposition and ion ranges may also be calculated by computer simulations. The best known of these is TRIM, a Monte Carlo simulation for amorphous targets [13,14] as well as to calculate sputter yields of rough surfaces [15]

The depth scale across an interface between two pure elements is, to a first approximation, given by the separate depth scales in the individual layers, where the interface is determined by the point at which the concentration falls to 50%. This analysis can, however, be difficult for thin layers or layers where the concentrations cannot be accurately determined. In Fig. 4.2a the

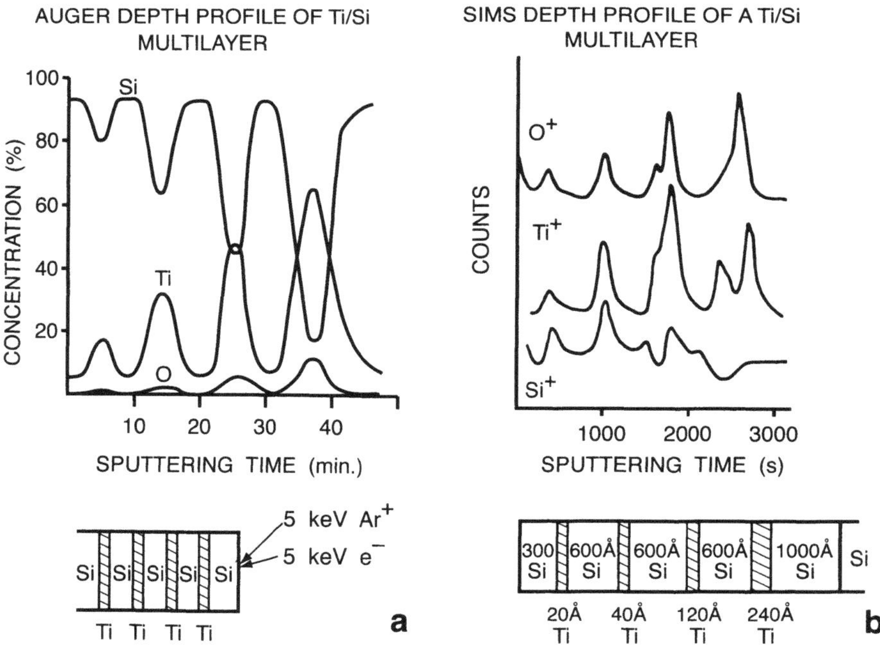

Fig. 4.2. Sputter depth profiles of the Ti/Si multilayer shown using (**a**) AES and (**b**) SIMS

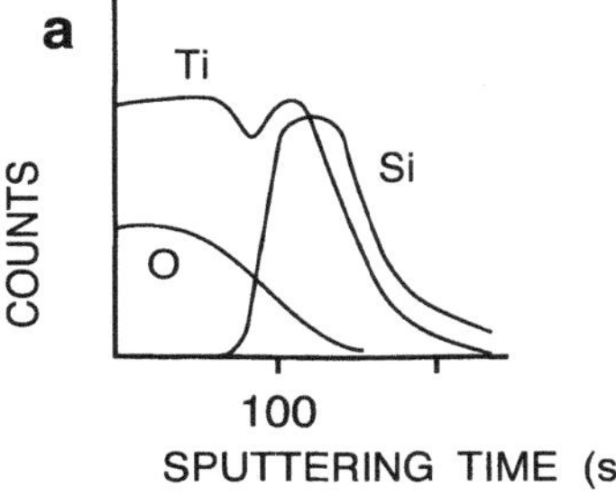

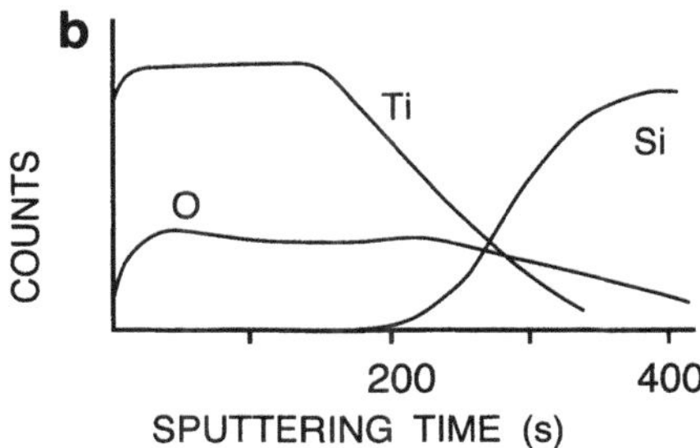

Fig. 4.3. SIMS sputter depth profiles of a Ti layer deposited onto a Si substrate. The ion beam was 5 keV Ar^+ at an incidence angle of 45°. The analyses were performed at oxygen partial pressures of (**a**) 10^{-8} Pa and (**b**) 2×10^{-6} Pa

Ti concentration, as measured by AES, never reaches 100% because the depth resolution of the depth profiling is comparable to the thickness of the individual Ti layers. A first depth calibration would be found by assuming that the sample was pure Si and then estimating a total Ti thickness in each layer from the integrated Ti concentration in each layer. It is difficult to similarly treat a SIMS depth profile of the same structure (Fig. 4.2b) since unambiguous Ti and Si concentration scales cannot be assigned due to changes in ion yield due to the changing oxygen concentration.

Sputter yields of pure targets may also alter during depth profiling due to reactions between the ion beam and target. For example, oxides are generally formed when oxygen ion beams or when oxyten flooding, namely ion bombardment at elevated oxygen partial pressures, are used during profiling, with a consequent drop in sputter yield. Figure 4.3 shows that the time to sputter through a Ti film increases by about factor of 3 when oxygen flooding is used during Ar^+ sputtering, due to the formation of TiO_2. In this case the sputter yield of TiO_2 is approximately equal to that of Ti so that the decrease in sputter rate is due the 3 fold increase in the number of atoms to be sputtered. Sputter yields for oxides [16,17] and nitrides are in most cases similar to (e.g. SiO_2) or higher than for the corresponding metals, but there are exceptions (e.g. Al_2O_3 and MgO). These changes are particularly important for depth quantification in the initial stages of SIMS profiling of shallow dopant profiles i.e. in the pre-equilibrium region, where the oxygen or caesium concentrations are changing with depth, and result in the apparent movement of peak positions. For example, the peaks in 1 keV O_2^+ depth profiles of a shallow 0.5 keV B_{11} implant in Si are heavily distorted at depths below 10 nm. The peak shift due to the initial short-term change in erosion rate for 1 keV O_2^+ was 3.5 nm at 34° incidence and 1.3 nm at 2° incidence. Hence it is mandatory to use O_2^+ energies that are perhaps half the implantation energy [18]. Good profiling conditions can be found which minimise this effect. For example, sputtering of B delta layers in Si with 500 eV O_2^+ at an angle of 50° while flooding with oxygen produced no measurable change in sputtering rate and resulted in no unexpected shift towards the surface of the B delta layers [19].

Sputtering of multicomponent targets presents even more difficulties in quantitation due to the uncertainty in Y and its variation with ion dose. A good review of the sputtering of multicomponent materials has been given by *Betz* and *Wehner* [20]. Briefly, the sputter yield, Y, of a metal alloy A_xB_{1-x} or multilayer of the same average composition is assumed to be related to the elemental yields Y_A and Y_B (with $Y_B > Y_A$) by

$$xY_A + (1-x)Y_B \; < \; Y \; < \; Y_B \, . \tag{4.7}$$

For more detail on many of the above aspects, the reader is directed to an extensive review of quantitative sputtering [21].

4.1.2 Calibration of the Concentration Scale

The signal measured in any one of the surface techniques used with sputter profiling is related to target elemental concentrations by factors which have been discussed in other chapters. In summary the counts, n_i, measured in time t for species i are given as

$$n_i = A\, t\, \gamma\, c_i\, \chi_i \tag{4.8}$$

where A is the analysed area (cm^{-2}), γ is the instrument transmission, c_i is the species concentration (cm^{-3}) and χ_i is the cross-section for production of the species i above the surface

$$\chi_i = JY\chi_{\text{ion}}/N \qquad \text{for SIMS} \tag{4.9}$$

$$\chi_i = dI\chi_{\text{el}} \qquad \text{for AES or XPS} \tag{4.10}$$

where J is the sputter ion current density (cm^{-2}), Y is the sputter yield, N is the atomic density (cm^{-3}), χ_{ion} is the cross-section for secondary ion production, d is the Auger or XPS information depth, I is the primary electron or photon flux (s^{-1}) and χ_{el} is the cross-section for Auger electron or photoelectron production.

In sputter depth profiling, quantification is often difficult since the sensitivity factors, Y, χ_{ion}, d and χ_{el}, generally change with sputter time. For example, in Fig. 4.2, depth profiling using AES shows four peaks in the Ti^+ concentration whereas a SIMS profile shows 5 Ti peaks which do not correspond to dips in the Si^+ signal. The two peaked structure from the deepest Ti layer is characteristic of SIMS profiles of concentrated layers and is due to the dependence of secondary ion yields on the matrix. In the above case, the Ti rich layer has been found to contain $TiSi_2$ by Auger lineshape analysis. The double peaked structure then results from the difference in yield of Ti^+ sputtered from a $TiSi_2$ matrix or from a predominately Si matrix. The presence of peaks in Si^+ signal at the position of the shallower Ti layers is probably due to oxygen enhancement of the Si^+ yield. Oxygen ion bombardment or oxygen flooding during inert gas bombardment would reduce these SIMS matrix effects by maintaining a relatively constant oxygen concentration throughout the depth profile. For example, in Fig. 4.3, the Ti^+ profile shows a dip then peak at the Ti-Si interface when profiled by Ar^+. When oxygen flooding is used the Ti^+ is shown to vary smoothly across the deposited film as expected. The Si^+ profile also corresponds better to the known distribution when flooding is used.

Quantification of the concentration is not only related to the matrix effects and interferences referred to above and in other chapters but is also determined by factors related to ion sputtering. The most important are ion beam induced compositional changes, sample contamination and preferential sputtering. For example, ion beam induced loss or gain of elements at the surface or in the bulk of the target can markedly affect bulk quantification. For example, ion irradiation causes loss of O from Ta_2O_5 to form a mean composition of Ta_2O [22]. On the other hand, elements may be gained from contamination of the sample from i) impurities in the ion beam, ii) adsorption from the residual gas in the analysis chamber and iii) redeposition of previously sputtered material onto the target during profiling.

Implantation into the target of impurities in the ion beam may cause erroneous depth profiles. These impurities may be ions which pass magnetic mass separation in the ion beam line, e.g. ArH^+ during depth profiling using Ar ion beams, or neutrals formed in the beam line and transported with the ion beam. To avoid

$$C = 2.63 \times 10^{24}\, P/MT \qquad (\text{particles cm}^{-2}\,\text{s}^{-1}) \tag{4.11}$$

and P, M and T are the pressure (in Pa), molecular mass of the gas and temperature respectively and γ and Y are the sticking probability and resputtering yield of the adsorbate [23]. A flux of $6 \times 10^{-6}\,\text{A cm}^{-2}\,\text{s}^{-1}$ is then needed to avoid target contamination for a chamber pressure of 10^{-7} Pa with unity sticking probability. This flux corresponds to an average erosion rate of $1\,\text{Å s}^{-1}$.

Redeposition of sputtered atoms into the analysis crater occurs due to gas scattering or, more commonly, resputtering from nearby surfaces. The high chamber pressures (10^{-3}–10^{-4} Pa) associated with oxygen flooding for SIMS as well as the use of non-differentially pumped ion guns in AES and XPS enhances these memory effects with ions or atoms being scattered as they leave the surface and subsequently redeposited into the crater. Such memory effects also occur if surfaces of ion lenses, for example, are within about one cm of the sample so that sputtered material is collected on the surface and redeposited into the crater. The latter effect is particularly severe for quantitative trace element analysis in semiconductors and can only be overcome by moving the analyser further from the target or having a method of replacing contaminated surfaces.

Preferential sputtering is defined as the difference between the composition of the flux of the sputtered particles and the composition of the outermost layers of the sample. Consider a homogenous binary alloy sample with bulk compositions, c_A^{b} and c_B^{b} for the constituents A and B. The equilibrium sputtered fluxes of A and B from such a target, which are proportional to $Y_A c_A^{\mathrm{s}}$ and $Y_B c_B^{\mathrm{s}}$ respectively, must have proportions of A and B identical to the bulk. Then

$$c_A^{\mathrm{s}}/c_B^{\mathrm{s}} = \left(c_A^{\mathrm{b}}/c_B^{\mathrm{b}}\right) Y_A/Y_B \tag{4.12}$$

at equilibrium. As a result the surface composition of A would change from the bulk value to the value in (4.12) as sputtering proceeded. The result of preferential sputtering is that transients occur in measured profiles of surfaces and interfaces. In Fig. 4.4 $Y_{\mathrm{Ag}} > Y_{\mathrm{Ta}}$ so that Ag is depleted at the surface with sputtering as Ta is enriched. These transients may however also be caused by other processes – preferential recoil implantation of light elements or radiation enhanced surface segregation. Preferential recoil implantation is seen for alloys with large mass differences between the components, i.e. $m_A/m_B > 5$. The depth over which this effect is evident is of the order of twice R_{p}. Surface segregation will be further discussed in a later section.

The above artifacts can be mostly overcome by the use of standards for calibration of the concentration and depth scales. For SIMS depth profiles implanted standards are used. If the implant is laterally homogeneous, both in implant dose and depth distribution agreement to within 10% can be obtained for analysis of B in Si by different SIMS instruments [24]. Thin film standards for calibration of depth scales in AES have also been developed, e.g. by National Physical Laboratory [25] and National Bureau of Standards [26,27].

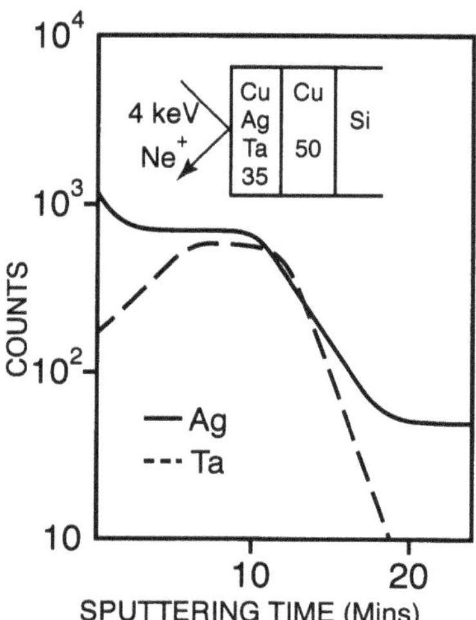

Fig. 4.4. LEIS sputter depth profile using 4 keV Ne^+ at 45° incidence of a Cu/Ta/Ag thin film deposited onto a Si substrate. There is an initial transient in the signal from Ag and Ta due to preferential sputtering. The width of the transient is about twice the projected range of the sputtering ion beam

4.2 The Depth Resolution of Sputter Profiling

Optimisation of the depth resolution is of great importance in accurate profiling so that, in particular, artifacts can be identified. The depth resolution of sputter profiling is determined by a combination of instrumental effects, effects related to the analysis technique (information depth), ion sputtering surface effects (preferential sputtering, surface topography) and bulk effects (ion mixing and enhanced diffusion and segregation).

4.2.1 Specification of the Depth Resolution

The depth resolution is generally described in terms of the measured width of an interface between two dissimilar layers. Figure 4.5a compares the ideal profile, which is rectangular in shape, with the measured profile which is generally represented as an error function. The interface width, Δz, is usually defined as the interval where the intensity drops from 84% to 16% (2σ, two standard deviations) of the maximum signal. If the layers are not thick then Δz will be of the same order of magnitude as the layer thickness, d (Fig. 4.5b). In the case illustrated, the variation in concentration, ΔC, is about half the peak height. This corresponds to Δz equal to 80% of the layer thickness. As Δz is increased ΔC will decrease [28] until effectively all resolution is lost at $\Delta z = 1.5d$. If the layer thickness is further reduced we have a marker geometry (Fig. 4.5c). Calculated [29] and measured [30] sputter profiles indicate that the measured profile of B is not gaussian but is asymmetric about the original interface with an extended tail at larger depths. The peak position is moved towards the surface with respect to the original marker position. In this case, the Δz cannot be strictly used to specify the resolution, although it is a useful approximation. Rather, beyond the marker depth, the concentration decreases exponentially with distance

$$C \propto \mathrm{e}^{-x/\delta} . \qquad (4.13)$$

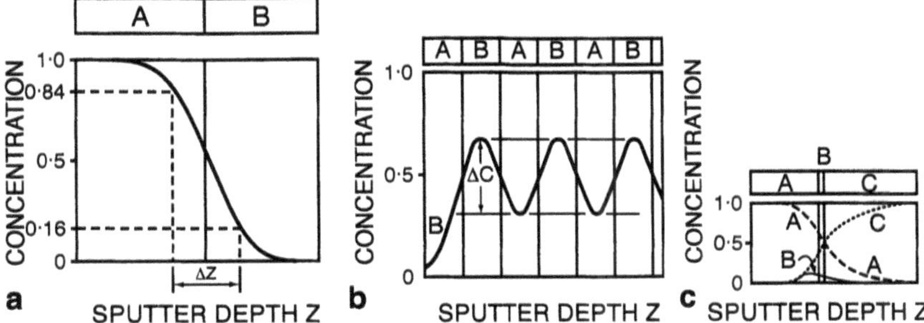

Fig. 4.5. Schematic representation of the depth profiles of elements A, B and C for (**a**) an AB bilayer (**b**) an AB multilayer and (**c**) a B marker layer sandwiched between A and C

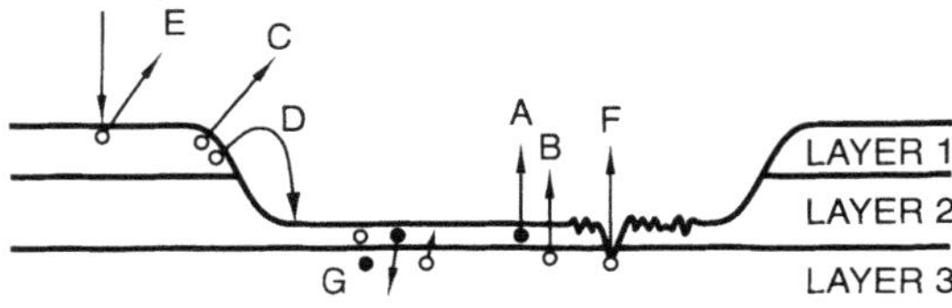

Fig. 4.6. Summary of the mechanisms which can lead to a loss of depth resolution in sputter profiling. (*A*) ejection of particles from the surface of a smooth crater; (*B*) subsurface ejection due to a larger escape depth; (*C*) analysis of material sputtered from the crater walls; (*D*) redeposition from crater walls into the analyzed crater; (*E*) analysis of material sputtered by neutrals in the ion beam; (*F*) ejection of particles from a rough crater; (*G*) atomic mixing within the target

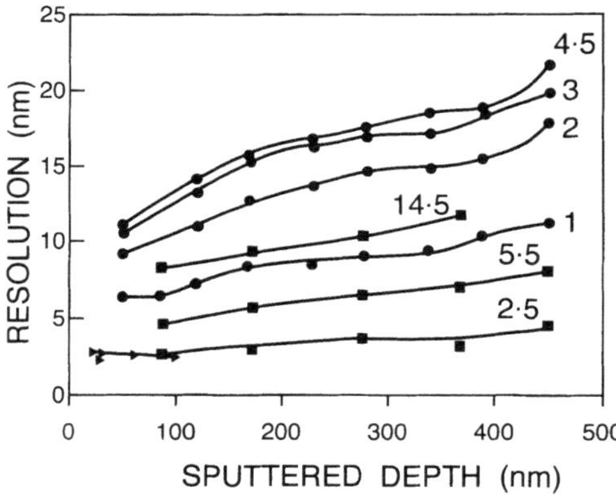

Fig. 4.7. The depth resolution Δz as a function of sputtered depth z for (•) 1, 2, 3 and 4.5 keV Ar^+ irradiated NiCr mulitlayers [32] (■) 2.5, 5.5 and 14.5 keV Cs^+ irradiated GaAs [33] (▶) SiO_2/Si [34]

Optimisation of the depth resolution requires a knowledge of all the factors which influence this quantity, illustrated in Fig. 4.6. They include instrumental effects (C, D, E), redistribution of material onto the surface (B) and within the bulk (G), and, finally, the formation and analysis of nonuniform surfaces (F).

Assuming that the individual contributions Δz_i are mutually independent and each results in broadening of a delta function initial distribution to a gaussian, then the total depth resolution is approximated by $\Delta z^2 = \sum(\Delta z_i^2)$. Compilations of many Δz values as a function of sputtered depth, z, have been presented [31] for argon ion sputtering using AES and SIMS. The depth resolution, Δz, ranges from less than 1 nm for $z = 2$ nm to about 50 nm at $z = 5$ microns. The trend of the data indicates that $\Delta z \propto z^{0.5}$ for a wide range of z. However, differences in Δz for different matrices indicate changes in the relative importance of different factors affecting depth resolution. Data for Ni/Cr, GaAs/GaAlAs and Si/SiO_2 are presented in Fig. 4.7. For 3 keV Ar^+ irradiation of Ni/Cr multilayers, for example, Δz is much larger than the ion projected range, R_p, of 26 Å. In contrast, Δz for 14.5 keV Cs^+ irradiation of GaAs and 1 keV Ar^+ irradiation of Si are approximately equal to their respective values of R_p, 80 Å and 20 Å.

However, despite the above comments, sputter profiling is capable of extremely good depth resolution. For example, values of δ for 500 eV Ar, Xe and SF_6 at 80° incidence in GaAs/AlAs multilayers were 1.4, 0.8 and 0.4 nm respectively [35] whereas B delta distributions in Si could be measured by normal incidence 1 keV O_2^+ with a FWHM of 2.9 nm and δ of 1.4 nm [18].

4.2.2 Instrumental Factors Determining the Depth Resolution

Depth resolution can be lost due to instrumental problems including those due to (i) the ion beam, (ii) the instrument geometry and stability and (iii) sample contamination.

Ion beam effects can be further subdivided into crater edge effects and surface charging of insulating targets. By far the most important of these factors in determining the depth resolution is the flatness of the ion beam crater over the analysis area. Normally the ion beam is rastered and the signal electronically gated to accept only the signal from the central 4–20% of the crater so as to avoid analysing the crater edges. Checkerboard gating is also used in more recent SIMS instruments. In this technique, the signal from the central crater region is divided into 8×8 or more pixels. The depth profiles from individual pixels can be excluded from the total profile if they correspond to contaminated regions within the crater [36]. In conjunction with the minichip technique, where the ion beam is rastered over an entire small sample (typically $0.5 \times 0.5\,\mathrm{mm}^2$) to avoid crater edges, dynamic ranges of over 7 decades can be achieved [24]. In AES and XPS, in contrast to SIMS, the analysis area is determined by the size of electron or X-ray beams. For AES, since the analysed area determined by the incident electron beam is typically only a few microns in diameter, the crater depth is generally uniform over the analysed area so crater edge effects are small. They can, however, be significant in XPS profiling since the analysed area is typically much larger. In general the depth resolution due to crater formation is gaussian with an interface broadening typically given by $\Delta z = az$ where a is 0.2–5% for SIMS data and 1% for AES.

Non-uniform craters may also result from the impact of neutrals which are formed along the beam line and hit the sample as an unfocussed beam which is not affected by the rastering of the final deflection plates. Formation of neutrals may occur by ion scattering at apertures. However, the largest contribution in HV systems is found to come from the interaction of the primary ion beam with the residual gas [37]. Neutral beam contamination is particularly important in SIMS profiling since the neutrals can produce a signal and degrade the dynamic range which is a feature of SIMS depth profiles. Secondary ions liberated from outside the raster scanned area by the energetic neutrals will be recorded if they originate from within the field of view of the analysis optics [38]. The field of view then is generally minimised by the use of lenses in the analysis optics. The region around the crater is also likely to contain higher oxygen concentrations than inside the crater so the ionisation probability and hence SIMS yield would increase. *Wittmaack* [39] has estimated that to achieve a dynamic range of 10^6 in depth profiling, a pressure lower than about 10^{-7} Pa is required for profiling with a high dynamic range in most analysis systems. Figure 4.8 shows an amalgamation of different experimental results for the influence of instrumental effects on SIMS depth profiles of B implanted into Si.

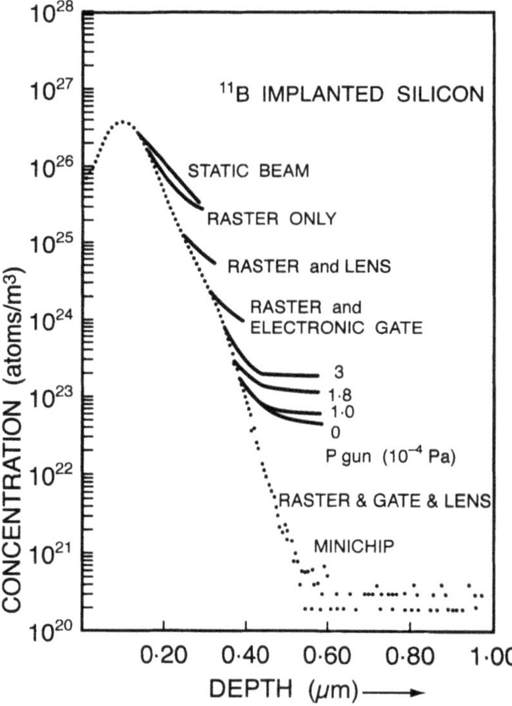

Fig. 4.8. Combination of various depth profiles of B implanted into Si showing the decrease in the background signal caused by (*i*) ion rastering, electronic gating and extraction lenses [40] (*ii*) residual gas pressure [37] and (*iii*) minichip sample preparation [36]

The other beam effect, induced charging, are most important for insulating samples. Ion beam induced sample charging causes dramatic changes in the energies of secondary ions and so in apparent yield of secondary or scattered ions and electrons. It can also cause field assisted migration of mobile species like Na within the sample. Surface charging is usually overcome by flooding the target with low energy electrons.

One of the other instrumental contributors to a loss in depth resolution, sample contamination, has been previously discussed. The final factor, instrument geometry and stability, is small for properly designed instruments. For example, only when the misalignment between the ion and electron beams in an AES instrument is greater than 0.8 times the variance of a static gaussian ion beam does the loss in depth resolution become significant in comparison to other factors [41].

4.2.3 Surface Effects Determining the Depth Resolution

These include the development of surface topography, preferential sputtering and the escape depth of detected particles. Of these, the development of surface topography is the most important in determining the overall depth resolution [42] especially for polycrystalline metals after long sputter times.

Initially surface roughness occurs due to the statistical process of removing atoms by sputtering. This effect stabilises rapidly to give a resolution of about 1 nm. For higher ion doses roughening increases even in almost contaminant free conditions. For example, conical forests a few microns high develop on the (11, 3, 1) Cu surface on extended bombardment [43]. For this surface it is thought that local variations in Y, caused by subsurface defects and dislocations initiated by the ion cascade, are responsible for the initiation of roughness. The roughness increases due to this processes and the variation in Y along different ejection directions [44]. For polycrystalline metals the resolution Δz has been empirically found to be

$$\Delta z = a z^{0.5} \tag{4.14}$$

where $a = 0.86 \pm 0.22$ and Δz and z are in nm [28]. Although there is a large variety of surface structures cones are generally aligned with the ion direction so that sample rotation during profiling (Zalar rotation) lessens the loss of resolution due to roughening [45]. The use of reactive gas ion beams can also improve resolution by forming amorphous compounds on the surface which exhibit no preferential ejection directions. Topographic effects are also seen on semiconductors. For example, 20keV Ar^+ bombardment of Si, initially produces transverse waves which transform, with increasing erosion, into larger amplitude corrugated and facetted wavelike structures [4.47]. Oxygen implantation also causes rippling, since a rapid onset of ripple formation, starts at depths as low as about 15 nm for 1 keV O_2^+ bombardment of boron delta distributions in Si [47]. The ripple formation can then cause the apparent shift of peak positions, e.g., a 41 nm boron delta appeared to be shifted towards the surface by as much 6 nm in the previous example.

The ultimate depth resolution of sputter profiling is determined by the information depth of the particular surface analysis technique used. This contribution to the depth resolution, typically up to a few monolayers, is however usually small in comparison to other sources even at the start of a depth profile. For LEIS, the signal comes from the top monolayer if inert gas ions are scattered, or a few monolayers for electropositive ions like Li^+. The majority of secondary ions detected in SIMS escape from the top 1–2 ML [48]. Typically the attenuation lengths of Auger electrons are 5–50 Å [49].

4.2.4 Bulk Effects Affecting the Depth Resolution

These are processes driven by the energy deposited into the sample by the ion beam as well as the implanted ions and include atomic mixing and radiation enhanced thermal diffusion and segregation.

All the processes which affect depth resolution except ion beam mixing may be circumvented by choice of target and ion species or by correct experimental procedure. However, the signal observed during sputter profiling is always affected by the radiation damage caused by the sputtering particle beam, since the target atom will be successively displaced, typically a few Å at a time, by many incident ion cascades before it is sputtered. The concentration, $C(x, z)$ of a species at a depth x below the surface after sputtering to a depth z is then different to the initial distribution of the species $C(x, 0)$. The measured depth profile, assumed proportional to the surface concentration, $C(0, z)$, is then a convolution

$$C(0, z) = \int C(x, 0) g(z, x)\, dx \tag{4.15}$$

of the original depth distribution and a function $g(z, x)$ which depends on the amount of ion mixing. The effect of ion mixing is to smooth the profile where concentrations change rapidly and to bring the peaks toward the surface with respect to their original positions (Fig. 4.5c). The redistribution of atoms during sputter profiling leads to a broadening of a delta function impurity distribution into a gaussian impurity distribution at a certain depth x (if sputtering and matrix relocation is ignored) with variance [50]

$$\sigma^2 = 2D(x)t = 0.608\, \phi F_{\mathrm{d}}(x) R_{\mathrm{c}}^2 / N E_{\mathrm{d}} \tag{4.16}$$

where $F_{\mathrm{d}}(x)(eV/\text{Å})$ is the energy deposited into collisional processes at depth x, N is the atomic density (atoms/Å^3), $(R_{\mathrm{c}}^2/E_{\mathrm{c}})$ is a factor which only depends on the target (Å^2/eV), ϕ is the ion fluence (ions/Å^2) and $D(x)$ is the diffusion constant (Å^2/s) over an irradiation time t at depth x. The diffusion formalism allows a parameter $(Dt/\phi F_{\mathrm{d}})$ to be used as a measure of mixing efficiency – the rate of mixing per unit of deposited energy. Experimental values for $(Dt/\phi F_{\mathrm{d}})$ typically are in the range 1–200 Å^5/eV. Low values are found for targets of high cohesive energy in which the cascade energy density is low [51]. The lower limit also corresponds to results from analytic theories of ballistic mixing [50,52]. The depth resolution Δz due to ion mixing may be found from $(Dt/\phi F_{\mathrm{d}})$ by solution of a diffusion equation together with associated boundary conditions [53]. It is found that Δz depends on the degree of preferential sputtering as well as $(Dt/\phi F_{\mathrm{d}})$ but that the depth resolution is not a sensitive function of $(Dt/\phi F_{\mathrm{d}})$. For the intermixing of Ge and Si, *Kirschner* and *Etzhorn* [54] also estimate Δz to be about 1.5 R_{p}. As a guide then, if ion mixing is dominant, Δz is about 1–2 R_{p}.

Since the interaction between the inert gas and the target atoms is small, it is reasonable to disregard the influence of inert gas implantation on measured depth profiles. Chemical reactions between different target elements and between implanted ions and target atoms will however alter depth profiles by either aiding or negating diffusional mixing. In general impurities with the high heats of reaction will preferentially segregate. *Hues* and *Williams* [55] in a study of the effect of an O_2 jet on ion mixing, found that impurities, like Ca, with higher oxygen affinities than Si, segregated to the surface whilst Ag with lower oxygen affinity than Si were displaced further into the bulk.

4.2.5 Minimisation of the Depth Resolution

Strategies for minimising the depth resolution depend on which factors predominately determine the resolution. Figure 4.9 shows typical values of Δz for keV Ar sputtering of the target types shown. For amorphous materials, where topographical effects are minimized, Δz is determined by information depth and atomic mixing for small values of z. The information depth can be minimised by the use of LEIS or SIMS. Δz for atomic mixing is primarily determined by the ion projected range as discussed above. In this case, the depth resolution Δz decreases with decreasing ion energy and increasing ion mass as a result of the decrease in the projected range. The deposited energy density, which is higher for heavier ions of the same energy, is of secondary importance, at least over the keV energy range typically used in sputtering. Lower values of R_p and thus depth resolution can also be obtained by using glancing incidence ($\theta < 70°$) ion irradiation. This has been seen, for example, in 1 keV Ar^+ (Fig. 4.10) profiling.

The latter result also illustrates that oblique incidence can reduce surface roughening since Fig. 4.9 indicates that this factor should dominate the resolution for polycrystalline metal films.

If the resolution function $g(z,x)$ in (4.15) is known then the original depth distribution $C(x,0)$ can, in principle, be found from the measured depth profile $C(0,z)$. When the initial impurity distribution is constant in the near surface regions where $g(z,x)$ is appreciable then (4.15) can be written as a convolution integral

$$C(0,z) = \int C(x,0)g(z-x)\,dx\,. \tag{4.17}$$

The resolution function depends in general on all the contributions to the depth resolution i.e. mixing, escape depth, topography etc. However, it can be experimentally determined with appropriate reference materials, e.g., samples containing delta layers or sharp interfaces between two media, and it can be theoretically predicted [57]. If the resolution function can be approximated by a gaussian then the above equation can be solved by Laplace transforms [58]. The procedure is simplified if $C(0,z)$ has the shape of an error function, i.e. $C(0,z) = \mathrm{erf}(z/\Delta z)$. In that case $C(x,0) = \mathrm{erf}(z/\Delta z_0)$ where $\Delta z_0^2 = \Delta z^2 - \Delta z_g^2$ and Δz_g is appropriate parameter in the gaussian resolution function. To obtain the original profile the broadening, Δz, of the measured profile is then corrected using the value Δz_g from the resolution function. The convolution

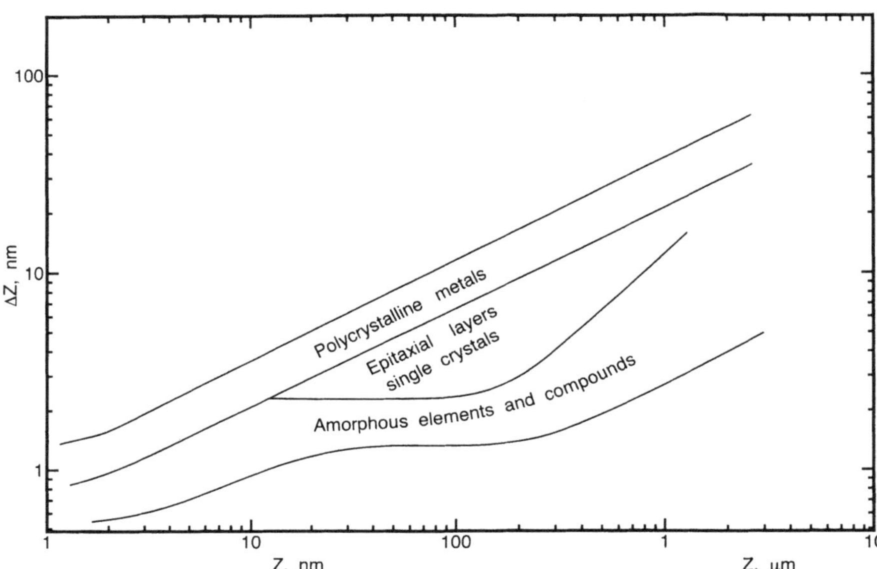

Fig. 4.9. Typical values of Δz for different targets as a function of the depth, z, eroded by keV Ar^+ [56]

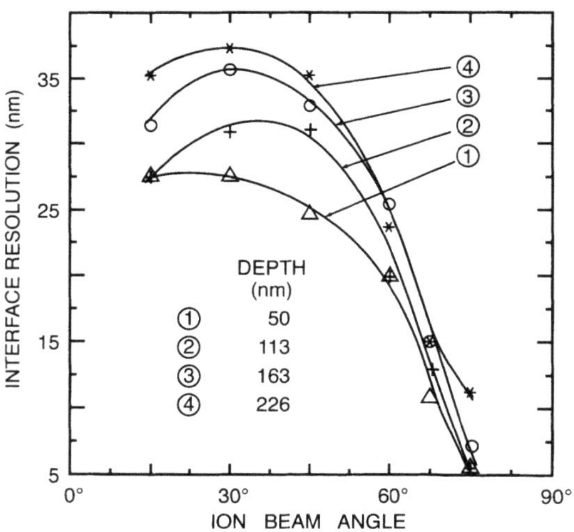

Fig. 4.10. Variation of Δz with ion beam angle to the target normal after sputtering to the indicated depths using 4 keV Ar^+ [32]

model is however only valid in the low concentration limit and alternatives exist, Fourier transform and maximum entropy methods, for deconvolution [59].

4.3 Conclusion

The theoretical developments associated with the study of effects due to the ion beam probe and also effects due to ion irradiation in general suggest that a reasonable understanding of the processes is emerging. Ion irradiation with the deliberate aim of inducing mixing and recoil implantation is likely to be a very significant, practical form of surface modification technology. As a consequence, there is a need to develop the available models in greater detail, for subsequent use in systems of practical importance.

References

1. H. Oechsner: In *Thin Film and Depth Profile Analysis*, ed. by H. Oechsner, Topics Curr. Phys. Vol. 37 (Springer, Berlin, Heidelberg 1984) p. 63
2. J. Kirschner, H.W. Etzhorn: Appl. Surf. Sci. **3**, 251 (1979)
3. M.P. Seah: J. Vac. Sci. Technol. A **3**, 1330 (1979)
4. H.H. Andersen, H.L. Bay: In *Sputtering by Particle Bombardment I*, ed. by R. Behrisch, Topics Appl. Phys. Vol. 47 (Springer, Berlin, Heidelberg 1981) p. 145
5. Y. Yamamura, H. Tawara: Atomic Data and Nuclear Data Tables **62**, 149–253 (1996)
6. G. P. Chambers and J. Fine: In *Practical Surface Analysis*, 2nd ed., Vol. 2, (ed. D. Briggs and M. P. Seah, Wiley 1992)
7. P. Sigmund: Phys. Rev. **184**, 383 (1969)
8. P. Sigmund: Phys. Rev. **187**, 768 (1969)
9. P.C. Zalm: J. Appl. Phys. **54**, 2660 (1983)
10. K.A. Gschneidner, Jr.: Solid State Phys. **16**, 344 (1964)
11. W.D. Wilson, L.G. Haggmark, J.P. Biersack: Phys. Rev. B **15**, 2458 (1977)
12. J. Schou: Nucl. Instr. Meth. **B 27**, 188 (1987)
13. J.P. Biersack, W. Eckstein: Appl. Phys. **A34**, (1984) 73
14. J.F. Ziegler, J.P. Biersack, U. Littmark: In *The Stopping and Range of Ions in Solids* (Pergamon, Oxford 1985)
15. M. Kustner, W. Eckstein, V. Dose, J. Roth: Nucl. Instrum. Meth. Phys. Res. **B145**, 320 (1998)
16. R. Kelly, N.Q. Lam: Radiat. Eff. **19**, 39 (1973)
17. C. Tian, W. Vandervorst: J. Vac. Sci. Technol. **A 15**, 452–459 (1997)
18. K. Wittmaack: J. Vac. Sci. Technol. **B. 16**, 2776–2785 (1998)
19. C.W. Magee, G.R. Mount, S.P. Smith, B. Herner, H.J. Gossmann: J. Vac. Sci. Technol. B16, 3099-3104 (1998).
20. G. Betz, G.K. Wehner: *Sputtering by Particle Bombardment II*, ed. by R. Behrisch, Topics Appl. Phys. Vol. 52 (Springer, Berlin, Heidelberg 1983) p. 11

21. P. Zalm: Surf. Interface Anal. **11**, 1 (1988)
22. E. Taglauer: Appl. Surf. Sci. **13**, 1980 (1982)
23. J.F. O'Hanlon: *A Users Guide to Vacuum Technology* (Wiley, New York 1980) p. 125
24. S.M. Hues, R.J. Colton: Surf. Int. Anal. **14**, 101 (1989)
25. M.P. Seah, C.P. Hunt, M.T. Anthony: Surf. Int. Anal. **6**, 92 (1984)
26. J. Fine, B. Navinsek: J. Vac. Sci. Technol. A **3**, 1408 (1985)
27. D.E. Newbury, D. Simons: In *Secondary Ion Mass Spectrometry SIMS IV*, ed. by A. Benninghoven, J. Okano, R. Shimizu, H.W. Werner, Springer Ser. Chem. Phys. Vol. 36 (Springer, Berlin, Heidelberg 1984) p. 101
28. S. Hofmann, J.M. Sanz: In *Thin Films and Depth Profile Analysis*, ed. by H. Oechsner, Topics Curr. Phys. Vol. 37 (Springer, Berlin, Heidelberg 1984) p. 141
29. U. Littmark, W.O. Hofer: Nucl. Instr. Meth. **168**, 329 (1980)
30. B.V. King, D.G. Tonn, I.S.T. Tsong, J.A. Leavitt: Mater. Res. Soc. Symp. Proc. **27**, 103 (1984)
31. M.P. Seah, C.P. Hunt: Surf. Int. Anal. **5**, 33 (1983)
32. J. Fine, P.A. Lindfors, M.E. Gorman, R.L. Gerlach, B. Navinsek, D.F. Mitchell, G.P. Chambers: J. Vac. Sci. Tech. A **3**, 1413 (1985)
33. M. Gauneau, R. Chaplain, A. Regreny, M. Salvi, C. Guillemot, R. Azoulay, N. Duhamel: Surf. Int. Anal. **11**, 545 (1988)
34. R. Helms, N.M. Johnson, S.A. Schwarz, W.E. Spicer: J. Appl. Phys. **50**, 7007 (1979)
35. A. Rar, S. Hofmann, K. Yoshihara, K. Kajiwara: Applied Surface Science **145**, 310–314 (1999)
36. R.v. Criegern, I. Weitzel, J. Fottner: In *Secondary Ion Mass Spectrometry SIMS IV*, ed. by A. Benninghoven, J. Okano, R. Shimizu, Springer Ser. Chem. Phys. Vol. 37 (Springer, Berlin, Heidelberg 1984) p. 308
37. K. Wittmaack, J.B. Clegg: Appl. Phys. Lett. **37**, 283 (1980)
38. P. Williams, C.A. Evans, Jr.: Int. J. Mass. Spectrom. Ion Phys. **22**, 327 (1976)
39. K. Wittmaack: Radiat. Effects **63**, 205 (1982)
40. C.W. Magee, W.L. Harrington, R.E. Honig: Rev. Sci. Instr. **49**, 477 (1978)
41. S. Duncan, R. Smith, D.E. Sykes, J.M. Walls: Surf. Int. Anal. **5**, 71 (1979)
42. T. Wohner, G. Ecke, H. Rossler, S. Hofmann: Fresenius Journal of Analytical Chemistry **353(3–4)**, 447–449 (1995)
43. G. Carter, B. Navinsek, J.L. Whitton: In *Sputtering by Particle Bombardment I*, ed. by R. Behrisch, Topics Appl. Phys. Vol. 47 (Springer, Berlin, Heidelberg 1981)
44. H.E. Roosendaal: In *Sputtering by Particle Bombardment I*, ed. by R. Behrisch, Topics Appl. Phys. Vol. 47 (Springer, Berlin, Heidelberg 1981)
45. S. Hofmann, A. Zalar, E.H. Cirlin, J.J. Vajo, H.J. Mathieu, P. Panjan: Surface & Interface Analysis **20**, 621–626 (1993)
46. V. Vishnyakov, G. Carter: Nuclear Instruments & Methods in Physics Research Section B – Beam Interactions with Materials & Atoms **106**, 174–178 (1995)
47. K. Wittmaack: Journal of Vacuum Science & Technology **B. 16**, 2776–2785 (1998)
48. G. Falcone, P. Sigmund: Appl. Phys. **25**, 307 (1981)
49. M.P. Seah, W.A. Dench: Surf. Interface Anal. **1**, 2 (1979)
50. P. Sigmund, A. Gras-Marti: Nucl. Instr. Meth. **182/3**, 5 (1981)

51. R.S. Averback: Nucl. Instr. Meth. B **15**, 675 (1986)
52. U. Littmark: Nucl. Instr. Meth. B **7/8**, 684 (1985)
53. B.V. King, I.S.T. Tsong: J. Vac. Sci. Technol. A **2**, 1443 (1984)
54. J. Kirschner, H.-W. Etzkorn: in "Thin Film and Depth Profile Analysis" (ed. H. Oechsner, Springer-Verlag 1984) 103
55. S.M. Hues, P. Williams: Nucl. Instr. Meth. B **15**, 206 (1986)
56. M.P. Seah: Vacuum **34**, 463 (1984)
57. S. Hofmann, J. Schubert: Journal of Vacuum Science & Technology A-Vacuum Surfaces & Films **16**, 1096–1102, (1998)
58. B.V. King, I.S.T. Tsong: Nucl. Instr. Meth. B **7/8**, 793 (1985)
59. M.G. Dowsett, R. Collins: Phil. Trans.Royal Soc. **A 354**, 2713–2729 (1996)
60. N.S. Smith, M.G. Dowsett, B. McGregor, P.Phillips: in "Secondary Ion Mass Spectrometry SIMS X" (ed. A. Benninghoven, B. Hagenhoff and H.W.Werner, Springer Verlag, Berlin 1995) 363

5 SIMS – Secondary Ion Mass Spectrometry

R.J. MacDonald and B.V. King

Secondary ion mass spectroscopy (SIMS) is an ion beam analysis technique useful for characterising the top few micrometres of samples. Primary ions of energy 0.5–20 keV, commonly O_2^+, Cs^+, Ar^+ but also ions such as Ga^+, Xe^+, O^-, $C_nH_m^+$ and SF_6^+ are used to erode the sample surface and the secondary elemental or cluster ions formed from the target atoms by the impact are extracted from the surface by an electric field and then energy and mass analyzed. The ions are then detected by a Faraday cup or electron multiplier and the resulting secondary ion distribution displayed as a function of mass, surface location or depth into the sample (Fig. 5.1).

SIMS has been available for thirty years or more, but in the last decade or so, recognition of the importance of the surface in many aspects of materials science and the growing importance of thin film processes and solid state electronics have all lead to a dramatic increase in the sophistication of surface analysis instrumentation, including SIMS. SIMS is probably the most

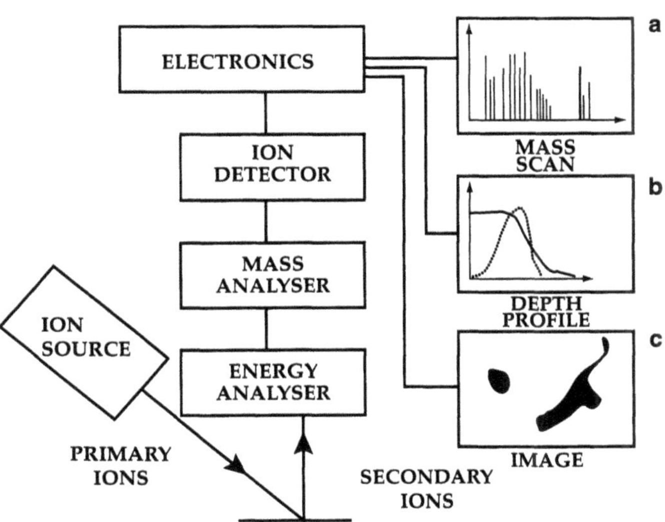

Fig. 5.1. Schematic representation of the main components of a SIMS experiment. The signal due to the detected secondary ions may be presented as a function of (**a**) the ion mass, (**b**) the duration of ion bombardment or (**c**) the location of the surface

sensitive technique currently commonly available for surface compositional analysis, with detection limits are 10^{13}–10^{18} atoms cm^{-3} for impurities in semiconductors. SIMS provides a large amount of information which at first sight is straightforward to interpret. It is however frustratingly difficult to relate the strength of the SIMS signal to the composition of the surface since SIMS has not yet been provided with a theory allowing absolute or relative quantification and so most quantitative analysis using SIMS will involve the use of well-characterised standards.

This review will not be exhaustive. For a fuller treatment the reader is referred to recent books dedicated to SIMS [1,2], SIMS depth profiling [3] and static SIMS [4]. There are also proceedings of biennial conferences devoted to SIMS which contain papers on both the theory and practice of SIMS [5] as well as web pages devoted to SIMS – e.g. [6].

5.1 The Practice of SIMS

5.1.1 Overview

Secondary ions are ejected from a surface subject to bombardment by a primary ion by a process known as sputtering. The complex nature of the interaction between an incident energetic ion or neutral and a solid surface is shown schematically in Fig. 5.2. Incident keV ionsgenerate collision cascades in the region of the surface of the solid. Within the cascade a small part of the momentum may be redirected towards the surface, providing sufficient energy for the ejection of atoms and molecules from the outermost surface layers. A proportion (10^{-3}–10 %) of the ejected atoms or molecules may be in a charged state. These ions provide the signal for SIMS.

SIMS information may be presented in three ways. Firstly, secondary ion intensities may be collected as the mass of the secondary ions which are accepted by the spectrometer is scanned. The resulting mass spectrum (Fig. 5.3) is typically obtained for low primary ion doses and may be representative of the top monolayer of the surface (static SIMS). Secondly, the secondary ions from selected elements may be monitored as a function of sputter time (Fig. 5.4) as the primary ion beam erodes up to a few microns into the sample (dynamic SIMS). The secondary ions detected in SIMS generally come from the top monolayer of the surface. The primary ion beam is then used for two purposes in dynamic SIMS: to eject secondary ions from the surface and to sputter the target, continuously exposing a new surface for SIMS analysis. In that way a profile of the target elemental concentrations is found as a function of depth into the target. The practice of depth profiling is considered further in another chapter. Thirdly, the secondary ion signal may be presented as a function of position on the surface in a similar way to scanning electron microscopy (Fig. 5.19).

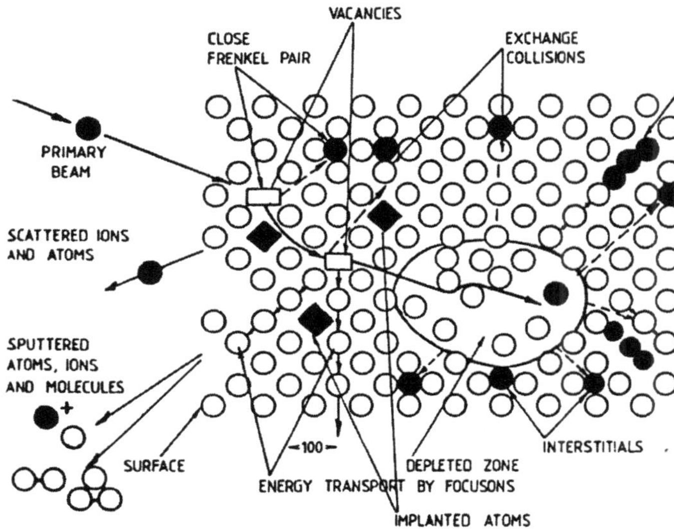

Fig. 5.2. Schematic diagram of the processes which take place after ion impact onto a surface. SIMS is concerned with analysis of the sputtered ions. The impact of one primary ion can cause many target atom displacements which affect the surface seen by subsequent ion impacts

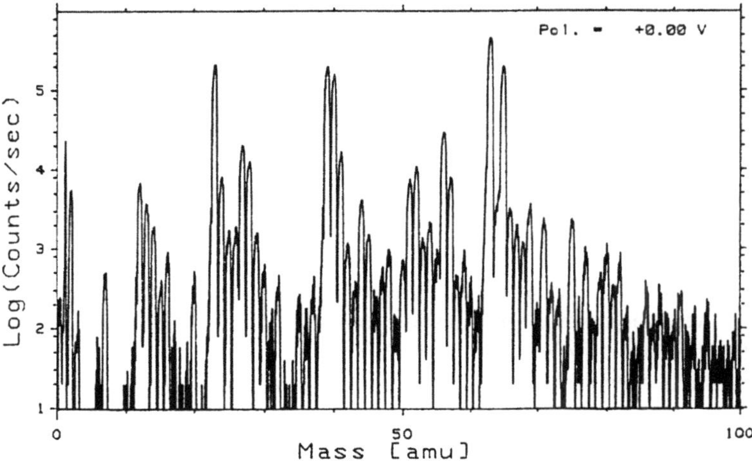

Fig. 5.3. Positive ion mass spectrum for Cu bombarded with 12 keV Ar^+. Prominent peaks are seen for Na, K, Ca, and the two isotopes of Cu

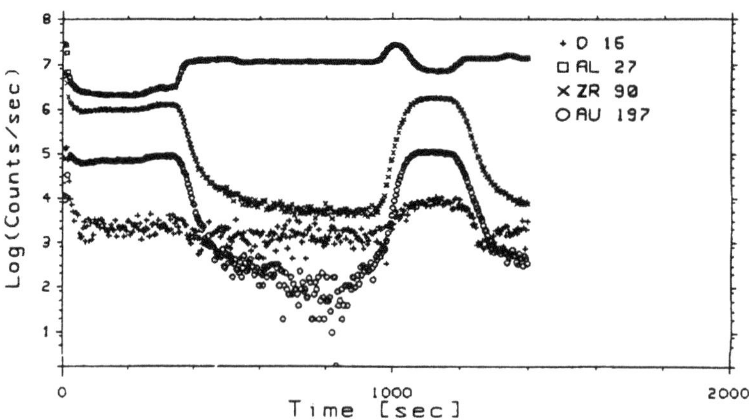

Fig. 5.4. SIMS depth profile of O, Al, Zr and Au as a function of primary ion bombardment time for a thin film sample comprising two layers of mixed Zr, Au and Al separated by a layer of "pure" Al. Note that all signals are high initially, due to enhancement of the ion yields by an increased O concentration at the surface

5.1.2 Advantages and Disadvantages of SIMS

The major advantages of the SIMS technique are:

i) excellent sensitivity (ppb–ppm) and a wide dynamic range (10^6) are found for most elements;
ii) the ability to distinguish different isotopes of the same element;
iii) the ability to obtain information on chemical bonding at the surface by monitoring molecular ions.

The primary disadvantage is the variation of detection sensitivities over orders of magnitude for different elements in one substrate and for the same element in different substrates. This may be overcome to a certain extent, as will be discussed later, by the use of reactive primary ions (O_2^+ or Cs^+) which enhance the yields of positive and negative ions respectively. In spite of this sensitivity variation, semi-quantitative analysis may be performed using standards with low primary ion doses. In that case for certain systems, SIMS can provide an accuracy of about 6 % and a precision less than 0.5 %.

Other disadvantages of SIMS analysis include

i) the requirement in some SIMS analysers that only one mass peak may be measured at any one time. Multi-element analysis then must be done sequentially by cycling the mass analyser through the different peaks to be measured. This procedure is time consuming, especially if static conditions are required;
ii) the alteration of the target by radiation-induced processes during dynamic SIMS measurements. These include ion-beam-induced intermixing of target elements, radiation-induced thermal segregation, surface

topography changes and redeposition of other material into the analysed region. These aspects are dealt with more fully in Chap. 4 on depth profiling.

5.1.3 The Yield of Secondary Ions

The yield of secondary ions of a given element, ejected from a solid surface, will depend on a number of parameters. We could write

$$Y_A\, dE\, d\Theta = J_A S_A N_A \sigma_A P_A T\, dE\, d\Theta\ , \tag{5.1}$$

where the symbols have the following meanings

$Y_A\, dE\, d\Theta$: yield of secondary ions of element A detected by the spectrometer
J_A: incident ion beam density (ions cm^{-2})
S_A: sputter yield of atoms of element A in the matrix comprising the surface layer of the target
N_A: surface density of atoms of element A (atoms cm^{-2})
σ_A: cross-section for ionisation of an atom of element A as it is ejected from the surface
P_A: probability that the ion of element A, once formed, survives in that ionised state to be detected
T: transmission of the spectrometer system for ions of element A
dE: acceptance width in energy of the spectrometer system
$d\Theta$: acceptance angle of the spectrometer system.

We can consider each of these factors in turn.

a) The Incident Ion Beam Density. This is a function of the ion source and the experimental measurement to be made. So-called static SIMS requires a very low primary beam density, perhaps 10^9 particles $cm^{-2}\,s^{-1}$ deposited over a few mm^2 to give a total dose in any experiment below $10^{13}\,cm^{-2}$, so that each incoming primary ion "sees" an undamaged surface If each ion impact affects an area of $10\,nm^2$ then only 10^{12} impacts cm^{-2} will cause 10% of the top monolayer to be damaged. For a particle beam density of $1\,nA\,cm^{-2}$, this damage threshold would be reached after 160 s.

In dynamic SIMS, beam densities of 0.1–$10\,mA\,cm^{-2}$ are typical although the exact value used depends on the time required to profile the complete structure. For example, to profile Si to a depth of $1\,\mu m$ in 20 min requires a current density of about $1\,mA\,cm^{-2}$. Typically O_2^+, Cs^+, or Ar^+ beams are used with beam diameters of 1–10 µm. The best depth resolution however is obtained when the detected ions all come from the same depth i.e. when the eroded crater is perfectly flat. Typically this is achieved by scanning the primary ion beam over areas of about $250\,\mu m \times 250\,\mu m$ on the sample. The requirement of rapid depth profiling places constraints on the size of the ion beam used. For example, if the ion beam used in the above Si depth profile was scanned in this way, a beam current of 0.6 μA would be required to achieve the desired current density of 1 mA cm^{-2}. Such a beam current could only be typically achieved for Ar^+, O_2^+ or Cs^+ ions by using a beam of at least 50 μm diameter. More details on dynamic SIMS are found in the chapter on depth profiling.

Imaging SIMS may be performed using an ion microscope or imaging microprobe. In microscope mode, the spatial distribution of ions emitted from the surface is retained through the secondary ion transport optics and measured with about 1 μm resolution by a position sensitive detector or on a fluorescent screen. In microprobe mode, a well focussed primary ion beam is rastered over the surface, as in the dynamic SIMS mode discussed above, and the secondary ion emission from the small bombarded area measured as a function of the beam position. This mode is inherently inefficient compared to the microscope mode since only a small part of the surface is being detected at any one time.

In most circumstances encountered, the density of the incident ion beam is such that the collision cascades initiated are not overlapping so that the sputtered secondary ion signal is linear with incident ion beam density. If, however, primary beams of molecular ions like O_2^+ are used, the molecule splits upon impact. The energetic atoms produced then initiate collision cascades in the target which can overlap in space and time. This will lead to nonlinear relationships between the incident ion beam density and the yield of secondary ions.

b) The Sputter Yield of Elements. The sputter yield of an element A, S_A, is related to the erosion rate, U, of its ion bombarded surface by

$$U = \frac{J_A S_A}{N_A} , \tag{5.2}$$

where J_A is the ion current density and N_A is the elemental atomic density. The sputter yield of a target depends on the ion energy and species as well as the target atomic number, surface crystallinity and topography. However, theoretical sputter yields calculated for smooth amorphous targets generally give good agreement with values found from experimental depth profiles of single element targets. Therefore theoretical sputter yields can be profitably used to calculate erosion rates in dynamic SIMS measurements. This then allows the SIMS depth profile which is given in terms of sputter time (Fig. 5.4) to be recast in terms of sputtered depth.

A theory of the sputtering of atoms from the smooth surface of an amorphous solid consisting of a single atomic species has been well developed by *Sigmund* [7]. In this case, the sputter yield of atoms, S, from a target bombarded by ions of mass M_1 and energy E is proportional to the nuclear stopping power, $F_d(E)$ (i.e. the energy spent in elastic collisions between atoms.)

$$S = 0.042 \, \frac{\alpha(M_2/M_1)F_d(E)}{E_b} \, , \tag{5.3}$$

where α is a function only of the ratio of the incident ion mass, M_1, to target atom mass, M_2 and E_b is the surface binding energy of a target atom to the surface. E_b is typically taken equal to the heat of sublimation of the target element. There are compilations of sputtering yield data for amorphous single element solids and for some compounds and alloys, e.g. *Matsunami* et al [8], *Andersen and Bay* [9]. Empirical relationships and computer codes [9] also exist as discussed in Chap. 4. Their accuracy in reproducing experimental values for the sputtering yields of elemental targets is good, but there exists little experimental data for comparison in the case of multi-element matrices. However we can predict the sputtering yield of a given element in almost any target matrix to better accuracy than we can account for some other parameters in (5.1). In general, S increases with primary ion energy and mass to a maximum of 0.1–10 at a primary ion energy of about 10 keV or more. S also increases with increasing ion incidence angle, θ, up to about 70° to the surface normal. Over this range S is approximately proportional to $\cos^{-1}\theta$ for inert gas bombardment, but not necessarily for active ion bombardment.

In general, the sputtering yield of an alloy may be approximated as the sputtering yield of the pure element A, multiplied by the fractional composition of A in the matrix. So for an alloy target $A_a B_{(1-a)}$ a first approximation to the total sputter yield is

$$S = aS_A + (1-a)S_B \, . \tag{5.4}$$

More details on sputter yields are included in Chap. 4.

c) The Surface Density of Atoms. We expect the ion yield of a given species of ion to be related to the surface density of atoms of that type. However the surface density of a given element may not correspond to the bulk composition and so may not be well known. The alteration may or may not be caused by the primary ion beam. For example, the primary ion irradiation process may modify the surface concentrations by ion implantation or ion-induced segregation, diffusion or mixing. In all cases the composition of the top monolayer will differ from the subsurface composition. This will lead to a time-dependent secondary ion yield in a dynamic SIMS experiment which complicates the determination of the bulk elemental composition from the measured SIMS signal of a particular element.

The surface density of atoms of a given element will be particularly affected by the process of preferential sputtering [11]. If, in a sample of two or more elements, the probabilities of sputtering atoms of different elements are different, there will be a modification of the surface layer concentrations as a result of sputtering. Atoms with higher probabilities of sputtering will be removed preferentially and hence the surface concentration of the atoms with lower probability of sputtering will increase with time. This process of modification of the surface layer concentration by preferential sputtering will continue until the sputtered particle flux has the bulk concentration and a concentration gradient with depth into the target will be established. An understanding of the modification in the so-called pre-equilibrium region is very important for accurate analysis of the sub keV implantations used in modern semiconductor devices.

d) The Cross-Section for Ionisation of a Sputtered Atom. The cross-section for ionisation of a sputtered atom is often expressed as a probability that an atom will be ejected from the surface in an ionised state. There are a variety of models for the cross-section, but also some facts we can state with certainty as a result of experimentation, e.g. the secondary ion yield is a function of the work function of the surface. *Ming Yu* and *Lang* have recently reviewed this evidence in a very precise way [12] for both negative and positive ions. The favoured model [13] considers a single-stage process which can be represented by the energy level diagram shown in Fig. 5.5. This represents an energy level on an atom leaving a solid as a function of distance from the surface of the solid. For positive ion formation this level would correspond in depth below the vacuum level to the ionisation energy of a given state of the

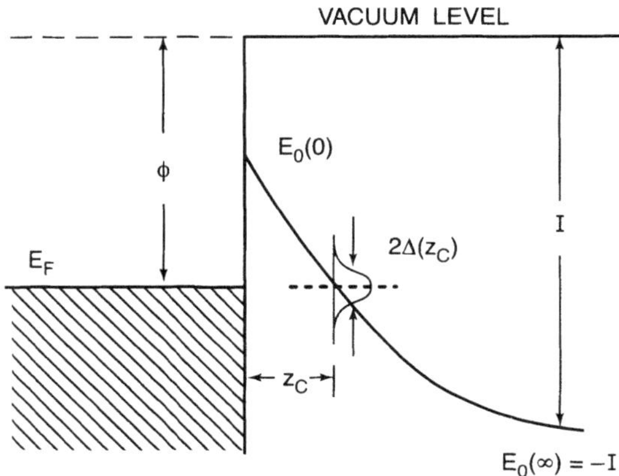

Fig. 5.5. Schematic energy diagram of an atom leaving a metal surface. E_F is the Fermi level and ϕ is the work function. Initially the atomic level E_i is broad and may lie above E_F before the atom is sputtered. The variation of the image potential causes E_a to lower with separation until $E_a = E_F$ at the crossing point. Electrons in the metal can tunnel out to fill the atomic level once $E_a < E_F$ beyond the crossing point

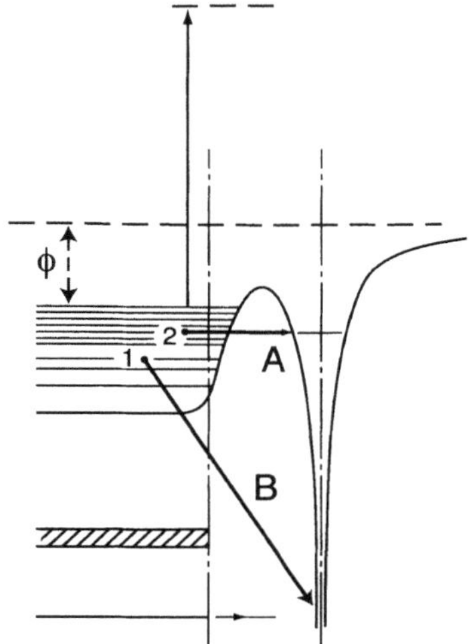

Fig. 5.6. Possible charge exchange processes involving an atom or ion moving away from a surface. (*A*) Resonant charge exchange (*B*) one type of Auger process

isolated atom, while for formation of a negative ion it would correspond to the electron affinity. The depth of the level below the vacuum level would be a function of the distance from the surface, because of the existence of the image potential and possible polarisation effects. This model neglects Auger processes, so only resonant transitions of electrons between the metal and the ion/atom are possible (Fig. 5.6).

According to Fig. 5.5, the sputtered atom will be ejected in an ionised state, so giving rise to the image potential, and the probability of neutralisation along the outgoing trajectory will be associated with resonance of the ionisation level and occupied states of the electron energy distributions in the solid, i.e. with states below the Fermi energy. The maximum chance of exchange will then occur when resonance occurs between the ionisation level and states near the Fermi level. The ionisation level is broadened as a result of the finite lifetime of the level and, since the wave function of the electron in the metal decays exponentially with distance z from the surface, we can approximate the level width by:

$$\Delta(z) = \Delta_0 \, \mathrm{e}^{-\gamma z} \, , \tag{5.5}$$

where γ is a characteristic length.

This model can be quantified, and for the case in which a crossing of the ionisation level E_a with the Fermi level E_F occurs at a distance z_c from the surface,

$$P^{\pm} = \exp\left[-2\Delta(z_\mathrm{c})/\hbar\gamma V_\perp\right] \tag{5.6}$$

where $\Delta(z_\mathrm{c})$ is the width of the ionisation level at the distance z_c at which the level E_a crosses E_F. $V_\perp$ is the ejected atom or ion velocity component

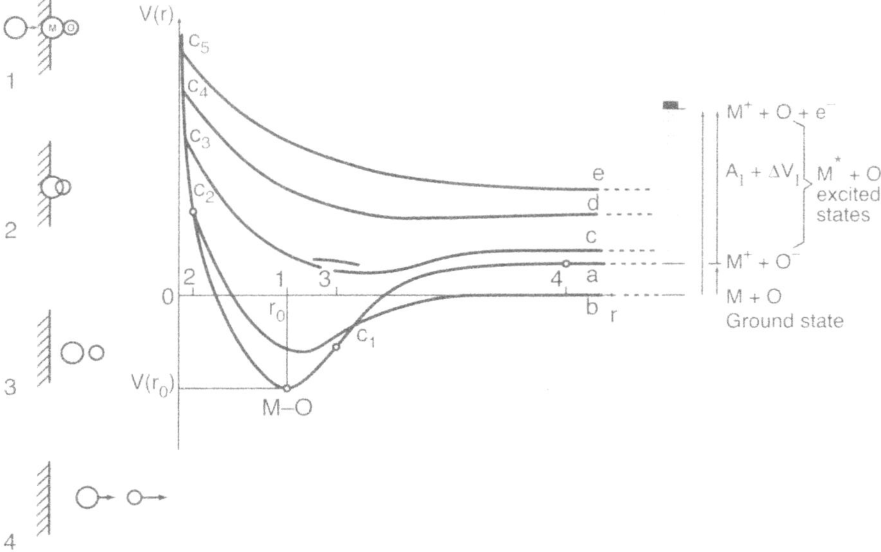

Fig. 5.7. Schematic diagram of the M–O system during its dissociation with increasing distance, r, from the surface. In this model of ionisation of sputtered atoms, energy is transferred from the collision cascade to a quasi-molecule comprising a substrate metal atom with an adsorbed oxygen atom (inset *1*). As the metal and oxygen atoms move together (inset *2*) the system moves to the left up the repulsive side of the potential energy curve. If the energy imparted in the collision is great enough the system may cross onto curves c, d or e at points c_3, c_4 or c_5 respectively, and excited metal atoms or $M^+ + \mathrm{O}$ will be sputtered. For the more probable low energy collisions, the system will return to the right along curves a or b and so metal and oxygen ions or atoms will be sputtered (inset *4*)

normal to the surface. This dependence of P^+ on $V_\perp$ has been observed experimentally [14].

The situation when an active gas is present, either in the form of the incident beam (e.g. O^-, O_2^+) or as a background gas in the chamber, is more complex. Secondary ion yields increase by one to two orders of magnitude, and cannot be explained by changes in the work function. The enhancement of the ion yield has, however, been shown to be linearly dependent on the surface coverage of adsorbate, down to coverages where the modification of the surface electronic energy levels would be minimal. The enhancement is then realised on the adsorbate site. To explain this, a bond-breaking model has been proposed [12]. The oxygen-surface atom bond is broken in the collision process resulting in ejection of the surface atom. The possibilities may be represented on a potential energy diagram represented schematically (Fig. 5.7). The path in this diagram to be followed will depend on the energy loss in the collisions, but it may give rise to the ejection of ionised particles. The logical suggestion is that, for a surface coverage which is small so that there is essentially M^+O^- bonding, the number of M^+ ions ought to equal the number of O^- ions.

It is obvious from the above that the general picture of secondary ion formation is known, but that an accurate analytical model allowing calculation of a cross-section for ionisation in (5.1) is not available. This remains the major barrier to quantification of SIMS.

e) Transmission of the Mass Spectrometer. This is an instrument function which can in principle be modelled by computer simulation or measured experimentally. Typically the transmission of a quadrupole analyser is 0.1% or less whilst a magnetic sector instrument has a transmission of 1–10% and a time-of-flight instrument 10% or more. This will be discussed in a later instrumental section.

f) The Energy and Angular Distributions of Secondary Ions. Equation (5.1) contains terms in dE and $d\theta$, related to the energy distribution and the angular distribution of the sputtered ions. The energy distribution will also enter the equation because the cross-section for ionisation and the probability of neutralisation are both likely to be energy dependent. The transmission of the instrument may also be energy dependent.

The angular distribution is important in that maximum sensitivity for measurement will occur at the maximum in the angular distribution of the secondary ion yield. The acceptance angle of the mass spectrometer should clearly be related to the half width of the sputtered ion angular distribution.

The Energy Spectrum of Sputtered Ions. The energy spectrum of sputtered atoms is fairly well understood. *Thompson* [16] showed that the energy spectrum of atoms sputtered from a random target is given by

$$N(E)\,dE \propto \frac{E}{(E+E_\mathrm{b})^3}\,dE\,, \tag{5.7}$$

where $N(E)dE$ is the yield of ions in the energy range dE at E and E_b is the binding energy of the atom to the surface.

The energy spectrum will peak at an energy of a few eV and will fall off as $1/E^2$ at higher energies, independent of the primary ion energy. It will be related to the ion yield detected, through the cross-section for ionisation and the probability of subsequent neutralisation. These cross-sections have the form of an exponential of $(-V_\mathrm{c}/V_\perp)$ so the energy spectrum of the secondary ions will be of the form

$$N(E)\,dE \propto \frac{E}{(E+E_\mathrm{b})^3}\,\mathrm{e}^{-V_\mathrm{c}/V_\perp}\,. \tag{5.8}$$

The neutralisation parameter will favour slower ions, hence it is likely that the energy spectrum will have its peak shifted to higher energies relative to the energy spectrum of sputtered neutrals, and a tail which shows a dependence close to $1/E^2$.

Angular Distribution of Secondary Ions. The angular distribution of secondary ions due to sputtering has not been well studied. In the case of sputtered neutrals, under circumstances in which the irradiation is normal to the surface, the angular distribution is close to cosinal. Taking into account the effect of the normal component of the secondary ion velocity, one would expect a removal of the ion by neutralisation to be occurring preferentially along trajectories close to the surface. Thus the distribution is likely to be of the form $\cos^n \theta$, where θ is the angle from the surface normal.

For non-normally incident ions, the angular distribution of sputtered atoms is a little more complex. It is approximately a cosine distribution about the specular direction. Secondary ion distributions due to sputtering by non-normal incident ions will be more complex, due to the neutralisation or ionisation cross-section dependence on the perpendicular component of the exit velocity. As the exit trajectory approaches the surface, the dependence on $V_\perp$ will lead to strong preferential neutralisation of the seondary ions. There has been little reported in the literature on such angular distributions.

5.2 Construction of a Secondary Ion Mass Spectrometer

The main elements of a SIMS analyser are (i) a primary ion source (ii) optics to transfer these ions to the target surface and to collect secondary ions from the surface and direct them to (iii) a mass spectrometer.

All SIMS machines must have a source of ions available to initiate the sputtering process. In general, the ions will have energy in the range 1–20 keV, but other characteristics of the incident beam will depend on the type of application. Many add-on SIMS analysers will use a reasonably simple ion beam, with a beam diameter of order 0.1–1 mm and with the facility to raster the beam across larger areas. Dedicated SIMS analysers on the other hand often have sophisticated ion beam systems, particularly if the analyser is likely to be used in microscope mode.

All ion guns comprise an ion source and a method of transferring the ions to the sample. This is normally done by floating the ion source at the accelerating voltage and leaving the target at ground potential. In some instruments, e.g. the Cameca IMS6F, the target may also be floated at a high potential. There are four common types of ion sources used, namely electron bombardment ion sources with and without plasma formation (duoplasmatron, hollow cathode), surface ionisation sources and field ionisation sources (liquid metal ion sources).

In duoplasmatron sources, ions are extracted in a strong external electrostatic field from a high density plasma which is magnetically confined

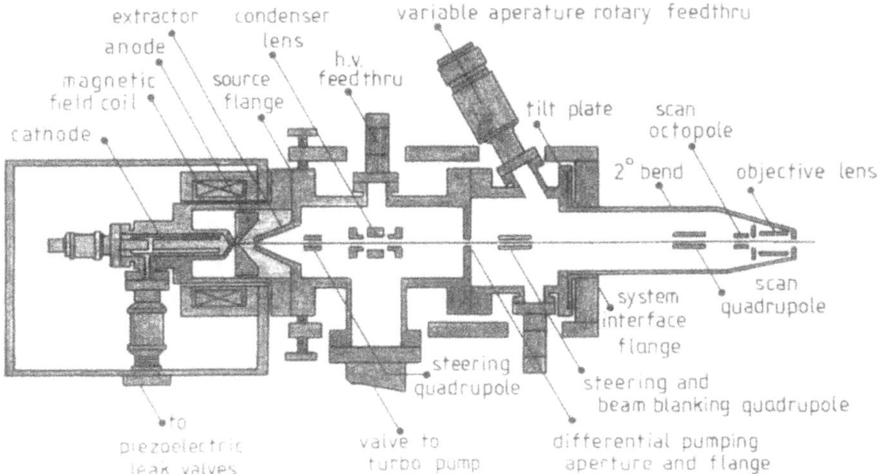

Fig. 5.8. Schematic diagram of a PHI duoplasmatron source and primary ion optics

between an anode and intermediate electrode. These sources are bright (10^6–$10^7\,\mathrm{A\,m^{-2}\,sr^{-1}}$) with an energy spread of about 10 eV, and so allow current densities of about $10\,\mathrm{mA\,cm^{-2}}$ into beams of diameter 1–50 μm. Both inert and active gas beams can be produced by this source. Figure 5.8 shows a schematic diagram of the PHI duoplasmatron source and ion transport optics.

In a surface ionisation source, alkali metals like caesium are varporised from a porous metal surface which has a high work function. When the temperature of the surface is high so that the caesium surface concentration is low, then the caesium is almost all desorbed as ions. The brightness of the source is about $10^6\,\mathrm{A\,m^{-2}\,sr^{-1}}$ but the source has a very low energy spread (< 1 eV) and so current densities and spot sizes comparable to duoplasmatrons can be achieved.

In liquid metal ion sources a low melting point metal (Ga, In, Cs) is supplied as a liquid over a metal tip which is subject to a high electric field. Electrons tunnel from surface atoms to the tip, leaving the metal ions to be extracted. These sources have high brightness ($10^{10}\,\mathrm{A\,m^{-2}\,sr^{-1}}$) [14] and allow nA currents to be delivered into spots of diameter less than 50 nm or pA to be delivered into a minimum spot size of 10 nm. Figure 5.9 shows a schematic diagram of the layout of the FEI single lens liquid metal ion source.

Duoplasmatron sources are mainly used for production of O_2^+ and Ar^+ for depth profiling and mass scans. Surface ionisation sources allow depth profiling using Cs^+ with the attendant advantages of increased negative ion yields and decreased matrix effects in positive ion depth profiling. Liquid metal ion sources are exclusively used for imaging SIMS since currents are normally too small for rapid depth profiling. However, for high spatial resolution imaging,

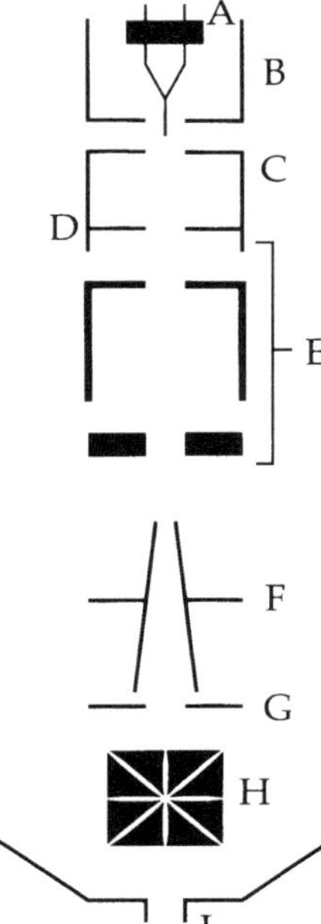

Fig. 5.9. Schematic diagram of the FEI liquid metal ion source. *A*–filament, *B*–suppressor, *C*–extractor, *D*–beam defining aperture, *E*–lens, *F*–blanking plates, *G*–blanking aperture, *H*–octapole, *I*–differential pumping aperture

the gun must be matched with a high transmission spectrometer since, for example, a volume of 20 nm × 20 nm × 5 nm contains only 50000 atoms. For a dilute impurity, e.g. at a concentration of 20 ppm, there will only be one atom in the volume. The impurity can then only be detected if the spectrometer transmission is high.

Primary ion optics, for transferring ions onto the sample consists of usually two electrostatic lenses for beam focussing, xy deflection for beam positioning, rastering and neutral rejection, a stigmator for focus correction independently in x and y directions when small spot sizes are used, and a mass filter to remove beam contamination, either multiply charged ions, clusters or impurity ions. The most commonly used mass filters are a Wien filter or a magnetic sector. In a Wien filter crossed magnetic and electric fields produce equal and opposite forces on ions of a certain M/Z ratio which then pass through the filter undeflected. Ions of other M/Z ratios do not pass through the filter. The electric fields may be shaped by shims to produce additional focussing. In a magnetic sector, ions are bent in a uniform magnetic field. Ions of energy eV and a specific M/Z ratio are bent in the horizontal plane with a radius of curvature

$$R = \sqrt{\frac{2V_0}{eB^2}\frac{M}{Z}}\,, \tag{5.9}$$

which, if it matches the radius of curvature of the instrument, allows the ions to be transmitted through the magnet. Focussing in the vertical plane can be achieved by rotating the magnetic boundaries with respect to the particle trajectory.

The dynamic range and depth resolution achievable in depth profiling is dependent on the crater flatness and the background signal reaching the detector. Flat craters are achieved by scanning the ion beam typically over five beam diameters. Electronic gating of the detected ions then ensures that only the ions emitted from the centre of the crater are counted. For this procedure to work correctly, neutrals arriving with the ion beam must be eliminated since they will generate secondary ions independent of any raster gating. A bend of at least 1° followed by apertures must be included in the section of the ion optical path to reject neutrals from the beam striking the sample.

SIMS analyses of insulating samples, common in static SIMS, creates problems due to surface charging. The injection of positive ions and ejection of secondary electrons (with an efficiency of almost unity at SIMS energies) can cause surface potentials to vary with time to a few hundred volts which can completely stop secondary negative ion emission and alter the energy spectrum of emitted positive ions making optimisation of ion optics difficult. Charging may be reduced by using thin samples, low ion current densities, electron flooding or fast atom bombardment. Electrons with energies of a few eV to a few hundred eV can be generated from just a filament or a simple gun. If the electron current is matched to the ion current, then the surface charging can be reduced. This process is self-stabilising since a positive surface potential will reduce the loss of secondary electrons which will in turn decrease the surface potential.

In the application of SIMS to analysis, the information on surface composition is contained in a range of secondary ions of variable mass, distributed over a broad angle and energy range. In order for the equipment to have a reasonable mass resolution, the energy range of particles accepted into the mass analysis section of the spectrometer must be narrow. This energy range is given by ΔE which is the spread in the ion energy E seen by the mass spectrometer. The energy E is equal to $qV + E_0$, where q and E_0 are the charge and energy, respectively, of the secondary ion and V the extraction potential. A spread in E then results from the spread in the initial energy of the secondary ions and from any drift in V. This energy resolution must be smaller if a larger mass resolution $M/\Delta M$ is desired. A SIMS mass spectrometer must therefore consist of an energy analyser preceding the mass analyser. The secondary ions may be extracted into this energy analyser by a high ($> 1\,\mathrm{kV\,mm^{-1}}$), or low ($< 10\,\mathrm{V\,mm^{-1}}$) electric field. Higher transmission results from the use of a high extraction field since more ions are collected into the analyser and the relative energy spread of the secondary ions ($\Delta E/qV$) is less. Typical electrostatic energy analysers have high transmission over a

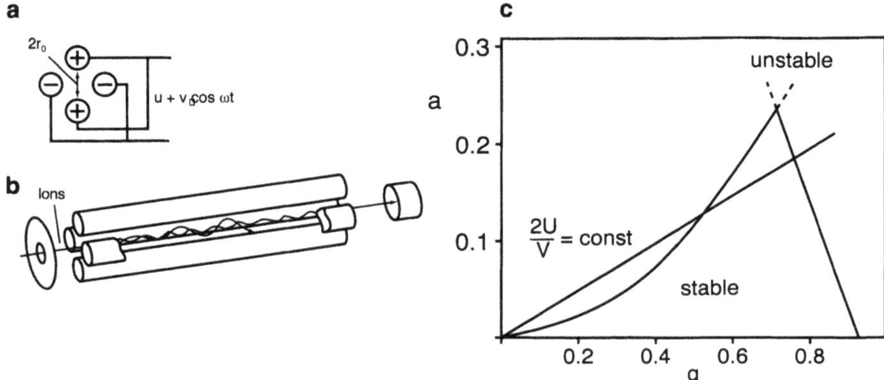

Fig. 5.10. Diagram of a rf quadrupole. (**a**) A combination of a dc voltage U and ac voltage V is applied to opposite pairs of rods. (**b**) The parameters a and q are given by $(4eU/m\omega^2 r_0^2)$ and $(2eV/m\omega^2 r_0^2)$ respectively. The ratio a/q is independent of m/e, the ion's mass to charge ratio. Therefore the orbits of all ions will lie on the straight line going through the origin. However, only ions lying within a narrow region of that line near $q = 0.7$ have stable trajectories. The width of this stable region may be changed by altering U/V. The value of m/e corresponding to this stable region is adjusted by altering U and V whilst keeping their ratio constant. (**c**) The ion will then travel through the rods and be detected. Ions with different mass-to-charge ratios will have diverging trajectories and strike rods or aperture

relatively narrow energy spread, typically 1–2%. To match this transmission to an energy spread ΔE of about 100 eV requires an extraction potential of 5–10 kV. If lower potentials are used, only a fraction of the total secondary ion spectrum can be collected and analyzed.

There are three main types of mass spectrometer: rf quadrupole, magnetic sector and time of flight. In an rf quadrupole analyser, low energy (< 50 eV) secondary ions travel down through the centre of four circular rods (Fig. 5.10) which are electrically connected in opposite pairs. A combination of dc and rf (~ 1 MHz) voltages is applied to one set of rods and an equal but opposite combination is applied to the other pair. Only ions of a certain M/Z ratio, where the ion charge q is given by $q = Ze$ with e the electronic charge, have a stable trajectory and are transmitted through the quadrupole. Ions of different M/Z ratios move in unstable trajectories and strike the rods or apertures. The mass of the transmitted ion is selected by increasing the rf and dc voltages whilst keeping their ratio constant. Only one M/Z ratio may be passed at any one time. Mass analysis of many elements must therefore be performed sequentially. The mass resolution, $M/\Delta M$ of the quadrupole is normally about 1000 over a restricted mass range. This resolution can be increased by using larger diameter rods at the expense of a lower obtainable mass range. Transmissions of the order of 1% and mass analyzes up to $M/Z \sim$

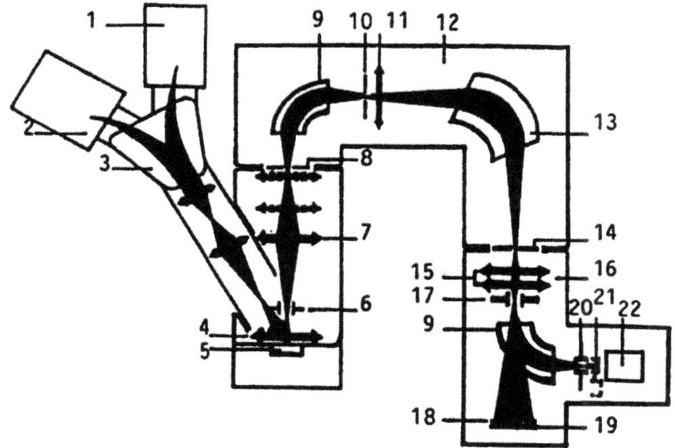

Fig. 5.11. Schematic diagram of the ion optics of a Cameca IMS-4F ion microscope. The electrostatic sector selects the ion energies passed to the double focussing magnetic sector. The ions may then be displayed on a fluorescent screen in microscope mode or collected in a Faraday cup or electron multiplier for quantitative static or dynamic analysis. The elements of the analyser are (*1,2*) – ion sources, (*3*) – primary beam mass filter, (*4*) – immersion lens, (*5*) – specimen, (*6,17,20*) – deflectors, (*7,11,15*) – lenses, (*8,10,14*) – slits, (*9*) – electrostatic sector, (*12*) – spectrometer, (*13*) – electromagnet, (*16*) – projection display and detection, (*18*) – channel plate, (*19*) – fluorescent screen, (*21*) – Faraday cup, (*22*) – electron multiplier

2000 are typical for rf quadrupoles. They are useful for adding SIMS analysis onto other analysis systems or where space is critical. Further discussion on rf quadrupole systems is given by *Dawson* [18].

Magnetic sector spectrometers send ions through a perpendicular magnetic field. The radius of curvature of the ions then depends on $(M/Z)^{1/2}$ as found previously. The mass resolution of the magnetic sector is proportional to the radius of curvature of the path but is sensitive to angular and energy spread in the incoming secondary ions. The loss in resolution may be alleviated by energy analysis in an electrostatic sector. The transmission of the instruments is high ($> 10\%$) and the mass resolution is high ($M/\Delta M > 10^4$) over a large mass range. The high mass resolution is important for many types of analysis. For example, a typical problem in semiconductors is to analyze P in Si. The mass of P^{31} is 30.9859 amu whereas the mass of $Si^{30}H$ is 30.9736 amu. If the analyser has a mass resolution $M/\Delta M$ less than 4000, the two peaks will overlap and unambiguous determination of low concentrations of P in Si will be made impossible.

Figure 5.11 shows a schematic diagram of a common magnetic sector instrument, the CAMECA IMS4F system. This is a SIMS microprobe which has been available for a number of years and has progressed through several generations of different analysers. It is a microscope system capable of producing images of about 1 μm spatial resolution. It has a high mass resolution (up to 10^4 in spectrometer mode) which is achieved by the use of a high extraction potential (about 5 kV), secondary ion energy and mass selection in 90° electrostatic and magnetic sectors. The combination of transfer lenses and sectors forms a stigmatic image of up to a 400 μm × 400 μm area of the target onto the fluorescent screen. Alternatively, the secondary ions can be directed into an electron multiplier for static and dynamic SIMS analysis.

Time of flight instruments rely on the measurement of the drift time of secondary ions along a flight tube. If the energy of the ions is known, eV_0, by accelerating them through a high potential difference, then the mass to charge ratio of the ion is related to the time taken to drift a distance d by

$$M/z = 2eV_0 t^2/d^2 \; . \tag{5.10}$$

Typical flight times are 10 μs–1 ms depending on the ion mass and the flight path. The secondary ions must, however, all start at the same time. This requires that the formation of the secondary ions must occur in short pulses, typically sub-nanosecond in duration for mass resolutions up to 10000. The time between pulses is set by the maximum flight time plus the data processing time. Typically the primary ion beam is pulsed by rapid deflection past slits in the primary ion column. Alternatively, secondary ion formation above the surface from sputtered neutrals could be pulsed, for example, by laser ionisation. The transmission of time of flight instruments is very high ($> 50\%$) but is dependent on the energy spread of the secondary ions. This may be circumvented by incorporating an energy compensation section or an energy filtering stage in the flight tube so that higher energy ions travel further than low energy ions of the same mass to charge ratio and ions of different energy arrive at the detector simultaneously [19]. For example, a 270° spherical sector energy filter has been used in the TFS surface analyser from Charles Evans & Associates (Fig. 5.12).

A detector system will then follow the mass filter, with the type of detector depending on the likely signal strength, i.e. it will consist of an ion counting system for low count rates or a low current amplifier if the count rate is above about 10^7 counts per second. Postacceleration of the secondary ions into the detector may also be used to improve detection limits.

There have been major advances in instrument design in the last few years, embodied in the new machines from the major manufacturers (e.g. ADEPT-1010 from PHI, IMS 6f, IMS1270 and NANOSIMS 50 from Cameca, Ion-TOF from the group of Prof. Benninghoven at the University of Muenster, SIMS 4500 from Atomika and Ionoptikas FLIG Floating Low Energy Ion Guns). The trends in instrumentation are to reduce the primary ion energies to sub keV for depth profiling, and to have increasing automation, especially for wafer-sized samples, in sample analysis and data reduction.

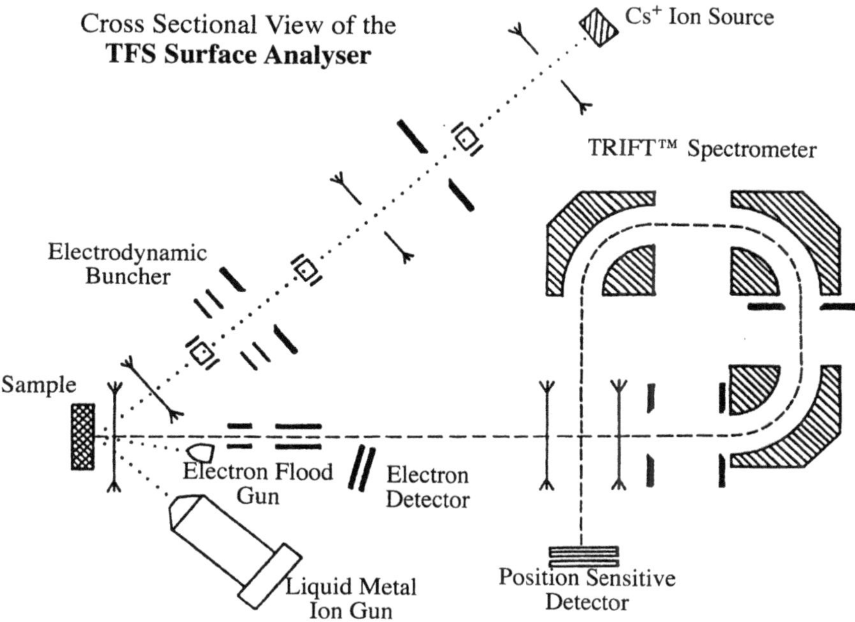

Fig. 5.12. Schematic diagram of a direct imaging time-of-flight SIMS analyser. The 270° analyser filters the energy of the secondary ions, allowing a mass resolution of more than 3000. The analyser can be used in microscope mode, with a lateral resolution of 1–2 μm or in microprobe mode (using a liquid metal ion source) with a 0.1 μm resolution

5.3 Topics in SIMS Analysis

5.3.1 Signal Enhancement by Surface Adsorption

It was indicated in Sect. 5.1.3d that the magnitude of the SIMS signal for a given ion was significantly affected by the presence of active gas adsorbates on the surface of the target. This is because the cross-section for ionisation is very dependent on the chemical state of the surface. The signal enhancement factor due to the presence of an active gas layer can be more than 100 times that of the ion yield from a clean surface. For example, the yield of positive secondary ions under 8 keV Ar^+ bombardment [20] increases by up to three orders of magnitude when the target surface is saturated with oxygen (Fig. 5.13). This change is not uniform even between cluster and multiply charged ions from the same element since, as the oxygen coverage of a silicon surface is changed from zero to a saturation coverage, the yield of Si^+ increases by 2.3 orders of magnitude, but SiO^{2+}, Si^{2+} and Si_2^+ remain the same or decrease [21] (Fig. 5.14). Negative ion yields can be similarly increased by introducing an electropositive element like Cs^+ onto the surface,

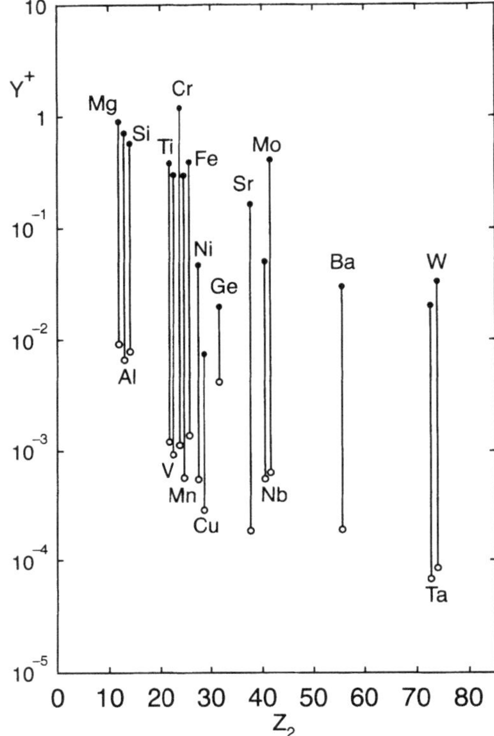

Fig. 5.13. Secondary ion yields for the indicated elements with (*filled circles*) or without (*open circles*) oxygen

typically by Cs^+ bombardment [22] (Fig. 5.15). However, it is still difficult to quantify concentrations from SIMS yields enhanced by reactive elements because

i) the enhanced yield is still a widely varying number from element to element as shown in Fig. 5.15. Systematic studies of the secondary ion yield from a wide range of elements have not been undertaken, in part because of the sensitivity of such yields to experimental parameters such as the cleanliness of the surface, the background gas concentration, the transmission characteristics of the SIMS analyser, etc.

ii) when an active gas is used to enhance the ion yield and hence the sensitivity, the measured ion yield is a function of the active gas concentration at the surface. This concentration of active gas may be established by irradiating the surface in a background environment of the active gas or alternatively, the incident ion beam may consist of ions of the active gas itself.

The use of primary Cs^+ beams and the subsequent detection of MCs^+ species or, for elements M with high electron affinity, MCs^{2+} [23] has provided a powerful alternative to O_2^+ beams for reduction of variations of the

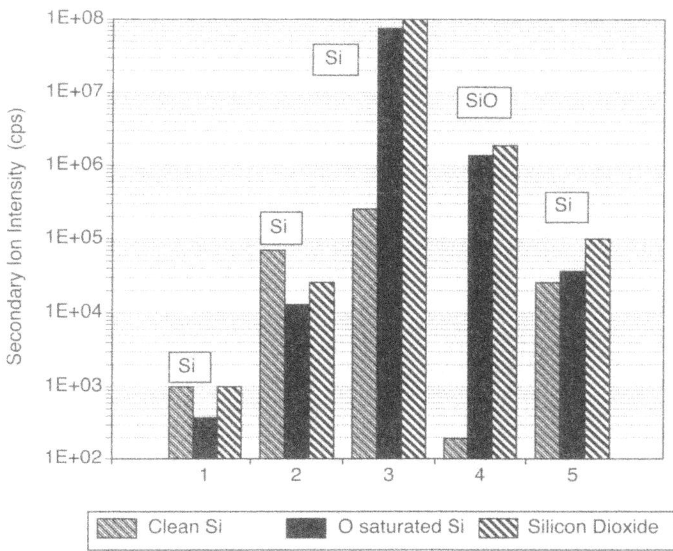

Fig. 5.14. Ion yields of silicon secondary ions for clean Si, oxygen-saturated silicon or silicon dioxide surfaces

ionisation cross-section [24]. The reduction is largely due to the formation of $Cs^+ + M$ in double collision sputtering events [25]. But some matrix effects remain. For example, the accumulation of Cs lowers the surface work function, giving a depression in the MCs^+ yield. Ratioing the MCs^+ to the Cs^+ signal can alleviate this problem [23]. In addition, there is a dependence of the MCs^+ yield on the surface oxygen concentration. For example, the SCs^+ signal from SiO_2 is 80 times that from Si [26]. The oxygen causes these changes due to the ejection of a substantial number of MO clusters. This reduces the amount of M or O available for Cs recombination to form MCs^+ or OCs^+.

5.3.2 Using Secondary Ion Energies in SIMS Analysis

The energy of the secondary ions which generate the SIMS signal is important for SIMS analysis in two ways. First, it allows molecular mass interferences to be overcome by tuning the acceptance energy in quadrupole systems. Second, it may be used to overcome matrix effects in the so-called infinite velocity method [27].

Many SIMS experiments do not require mass resolution equivalent to isotope separation, but it is often necessary to detect the signals due to atomic and molecular ions. This can be done on the basis of the energy spectrum of the secondary ions. The energy spectrum of molecular ions, e.g. M_n^+ or $M_nO_x^+$, maximise yield at a lower energy than that of the atomic species

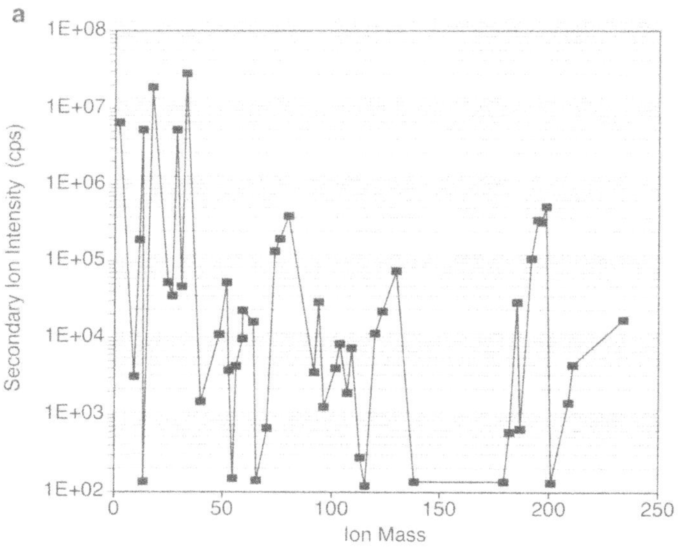

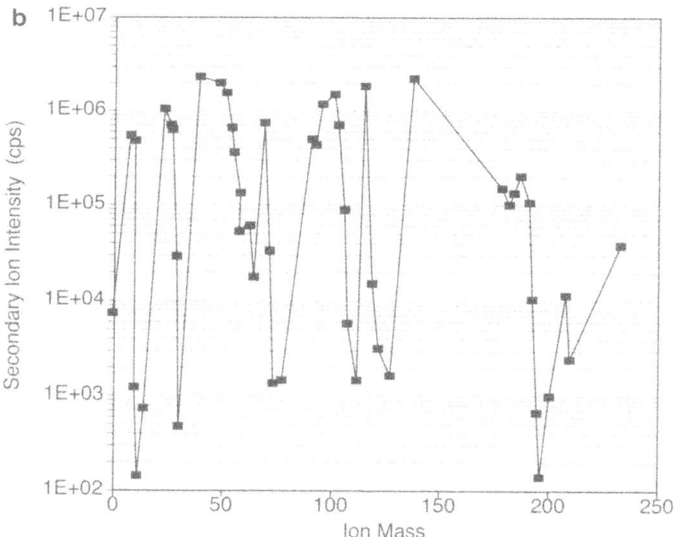

Fig. 5.15. (**a**) Positive ion yields for the indicated targets bombarded with 13.5 keV O^- at normal incidence. (**b**) Negative ion yields for the indicated targets bombarded with 16.5 keV Cs^+ at normal incidence

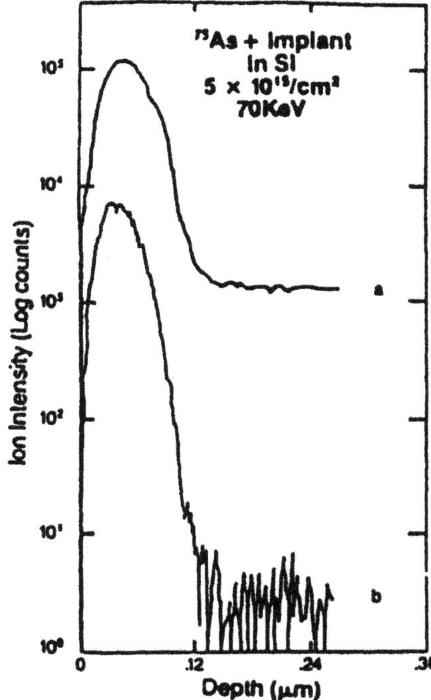

Fig. 5.16. Biassing the secondary ion optics to only accept high energy secondaries excludes molecular ions which interfere with and limit the dynamic range of dopant profiling.

M^+. In addition, the energy spectrum is substantially narrower than that of the atomic ion. In cases where overlap of an elemental peak of interest by a molecular ion peak is a possible problem, the case can often be resolved by changing the pass energy of the energy analyser preceding the mass analysis section of the instrument. The molecular ion contribution to any signal will be preferentially removed as the pass energy increases. Such a procedure is particularly useful in depth profiling dopants in semiconductors where molecular ion interferences cause constant backgrounds which limit the dynamic range available. For example, in Fig. 5.16, the detection limit for $^{75}As^+$ implanted in Si is much lower when a 40 V offset is put on the secondary ion optics so as to only accept high energy secondaries [28]. This is because the molecular ion $^{29}Si^{30}Si^{16}O^+$ also has mass of 75 amu and, since it comes from the bulk Si being sputtered by O_2^+, has constant intensity with depth. The energy spectrum of the molecular ion is much narrower than for the elemental ion and so the molecular ion is discriminated against when high energy secondaries are selected.

5.3.3 The Relative Sensitivity Factor

It is possible to circumvent some of the absolute uncertainties in ionisation cross-section to gain quantitative information of concentrations by the use of standards. If dilute concentrations are to be measured the easiest method is to implant a known small dose of the impurity into an identical matrix. Such implant standards are available from NIST, which offers certified reference materials, SRM 2137 (^{10}B implant in Si) and SRM2134 (As implant in Si) [29]. NPL also offers a Ta_2O_5 certified reference material CRM 261 [30].

The relative sensitivity factor for element E, RSF_E, is defined by the following equation

$$I_R/C_R = RSF_E I_E/C_E \tag{5.11}$$

where I_E and I_R are the secondary ion intensities for element E and reference (or matrix) element R and C_E and C_R are their respective concentrations. For trace element analysis, C_E is small so C_R doesn't change with a change in C_E. In this case, we rename the reference (R) as the matrix (M) and can combine the matrix concentration with the elemental RSF to give

$$RSF = RSF_E \, C_M \,. \tag{5.12}$$

Then

$$C_E = RSF \, \frac{I_E}{E_M} \,. \tag{5.13}$$

RSF values for positive impurity ions [3] from oxygen bombardment of silicon range from less than 10^{21} atoms cm^{-3} for Na to more than 10^{24} atoms cm^{-3} for common dopants like P. For Cs bombardment, negative ion RSFs range from less than 10^{22} atoms cm^{-3} for S to more than 10^{27} atoms cm^{-3} for Be. Tables of RSFs can be found in [3]. RSFs measured on the same machine vary by $< 50\%$ over time. Differences between different instruments are of the same magnitude, at least for GaAs [5.30], and are primarily ascribed to different primary ion impact angles.

5.4 Static SIMS Analysis

Static SIMS is ideally used to obtain the chemical composition of the surface without damaging the surface. A common SIMS problem is to identify differences in trace element concentrations between two targets, e.g. from contaminated and uncontaminated samples. In this case absolute concentration determinations are not required, thus avoiding SIMS ion yield uncertainties. In the following example, an epitaxial CdTe film had been grown on a CdTe(111) substrate by molecular beam epitaxy. The film was a single crystal, but measurement of optical properties indicated the presence of impurities in the film. Figures 5.17a, b shows SIMS mass spectra of a CdTe substrate and the CdTe film, respectively, which were taken on a Riber MIQ 156 RF quadrupole-based analyser. The level of oxygen and hydrocarbons (masses 12–16) in the CdTe substrate is much lower than in the CdTe film.

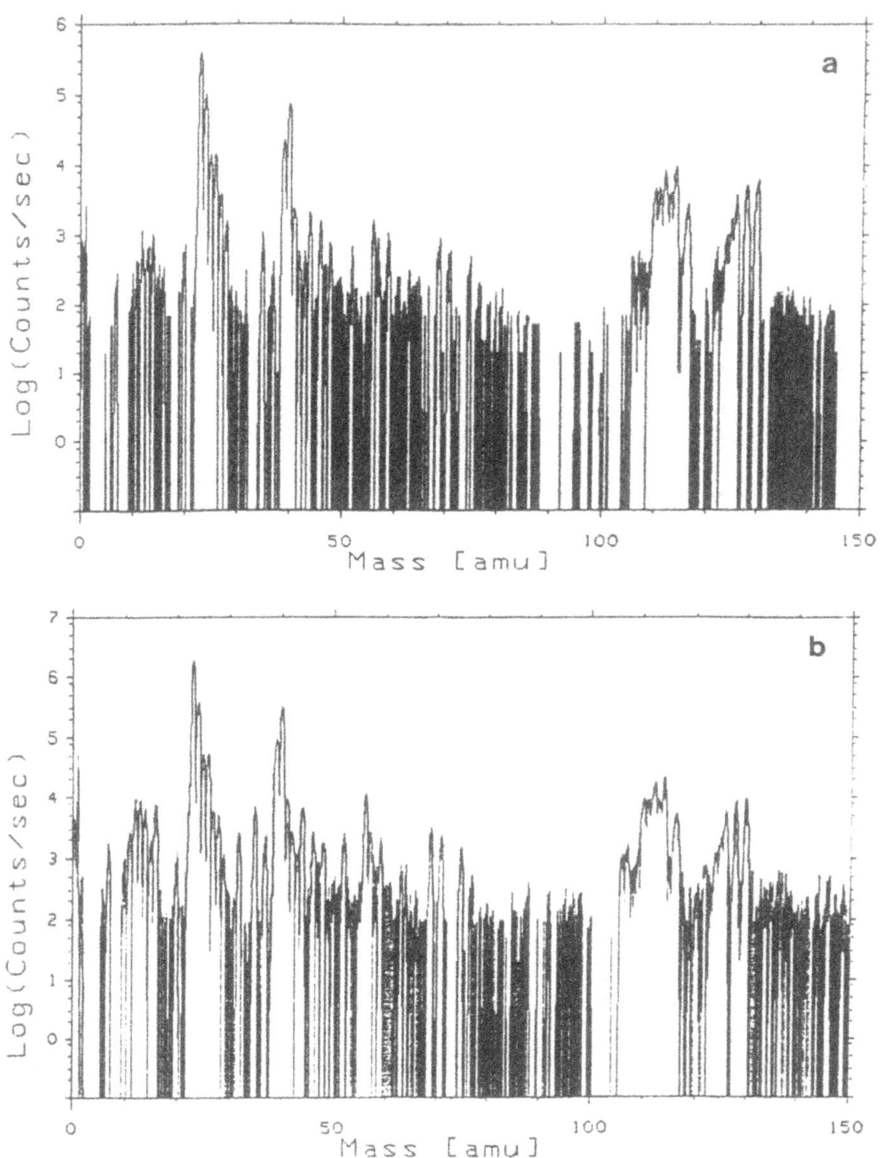

Fig. 5.17. (**a**) Positive ion spectrum from a clean CdTe substrate. (**b**) Positive ion spectrum for a MBE–grown CdTe thin film

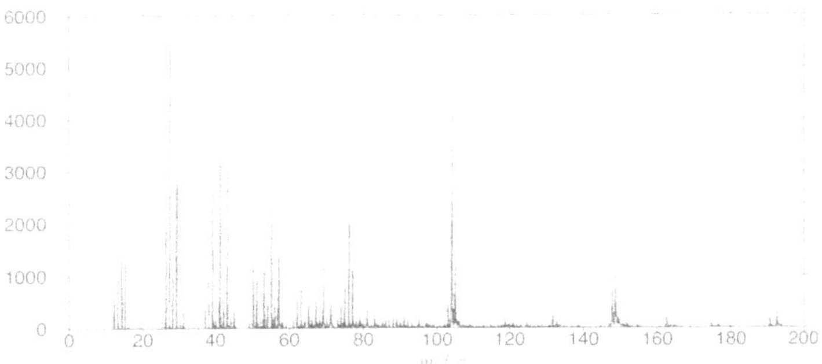

Fig. 5.18. SIMS mass spectra from a PET surface. The protonated monomer has a mass of 193 Daltons. The many lower mass peaks are due to contaminant molecules and molecular fragments.

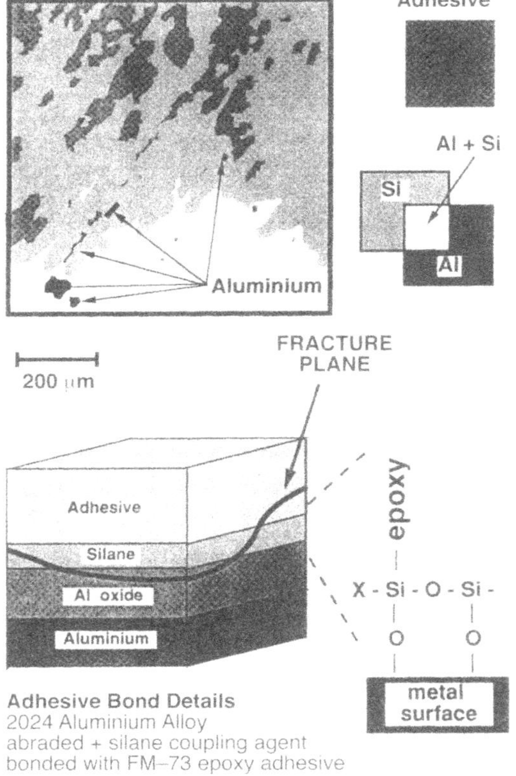

Fig. 5.19. SIMS composition map of an adhesive bond fracture on 2024 aluminium treated with silane, showing the locus of fracture based on the distribution of Si and Al. The dark grey areas in the map contain no Al or Si and so correspond to adhesive at the surface. The white areas contain both Al and Si so correspond to the surface fracturing at the silane-oxide interface

There is also increased Fe (mass 56) in the deposited film. These are probably the impurities responsible for the altered optical properties.

A more complicated static SIMS spectrum, from a polymer surface, is shown in Fig. 5.18 [32]. In this case many of the peaks may be due to fragment ions. To make sense of specific spectra may well require an understanding of molecular fragmentation and attachment processes. However, libraries of static SIMS spectra [33] will give a first guide to the interpretation of results. Principal factor analysis has also proved useful for static SIMS analysis [34].

References

1. A. Benninghoven, F.G. Ruedenauer, H.W. Werner: *Secondary Ion Mass Spectrometry* (Wiley, New York 1987)
2. J.V. Vickerman, A. Brown, N. Reed (eds.): *Secondary Ion Mass Spectrometry* (Oxford University Press, Oxford 1989)
3. R.A. Wilson, F.A. Stevie, C.W. Magee: *Secondary Ion Mass Spectrometry* (Wiley, New York 1989)
4. D. Briggs, A. Brown, J.C. Vickerman: *Handbook of Static Secondary Ion Mass Spectrometry (SIMS)* (Wiley, Chichester 1989)
5. The most recent proceedings is "SIMS XII" (Wiley, Chichester 1999)
6. www.simsworkshop.org
7. P. Sigmund: Phys. Rev. **184**, 383 (1969)
8. N. Matsunami, Y. Yamamura, Y. Itikawa, N. Itoh, Y. Kazamuta, S. Miyagawa, K. Morita, R. Shimizu, H. Tawara: Atomic Data and Nuclear Data Tables **31** (1984)
9. H.H. Andersen, H.L. Bay: In *Sputtering by Particle Bombardment I*, ed. by R. Behrisch (Springer, Berlin, Heidelberg 1981) p. 145
10. J.P. Biersack, L.G. Haggmark: Nucl. Instr. Meth. **174**, 250 (1980) More recent versions of the TRIM code can be found at http://www.research.ibm.com/ionbeams/
11. G. Betz, G.K. Wehner: In *Sputtering by Particle Bombardment II*, ed. by R. Behrisch (Springer, Berlin, Heidelberg 1983) p. 11
12. M.L. Yu, N.D. Lang: Nucl. Instr. Meth. B **14**, 403 (1986)
13. H.D. Hagstrum: In *Inelastic Ion Surface Interactions*, ed. by N.H. Tolk, J.C. Tully, W. Heiland, C.W. White (Academic, New York 1977)
14. R.J. MacDonald, E. Taglauer, W. Heiland: Appl. Surf. Sci. **5**, 197 (1980)
15. I.S.T. Tsong: In *Inelastic Particle-Surface Collisions*, ed. by E. Taglauer, W. Heiland (Springer, Berlin, Heidelberg 1981) p. 258
16. M.W. Thompson: Phil. Mag. **18**, 337 (1968)
17. J. Orloff: Rev. Sci. Instrum. **64**, 1105 (1995)
18. P.H. Dawson: In *Quadrupole Mass Spectrometry* (Elsevier, Amsterdam 1976) p. 9
19. W.P. Poschenrieder: Int. J. Mass. Spectrom. Ion Phys. **9**, 357 (1972)
20. A. Benninghoven: Surf. Sci. **53**, 596 (1975)
21. J.L. Maul, K. Wittmaack: Surf. Sci. **47**, 358 (1975)
22. H.A. Storms, K.F. Brown, J.D. Stein: Anal. Chem. **49**, 2023 (1977)

23. Y. Gao, J.W. Erickson and R.A. Hockett: In *Secondary Ion Mass Spectrometry X* eds A. Benninghoven, B. Hagenhoff, H.W. Werner (John Wiley, Chichester 1997) p. 339
24. K. Wittmaack: In *Secondary Ion Mass Spectrometry X* eds A. Benninghoven, B. Hagenhoff, H.W. Werner (John Wiley, Chichester 1997) p. 39
25. H. Gnaser: J. Vac. Sci. Technol. A **12**, 452 (1994)
26. Y. Homma, Y. Higashi, T. Maruo, C. Maekawa, S. Ochiai: In *Secondary Ion Mass Spectrometry IX* eds A. Benninghoven, B. Hagenhoff, H.W. Werner (John Wiley, Chichester 1995) p. 398
27. *Secondary Ion Mass Spectrometry X* eds A. Benninghoven, B. Hagenhoff, H.W. Werner (John Wiley, Chichester 1997) p. 131
28. Reference [3] page 1.8.5
29. http://ois.nist.gov/srmcatalog/
30. http://www.npl.co.uk/npl/cmmt/sis/refmat.html#Tantalum
31. *Secondary Ion Mass Spectrometry X* eds A. Benninghoven, B. Hagenhoff, H.W. Werner (John Wiley, Chichester 1997) p. 135
32. S. Bryan, F. Reich, B.W. Schueler, G. Marsh: In *Secondary Ion Mass Spectrometry X* eds A. Benninghoven, B. Hagenhoff, H.W. Werner (John Wiley, Chichester 1997) p. 939
33. http://www.SurfaceSpectra.com/
34. *Secondary Ion Mass Spectrometry X* eds A. Benninghoven, B. Hagenhoff, H.W. Werner (John Wiley, Chichester 1997) pp. 313, 887

6 Auger Electron Spectroscopy and Microscopy – Techniques and Applications

P.C. Dastoor

6.1 Introduction

Auger electron spectroscopy (AES) is one of the most commonly used surface analytical techniques available to the materials scientist. It has the ability to measure the chemical composition of the first few monolayers of a given surface with a sensitivity of the order of 0.1 atomic % and a spatial resolution of the order of 10 nm [1]. Its ease of interpretation means that AES is often the analysis technique of choice and has been used to study a wide range of different materials.

This chapter outlines the basic principles of Auger electron spectroscopy, the equipment used and the advantages and disadvantages of the technique. The various methods of performing an Auger analysis are described, with a view to providing the reader with expert insight into how AES may used to tackle a variety of surface materials analysis problems. The diverse capability of Auger electron spectroscopy is illustrated by a number of examples of its application in a wide range of fields. Finally, the possible future trends in the development of the technique are discussed, foreshadowing the future capabilities of the technique.

6.2 Fundamentals

The Auger effect was first independently discovered by *Meitner* [2] and *Auger* [3] who found that if an atom is ionised, the resulting electronic reorganisation can cause the ejection of an electron with an energy that is characteristic of the originating atom. The Auger process is essentially a three-electron process that is typically initiated by the ejection of a core electron by an incident high-energy electron. The core hole thus created is rapidly filled by an electron from a higher energy shell leaving the atom in an energetically excited state (Fig. 6.3). The excited atom can lose energy in one of two ways. Firstly, an X-ray photon of the appropriate energy can be ejected. The measurement of the energy of the energy of this X-ray, which is characteristic of the originating atom, forms the basis of the analytical technique known as X-ray fluorescence (XRF). The second way that the excited atom may lose its excess energy is through the ejection of another electron

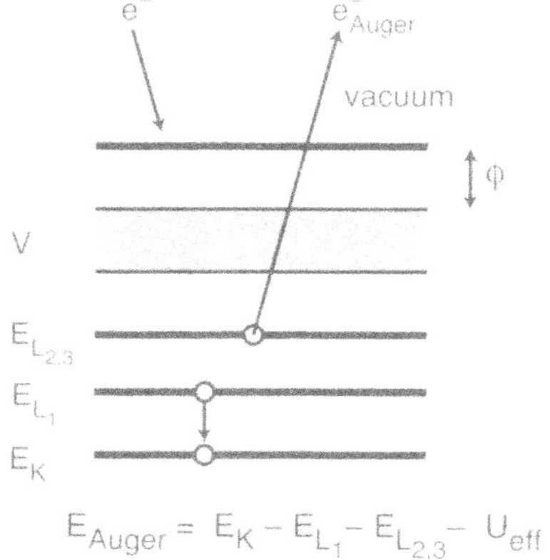

Fig. 6.1. Schematic diagram of the electronic energy levels involved in the Auger effect. Ionisation of a core hole (E_K) by the primary electron beam, results in the decay of an electron from a higher energy level (E_{L_1}). The energy thus released causes the ejection of an electron from a third energy level ($E_{L_{2,3}}$), which is termed the Auger electron (E_{Auger})

with a certain kinetic energy. The energy of this so-called Auger electron is also characteristic of the source atom and is the measurement of the energy distribution of these Auger electrons that forms the basis of Auger electron spectroscopy. This relaxation involves both the electrons in the atom itself [5] and the surrounding material [6].

So, what is it that determines whether an atom de-excites through the emission of an X-ray photon or through the ejection of an Auger electron? It turns out that the probability, or cross-section, of these two processes is highly dependent upon the atomic number (Z) of the excited atom. For low Z elements ($Z < 40$), the de-excitation is dominated by the Auger process and thus for many of the more abundant elements a strong Auger signal is observed.

If the Auger electron energy (E_{Auger}) is written in terms of experimental photoelectron energies (referenced to the Fermi level) then the energy can be written as:

$$E_{\mathrm{Auger}} = E_x - E_y - E_z - U_{\mathrm{eff}} \tag{6.1}$$

where E_x, E_y and E_z are the binding energies of the three participating electrons and U_{eff} is the extra energy needed to remove an electron from a doubly ionised atom [7].

The characteristic energy distribution of electrons emitted from a solid surface upon excitation by a primary electron beam is shown in Fig. 6.2. This spectrum can be divided into main three regions. The large peak at low energies ($\approx 50\,\mathrm{eV}$) in region 1 is due to the so-called "true secondary cascade". In region 2 the presence of Auger peaks can be observed superimposed upon a background of rediffused primary and backscattered electrons. In region 3, the elastically scattered electrons can be observed in the form of the elastic peak. Indeed, it is these electrons that are used in low energy electron diffraction

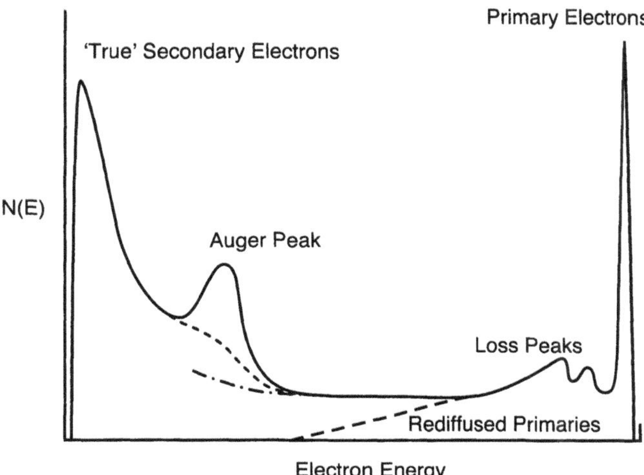

Fig. 6.2. Schematic diagram of the characteristic secondary electron spectrum from a solid surface. The peak at high energy corresponds to the elastically scattered electrons and thus occurs at the energy of the primary beam. Electron energy loss peaks are observed on the low energy tail of the elastic peak and correspond to electrons that have undergone inelastic interactions with the material (such as generating surface and bulk plasmons). The peak at low energy corresponds to the so-called "true secondaries", which have undergone multiple inelastic collisions within the solid surface, and is typically located at an energy of 50–100 eV. The Auger peaks (shown exaggerated in the figure) are superimposed upon the long exponential tail from the low energy peak and these electrons initiate further contributions to the background

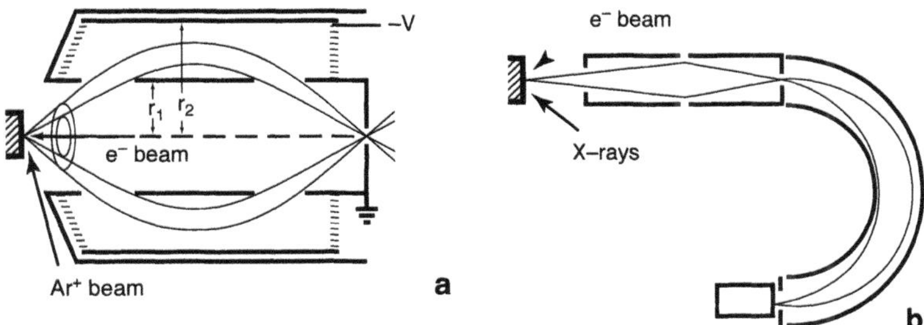

Fig. 6.3. Schematic diagram of the common electron analysers used for Auger electron spectroscopy. **(a)** A single pass cylindrical mirror analyser (CMA). **(b)** A concentric hemispherical analyser (CHA)

(LEED) studies. On the low energy side of this peak there are a number of features associated with electron loss processes in the solid, including losses due to both bulk and surface plasmons. These electrons are the ones that are probed during electron energy loss spectroscopy (EELS).

The surface sensitivity of the technique arises from the energy range of the Auger electrons that are ejected. When an energetic electron passes though a solid, it continually has inelastic collisions with the electrons in the solid. These collisions ultimately result in the electron becoming incorporated in the electronic structure of the solid itself. The number of collisions that occur, and hence the range of the electron in the solid, is known as the inelastic mean free path (imfp) and is a strong function of the electron's kinetic energy. Early studies suggested that the imfp of most materials lay on a universal curve [4]. Indeed, for the usual kinetic energies encountered in Auger spectroscopy (20–2000 eV), the imfp of the Auger electrons is very low; typically between 5 and 20 Å. Therefore, only electrons that originate within the first 5–20 Å(or 3 to 10 atomic layers) of the surface will be ejected from the surface and hence detected. However, as will be discussed later in this chapter, recent work has shown the concept of a "universal curve" for imfp is inappropriate [8].

6.3 Instrumentation

Although Auger spectra can be stimulated using incident x-rays or ion beams, most dedicated Auger systems employ an electron beam to irradiate the surface. Historically, the incident electrons typically have a kinetic energy of either 3 keV or 5 keV, although increasingly beam energies of 10 keV are being employed. The choice of beam energy actually influences the yield of secondary electrons emitted from a solid surface. Typically, the yield of secondary electrons of a given energy, E, is maximised when the beam energy is $3E$. The use of an electron beam provides a number of further advantages when it comes to using Auger spectroscopy to provide chemical images of surfaces. In this application, the incident electron beam can be readily focussed to produce a small spot size upon the surface and subsequently rastered over the surface. Many of the early Auger systems used electron sources that were based around a thermionic electron gun. Although relatively low cost, the inherent electron-optical aberrations associated with these devices meant that they were only capable of minimum beam spot sizes of greater than 0.2 µm. For applications requiring a high brightness, small spot size electron emitter, single crystal sources (such as lanthanum hexaboride, LaB_6) or field emission sources are used. The spot size of the electron beam governs the ultimate resolution of the Auger imaging system and current systems are capable of spot sizes of the order of 10 nm [9].

The second main component in any Auger system is the electron energy analyser that allows the energy distribution of secondary electrons to be collected. There are three main designs of electron energy analyser used in conventional Auger spectroscopy. The earliest analysers were based upon an adaptation of the 3-grid low energy electron diffraction apparatus used to characterise surface structures. In a LEED system, (Fig. 6.1), elastically scattered electrons are imaged on a hemispherical phosphor screen. The addition of a 4th grid permits the energy of the collected electrons to be selected and by counting all of the electrons that are received at the final screen a simple electron energy analyser is formed [10]. However, this so-called retarding field analyser (RFA) suffers from relatively low collection efficiency and poor energy resolution.

The cylindrical mirror analyser (CMA) incorporates two concentric cylinders as shown in Fig. 6.3. Electrons ejected from the sample surface with the correct kinetic energy follow a curved trajectory through the annular gap between the two cylinders before being collected at an electron detector. The voltages on the inner and outer cylinders determine the kinetic energy of the electrons that can pass through the analyser. In its usual mode of operation for Auger electron spectroscopy, the inner cylinder is held at ground while the voltage on the outer cylinder is ramped. An aperture at the front of the analyser defines the analysis area while the transmitted electrons are detected by an electron multiplier or channeltron.

The concentric hemispherical analyser (CHA) was initially used for X-ray photoelectron spectroscopy. The analyser (Fig. 6.3) consists of two hemispherical surfaces between which is maintained a $1/r^2$ electric field via the application of suitable voltages to the hemispheres. Although CMA systems are a popular choice for AES analysis, CHA based systems do offer certain advantages with regards to quantification [11,12].

6.4 Quantification

The functional form of the true secondary electron cascade (region 1 in Fig. 6.2) is important for quantitative analysis in Auger spectroscopy since most of the Auger peaks are superimposed upon this background. This slowly decaying electron signal arises from inelastically scattered Auger electrons, and backscattered primary electrons, that in turn generate secondary electrons in the solid. Using a diffusion-based argument, Sickafus showed that this cascade of electrons has the general form:

$$N(E) = AE^{-m} \tag{6.2}$$

where $N(E)$ is the energy distribution of electrons and A and m are constants related to the material properties of the solid [13–15]. This exponential form for the background has been shown to be suitable for many materials in the region $E < \frac{1}{2} E_{\text{primary}}$ (where the contribution of losses from primary electrons can be neglected). In an improvement to this approach, Matthew et al. suggested that a more appropriate form for the background function was:

$$N(E) = A(E + J)^{-m} \tag{6.3}$$

where J is the mean ionisation energy of the surface atoms [16].

One of the major requirements of any surface analytical technique is the quantitative determination of the chemical composition of a surface. Indeed, the application of AES to quantitative surface analysis has a long history, with a number of reviews published over the last 20 years [17–19]. As discussed by *Seah* [19], there are three main approaches to quantitative analysis:

1. Calculate the relevant terms from first principles. Unfortunately, this approach is not practical since the origin of the secondary electron spectrum is still not sufficiently well understood.
2. Calculate the surface concentrations from published databases. While more practical, this approach does not take into account the effect of the measuring instrument.
3. Calibrate the concentration using locally produced standards and databases. This is the most desirable approach since it compensates for the transmission function of each individual apparatus.

It should be noted that there are a number of different approaches that can be used for quantifying Auger spectra in different circumstances, such as dilute alloys, thin films or multi-component systems [20]. The most common method for quantification of surface concentration involves the use of sensitivity factors from pure elemental standards [21,22]. These sensitivity factors will differ from one apparatus to another and thus each operator needs to generate their own set of standard spectra. Using bulk elemental standards, the surface concentration, C_i, of element i, is given by:

$$C_i = \frac{I_i/I_i^{\infty}}{\sum_j F_{ij} I_j/I_i^{\infty}} \tag{6.4}$$

where j is the number of elements on the surface, I_i is the Auger signal of element i, is the intensity of the pure element standard and F_{ij} is the matrix correction factor. The matrix correction factor includes all of the effects associated with the interaction of the primary beam with the target, such as; changes in backscattering [23], atomic density and attenuation length between the pure elemental standard and the unknown sample, microtopography [24] and primary electron diffraction [25,26].

6.5 Techniques

Generally speaking there are four main modes of AES operation: point analysis, line scan, profiling and scanning. The four different modes originate from the typical surface regions that are analysed.

6.5.1 Spot Analysis Mode

In this mode, the incident electron beam spot remains stationary on the surface and the Auger electron spectrum is measured. Figure 6.4 shows the direct form of a typical survey spectrum for a copper surface, with the number of detected electrons N(E)/E plotted as a function of electron kinetic energy. The characteristic copper peaks between 700 and 1000 eV are clearly observed, together with peaks associated with surface contamination. Figure 6.4 also shows the differentiated form of the same survey scan, illustrating how this method tends to emphasise the fine structure in the direct spectrum [19]. The point analysis mode is the most common analysis mode for Auger electron spectroscopy since it allows for the rapid determination of the elemental composition of a small region of the material surface. Although

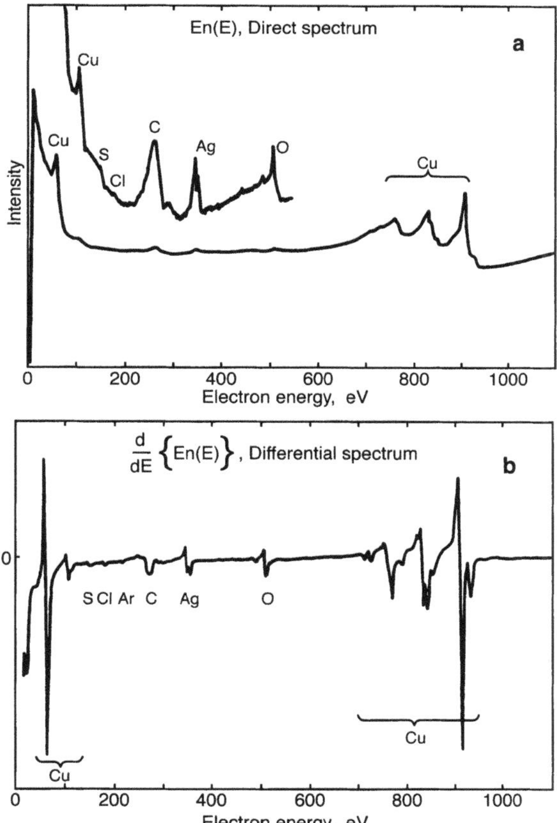

Fig. 6.4. Auger surface obtained from a contaminated copper surface. **(a)** The direct spectrum $EN(E)$. **(b)** The differential spectrum, $\frac{d}{dE}\{EN(E)\}$, obtained by sinusoidal energy modulation

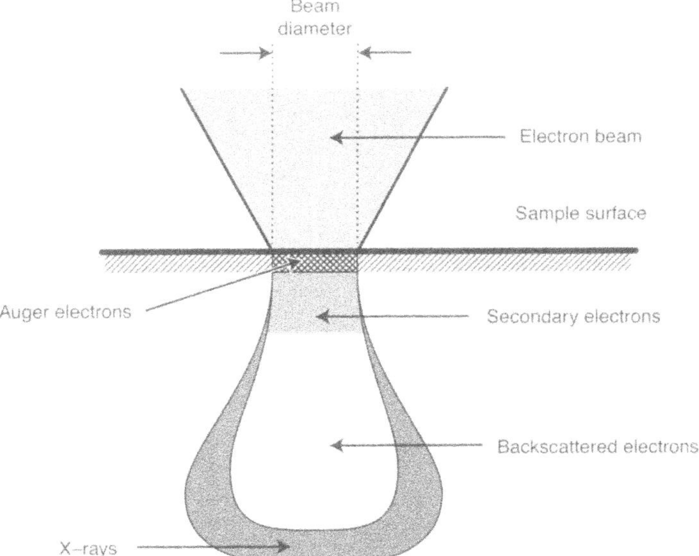

Fig. 6.5. Schematic diagram showing the interaction volume produced by an incident high-energy electron beam on a solid surface. The incident electron beam induces a pear-shaped interaction volume and generates both electrons and X-rays to be ejected from the surface. The short inelastic mean free path of the relatively low energy Auger electrons means that AES information is obtained only from the uppermost atomic layers. As a consequence the lateral resolution of AES is close to the diameter of the primary beam. The secondary electron escape depth is larger than that of the Auger electrons whereas the backscattered electrons, whose energy is close to that of the primary beam, escape from depths of 100–1000 nm. The X-ray escape depth and lateral resolution correspond to the dimensions of the entire interaction volume

the interaction volume of the high energy incident electrons is of the order of a micron, the short imfp of the escaping Auger electrons means that the lateral extent of Auger emission is determined by the minimum spot size of the electron gun (Fig. 6.5). For modern electron guns, the size of this spot (and hence the lateral resolution of the technique) is typically of the order of 10 nm.

6.5.2 Line Scan Mode

In this mode the electron beam is rastered laterally over the surface in one dimension while simultaneously collecting the ejected Auger electrons. The main application for this analysis mode is actually in determining the depth variation of a material's chemical composition. For materials that consist of many layers, or perhaps possess a relatively thick surface coating, the long ion

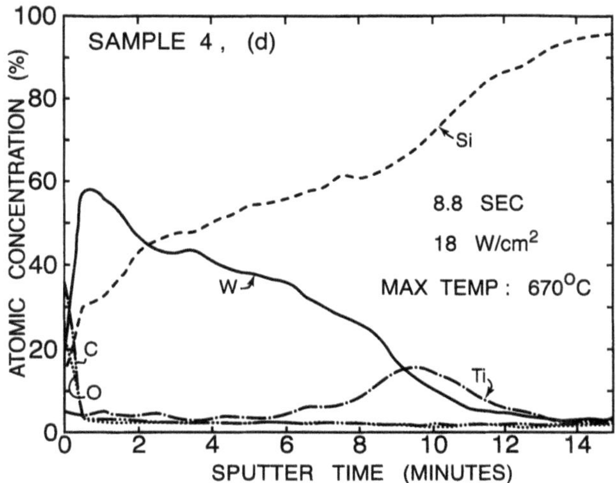

Fig. 6.6. Auger depth profile from a study of the formation of tungsten silicide on silicon. A titanium film has been interposed between the tungsten surface film and the silicon substrate. The study showed that upon annealing, the formation of tungsten silicide was enhanced by the presence of the titanium film with a corresponding decrease in surface roughness and improvement in adhesion

sputtering times required for in-situ depth profiling may well be impractical. However, the variation of chemical composition with depth can be obtained for such materials by collecting a line scan over their cross-section. This cross section may be obtained by ball-cratering or by fracturing the sample and analysing the cross-section of the fracture cross-section [27–30]. Indeed, a number of surface analysis systems that are designed for industrial use possess facilities to allow samples to be fractured in-vacuo. The relatively long times required for a complete survey scan mean that typically a small number of Auger peaks are collected during each line scan. Each monitored element is determined by selecting an energy window that covers the appropriate Auger electron energy. The intensity of the Auger electron peak that corresponds to the element of interest can thus be automatically measured as a function of lateral distance along the cross-section, which in turn corresponds to a depth.

6.5.3 Scanning Mode

Surface compositional maps can be obtained by performing Auger electron spectroscopy in the scanning mode. In this mode, the electron beam is rastered over a selected area of the surface. By tuning the analyser to detect Auger electrons from the elements of interest a compositional map of the surface elemental concentrations can be obtained. If the target current to ground is measured while the electron beam is rastered across the surface it is possible to display a low resolution electron micrograph of the surface. The contrast is due to changes in the secondary electron emission that occur and provides an image of the surface morphology. This image, which is akin to an electron microscope image, can be superimposed on the Auger compositional map to provide a measure of the chemical composition of structural features on the surface.

6.5.4 Scanning Auger Microscopy

The logical application of the scanning Auger mode is the development of a microscope that can routinely provide images of the chemistry of a surface. Indeed, the development of just such an instrument was pioneered by the work of *Prutton* and co-workers over the last thirty years [31]. *Macdonald* and *Waldrop* first demonstrated the principle of Auger microscopy in 1971 [32].

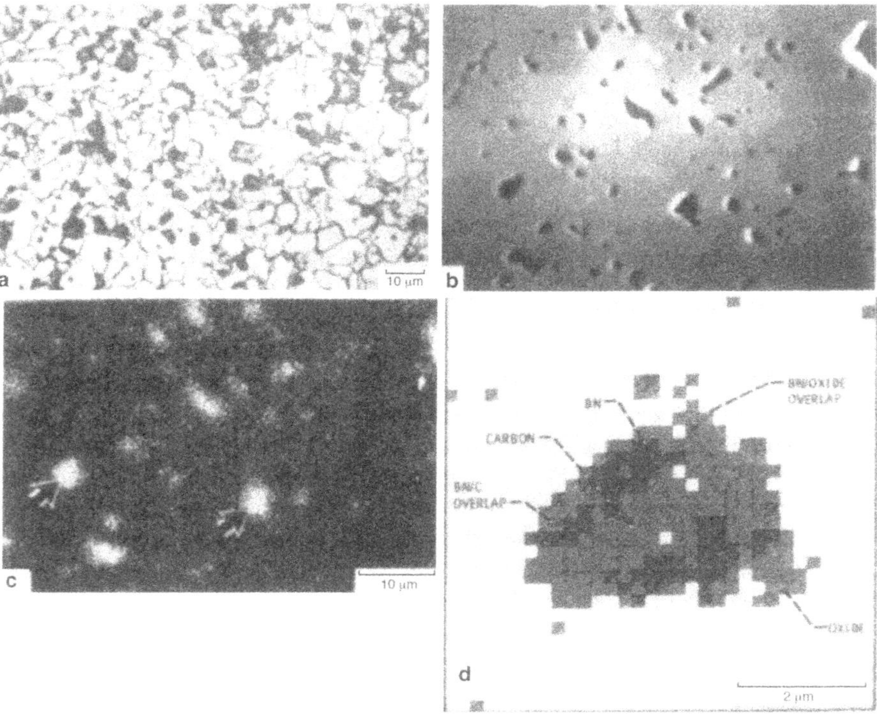

Fig. 6.7. Micrographs of SiC sintered in a nitrogen atmosphere. **(a)** Optical micrograph of polished and etched microstructure. **(b)** 15 keV backscatter electron image. **(c)** Carbon X-ray map by wavelength dispersive analysis. **(d)** Monochrome reproduction of chemical phase image derived from Auger imaging. The phase image is of one particle at the triple point between SiC grains. Note the different scales between this and the X-ray image in (c)

The first instruments were analogue systems that involved the combination of a UHV electron microscope with a CHA [33]. The images were stored on magnetic tape and the system was relatively slow and cumbersome. There followed the development of a digital instrument [34] and shortly afterwards the development of commercial instruments.

The spatial resolution of state-of-the-art commercial instruments [36] is currently of the order of 5–10 nm, with the electron beam generated using a Schottky field emission electron gun [37]. This resolution is far superior to that commonly available with other surface microprobe techniques, such as imaging X-ray photoelectron spectroscopy with a typical working resolution of the order of a few microns. The versatility of the SAM technique means that it has found applications in studying a wide variety of material surfaces including geological specimens [38–43], semiconductors [44], superconducting oxides [45,46] and engineering ceramics [47,48].

As an example, Fig. 6.7 shows a comparison of the optical, backscattered, X-ray wavelength dispersive and Auger microprobe images of a SiC surface sintered in a nitrogen atmosphere [49].

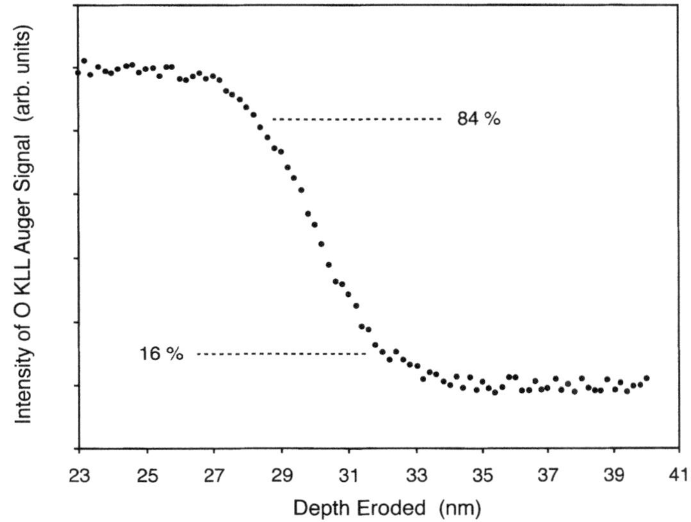

Fig. 6.8. Schematic diagram of the Auger depth profile obtained from a 30 nm Ta_2O_5/Ta standard. The resolution of the interface (Δz) is taken as the depth between the 84% and 16% levels of the O KLL Auger signal. The 50% point is taken as the position of the interface

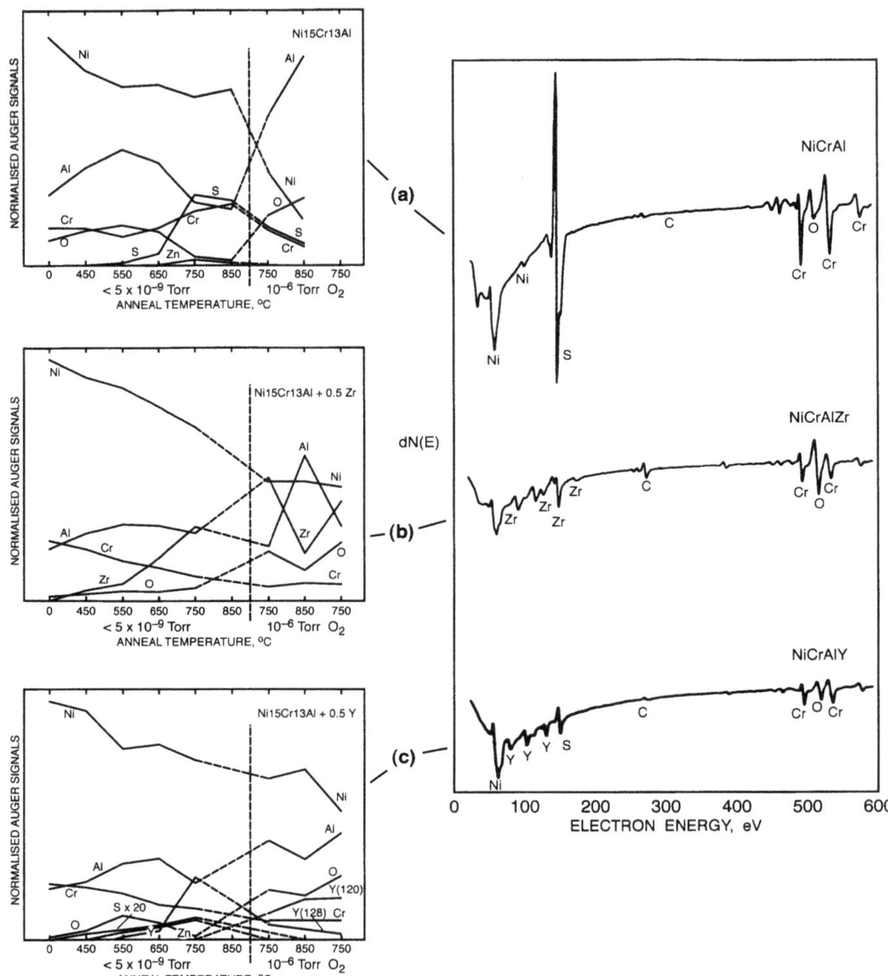

Fig. 6.9. Competitive segregation of sulphur, zirconium and yttrium on doped $Ni_{15}Cr_{13}Al$ high temperature superalloy surfaces as a function of temperature under different oxidation conditions as monitored by AES. Each annealing profile (left hand panels) consists of two regions. The region to the left of the vertical dashed line corresponds to annealing under UHV conditions whereas the region to the right of the dashed line corresponds to an oxidising anneal at a partial oxygen pressure of 10^{-6} Torr. The AES spectra shown in the panel on the right hand side of the figure correspond to the 705°C anneal under UHV conditions. **(a)** Auger annealing profile and Auger signal intensity for an undoped $Ni_{15}Cr_{13}Al$ surface. **(b)** Auger annealing profile and Auger signal intensity for a Ni15Cr13Al surface doped with 0.5 % zirconium. **(c)** Auger annealing profile and Auger signal intensity for a Ni15Cr13Al surface doped with 0.5 % yttrium

6.5.5 Depth Profiling Mode

In this mode AES is conducted simultaneously with ion bombardment of the surface. In this way, a material's surface can be analysed as it is eroded by the incident ion beam, thus providing a measure of the variation of chemical composition with depth. A calibration of the depth as a function of erosion time can be obtained through the use of internationally recognised standard materials. Tantalum oxide (Ta_2O_5) films are commonly used as depth standards, since they can be made reproducibly with a known oxide thickness and possess a sharp metal/oxide interface [50]. The thickness of the standard oxide layer is separately obtained with nuclear reaction analysis (NRA). Figure 6.8 shows the depth profile for a 100 nm Ta_2O_5 film obtained by bombarding the surface with 5 keV Ar ions. In practice, the calibration of a depth profile involves several distinct steps:

1. The time required to reach the interface is measured.
2. The equilibrium Auger amplitudes in the film are determined.
3. The sputtering time is converted to a depth from knowledge of the thickness of the oxide standard.

The depth profile of the Ta_2O_5 standard also provides a measure of the resolution of the depth profiling technique. The depth resolution is defined by the width of the measured interface, which itself is defined by the distance between the 84% and 16% steady state oxygen composition levels across the interface [51]. For convenience, the position of the interface is taken as the point at which the oxygen signal has reached 50 % of its equilibrium value in the oxide film.

However, even with the use of independently calibrated standard materials there are a number of additional factors that need to be considered during the measurement of a compositional depth profile. A typical depth profile is shown in Fig. 6.5, which is from a study of the effect of a titanium barrier layer on the reaction between a tungsten film and silicon substrate. The depth distribution of the reactants shows that the titanium film remains after annealing and enhances the formation of an interfacial layer of tungsten silicide [52].

6.5.6 Preferential Sputtering

One of the complicating factors associated with ion sputtering is the phenomenon of preferential sputtering, whereby certain elements in the matrix are sputtered faster than others [53]. Thus, the surface layer will be depleted in those elements with the higher erosion rate and consequently the composition of the profiled material can be altered under the action of the eroding ion beam [53,54]. Furthermore, the incident ion beam can chemically reduce many surface oxides resulting in inaccurate measurements of the surface oxide to metal ratio [55].

6.5.7 Attenuation Length

As discussed earlier in this chapter, the surface sensitivity of Auger electron spectroscopy arises from the very short distance that Auger electrons can travel in a solid material before they are reabsorbed through inelastic scattering processes. This distance is typically of the order of a few tens of nanometres and as a consequence only electrons that are ejected from within this distance of the sample surface will be collected by an external electron analyser. Historically, attempts to determine quantitatively the surface sensitivity of AES (and the allied technique of X-ray photoelectron spectroscopy) lead to the development of the concept of the "universal curve" [56]. This approach, which assumed a negligible material dependence of the measured attenuation length of electrons in solid materials, was useful as a rough indication of the surface sensitivity and has been widely used and reproduced. The historical development of the "universal curve" has been clearly outlined by *Jablonski* and *Powell* in their excellent review [57]. Suffice to say that the increasing accuracy of experimental measurements of electron attenuation lengths together with more sophisticated numerical analysis techniques have lead to the gradual acceptance that the "universal curve" is no longer an acceptable guide to the surface sensitivity of AES [58]. For example, elastic scattering of electrons means that the effective attenuation length (EAL), upon which the universal curve, is not necessarily equivalent to the inelastic mean free path (IMFP) of the Auger electrons [59].

Consequently, a predictive formula for Tanuma and co-workers have developed the calculation of IMFP based upon the modified Bethe equation [60] for inelastic electron scattering in matter [60–68]. By fitting the results of their calculations of IMFP for more than 60 materials (including elements, inorganic and organic compounds) to this modified Bethe equation it is now possible to more accurately determine the IMFP of electrons with energy between 50 and 2000 eV for virtually any material. A comparison of the imfps calculated from the predictive formula compared with those calculated directly from optical data shows that the root mean square deviations between the two datasets are [65]:

Material Group	Elements	Organic compounds	Inorganic compounds
RMS deviation	10.2 %	8.5 %	18.9 %

It is now possible to obtain the most up to date measurements for the imfps for a variety of materials using the National Institute of Standards and Technology (NIST) Electron Inelastic-Mean-Free-Path Database [65].

6.5.8 Chemical Effects

The shape and position of a given Auger peak can also be influenced by the presence of differing neighbour atoms. These so-called chemical effects can include [66–69]:

- the appearance of new peaks
- the disappearance of existing peaks
- changes in peak width and peak position
- changes in relative intensities of Auger features.

These chemical effects have been extensively reviewed by *Madden* [70]. Transitions that involve valence electrons are more likely to exhibit these chemically induced changes since the effect of chemical bonding typically results in changes to the energy levels and density of states of the valence band.

The presence of chemically induced energy shifts in the Auger peak position is an important feature of Auger spectroscopy, since it allows, in theory, the extraction of chemical state information from the Auger spectrum [71]. In this way it is similar to the well-established chemical shifts observed in X-ray photoelectron spectroscopy (XPS). These Auger chemical shifts actually arise from a variety of mechanisms including relation effects, changes in the hole-hole interaction energy as well as the changes in core levels due to chemical environments. However, these chemical shifts are often more complicated to interpret in Auger spectroscopy (as compared with XPS) because of the increased number of contributing factors and processes.

A valuable indicator of the Auger chemical state is the so-called Auger parameter (α) that was introduced by *Wagner* [71]. This parameter combines information from both the photoelectron and Auger lines and is defined as:

$$\alpha = KE_{\mathrm{Auger}} + BE_{\mathrm{photoelectron}} \quad (6.5)$$

A common method of presenting Auger parameter data is in the form of a chemical state plot whereby Auger kinetic energy is plotted against photoelectron binding energy (plotted in reverse). In such a two-dimensional diagram, each chemical state has a unique position.

6.6 Applications

6.6.1 Thin Film Analysis

When used in conjunction with ion sputtering, Auger electron spectroscopy is a powerful technique for obtaining depth profiles of the elemental distribution in thin films. This technique is widely used for materials analysis applications in, for example the semiconductor industry [72], the investigation of oxidation layers [77] as well as being applied to the study of passivation at surfaces [78].

Recent developments in depth profiling have involved the application of modern numerical techniques, such as target factor analysis (TFA) to the analysis and interpretation of the measured spectra [79]. TFA, as applied to Auger sputter depth profiles, is used to determine and distinguish between the different chemical states that are present at different depths of a given sample. The analysis process involves storing each Auger spectrum from a given depth as a column in a data matrix, D. The objective of the analysis is to determine the two factors R and C that express the measured spectra (D) in the form:

$$D_{ij} = R_i * C_j \tag{6.6}$$

where R is a column vector containing representative spectra for the chemical components in the profile and C is a row vector containing the concentrations of these spectra for each sputter depth. In order to accurately decompose the acquired depth profile into the factors R and C, the number of significant spectral components (or eigenvectors) needs to be determined. Typically this involves a decision by the spectroscopist as to which elements are present on the surface, each species then has a characteristic Auger spectrum corresponding to a significant eigenvector for the purposes of the analysis.

Once R and C have been determined it is possible, in principle, to split the original data matrix into terms corresponding to different spectral features of the original matrix, such that:

$$D \approx \sum_i D_i \tag{6.7}$$

where the matrices D_i represent different spectral features, such as oxidation states. The matrices can then be quantified, together with matrices corresponding to the other elements in the depth profile, using the appropriate sensitivity factors.

6.6.2 Surface Diffusion and Segregation

AES has proven to be a powerful tool in furthering the understanding of kinetic processes at surfaces (e.g. solid state diffusion [80,81]). In particular, its surface sensitivity means that it is well suited to studying the effects of temperature and dopants upon grain boundary segregation and the role that this phenomenon plays in determining material fracture toughness [82,83]. Indeed, the effects of heating time, temperature and surface concentration can all be measured dynamically [82]. The distribution of different metallic phases on the surface can also be measured [84–87]. The use of the Auger microprobe for these studies allows surface composition to be directly related to surface structure and a number of systems have been studied including high temperature alloys [88], magnetic alloys [89] and ceramics [90]. A good example of this type of application is the study of scale formation on high temperature superalloys [91–93]. These materials (such as NiCrAl) are commonly used in aircraft turbine blades and form a protective alumina scale at high temperatures [94,95]. However, in order to ensure that this scale does not spall off during thermal cycling then rare earth elements (such as yttrium and zirconium) are added to the superalloy [92]. Without these additives turbine life is dramatically reduced. Figure 6.8 shows how AES can be used to monitor the segregation of these dopants and the suppression of sulphur segregation on the NiCrAl surface in vacuo.

6.7 Future

Auger electron spectroscopy as a technique is continually being developed and improved, both from the point of view of experimental practice and instrumentation. Increasingly, there are international efforts to standardise the practice of AES and the quantification of spectra. Initial efforts were focussed around the Versailles Project on Advanced Materials and Standards (VAMAS), with the establishment of a working party and the determination of a standard reference spectrum for copper [96]. Since then a committee of the International Standards Organisation (ISO) has been established (ISO/TC 201 for Surface Chemical Analysis) with the purpose of establishing internationally recognised standards for Surface Chemical Analysis. Indeed, a number of standards are being developed that are directly related to AES. For example, ISO 17973 - Surface chemical analysis - Medium resolution Auger electron spectrometers - Calibration of energy scales for elemental analysis specifies a method for calibrating the kinetic energy scale of Auger electron spectrometers with an uncertainty of 3 eV for general analytical use for identifying elements at surfaces and specifies a method to establish a calibration schedule [97,98].

In terms of instrumentation, one of the major trends in Auger spectroscopy has been the continual advance in the ability to produce Auger microscopic images of surfaces. The study of multichannel image collection and processing techniques have resulted in the development of the multi-imaging scanning analytical microscope (MULSAM) [35,99]. The development of the hyperbolic field analyser (HFA) offers the possibility of collecting an entire energy spectrum in parallel, thus allowing the very rapid acquisition of energy spectra [100]. Indeed, a new instrument, the so-called spectrum imaging scanning analytical microscope (SISAM), based on the implementation of the HFA [101]. These, and other developments, mean that the study and application of Auger electron spectroscopy is likely to continue for many years to come.

References

1. H. Todokoro, Y. Sakitani, S. Fukuhara, Y. Okajima: J. Electron Microsc. **30**, 107 (1981)
2. L. Meitner: Z Phys. **17**, 54 (1923)
3. P. Auger: Comptes Rendus **177**, 169 (1923)
4. M.P. Seah, W.A. Dench: Surf. Int. Anal. **1**, 2 (1979)
5. D.A. Shirley: Phys. Rev. A. **7**, 1520 (1973)
6. R. Hoogewijs, L. Fiermans, J. Vennik: Surf. Sci. **69**, 273 (1977)
7. R. Weissmann, K. Muller: Surf. Sci. Rep. **105**, 251 (1981)
8. S. Tanuma, C.J. Powell, D.R. Penn: Surf. Interface Anal. **25**, 25 (1997)
9. C.-O. A. Olsson, S.E. Hrnstrm, S. Hogmark: In Surface Characterisation: a user's sourcebook, ed. by D. Brune, R. Hellborg, H.J. Whitlow, O. Hunderei, (Wiley, New York 1997)

10. C.C. Chang: Surf. Sci. **25**, 53 (1971)
11. M. Prutton, M.M. El Gomati: Surf. Int. Anal. **9**, 99 (1986)
12. C.G.H. Walker, D.C. Peacock, M. Prutton, M.M. El Gomati: Surf. Int. Anal. **11**, 266 (1988)
13. E.N. Sickafus: Phys. Rev. B **16**, 1436 (1977)
14. E.N. Sickafus: Phys. Rev. B **16**, 1448 (1977)
15. E.N. Sickafus, B. Kukla: Phys. Rev. B **19**, 4056 (1977)
16. J.A.D. Matthew, M. Prutton, M.M. El Gomati, D.C. Peacock: Surf. Int. Anal. **11**, 173 (1988)
17. M.P. Seah: Surf. Int. Anal. **9**, 85 (1986)
18. J.T. Grant: Surf. Int. Anal. **14**, 271 (1989)
19. M.P. Seah: SEM 83, SEM Inc, AMF O'Hare, Chicago (1983), p.521-536
20. P. Lejcek: Surf. Sci. **202**, 493 (1988)
21. L.E. Davies, .C. MacDonald, P.W. Palmberg, G.E. Riach, R.E. Weber: Handbook of AugerElectron Spectroscopy (Physical Electronics, Eden Prairie, MN)
22. T. Sekine, A. Mogami, M. Kudoh, K. Hirata: Vacuum **34**, 631 (1984)
23. S. Ichimura, R. Shimizu: Surf. Sci. **112**, 386 (1981)
24. P.H. Holloway: J. Electron Spectrosc. Relat. Phenom. **7**, 215 (1975)
25. H.E. Bishop, B, Chornik, C. LeGressus, A. LeMoel: Surf. Int. Anal., **6**, 116 (1984)
26. F.E. Doern, L. Kover, N.S. McIntyre: Surf. Int. Anal. **6**, 282 (1984)
27. C. Lea: Metal Sci. **17**, 357 (1983)
28. J.M. Walls, I.K. Brown, D.D. Hall: Appl. Surf. Sci. **15**, 93 (1983)
29. I. Lhermitte-Sebire, M. Lahaye, R. Colmet, R. Naslain: Thin Solid Films **138**, 209 (1986)
30. J.F. Bresse: Surf. Sci. **168**, 810 (1986)
31. M. Prutton: Microsc. Microanal. Microstruct. **6**, 289 (1995)
32. N.C. MacDonald and J.R. Waldrop: Appl. Phys. Lett. **19**, 315 (1971)
33. R. Browning, P.J. Basset, M.M. El Gomati, and M. Prutton: Proc. Roy. Soc. London **A357**, 213 (1977)
34. R. Browning, M.M. El Gomati, M. Prutton: Surf. Sci. **68**, 328 (1977)
35. M. Prutton, C.G.H. Walker, J.C. Greenwood, P.G. Kenny, J.C. Dee, I.R. Barkshire, R.H. Roberts, M.M. El Gomati: Surf. Int. Anal. **17**, 71 (1991)
36. www.phi.com
37. R.H. Roberts, M.M. El Gomati, J. Kudhjoe, I.R. Barkshire, S.J. Bean, M. Prutton: Meas. Sci. Technol. **8**, 536 (1997)
38. M.H. Hochella, M.F. Turner, D.W. Harris: SEM II (1986) p.337
39. T. Gold, E. Bilson, R.L. Baron: Proc. 5th Lunar Sci. Conf. (1975) p2413
40. T. Gold, E. Bilson, R.L. Baron: Proc. 6th Lunar Sci. Conf. (1975) p3285
41. J.W. Morse, A. Mucci, L.M. Walter, M.S. Kaminsky: Science **205**, 904 (1979)
42. A. Mucci, J.W. Morse, M.S. Kaminsky: Am. J. Sci. **285**, 289 (1985)
43. A. Mucci, J.W. Morse, M.S. Kaminsky: Am. J. Sci. **285**, 306 (1985)
44. J.F. Jongste, F.E. Prins, G.C.A.M. Janssen: Matt. Lett. **8**, 273 (1989)
45. M.G. Ransey, F.P. Netzer: Mat. Sci. Eng. B **2**, 269 (1989)
46. A. Roshko, Y.M. Chiang: J. Appl. Phys. **66**, 3710 (1989)
47. H.H. Madden, W.O. Wallace: Surf. Sci. **172**, 641 (1986)
48. R. Sherman: J. Am. Ceram. Soc. **68**, C7 (1985)
49. R. Browning, J.L. Smialek, N.S. Jacobsen,: Advanced Ceramic Materials, **2**, 773 (1987)

50. BCR reference material no. CRM 261 available from Institute for Reference Materials and Measurements (IRMM), Retieseweg, B-2440, Geel, Belgium. (www.irmm.jrc.be)
51. C.P. Hunt, H.J. Mathieu and M.P. Seah: Surf. Sci. **139**, 549 (1984)
52. C.S. Wei, M. Setton, J. Van der Spiegel, J. Santiago: J. Appl. Phys **61**, 1429 (1987)
53. P.H. Holloway: Surf. Sci. **66**, 479 (1977)
54. R. Li, L. Tu, Y. Sun: Surf. Sci. **163**, 67 (1985)
55. T. Sekine, A. Mogami, J.D. Geller, SEM, SEM Inc, AMF O'Hare, Chicago (1981), p.245
56. H.J. Mathieu: In *Surface Analysis - Principal Techniques*, ed. by J.C. Vickerman (Wiley, New York 1997), Chap. 4
57. A. Jablonski, C.J. Powell: J. Electron Spectrosc. Relat. Phenom. **100**, 137 (1999)
58. C.J. Powell, A. Jablonski, I.S. Tilinin, S. Tanuma, D.R. Penn: J. Electron Spectrosc. Relat. Phenom. **98–99**, 1 (1999)
59. C.J. Powell, A. Jablonski: Surf. Interface Anal. **29**, 108 (2000)
60. S. Tanuma, C.J. Powell, D.R. Penn: Surf. Interface Anal. **11**, 577 (1988)
61. S. Tanuma, C.J. Powell, D.R. Penn: Surf. Interface Anal. **17**, 911 (1991)
62. S. Tanuma, C.J. Powell, D.R. Penn: Surf. Interface Anal. **17**, 929 (1991)
63. S. Tanuma, C.J. Powell, D.R. Penn: Surf. Interface Anal. **20**, 77 (1988)
64. S. Tanuma, C.J. Powell, D.R. Penn: Surf. Interface Anal. **21**, 165 (1994)
65. C.J. Powell, A. Jablonski: NIST Electron Inelastic-Mean-Free-Path Database (Version 1.0), National Institute of Standards and Technology: Gaithersburg, MD, 1999
66. P. Morgan, B. Jorgensen: Surf. Sci. **208**, 306 (1989)
67. P.J. Lurie, J.M. Wilson: Surf. Sci. **65**, 476 (1977)
68. T.W. Hass, J.T. Grant, G.J. Dooley III: J. Appl. Phys. **43**, 1853 (1972)
69. Y. Mizokawa, T. Miyasato, S. Nakamura, K.M. Geib, C.W. Wilmsen: Surf. Sci. **182**, 431 (1987)
70. H.H. Madden: Surf. Sci. **126**, 80 (1983)
71. C.D. Wagner: Anal. Chem. **47**, 1201 (1975)
72. L.P. Erickson, B.F. Phillips: J. Vac. Sci. Technol. B **1**, 158 (1983)
73. H.H. Busta, C.H. Tang: J. Electrochem. Soc. **133**, 1195 (1986)
74. F. Marchetti, M. Dapor, S. Girardi, F. Giacomozzi, A. Cavalleri: Mater. Sci. Eng. A **115**, 217 (1989)
75. R. Pantel, F.A. D'Avitaya: Thin Solid Films **140**, 177 (1986)
76. S.J. Pearton, A.B. Emerson, U.K. Chakrabarti, E. Lane, K.S. Jones, K.T. Short, A.E. White, T.R. Fullowan: J. Appl. Phys **66**, 3839 (1989)
77. T.T. Huang, B. Peterson, D.A. Shores, E. Pender: Corrosion Science, **24**, 167 (1984)
78. S. Mathieu, La Revue de Metallurgie, Jan 1989, p. 73
79. E.R. Malinowski, Factor Analysis in Chemistry, 2nd ed. (Wiley, New York 1991)
80. W. Palmer: Appl. Phys. A **42**, 219 (1987)
81. B.M. Clemens: J. Non Cryst. Solids **61–62**, 817 (1984)
82. H.J. Grabke: ISIJ **29**, 529 (1989)
83. M. Militzer, J. Wieting: Acta Metall. **37**, 2585 (1989)
84. D.C. Peacock: Appl. Surf. Sci. **26**, 306 (1986)

85. D.C. Peacock: Appl. Surf. Sci. **27**, 58 (1986)
86. P. Humbert, A. Mosser: Surf. Sci. **126**, 708 (1983)
87. J. du Plessis, P.E. Vijoen, F. Bezuidenhout: Surf. Sci. **138**, 26 (1984)
88. M. Takeyama, C.T. Liu: Acta Metall. **37**, 2681 (1989)
89. W.E. Wallace, S.G. Sankar, J.M. Elbiki, S.F. Cheng: Mat. Sci. Eng. B **3**, 351 (1989)
90. R. Hamminger, G. Grathwohl, F. Thummler: J. Mater. Sci. **18**, 3154 (1983)
91. J.G. Smeggil, A.W. Funkenbush, N.S. Bornstein: Metall. Trans. **17A**, 923 (1986)
92. J.L. Smialek, R. Browning: Proc. Symp. High Temp. Material Chemistry-III, Electrochem., Soc., (1986) p.258
93. K.L. Luthra, C.L. Briant: Mater. Sci. Forum **43**, 299 (1989)
94. K. Sato, Y. Inoue: ISIJ **29**, 246 (1989)
95. M. Tomita, T. Tanabe, S. Imoto: Surf. Sci. **209**, 173 (1989)
96. G.C. Smith, M.P. Seah: Surf. Int. Anal. **12**, 105 (1988)
97. M. P. Seah and I. S. Gilmore: J. Electron Spectrosc. Relat. Phenom., **83**, 197 (1997)
98. M. P. Seah: J. Electron Spectrosc. Relat. Phenom. **97**, 235 (1997)
99. R. Browning: MRS Bull. **12**, 75 (1987)
100. M. Jacka, M. Kirk, M.M. El Gomati, M. Prutton: Rev. Sci. Instrum., **70**, 2282 (1999)
101. M. Prutton: Surf. Interface Anal. **29**, 561 (2000)

7 X-Ray Photoelectron Spectroscopy

M.H. Kibel

The detection and energy analysis of photoelectrons produced by radiation whose energy exceeds their binding energies is the subject of an extensively-used technique known as Photoelectron (PE) Spectroscopy. This technique can be conveniently divided into two broad areas, the first employing ultraviolet radiation, hence called Ultraviolet Photoelectron Spectroscopy (UPS), and the second using X-rays, termed X-ray Photoelectron Spectroscopy (XPS). The latter spectroscopy is the subject of this present chapter, while UPS is discussed in Chap. 14.

The history of the development of XPS is a fascinating topic by itself, but involved discussion is not appropriate here. For a detailed account, the reader is referred to the excellent work of Jenkin et. al. [1–4]. XPS is also known as electron spectroscopy for chemical analysis (ESCA), a name coined by *Siegbahn* in order to emphasise the presence of both photo- and Auger electron peaks in the XP spectrum [5,6].

7.1 Basic Principles

7.1.1 Theory

Presented schematically in Fig. 7.1 are the related processes that are involved in the ejection of a photo- or Auger electron. XPS involves the removal of a single core electron, while AES is a two-electron process subsequent to the removal of the core electron, with the Auger electron ejected following reorganisation within the atom. Auger electrons are produced in XPS along with photoelectrons, and these can complicate the interpretation of the subsequent spectra as will be discussed later in this chapter.

The photoemission process is shown on an energy level diagram in Fig. 7.2. The sample is irradiated with X-rays of known energy, $h\nu$, and electrons of binding energy (BE) E_b are ejected, where $E_\mathrm{b} < h\nu$. These electrons have a kinetic energy (KE) E_k which can be measured in the spectrometer, and is given by

$$E_\mathrm{k} = h\nu - E_\mathrm{b} - \Phi_\mathrm{sp} \, , \tag{7.1}$$

where Φ_sp is the spectrometer work function, and is the combination of the sample work function, Φ_s, and the work function induced by the analyser.

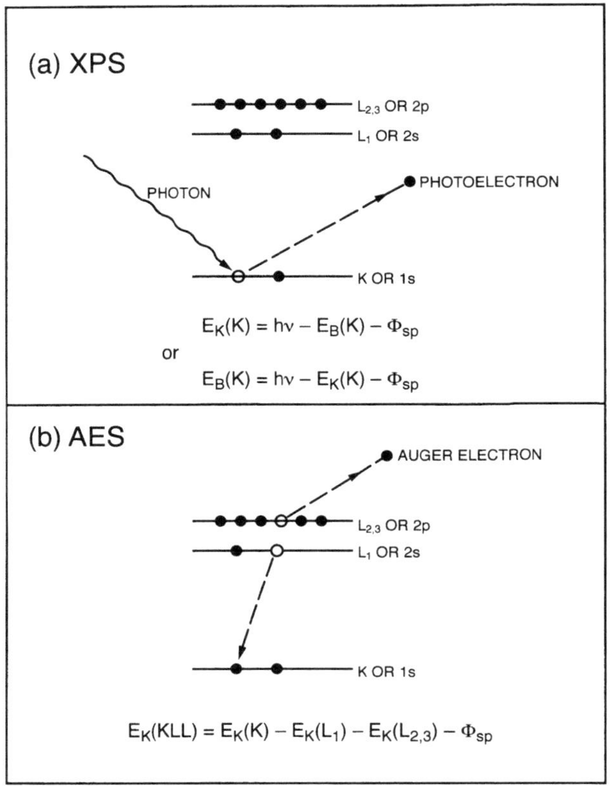

Fig. 7.1. Representation of the processes involved in XPS and AES. **(a)** XPS involves the removal of a single core electron while **(b)** AES is a two-electron process following the removal of the core electron

Since we can compensate for the work function term electronically, it can be eliminated, leaving

$$E_{\mathrm{k}} = h\nu - E_{\mathrm{b}} \tag{7.2}$$

or

$$E_{\mathrm{b}} = h\nu - E_{\mathrm{k}} \, . \tag{7.3}$$

Thus by measuring the KE of the photoelectrons, (7.3) can be used to translate this energy into the BE of the electrons.

7.1.2 Typical Spectrum

An XP spectrum is generated by plotting the measured photoelectron intensity as a function of BE, as shown in Fig. 7.3, which is the Mg $K\alpha$

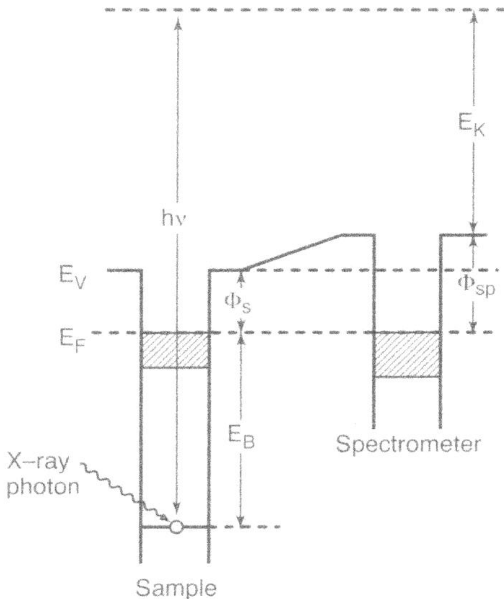

Fig. 7.2. Schematic diagram of the photoemision process. The sample is irradiated with X-rays of known energy, $h\nu$, and electrons of binding energy E_b are ejected. These electrons have a kinetic energy, E_k, which can be measured in the spectrometer, and is given by (7.1). Φ_s is the work function of the sample, and Φ_{sp} is the spectrometer work function

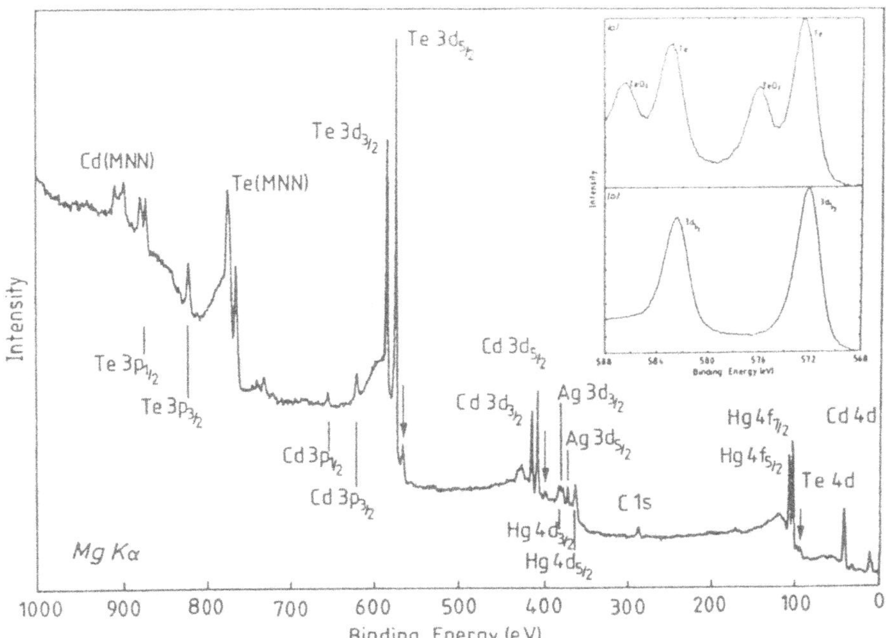

Fig. 7.3. Mg $K\alpha$ XP spectrum of $Hg_{0.6}Cd_{0.4}Te$ grown via Metal Organic Chemical Vapour Deposition (MOCVD). Inset: (**a**) Te 3*d* region of full spectrum before cleaning, showing the presence of oxide. (**b**) Te 3*d* region after ion etching, which has removed the oxide

($h\nu = 1253.6\,\mathrm{eV}$) XP spectrum of a HgCdTe film [7]. The resulting series of lines are superimposed on a background caused by the Bremsstrahlung radiation inherent in non-monochromatic X-ray sources. The BEs of these lines are characteristic for each element, and are a direct representation of the atomic orbital energies. Published tables of photoelectron BEs for all elements can be used to assist in the assignment of peaks in XP spectra [8,9]. Note that the carbon 1*s* orbital is represented by a single line in the spectrum, while the various *p*, *d* and *f* orbitals produce two lines. These doublets are due to spin-orbit splitting, and their occurrence will be explained in Sect. 7.3.1. An expansion, of the Te 3*d* region is presented in Fig. 7.3a (inset), which shows that there are actually four components apparent for Te. The higher intensity peaks represent elemental Te in the zero oxidation state, with the other two peaks representing TeO_2, i.e. Te in an oxidation state of +4. This is a typical example of a chemical shift in the BE, which will be the topic of Sect. 7.3.2. From Fig. 7.3b the effect of a light argon-ion bombardment can be seen, with the complete removal of the TeO_2, showing how XPS can be employed to monitor surface cleanliness.

Returning to Fig. 7.3, note also the presence of Cd and Te *MNN* Auger lines. While these groups of peaks can sometimes be considered a hindrance to spectral interpretation due to overlap with XP peaks, their presence can also be of substantial value. As will be discussed in Sect. 7.3.3, the use of these Auger lines can yield valuable chemical shift information.

7.1.3 Surface Specificity

As was discussed in Sect. 1.5.1, the surface sensitivity of electron spectroscopies is due to the low inelastic mean-free path, λ_m, of the electrons within the sample. For XPS, the main region of interest relates to electron energies from 100–1200 eV, which gives rise to a λ_m value of 0.5–2.0 nm (or 5–20Å). However the actual escape depth, λ, of the photoelectrons depends on the direction in which they are travelling within the solid, such that [10]

$$\lambda = \lambda_\mathrm{m} \cos\theta \qquad (7.4)$$

where θ is the angle of emission to the surface normal. Thus electrons emitted perpendicular to the surface ($\theta = 0°$) will arise from the maximum escape depth, and hence will carry information which may be indicative of the bulk material, whereas electrons emitted nearly parallel to the surface ($\theta \sim 90°$) will be purely from the outermost surface layers. Thus XPS is an extremely surface sensitive technique, and as we will see from Sect. 7.5.3, varying the angle of detection during an experiment, as described above, can enhance this surface sensitivity.

7.2 Instrumentation

As with most techniques, certain components are essential, while others, although desirable, can be optional. In this section we discuss the components used in modern XPS systems, and consider different alternatives available for many of these components.

7.2.1 Essential Components

Fig. 7.4 is a schematic diagram indicating the essential components necessary for performing XPS. These consist of an X-ray source, a sample/support system, an electron energy analyser and an electron detector/multiplier, all maintained under ultra-high vacuum (UHV), and suitable electronics to convert the detected current into a readable spectrum. These components will now be discussed individually in more detail.

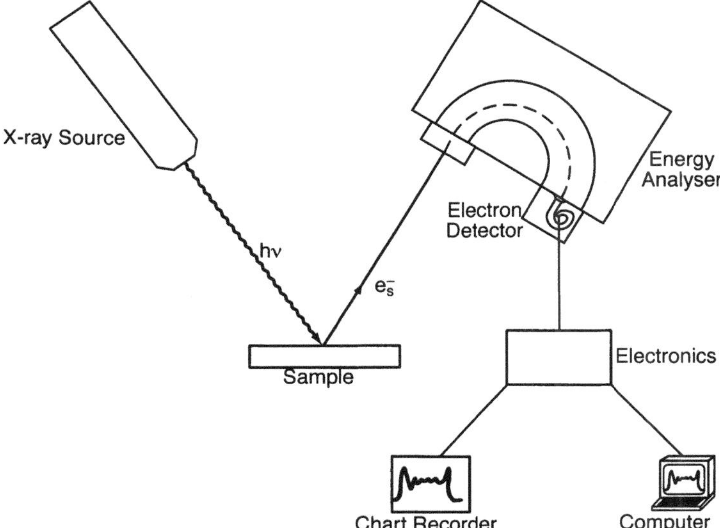

Fig. 7.4. Schematic representation of the components necessary for performing XPS

a) X-Ray Source. An ideal photon source must be sufficiently energetic to access core levels, intense enough to produce a detectable electron flux, have a narrow line width and be simple to use and maintain. Listed in Table 7.1 are some of the more common sources, together with their energies and line widths. The full-width-half-maximum (FWHM) of an XPS peak depends on several factors, but the major contribution comes from the line width of the

Table 7.1. Common x-ray sources

X-ray line	Energy [eV]	Line Width [eV]	Comments
Y $M\zeta$	132.3	0.47	Very low energy; very narrow line
Zr $M\zeta$	151.4	0.77	Very low energy; narrow line
Na $K\alpha$	1041.0	0.70	Very difficult to design
Mg $K\alpha$	1253.6	0.70	Requires 15 keV; narrow line; stable
Al $K\alpha$	1486.6	0.85	Requires 15 keV; narrow line; stable
Zr $L\alpha$	2042.4	1.7	Wide line
Ti $K\alpha$	4510.0	2.0	Requires 20 keV source; wide line
Cu $K\alpha$	8048.0	2.6	Requires 30 keV source; wide line

X-ray line. From Table 7.1, the only lines which are sufficiently energetic and narrow are the Na $K\alpha$, Mg $K\alpha$ and Al $K\alpha$ lines. However it is very difficult to design a suitable, stable sodium anode, thus the Mg and Al sources are most commonly used. Many manufacturers now provide X-ray sources with dual Mg/Al anodes as standard. Other dual-anode combinations are also available, such as Mg/Zr and Al/Zr, used since the Zr $L\alpha$ line has a particularly high sensivity to elements such as aluminium and silicon [11].

The sources described above do not, however, produce single X-ray lines, but a series of lines superimposed on the Bremsstrahlung continuum. One method of removing the unwanted components, and eliminating the continuum, is to monochromatise the radiation. This is most conveniently achieved by using a diffraction grating, with the line spacings for selecting the correct component determined from the Bragg relation [12]. Of course the resulting monochromatised radiation will be greatly reduced in intensity, due to dispersion.

b) Electron Energy Analyser and Detector. Once the photoelectrons have been produced, they must be separated according to their energy and subsequently converted into a spectrum. The electron energy analyser is thus the heart of any XPS system and is the critical component in determining sensitivity and resolution. There are several types of analysers that have been used in electron spectrometers, but the simplest is the Retarding Field Analyser (RFA) which although used mainly in LEED/AES systems, was occasionally used in UPS/XPS [13–15]. Of the electrostatic analysers, the simplest is the 127° cylindrical analyser, which consists of two concentric cylindrical deflectors focussing the electrons, which travel through an arc of 127° ($\pi/\sqrt{2}$), in one dimension. This type of analyser is relatively easy to construct, and hence appears often in home-built systems [16]. The two most common analysers encountered in XPS are the Cylindrical Mirror Analyser (CMA) and the Concentric Hemispherical Analyser (CHA).

The CHA is also known as the spherical deflection analyser, and the basic operating principle is depicted in Fig. 7.5 [17]. Two hemispherical surfaces

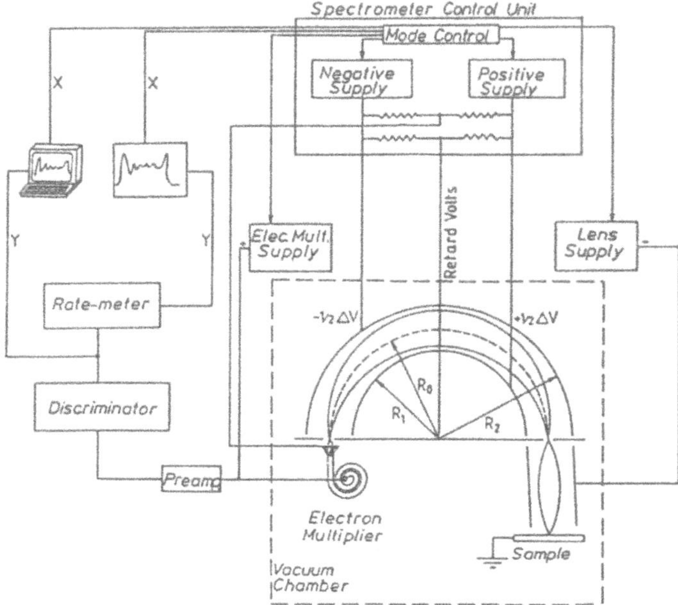

Fig. 7.5. Schematic diagram showing the operating principles of a concentric hemispherical analyser (CHA) (see text for explanation), together with the electronics used to produce XP spectra

of inner radius R_1 and outer radius R_2 are positioned concentrically, with a potential ΔV applied such that the outer sphere is negative and the inner positive (with respect to ΔV). The entrance and exit slits lie on the median radius, R_0, where

$$R_0 = (R_1 + R_2)/2 \tag{7.5}$$

and

$$eV = E_k(R_2/R_1 - R_1/R_2) \tag{7.6}$$

where the entrance and exit slits are separated by an angle of 180°. This results in focussing in two dimensions, i.e. at a point.

The resolution of an analyser can be defined in two ways. One is the *absolute* resolution ΔE measured as the full width at half-maximum height (FWHM) of a peak. The other is the *relative* resolution, ϱ, of a peak at KE E_0

$$\varrho = E_0/\Delta E\,. \tag{7.7}$$

Thus the absolute resolution is independent of peak position, but the relative resolution must be referred to the kinetic energy of the peak.

The current actually reaching the analyser exit slit following photoionisation is typically in the region of 10^{-16}–10^{-14} A, which is well below conventional current-measuring techniques. Thus pulse counting is required, and an electron multiplier is used as the detector. There are two types of electron multiplier currently in use in modern spectrometers: the traditional discrete-dynode type and the more common channel electron multiplier.

The dynode multiplier functions by allowing the electrons to strike the first electrode (or dynode), and then achieving current amplification via secondary electron multiplication of many stages (typically 10–20). A resistor chain is used to establish the separate dynode potentials. Multipliers of this type can achieve amplification of 10^4–10^7, depending on their design.

A channel electron multiplier consists of a small curved glass tube, the inside wall of which is coated with high resistance ($\sim 10^9\,\Omega$ full length) material. When a potential ($\sim 3\,\mathrm{kV}$) is applied between the ends of the tube the resistive surface becomes a continuous dynode. An electron entering the low potential (typically $+500\,\mathrm{V}$) end of the multiplier generates secondary electrons on collision with the wall of the tube. These are accelerated until they strike the wall again, giving an avalanche effect. The gain of these devices is of the order of 10^7–10^8.

Apart from single multipliers, substantial increase in performance is possible with multichannel acquisition via position-sensitive detection [32]. This involves use of multiple arrays of discrete dynodes or channeltrons, providing vastly increased signal-to-noise (S/N), thus reducing acquisition times and thus less sample exposure to X-rays.

As seen in Fig. 7.5, the multiplier output is normally taken through a pre-amplifier, an amplifier, a discriminator and a rate-meter system, with the spectrum being displayed on an X-Y recorder. More commonly, a computer interface is situated between the discriminator and rate-meter, with the spectra subsequently digitised and stored for future reference. Also shown in Fig. 7.5 is a schematic of the electronics required to operate the analyser/detector system.

7.2.2 Optional Components

There are many experiments performed via XPS that require further components, other than those discussed above. For example, the addition of an electron gun allows AES to be performed with the same analyser/electronics. The addition of a rare-gas ion gun allows samples to be cleaned in vacuum, as well as permitting depth-profiling studies to be performed. The ability to heat and cool a sample in situ can also be extremely useful, especially in adsorption/desorption and catalysis studies. While it is ideal to have a sample manipulator with these facilities, one can provide these options in a separate part of the instrument, e.g. a sample preparation chamber. It may also be necessary to install a low-energy electron flood gun for neutralising the charge on insulating materials, such as ceramics or polymers. A mass spectrometer is also extremely valuable for residual gas analysis and, if the correct geometry is chosen, for thermal desorption spectroscopy. Many of these options are discussed elsewhere in this book, and in review articles [18,19].

7.2.3 Synchrotron Radiation

All the X-ray sources discussed thus far are discrete line sources. Synchrotron radiation provides a continuous source of photons from 10eV to 10keV, which can allow the radiation to be tuned to attain the ideal photoinisation cross-section for a particular set of core levels [18]. Variation of the cross section permits the photoelectron intensity of one element to be enhanced compared to another, or enables resonance effects associated with ionisation thresholds to be followed. Synchrotron radiation is produced by accelerating electrons to near-relativistic velocities around an approximately circular torus by pulsed magnetic fields. Because the electrons are forced into curved paths they emit light in a continuous spectrum, with the intensity of the radiation proportional to the radius of curvature of the path, and inversely proportional to the cube of the electron energy. The radiation thus produced is concentrated in a very narrow cone tangential to the electron orbit, hence is ideal for transmission through a monochrometer, with intensity loss a minor concern since the X-ray flux is at least two orders of magnitude more intense than that from conventional line sources. The major limiting factor, however, is that synchrotron time is extremely limited and very expensive, hence not available to most people.

7.2.4 Imaging XPS

While imaging techniques have been used extensively in AES (i.e. Scanning Auger Microscopy, SAM), the routine application of these techniques to XPS is relatively new. XPS imaging systems are either defined by the source, i.e. the lateral resolution will be a consequence of narrowly focussed incident X-rays or defined by the analyser, i.e. the analyser is restricted to observing a small region of the surface. Several valuable review articles on imaging XPS and selected area XPS (SAXPS) have been published recently [20–24]. Unlike SAM, the difficulty in focussing X-rays limited early attempts at improved XPS spatial resolution to reliance on a restricting aperture in front of the analyser [21], with deflection plates introduced between the aperture and the sample, allowing a raster-scanned virtual true image of the surface to be produced [21,22]. The drawback of this method is a dramatic loss of S/N and long acquisition times. The introduction of a magnetic lens into an imaging system [25] has led to significant advances, including commercial instruments (e.g. Kratos AXIS HSi).

The development of the photoelectron spectromicroscope provided a direct imaging approach which involved placing the sample in a strong magnetic field, then illuminating it with UV- or X-radiation [26,27]. The exiting photoelectrons spiral out in tight trajectories along divergent magnetic field lines. Energy analysis is achieved by a simple RFA, and spatial detection is via a channel plate display. The image is always in focus as this is, essentially, a point projection method. A further development (MicroESCA) uses synchrotron radiation and acquires the images digitally, enabling images at different retarding potentials to be subtracted. This gave a lateral resolution of 5 mm and an energy resolution of 0.1 eV below 100 eV. However the low energy and the use of synchrotron radiation makes the method of limited application.

Rather than point by point imaging, as described above, an imaging method which can gather all the spatial information simultaneously has been developed by VG Scientific [28]. This uses a parallel imaging method for CHAs that overcomes the convolution of energy and spatial information in the output plane of the analyser [20,21]. Objective lens and aperture systems convert the spatial information into angular information, so that electrons leaving the surface at a given point, but over a range of collection angles, actually enter the analyser as a parallel beam at a given angle, effectively providing a Fourier transform of the suface image. This system is capable of $< 10\,\mu\text{m}$ resolution with a standard X-ray source. Further development followed with the ESCALAB 220i–XL (VG Scientific), which involves the incorporation of a magnetic lens close to the sample, which improves performance (similiar to the AXIS HSi), with a monochromator using a variable spot electron gun to irradiate the X-ray anode. This instrument is capable of field of view of 0.5 to $8\,\mu\text{m}$ at $10\,\mu\text{m}$ resolution with electrostatic lenses and 80 mm to 1 mm at < 2 mm resolution with the magnetic lens.

Several other methods of XPS imaging have also been used and discussed in the literature, such as the focussed X-ray probe using a focussing crystal monochromator [24,29], a focussed X-ray probe using synchrotron radiation [30], and the line-by-line or E-x approach [31], but these will not be discussed here; the reader is referred to appropriate reviews [20–24] and the original articles.

7.3 Spectral Information

Much important information can be gleaned from an XP spectrum, beyond simple BE data and elemental identification. In this section we look at what additional information the spectra can reveal.

7.3.1 Spin-Orbit Splitting

If an unpaired electron is in a degenerate orbital (i.e. $p, d, f, \ldots$) the spin angular momentum, S, and the orbital angular momentum, L, can combine in different ways and produce new states that are characterised by the total electronic angular momentum, J:

$$J = |L \pm S| \tag{7.8}$$

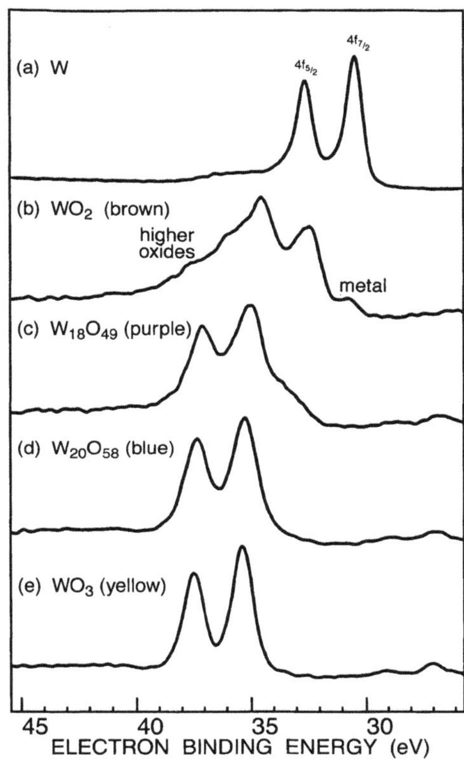

Fig. 7.6. W $4f$ XP spectra of tungsten and four tungsten oxides. (Reproduced with permission from [32])

where $L = 0, 1, 2, \ldots$; $S = 1/2$; $J = 1/2, 3/2, 5/2, \ldots$. The energies of these new states are thus different because the magnetic moments due to the electron spin and orbital motion may oppose or reinforce each other. The degeneracies of these states is $2J + 1$, and the relative intensities of these split peaks is given by the ratio of these degeneracies.

For example, for the Tungsten $4f$ orbital ($L = 3$) shown in Fig. 7.6a [49],

$$J = |3 \pm 1/2| = 7/2, 5/2 \, . \tag{7.9}$$

Thus we have two components, viz. $4f_{7/2}$ and $4f_{5/2}$. The relative peak intensities in this example are therefore $(2 \times 7/2 + 1) : (2 \times 5/2 + 1)$, which equates to $4 : 3$.

7.3.2 Chemical Shifts

As with a technique such as nuclear magnetic resonance (NMR), XPS can distinguish between a particular element in different environments. This is due to the fact that placing a particular atom in a different chemical environment, or a different oxidation state, or in a different lattice site, etc., gives rise to a change in BEs of the core-level electrons. This BE variation is called the chemical shift, and appears as a definite "movement" of the elemental peak in the XP spectrum. As an example, returning to Fig. 7.6, which shows the $4f$ peaks for Tungsten (W) and four common oxides of W [32], we can observe the BE of these peaks moving to higher binding energy as the oxidation number of the W increases. A great many of these shifts have been documented for instant reference, and an example for W can be found in the Perkin–Elmer Handbook [8]. These can be of substantial aid in determining the chemical environment of an element from a given XP spectrum.

7.3.3 Auger Chemical Shifts in XPS

Chemical information is also available from AES, although the effects on spectral features are not as well understood as in XPS. However since many core-type Auger lines are generated via Mg $K\alpha$ and Al $K\alpha$ X-rays, it is convenient to consider the combined chemical shifts in both the photoelectron and Auger lines. The difference between Auger and photoelectron chemical shifts results from the difference in final-state relaxation energies between chemical states. It is thus possible to define a parameter, α, called the Auger parameter, such that [33]

$$\alpha = E_{\mathrm{k}}(A) - E_{\mathrm{k}}(P) \tag{7.10}$$

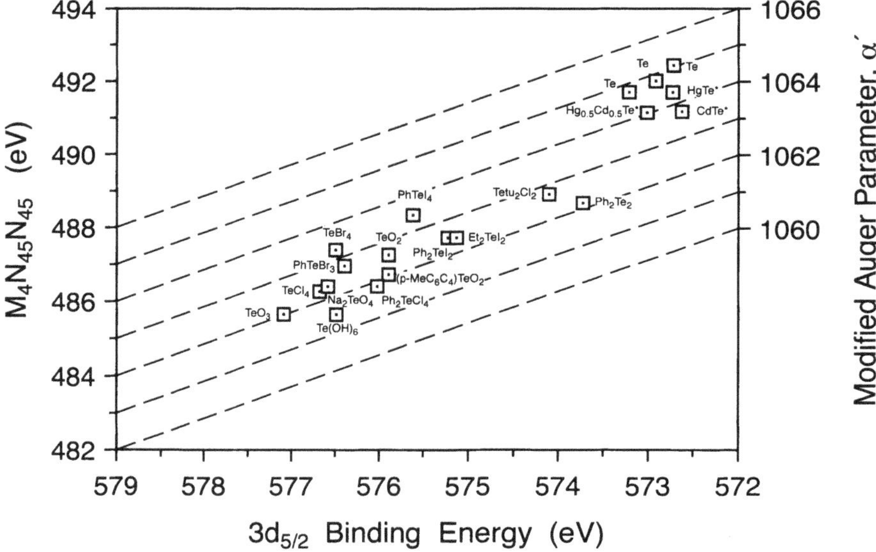

Fig. 7.7. Plot of Auger $M_4M_{45}M_{45}$ kinetic energy versus $3d_{5/2}$BE for Te. The diagonal lines represent the sum of these two quantities, and hence yields the modified Auger parameter, (see (7.12)). (Data obtained from [7,8])

where $E_\mathrm{k}(A)$ = Auger electron KE, and $E_\mathrm{k}(P)$ = photoelectron KE. One of the main advantages of α is that it is independent of static charging, and is characteristic of a particular chemical state. Unfortunately, for some systems α can have negative values, however (7.12) can be modified by using (7.2) as follows:

$$\alpha = E_\mathrm{k}(A) + E_\mathrm{b}(P) - h\nu \tag{7.11}$$

or, the "modified" Auger parameter,

$$\alpha' = \alpha + h\nu = E_\mathrm{k}(A) + E_\mathrm{b}(P) \,. \tag{7.12}$$

Thus a graph $E_\mathrm{k}(A)$ vs $E_\mathrm{b}(P)$ becomes independent of photon energy. Such two-dimensional plots have been generated for many elements, one of which is tellurium, Te, shown in Fig. 7.7 [8]. Many other examples of the use of the Auger parameter can also be found in the literature [35–41].

7.3.4 X-Ray Line Satellites

As mentioned in Sect. 7.2.1(a), X-ray sources do not produce single lines, but a series consisting of a main line with some minor components at higher photon energies. Those minor lines can cause small satellite peaks to occur for each major photoelectron peak, the intensities and spacings of which are characteristic of the particular X-ray anode material. An example of this is presented in Fig. 7.3, where the $K\alpha_{1,2}$ peaks for the Hg $4f$, Cd $3d$ and Te $3d$ orbitals have accompanying $K\alpha_3$ satellite peaks (arrowed), the latter being displaced by 8.4 eV. Table 7.2 lists the X-ray satellite energies and intensities for Mg and Al radiation.

Table 7.2. X-ray satellite energies and intensities

		$\alpha_{1,2}$	α_3	α_4	α_5	α_6	β
Mg	relative intensity	100 ($\overbrace{67, 33}$)	8.0	4.1	0.55	0.45	0.5
	displacement, eV	0	8.4	10.2	17.5	20.0	48.5
Al	relative intensity	100 ($\overbrace{67, 33}$)	6.4	3.2	0.4	0.3	0.55
	displacement, eV	0	9.8	11.8	20.1	23.4	69.7

7.3.5 "Shake-up" Lines

Following photoionisation there is often a finite probability that the resultant ion will remain in an excited state, a few electron volts above ground state.

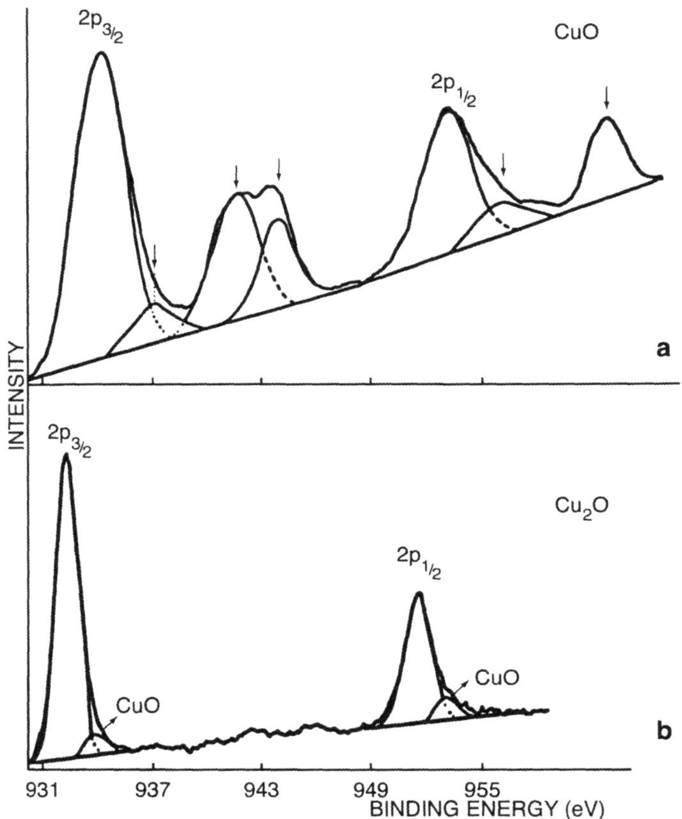

Fig. 7.8. The copper $2p_{1/2}$ and $2p_{3/2}$ XP spectra from (**a**) CuO, showing very prominent shake-up satellites, and (**b**) Cu_2O, where such satellites are absent. (Reproduced with permission from [42])

The resultant photoelectron thus suffers a loss in KE, corresponding to the energy difference between the ground and excited states. In the spectrum this appears as a satellite a few eV higher in BE from the main peak. An example of this phenomenon can be seen in the XP spectrum of CuO, reproduced in Fig. 7.8 [42], where several satellites occur. These satellites can often be useful in determining the chemical nature of an element, e.g. distinguishing between CuO and Cu_2O, as seen in Fig. 7.8.

7.3.6 Ghost Lines

Ghost lines are small peaks appearing in an XP spectrum which result from X-radiation from foreign material. For example, common ghost lines arise from Mg impurity in the Al source, or *vice versa* in a dual anode source.

Other sources can be Cu from the anode base structure or X-ray photons arising from the foil window. The positions of these ghost lines can be easily calculated, and Table 7.3 shows where such lines are expected to occur.

Table 7.3. Displacements of X-ray ghost lines

	Anode material	
Contaminant Radiation	Mg	Al
O ($K\alpha$)	728.7	961.7
Cu ($L\alpha$)	323.9	556.9
Mg ($K\alpha$)	–	233.0
Al ($K\alpha$)	−233.0	–

7.3.7 Plasmon Loss Lines

A plasmon loss occurs following interaction between photoelectrons and surface electrons in a solid, i.e. electrons passing through a solid can excite a collective oscillation of bulk electrons. This is called a bulk plasmon energy loss, and if the frequency of oscillation is ω_{b}, then the plasmon energy loss is $\hbar\omega_{\mathrm{b}}$. Subsequent plasmon losses can also occur for these photoelectrons, resulting in a series of lines in the spectrum, all equally spaced by $\hbar\omega_{\mathrm{b}}$, but of decreasing intensity.

At the surface a more localised oscillation can occur, and this is called a surface plasmon, with frequency ω_{s}, and hence energy loss $\hbar\omega_{\mathrm{s}}$. The surface plasmon loss can be related to the bulk plasmon loss as follows [43]:

$$\omega_{\mathrm{s}} = \omega_{\mathrm{b}}/\sqrt{2}\,. \tag{7.13}$$

For all elements the fundamental, or first, bulk plasmon will always be observable, with multiple plasmon loss peaks also visible in some cases. Depending on surface conditions, and the energy of the corresponding photoelectron peak, surface plasmons may also be seen, especially in the case of clean metal surfaces. An example of this can be seen in Fig. 7.9, for aluminium, showing plasmon loss peaks, both bulk and surface, for the $2s$ photoelectrons [44].

7.4 Quantitative Analysis

One of the major advantages of XPS is the ease with which quantitative data can be routinely obtained. This is usually performed by determining the area under the peaks in question and applying previously-determined sensitivity factors. For a homogeneous sample, the number of photoelectrons per second in a given peak, assuming constant photon flux and fixed geometry, is given by [45]:

$$I = KN\sigma\lambda AT \, , \tag{7.14}$$

where K = const, N = number of atoms of the element per cm^3, σ = photoionisation cross section for element, λ = inelastic mean-free path length for photoelectrons, A = area of the sample from which the photoelectrons eminate and T = analyser transmission function. If we define the sensitivity factor for element x as

$$S_x = K\sigma\lambda AT \, , \tag{7.15}$$

then

$$I_x = N_x S_x \tag{7.16}$$

or

$$N_x = I_x / S_x \, . \tag{7.17}$$

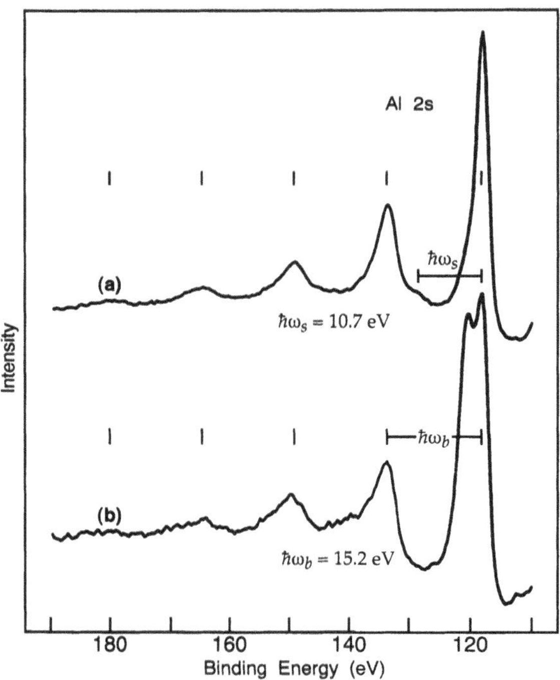

Fig. 7.9. XP spectrum of the Al 2s region, showing a succession of sequential bulk plasmon losses ($\hbar\omega_b$ = 15.2 eV). In spectrum (a), for clean Al, a surface plasmon loss can also be seen ($\hbar\omega_s$ = 10.7 eV), which disappears on oxidation, as in spectrum (b). (After [44])

If we wish to know the relative amounts of elements in a sample, it is thus necessary to know the sensitivity values for the elements and to measure their intensities (i.e. peak areas)

$$N_1 = I_1/S_1 \; ; \quad N_2 = I_2/S_2 \; . \tag{7.18}$$

This approach will provide semi-quantitative results for most situations, except where heterogeneous samples are involved, or where serious contamination layers obscure the underlying elements. Also, any peak interference (e.g. overlapping Auger lines or another XPS peak) must be avoided. Ideally elemental sensitivity factors should be determined specifically for each instrument.

A simple application of this type of quantitative analysis has been provided by *Leech* et al. in a study of the effect of etchants on II-VI semiconductor materials [46,47]. Their work used XPS to determine the relative amounts of Hg, Cd and Te in the samples before and after etching, and hence were able to follow the changes which the etchants induced.

7.5 Experimental Techniques

7.5.1 Variation of X-Ray Sources

As mentioned in Sect. 7.2.1(a), many X-ray sources are available for use in XPS, with the most common being a dual Mg/Al source. What is the value of the availability of two different X-ray lines? Firstly, the Mg $K\alpha$ line is narrower than the Al $K\alpha$, thus allowing a better resolution to be achieved. Secondly, some core levels may not be accessible with the Mg source, and require the slightly higher (by 233 eV) energy Al line. Thirdly, often in an XP spectrum Auger lines and XP peaks may overlap, and changing to another source can alleviate this problem, since the Auger electrons always have the same KE, whereas the photoelectron KE depends on the ionising radiation as the BE is constant. Thus the Auger lines appear to move relative to the XP peaks as the X-ray energy is varied. An example of this application is shown in Fig. 7.10, where spectra have been recorded for galvanised iron. In the Al $K\alpha$ spectrum (b), the Zn(LMM) lines overlap with the O 1s and Cr 2p orbitals, while the Cr(LMM) lines interfere with the Zn 2p peaks. Changing to Mg $K\alpha$ (a) solves this problem, with the Cr(LMM) lines clear of any other features, and the Zn(LMM) lines now overlapping with the C 1s peak.

7.5.2 Depth Profiles

This topic has been broached previously in this book, both generally and with specific application to AES (Chap. 6). Depth profiling has applications in XPS also, although it has been less utilised than in AES. The process

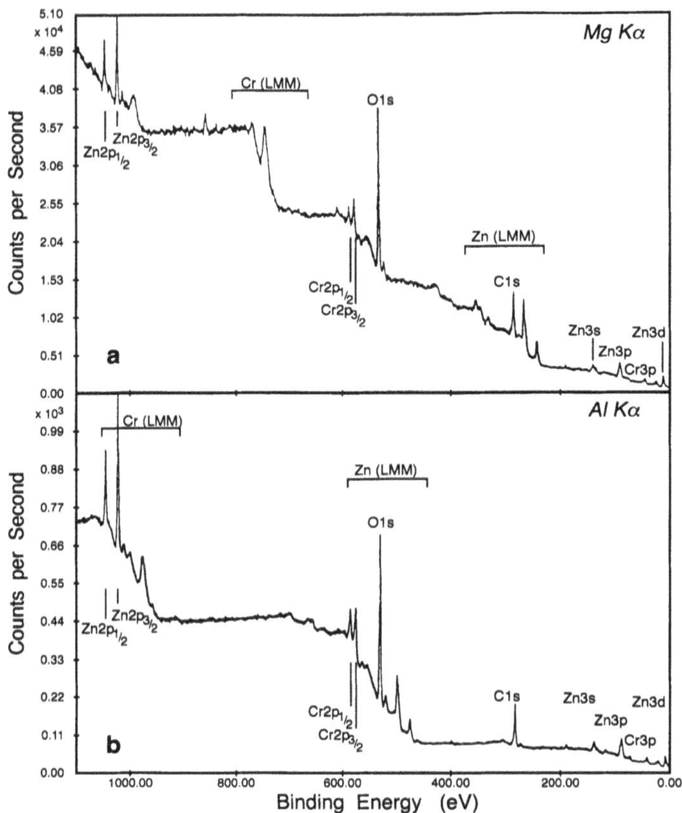

Fig. 7.10. XP spectra of galvanised steel obtained using (**a**) Mg $K\alpha$ and (**b**) Al $K\alpha$ X-ray sources. Note the relative positions of the Auger and Photoelectron lines

again involves etching, or sputtering, away the material using a rare-gas ion beam (usually Ar) and recording spectra as a function of depth. However for an XPS depth profile a relatively large surface area must be etched. This is usually necessary since the X-ray beam cannot easily be focussed to a small spot in most instruments (as discussed in Sect. 7.2.4).

Although ion sputtering can change the chemical states in a material, much information is still available from an XPS depth profile. An excellent example is the variation of the oxidation state of an element in a material. Figure 7.11 shows a simple depth profile for a CdTe film on a nickel substrate, where the tellurium has oxidised [37]. The first thing to notice is that Cd seems to be preferentially concentrated at the surface, whereas the Te which appears near the surface occurs largely as the oxide TeO_2, which disappears quickly on sputtering. The sequence of XP spectra recorded as a function of

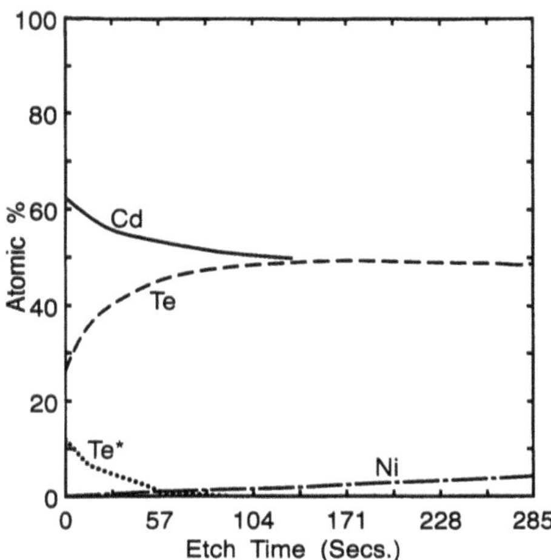

Fig. 7.11. XPS depth profile for a CdTe electrodeposited film. "Te*" represents Te arising from TeO_2

depth can be presented as a montage, as in Fig. 7.12, which highlights the relative variation of the peaks.

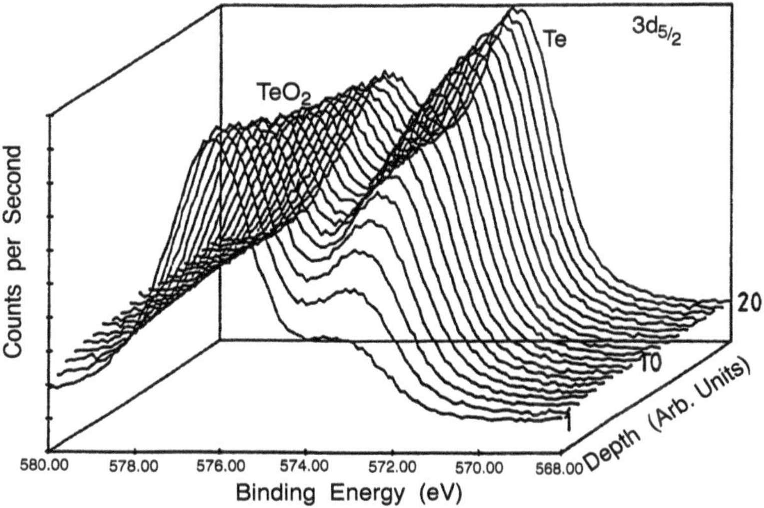

Fig. 7.12. Montage showing the variation of Te peaks at the TeO_2/Te interface as a function of depth

7.5.3 Angular Variations

While discussing surface specificity in Sect. 7.1.3, it was pointed out that the depth from which the photoelectrons emanated depended on the angle of detection, i.e. the angle of emission to the surface normal, θ. This is shown diagrammatically in Fig. 7.13, where it can be seen that detection close to the normal enhances the signal from the bulk relative to the surface, while detection close to the surface plane enhances the signal from the surface relative to the bulk. Thus varying the angle of detection can yield non-destructive depth information, an example of which is presented in Fig. 7.14, showing data for a thin film of SiO_2 on Si [48]. For low values of θ the main contribution to the spectrum is from the bulk Si, while at larger values of θ the contribution from the oxide layer becomes substantial. This approach is obviously preferable to the destructive ion etching, but is limited to very thin layers.

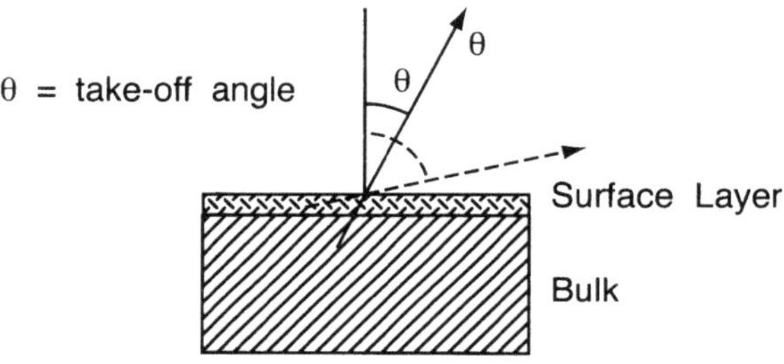

Fig. 7.13. Schematic showing surface sensitivity as a function of emission angle. Small θ enhances the signal from the bulk, while large θ enhances the signal from the surface

7.5.4 Sample Charging

When insulating materials are irradiated by X-rays they quite often develop a static charge due to their inability to replace the photoemitted electrons. This charging can be substantial, say 2–5 eV, but is always positive and quite small in comparison to electron-induced AES where charging can be of the order of several hundred eV. By using reference elements, such as a small amount of gold or silver (or, in some cases, adventitious carbon) the BE shifts can easily be calculated. In some instances, partial charging is possible, e.g. insulating domains on a conducting substrate (see Fig. 17.6 in Chapter 17). The most common approach to this problem is to use a neutralising source, such as a low energy electron flood gun, to compensate for the charge. Using a reference peak to observe the amount of BE shift, the flood gun can be tuned to provide just the right amount of current to shift the peaks back to their "uncharged" binding energies [32].

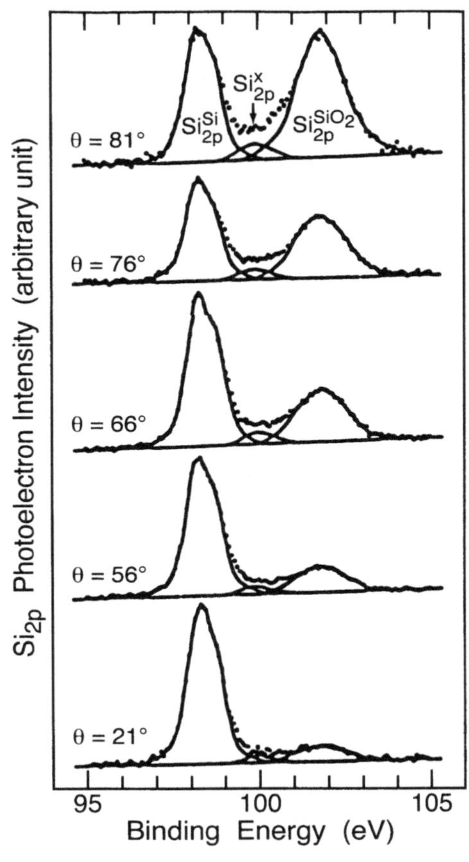

Fig. 7.14. Study by XPS of the interface between silicon and a thin (0.91 nm) film of SiO_2 on its surface. θ varies from 21° to 81° away from the normal to the surface. (Reproduced with permission from [48])

7.6 Comparison with Other Techniques

Obviously no one technique can ever solve every surface problem, and for most studies two or more techniques are combined. Compromises must be made in the use of a particular method, between, for example, resolution or sensitivity, or lateral resolution or chemical information. In order to place XPS in perspective with respect to other surface-science tools, we will compare its characteristic properties with those of two other frequently used techniques, viz. AES and secondary ion mass spectrometry (SIMS). (Refer to Tables 1.3, 1.4 for a list of the specific properties of these surface analytical techniques.)

All three spectroscopies have similar element detection capabilities, although only SIMS can detect hydrogen. XPS can detect all other elements, while AES can detect all but hydrogen and helium. Thus, except for special applications, the range of elemental detection is fairly comparable. Quantitative data can be obtained from all the methods, although standards are

Table 7.4. Applications of XPS

Area of Application	Information available	References
Corrosion and oxidation	Elemental Identification; chemical (oxidation) state of corrosion products; change of chemical composition of surface (or bulk: depth profiling) during process	[49–58]
Metallurgy / tribology	Elemental identification; analysis of alloys: effects of added lubricants; study of surface protective coatings	[10,59–66]
Thin films	Measure of film purity and thickness; details of bonding between film and substrate; modification of films during processing	[67–75]
Catalysis	Identification of intermediate species formed; oxidation state of active species; study of modification of catalyst and support material during reaction	[32,45,76–81]
Semiconductors	Characterisation of thin film coatings; identification of native oxides; interface characterisation	[82–90]
Fibres and Polymers	Elemental composition data without serious charging; information available on typical polymer groupings; shake-up satellites for indicating aromaticity; contamination identification	[91–101]
Chemisorption	Variation of chemical state of substrate and adsorbate during chemisorption; uptake curves	[102–110]
Superconductors	Valence states; stoichiometries; electron structure	[111–120]
Biomaterials	Details of bonding between biological and inorganic materials; identification of intermediate species	[121–126]

needed, however AES and XPS can provide semi-quantitative information directly from peak heights. In some situations SIMS can detect elements down to concentrations of less than 0.01 ppm (10^{-6} Atomic %), whereas XPS and AES are virtually identical in their detection ability of only 0.1 Atomic %. All three techniques are comparable when it comes to depth resolution, however AES and SIMS can achieve high lateral resolution unlike XPS. This arises because it is much easier to focus an electron or ion beam than an X-ray flux, although as discussed in Sect. 7.2.4, the ability to image with X-rays, and the resultant lateral resolution, has improved recently.

One of the major advantages of XPS is its ability to provide consistent chemical information. SIMS also provides chemical information, although more difficult to interpret, and AES chemical information is both difficult to obtain routinely and to interpret. It is difficult to study organic materials or many adsorbate systems, due mainly to electron-beam induced damage, although this is minor compared to the destructive nature of SIMS. XPS is by far the least destructive of the three techniques considered here, and X-ray damage is not often a problem. Finally, it is much simpler to eliminate static charging during XPS than when using AES or SIMS, since the latter methods rely on beams of charged particles.

In summary, the main advantages of XPS (in comparison to AES and SIMS) are its ability to provide vital chemical information, simple quantitative information, low sample damage and the fact that it can be used to study insulating materials. The only major disadvantage of XPS is its relatively poor lateral resolution which has been improved in recent times.

7.7 Applications

While there are a plethora of applications in which XPS has played a major role, the list is far too exhaustive to consider in any detail here. Instead the main categories in which XPS has been used are tabulated in Table 7.4 along with the information yielded by the technique in each case, and refcrences to some of the original, and more recent, work.

7.8 Conclusion

X-ray photoelectron spectroscopy has been used extensively over the past twenty years in many areas of surface and materials analysis. The disadvantages of the technique, such as the need for UHV and the poor spatial resolution of the X-ray source, are more than adequately outweighed by the advantages, viz. the ease of interpretation of spectra and the ability to derive chemical-state and simple quantitative information. Over the past few years the technical aspects of XPS have improved substantially, and this trend will continue into the next decade. The number of areas of application for XPS has also grown considerably, making this technique almost mandatory in modern analytical laboratories.

Acknowledgement. I would like to thank C.G. Kelly for valuable suggestions relating to this work. The permission of the Director, Telstra Research Laboratories, to publish this chapter is also acknowledged.

References

1. J.G. Jenkin, R.C.G. Leckey, J. Liesegang: J. Electron Spectrosc. **12**, 1 (1977)
2. J.G. Jenkin, J.D. Riley, J. Liesegang, R.C.G. Leckey: J. Electron Spectrosc. **14**, 477 (1978)
3. J.G. Jenkin, J. Liesegang, R.C.G. Leckey, J.D. Riley: J. Electron Spectrosc. **15**, 307 (1979)
4. J.G. Jenkin: J. Electron Spectrosc. **23**, 187 (1981)
5. K. Siegbahn, C. Nordling, A. Fahlman, R. Nordberg, K. Hamrin, J. Hedman, G. Johansson, T. Bergmark, S.-E. Karlsson, I. Lindgren, B. Lindgren: *ESCA: Atomic, Molecular and Solid State Structure Studied by Means of Electron Spectroscopy* (Almqvist & Wiksells, Stockholm 1967)
6. K. Siegbahn, G. Nordling, G. Johansson, J. Hedman, P.F. Heden, K. Hamrin, U. Gelius, T. Bergmark, L.O. Werme, R. Manne, Y. Baer: *ESCA Applied to Free Molecules* (North-Holland, Amsterdam 1969)
7. M.H. Kibel: *X-Ray Spectrometry* **19**, 73 (1990)
8. J. Chastien (Ed.): *Handbook of X-Ray Photoelectron Spectroscopy* (Perkin-Elmer Corp., 1992)
9. D. Briggs, M.P. Seah (Eds): *Practical Surface Analysis by Auger and Photoelectron Spectroscopy* (Wiley, New York 1983) Appendix 6
10. J.B. Lumsden: In *Metals Handbook*, 9th edn. Vol. **10** (1986) pp. 568–580
11. J.E. Castle, L.B. Hazell, R.H. West: J. Electron Spectrosc. **16**, 97 (1979)
12. M.A. Kelly, C.E. Tyler: Hewlett-Packard Journal **24**, 2 (1972)
13. D.E. Golden, A. Zecca: Rev. Sci. Instr. **42**, 210 (1971)
14. J.D. Lee: Rev. Sci. Instr. **43**, 1291 (1972)
15. F.J. Leng, G.L. Nyberg: J. Phys. E. **10**, 686 (1977)
16. R.G. Dromey, J.B. Peel: Aust. J. Chem. **28**, 2353 (1975)
17. D. Roy, J.-D. Carette: Canadian J. Phys. **49**, 2138 (1971)
18. J.C. Riviere: In *Practical Surface Analysis by Auger and X-Ray Photoelectron Spectroscopy*, ed. by D. Briggs, M.P. Seah (Wiley, New York 1983) Chap. 2
19. A. Barrie: In *Handbook of X-Ray and and Ultraviolet Photoelectron Spectroscopy* ed. by D. Briggs (Heyden & Sons, London 1977) Chap. 2
20. C. Klauber: Microbeam Analysis **4**, 341 (1995)
21. I.W. Drummond: Phil. Trans. R. Soc. Lond. A **354**, 2667 (1996)
22. M.P. Seah, G.C. Smith: Surf. Interface Anal. **11**, 69 (1988)
23. I.W. Drummond, T.A. Cooper, F.J. Street: Spectrochimica Acta **40B**, 801 (1985)
24. R.L. Chaney: Surf. Interface Anal. **10**, 36 (1987)
25. A.R. Walker: *A charged particle energy analyser*, European Patent Specification, Publication no. 0 243 060 B1, European Patent Office (1991)
26. G. Beamson, H.Q. Porter, D.W. Turner: Nature **290**, 556 (1981)
27. D.W. Turner, I.R. Plummer, H.Q. Porter: Phil. Trans. R. Soc. Lond. A **318**, 219 (1986)
28. P. Coxon, J. Krizek, J. Humpherson, I.R.M. Wardell: J. Electron Spectrosc. **52**, 821 (1990)
29. P.E. Larson, P.W. Palmberg: *Scanning and high resolution X-ray photo electron spectroscopy and imaging*, European Patent Specification, Publication no. 0 590 308 A2, European Patent Office (1994)

30. H.W. Ade, J. Kirz, S.L. Hulbert, E.D. Johnson, E. Anderson, D. Kern: Appl. Phys. Lett **56**, 1841 (1990)
31. U. Gelius, B. Wannberg, P. Baltzer, H. Fellner-Felegg, G. Carlsson, C.-G. Johansson, J. Larsson, P. Münger, G. Vegefors: J. Electron Spectrosc. **52**, 747 (1990)
32. P.J.C. Chappell, M.H. Kibel, B.G. Baker: J. Catalysis **110**, 139 (1988)
33. C.D. Wagner: Anal. Chem. **47**, 1201 (1975)
34. C.D. Wagner, L.H. Gale, R.H. Raymond: Anal. Chem. **51**, 466 (1979)
35. R.H. West, J.E. Castle: Surf. Interface Anal. **4**, 68 (1982)
36. J.C. Riviere, J.A. Crossley, B.A. Sexton: J. Appl. Phys. **64**, 4585 (1986)
37. M.H. Kibel, C.G. Kelly: Materials Aust. **19** (10), 15 (1987)
38. C.D. Wagner, D.E. Passoja, H.F. Hillery, T.G. Kinisky, H.J.A. Six, W.T. Jansen, J.A. Taylor: J. Vac. Sci. Technol. **21**, 933 (1982)
39. A. D'Agnano, M. Gargano, C. Malitesta, N. Ravasio, L. Sabbatini: J. Electron Spectrosc. **53**, 213 (1991)
40. R. Alfonsetti, G. De Simone, L. Lozzi, M. Passacantando, P. Picozzi, S. Santucci: Surf. Interface Anal. **22**, 89 (1994)
41. M.H. Kibel, P.W. Leech: Surf. Interface Anal. **24**, 605 (1996)
42. M. Scrocco: Chem. Phys. Lett. **63**, 52 (1979)
43. J.M. Ritchie: Phys. Rev. **106**, 874 (1957)
44. A. Barrie: Chem. Phys. Lett. **19**, 109 (1973)
45. B.A. Sexton: Materials Forum **10**, 134 (1987)
46. P.W. Leech, P.J. Gwynn, M.H. Kibel: Appl. Surf. Sci. **37**, 291 (1989)
47. P.W. Leech, M.H. Kibel, P.J. Gwynn: J. Electrochem. Soc. **137**, 705 (1990)
48. A. Ishizaka, S. Iwata: Appl. Phys. Lett. **36**, 71 (1980)
49. B.L. Maschhoff, K.R. Zavadil, K.W. Nebesny, N.R. Armstrong: J. Vac. Sci. Technol. A **6**, 907 (1988)
50. D.E. Fowler, J. Rogozik: J. Vac. Sci. Technol. A **6**, 928 (1988)
51. T. Czeppe, P. Nowak: Materials Sci. Forum **25/26**, 525 (1988)
52. T.M.H. Saber, A.A. El Warraky: J. Mater Sci. **23**, 1496 (1988)
53. E. Paparazzo: Surf. Interface Anal. **12**, 115 (1988)
54. E.W.A. Young, J.C. Riviere, L.S. Welch: Appl. Surf. Sci. **28**, 71 (1987)
55. N.S. McIntyre: In *Practical Surface Analysis by Auger and Photoelectron Spectroscopy*, ed. by D. Briggs, M.P. Seah (Wiley, New York 1983) Chap. 10, p. 397 and references therein
56. D. Landolt: Surf. Interface Anal. **15**, 395 (1990)
57. P.M.A. Sherwood: Anal. Chim. Acta **283**, 52 (1993)
58. M.F. Guimon, G. Pfister-Guillouzo, M. Bremont, W. Brockmann, C. Quet, J.Y. Chenard: Appl. Surf. Sci. **108**, 149 (1997)
59. M. Ben-Haim, U. Atzmony, N. Shamir: Corrosion Sci. **44**, 461 (1988)
60. K.L. Rhodes, P.C. Stair: J. Vac. Sci. Technol. A **6**, 971 (1988)
61. J.C. Langevoort, I. Sutherland, L.J. Hanekamp, P.J. Gellings: Appl. Surf. Sci. **28**, 167 (1987)
62. F.-M. Pan, P.C. Stair: J. Vac. Sci. Technol. A **5**, 1036 (1987)
63. R.J. Bird: Metal Sci. J. **7**, 109 (1973)
64. H.J. Brabke: Surf. Interface Anal. **14**, 686 (1989)
65. J. Jedlinski, A. Glazkov, M. Konopka, G. Borchardt, E. Tscherkasova, M. Bronfin, M. Nocun: Appl. Surf. Sci. **103**, 205 (1996)
66. I. Milosev, H.-H. Strehblow, B. Navinsek: Surf. Interface Anal. **26**, 242 (1998)

67. S. Akhter, X.-L. Zhon, J.M. White: Appl. Surf. Sci. **37**, 201 (1989)
68. J.G. Clabes, M.J. Goldberg, A. Viehbeck, C.A. Kovac: J. Vac. Sci. Technol. A **6**, 985 (1988)
69. H.M Meyer III, S.G. Anderson, Lj. Atanasoska, J.H. Weaver: J. Vac. Sci. Technol. A **6**, 1002 (1988)
70. J.L. Jordan, C.A. Kovac, J.F. Morar, R.A. Pollack: Phys. Rev. B **36**, 1369 (1987)
71. H. Ohno, T. Ichikawa, N. Shiokawa, S. Ino, H. Iwasaki: J. Mater Sci. **16**, 1381 (1981)
72. C. Argile, G.E. Rhead: Surface Sci. Rep. **10**, 277 (1989)
73. W.W. Zhao, F.J. Boerio: Surf. Interface Anal. **26**, 316 (1998)
74. A.J. Hartmann, R.N. Lamb: Current Opinion in Solid State & Mater. Sci. **2**, 511 (1997)
75. N. Bertrand, B. Drévillon, A. Gheorghiu, C. Sénémaud, L. Martinu, J.E. Klemberg-Sapieha: J. Vac. Sci. Technol. A **16**, 6 (1998)
76. G. Moretti, G. Fierro, M. Lo Jacono, P. Porta: Surf. Interface Anal. **6**, 188 (1980)
77. T.L. Barr: In *Practical Surface Analysis by Auger and X-Ray Photoelectron Spectroscopy*, ed. by D. Briggs, M.P. Seah (Wiley, New York 1983) Chap. 8 p. 283 and refences therein
78. J.S. Brinen: In *Applied Surface Analysis* ed. by T.L. Barr, L.E. Davis (ASTM, Philadelphia 1978) p. 24
79. J. Grimblot, L. Gengembre, A. D'Huysser: J. Electron Spectrosc. **52**, 485 (1990)
80. B.G. Baker, Chapter 17, and references therein
81. F. Gaillard, P. Artizzu, Y. Brullé, M. Primet: Surf. Interface Anal. **26**, 242 (1998)
82. J.H. Thomas III, G. Kaganowicaz, J.W. Robinson: J. Electrochem. Soc. **135**, 1201 (1988)
83. T. Wada: Appl. Phys. Lett. **52**, 1056 (1988)
84. R.N.S. Sodhi, W.M. Lan, S.I.J. Ingrey: Surf. Interface Anal. **12**, 321 (1988)
85. C.F. Yu, M.T. Schmidt, D.V. Podlesnik, R.M. Osgood, Jr.: J. Vac. Sci. Technol. B **5**, 1087 (1987)
86. M.D. Biedenbender, V.J. Kapoor, W.D. Williams: J. Vac. Sci. Technol. A **5**, 1437 (1987)
87. H.J. Kim, R.F. Davis, X.P. Cox, R.W. Linton: J. Electrochem. Soc. **134**, 2269 (1987)
88. T. Hattori: J. Vac. Sci. Technol. B **11**, 1528 (1993)
89. W. Palmer: Surf. Interface Anal. **22**, 331 (1994)
90. L.-Y. Chen, G.W. Hunter, P.G. Neudeck, D. Knight: J. Vac. Sci. Technol. A **16**, 2990 (1998)
91. A. Nelson, S. Glenin, A. Frank: J. Vac. Sci. Technol. A **6**, 954 (1988)
92. B. Mutel, O. Dessaux, P. Goudmand, J. Grimblot, A. Carpentier, A. Szarzbajnski: Rev. Phys. Appl. **23**, 1253 (1988)
93. D.T. Clark, D.R. Hutton: J. Polym. Sci. A **25**, 2643 (1987)
94. T.J. Hook, J.A. Cardella Jr., L. Salvati, Jr.: J. Mater Res. **2**, 117 (1987)
95. G. Gilberg: J. Adhes. **21**, 129 (1987)
96. D. Briggs: In *Practical Surface Analysis by Auger and Photoelectron Spectroscopy* ed. by D. Briggs, M.P. Seah (Wiley, New York 1983) Chap. 9, p. 359 and references therein

97. M.M. Millard: In *Industrial Applications of Surface Analysis*, ACS Symposium Series, ed. by L.A. Casper, C.J. Powell (Am. Chem. Soc. 1982) Chap. 8, p. 143
98. H.S. Munro, S. Singh: In *Polymer Characterisation* ed. by B.J. Hunt, M.I. James (Blackie, Glasgow, UK, 1993) pp 333–356
99. F.C. Loh, K.L. Tan, E.T. Kang, K. Kato, Y. Uyama, Y. Ikada: Surf. Interface Anal. **24**, 597 (1996)
100. R.W. Paynter: Surf. Interface Anal. **26**, 674 (1998)
101. E.A. Thomas, J.E. Fulghum: J. Vac. Sci. Technol. A **16**, 1106 (1998)
102. M.A. Barteau, E.I. Koh, R.J. Madix: Surf. Sci. **102**, 99 (1981)
103. G.A. Sormorjai: *Chemistry in Two Dimensions: Surfaces* (Cornell University Press, Ithaca 1981)
104. M.W. Roberts, C.S. McKee: *Chemistry of the Metal–Gas Interface* (Clarendon, Oxford 1981)
105. M.W. Roberts: Adv. Catal. **29**, 55 (1980)
106. R. Gomer (Ed.): *Interactions on Metal Surfaces* (Springer, Berlin, Heidelberg 1975)
107. C.R. Brundle, C.J. Todd (Eds.): *Adsorption at Solid Surfaces* (North-Holland, Amsterdam 1975)
108. C. Poncey, F. Rochet, G. Dufour, H. Roulet, F. Sirotti, G. Panaccione: Surf. Sci. **338**, 143 (1995)
109. S.R. Bare, K. Griffiths, W.N. Lennard, H.T. Tang: Surf. Sci. **342**, 185 (1995)
110. G.C. Allen, K.R. Hallam, J.R. Eastman, G.J. Graveling, V.K. Ragnarsdottir, D.R. Skuse: Surf. Interface Sci. **26**, 518 (1998)
111. H. Ihara, M. Jo, N. Terada, M. Hirabayashi, H. Oyanagi, K. Murata, Y. Kimura, R. Sugise, I. Hayashida: Physica C **153–155**, 131 (1988)
112. P.C. Healy, S. Myhra, A.M. Stewart: Philos. Mag. B **58**, 257 (1988)
113. A. Mogro-Campero, L.G. Turner, E.L. Hall, M.C. Burrell: Appl. Phys. Lett. **52**, 2068 (1988)
114. G.G. Peterson, B.R. Weinberger, L. Lynds, H.A. Krasinski: J. Mater. Res. **3**, 605 (1988)
115. S. Horn, J. Cai, S. Shaheen, C.L. Chang, M.L. Den Boer: Rev. Solid State Sci. **1**, 411 (1987)
116. P. Steiner, R. Courths, V. Kinsinger, I. Sander, B. Siegivart, S. Hüfner, C. Politis: Appl. Phys. A **44**, 75 (1987)
117. M.S. Golden, S.J. Golden, R.G. Edgell, W.R. Flavell: J. Mater. Chem. **1**, 63 (1991)
118. C.R. Brundle, D.E. Fowler: Surface Sci. Rep. **19**, 143 (1993)
119. D. Klissurski, V. Rives: Appl. Catal. A **109**, 1 (1994)
120. A. Hartmann, G.J. Russell, D.N. Matthews, J.W. Cochrane: Surf. Interface Anal. **24**, 657 (1996)
121. B.D. Ratner, Cardiovasc. Pathol. **2**, 87S (1993)
122. B.D. Ratner, B.J. Tyler, A. Chilkoti: Clin. Mater. **13**, 71 (1993)
123. J.C. Lin, S.L. Cooper: Polymer Prepr. (Am. Chem. Soc., Div. Polymer Chem.) **34**, 659 (1993)
124. D.S. Sutherland, P.D. Forshaw, G.C. Allen, I.T. Brown, K.R. Williams: Biomaterials **14**, 893 (1993)
125. J. Lausmaa, J. Electron Spectrosc. **81**, 343 (1996)
126. R.A. Brizzola, J.L. Boyd, A.E. Tate: J. Vac. Sci. Technol. A **15**, 773 (1997)

8 Vibrational Spectroscopy of Surfaces

R.L. Frost and N.K. Roberts

There are many forms of vibrational spectroscopies applied to surfaces but the most accessible techniques are Fourier transform infrared spectroscopy and Raman spectroscopy. They will be discussed in this chapter. Other vibrational spectroscopies (e.g. PAS, EELS) are listed in the Appendix and Bibliography.

8.1 Introduction

Infrared spectroscopy has been a widely used technique in industry for the structural and compositional analysis of organic, inorganic and polymeric samples and for quality control of raw materials and commercial products. With the advent of Fourier transform infrared spectroscopy (FTIR) the range of applications and the materials amenable to study has increased enormously, owing to its increased sensitivity, speed, wavenumber accuracy and stability. Oils, coal, shale, polymers, paints, catalysts, pharmaceuticals and industrial gases have been successfully analyzed [1]. More recently FTIR has been developed for the quality control of minerals and in exploration [2]. Cellulosic materials, wood, paper and cotton have also proved amenable [3]. It is also now possible to study electrode processes using FTIR [4,5]. Biochemical processes once excluded from the field of infrared spectroscopy owing to the presence of water now form one of the fastest growing areas of research in FTIR. Specific examples are discussed in the text.

Conventional infrared spectroscopy relies on the dispersion of an infrared beam via a grating into its monochromatic components, and slowly scanning through the entire spectral region of interest. When a sample is placed in the beam, various wavelengths of infrared radiation are absorbed by the sample as the beam is scanned, and the result recorded as the infrared spectrum of the sample. To maintain a resolution of 4 wavenumbers, this typically requires around 2–3 minutes for a single scan of the spectrum.

Fourier transform infrared (FTIR) spectroscopy relies on a totally different principle to record the same information – that of interferometry. A Michelson or Genzel interferometer forms the basis of the FTIR spectrometer. Figure 8.1 shows a comparison of the optics for dispersive and Michelson interferometric instruments. The FTIR instrument consists of a standard infrared source, collimation mirrors, a beamsplitter, a fixed mirror and moving mirror. 50% of the beam is passed through the beamsplitter and 50% is reflected both on the initial pass and from reflections from the two mirrors. This

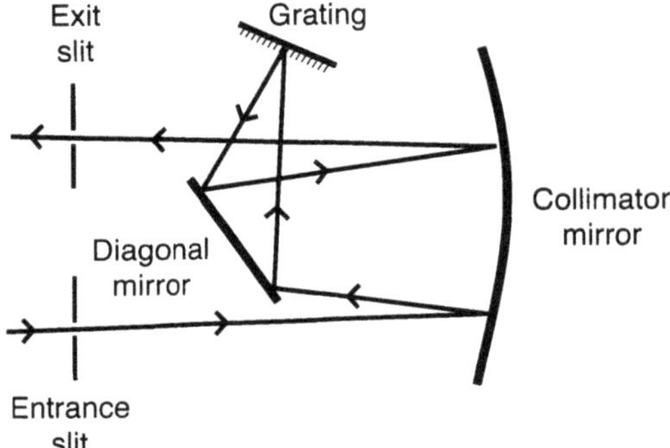

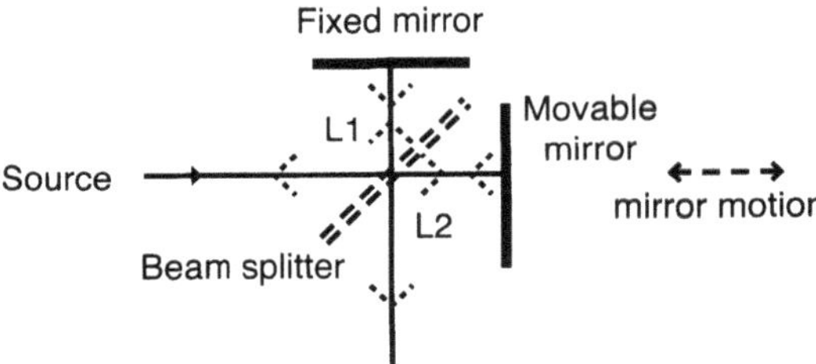

Fig. 8.1. Comparison of optics for dispersive and Michelson interferometric instruments

results in the potential for interference patterns to be generated by varying the path length that one portion of the beam travels before recombination at the beamsplitter.

When $L2 = L1 + n\lambda$, constructive interference occurs, while when $L2 = L1 + n\lambda/4$, destructive interference occurs. For a monochromatic light source this results in an "interferogram" of intensity versus distance travelled in the form of a cosine wave. A normal infrared source covers a wide range of wavelengths and the interferogram obtained is a combination of all the individual interference patterns. Only at the point $L1 = L2$ are all waves in phase together, giving a strong "centreburst". The further away from this point one travels in either direction the various wavelengths tend to cancel giving a reduced signal.

The interferogram is detected as a function of mirror travel at the detector. For example, two optical velocities, $0.31\,\mathrm{cm\,s^{-1}}$ and $1.2\,\mathrm{cm\,s^{-1}}$, are possible with Digilab FTS-2OE instrument. A choice of optical velocities is necessary for the photoacoustic technique described later, usually within this range. Any substance which absorbs infrared radiation modifies the interferogram according to the wavelengths that are absorbed. Thus all the absorption information is contained within a single interferogram (the "Fellgett" advantage) which at a resolution of 4 wavenumbers/cm can be obtained in about 2 seconds (corresponding to a moving mirror travel of 0.125 cm). However, this only becomes useful when the wavelength information is "unscrambled" from the interferogram by the mathematics of the Fourier transformation resulting in a "single beam" spectrum of intensity versus wavenumber (or wavelength if desired). The "single beam" spectrum represents the combination of the source and beamsplitter characteristics and any absorbances due to components in the air inside the spectrometer. In the FTS-2OE a purge of dry air is maintained to minimize the contribution from water vapour while carbon dioxide is not removed and hence a large absorbance is observed. The spectral information is not usually presented in this format. Instead, a "reference" spectrum is taken of the system without the sample, and the subsequent sample spectrum is ratioed to this producing either a transmission spectrum or an absorption spectrum.

One of the important features of FTIR spectroscopy is the ability to signal average a large number of scans in a relatively short amount of time. Thus spectra on very small quantities of material, or on highly absorbing materials, in which the signal to noise ratio of individual scans is very poor, can be achieved. The signal to noise ratio improves as the square root of the number of scans taken. It is thus possible to acquire 10 000 scans overnight to achieve a 100-fold improvement in the signal to noise ratio. In practice, the interferograms are signal averaged first, and a single Fourier transform then done, as the Fourier transform can take up quite significant amounts of computing time.

The second major advantage of FTIR over conventional dispersive instruments is the high throughput of infrared radiation, since narrow slits are no longer necessary to achieve resolution. This is the so-called "Jacquinot" advantage. In a dispersive instrument measuring in the range 4000–400 cm^{-1} at a resolution of 8 cm^{-1} only 8/3600 or 0.2% of the incident radiation reaches the detector. For 1 cm^{-1} resolution this is reduced to 0.25%. The FTIR spectrometer is also capable of very high resolution of absorption bands. Higher resolution is achieved by moving the mirror further while maintaining the same starting point. By allowing mirror movement of 5 cm, resolution of 0.1 wavenumbers is possible, taking 30 s per scan on the FTS-2OE, and over 30 minutes to compute the Fourier transform. More recent FTIR spectrometers are faster. Moreover, the interferometric derived frequencies are internally calibrated by a laser giving greater frequency reproducibility. This advantage is very important in multiple scanning needed for low energy signals. FTIR spectroscopy, therefore, has two main advantages over conventional IR spectroscopy, improved sensitivity and improved computational ability from having the data in digital form in a powerful computer [6,7].

8.2 Surface Techniques

The main techniques for studying surfaces with FTIR spectroscopy are,

a) Specular and grazing angle reflectance
b) Diffuse reflectance (DRIFT)
c) Attenuated total reflectance (ATR)
d) Photoacoustic spectroscopy (PAS)
e) Infrared emission spectroscopy

Of these techniques (a), (b) and (c) have been used successfully with dispersive IR instruments on solid samples. FTIR has extended infrared spectroscopy to aqueous solutions using (c) and enabled (d) to be carried out in the IR region; previously PAS was only possible in the UV-visible region. Both ATR and PAS allow depth profiling of the surface. (e) IES enables the FTIR spectra of surfaces and thin films to be obtained at elevated temperatures and in situ.

a) Specular and Grazing Angle Reflectance. This technique has rarely been used because of the usual requirement for a collimated beam, the strict requirement for sample flatness and the difficulty of interpreting spectra dominated by the effect of the absorbing species on the complex refractive index. Its main application is in the area of surface coatings, such as lubricants, adhesives and paints.

b) Diffuse Reflectance is used for powdered samples. Although little or no sample preparation is required, the technique has the disadvantage that specular reflection may give rise to spurious spectral features. It is important that the samples be highly scattering. Coal, pharmaceutical products, foodstuffs and mineral samples are amenable to this technique.

c) Attenuated total reflectance may be used to study

i) the surface of solid samples, provided good contact can be achieved between the sample and the ATR element. Depth profiling is also possible by varying the angle of incidence provided it does not exceed the critical angle. This technique is ideal for solid samples that can make good contact with the ATR element, in this class fall rubber, polymer films fabrics, coated and painted surfaces.
ii) ATR has been used with excellent results to investigate the solid water interface. The application of FTIR-ATR Spectroscopy to aqueous systems has opened up a whole new area of study, from the clotting mechanism of blood on foreign surfaces to the mode of action of flotation collectors on mineral surfaces.

d) Photoacoustic Spectroscopy has the great advantage of requiring very little sample preparation and in contrast to the ATR method the sample does not have to be in contact with an element. Higher surface/volume ratios of the sample increase the signal intensity. Thus powdered samples and rough surface morphologies are more favorable in PAS. The technique has been used with great success in the depth profiling of solid surfaces. Previously variable angle ATR was the only method for depth profiling and the maximum depth was of the order of a micron. PAS allows maximum depths of many microns by using low mirror velocities. As with the ATR method the sampling depth is wavelength dependent, making quantification of surface concentrations difficult. PAS is best suited to intractable solid samples. Of the four techniques only ATR has found any application in aqueous systems. Biological applications are now dominating the field of FTIR-ATR. Previously dispersive instruments make aqueous systems inaccessible to IR spectroscopy. A more detailed description of these techniques follows with some specific applications.

e) Fourier Transform Infrared Emission Spectroscopy. The technique of measurement of discrete vibrational frequencies *emitted* by thermally excited molecules, is known as **Fourier transform Infrared Emission Spectroscopy** (FTIR ES or simply IES) [18,19]. IES has the great advantage that no sample preparation is involved apart from micronising the sample to $< 1\,\mu m$ size. In the IES technique, the heated sample acts as a source of the infrared radiation and so no IR source is required. The technique is particularly useful for the study of, for example, thin films, polymers, catalyst and mineral surfaces. The major advantage of IES is that the samples are measured *in situ* at *elevated temperatures.* Hence the technique removes the uncertainties introduced by heating the sample to elevated temperatures and quenching before measurement, as IES measures the surface phenomena as it is actually taking place. Thus phase changes, dehydration, dehydroxylation, surface adsorption and desorption may be readily studied. The technique when combined with self-absorption is useful for the study of chemical changes on catalyst surfaces.

8.2.1 Diffuse Reflectance Infrared Fourier Transform (DRIFT)

The theory of diffuse reflectance at scattering layers within powdered samples has been developed by *Kubelka* and *Munk* [16,17]. For an "infinitely thick" layer, the Kubelka–Munk equation may be written as

$$f(R_\infty) = \frac{(1 - R_\infty)^2}{2R_\infty} = \frac{k}{s} \tag{8.1}$$

where R_∞ is the absolute reflectance of the layer, s is a scattering coefficient, and k is the molar absorption coefficient. In practice there is no perfect diffuse reflectance standard and R_∞ is replaced by R'_∞ where

$$R'_\infty = \frac{R'_\infty \text{ (sample)}}{R'_\infty \text{ (standard)}} . \tag{8.2}$$

R'_∞ (sample) represents the single-beam reflectance spectrum of the sample and R'_∞ (standard) is the single-beam reflectance spectrum of a selected non-absorbing standard exhibiting high diffuse reflectance throughout the wavelength regions being studied. Potassium chloride or potassium bromide with a particle size of less than 10 μm is often used as a standard. The depth of the KCl or KBr matrix required to yield samples of "infinite thickness", i.e where further increasing the thickness causes no change in the spectrum, is less than 5 mm. The Kubelka–Munk theory predicts a linear relationship between the molar absorption coefficient, k, and the peak value of $f(R_\infty)$ for each band, provided s remains constant. Since s is dependent on particle size and range, these parameters should be made as consistent as possible if quantitative data are needed.

Diffuse reflectance can be used with dispersive IR instruments, however the low signal level and the poor sensitivity limits the range of applications. The advent of FTIR spectroscopy has made the technique more accessible. The samples must be highly scattering, that is, the optical depth, $\mu\beta$, must be larger than the geometrically average path length, μD, through the sample. For scattering samples μD is a function of particle size. Light impinging on an opaque, nonscattering sample is completely absorbed, except for a small reflected component ($\sim 7\%$ at normal incidence). If the surface is rough, a diffuse reflectance accessory ($\sim 20\%$ efficient) is required to collect this small amount of light. The great advantage of DRIFT is the short preparation time and the ability to obtain a strong spectrum with less than 100 μg of sample in a suitable matrix. Diffuse Reflectance has been successfully applied to finely powdered samples, however it is not suitable for samples in the form of coarse powders, granules or intractable lumps. High pressure and temperature DRIFT cells are now available for studying in situ catalytic reactions at surfaces. Samples may also be used without being mixed with a matrix powder, but considerable distortions in relative peak heights within the spectrum may occur if the diffuse reflectance accessory does not completely eliminate the specularly reflected radiation.

A successful application of this technique is ex-situ investigation of the adsorption of the flotation collector, amylxanthate, on the mineral pyrite [8]. For the more interesting and relevant in-situ investigations this method cannot be used and the ATR method is the method of choice (see next section). Figure 8.2 shows the DRIFT spectra of pyrite after wet grinding at natural pH (5.1) and adsorption of the collector amylxanthate. The sample, of which 50% was less than 7 μm, was dispersed in KBr. Above monolayer coverage bands appear at 1026 cm^{-1} (CS stretching mode) 1259–1261 cm^{-1} (C-O-C

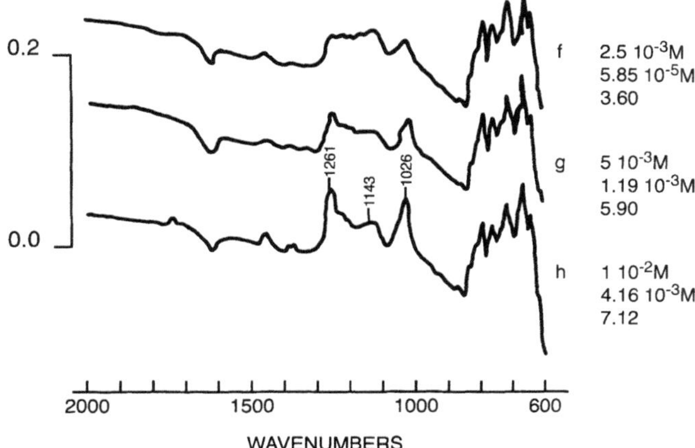

Fig. 8.2. Diffuse reflectance spetra of pyrite after wet grinding at natural pH (5.1) and adsorption of amylxanthate with a conditioning time of 10 min. For each spectrum are given the initial and final concentration, and the statistical surface coverage

stretching in dixanthogen) and bands due the hydrocarbon chain at 1347 and 1465 cm^{-1}. The bands at 1746 and 1716 cm^{-1} indicate the presence in the adsorbed layer of the dimer of amylmonothiocarbonate from an oxydegradation reaction. The bands between 600 and 900 cm^{-1} are due to the oxidation products on pyrite surface after grinding and before contact with the xanthate solutions.

8.2.2 Attenuated Total Reflectance Spectroscopy (ATR)

Attenuated total reflectance (ATR) spectroscopy, sometimes known as internal reflection spectroscopy (IRS), was developed independently by *Harrick* and *Fahrenfort* [9] in 1960. The theory and practice of the technique is described extensively by *Harrick* (1967) [10]. If a beam of radiation strikes the interface of two media of refractive indices n_1 and n_2 from the side of an IRS crystal (n_1) at any angle greater than the critical angle, it is totally reflected and a standing wave is established at the crystal-sample interface. The intensity of the radiation that penetrates into the rarer medium (n_2) falls off exponentially from its value I_0 at the surface, and the total reflection is attenuated (Fig. 8.3). The depth of penetration of the sample is given by the relation,

$$d_{\mathrm{p}} = \frac{\lambda}{2\pi}(\sin^2\theta - n_{21}^2)^{1/2} \tag{8.3}$$

where d_{p} = depth at which the intensity is reduced to 1/e of its original value. θ is the angle of incidence and n_{21} the ratio of the refractive indices of

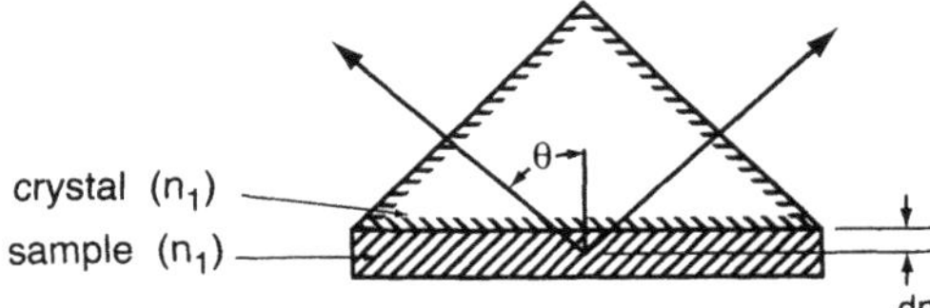

Fig. 8.3. Schematic of the attenuated total internal reflection process

the sample (n_2) and the internal reflection element (n_1). Consequently the penetration is greater at longer wavelengths. Over the mid-IR range (2.5–25 μm) there is a factor of 10 difference in sampling depth. Furthermore as θ gets larger, d_p gets smaller and approaches $0.1\,\lambda$ at grazing incidence ($\theta = 90°$) for high index media. Theoretically d_p is infinitely large at the critical angle. It is not possible to vary the angle continuously and several elements of fixed angle, usually 30°, 45° and 60°, are employed, provided they do not exceed the critical angles. The sensitivity of the ATR technique can be markedly increased by using multiple internal reflection elements. A typical crystal 50 mm long with its end faces cut at the crystal angle can produce between 20 and 30 reflections. If l is the crystal length and d its thickness, the number of reflections is given by

$$N = l/d \cot\theta \,. \tag{8.4}$$

The most common materials employed as ATR elements are thallium bromoiodide-5, germanium, zinc selenide, silicon, silver bromide, synthetic sapphire and diamond (used mainly in the far infrared region). Table 8.1 lists typical depths of penetration of the infrared beam for ATR elements of germanium and KRS-5 of different crystal angles in contact with polyethylene [1].

Thus one of the major differences between transmission and ATR spectra is the dependence of the intensity of the spectral bands on wavelength in ATR spectra since d_p is larger at longer wavelengths. There is also distortion on the longer wavelength side. Programs are available for correcting for these effects.

Table 8.1. Penetration depths (d_p) in μm for polyethylene for various ATR elements

		Wavenumber [cm^{-1}]			
ATR element	Angle	2000	1700	1000	500
KRS-5	45	1.00	1.18	2.00	4.00
KRS-5	60	0.55	0.65	1.11	2.22
Ge	30	0.60	0.71	1.20	2.40
Ge	45	0.33	0.39	0.66	1.32
Ge	60	0.25	0.30	0.51	1.02

ATR spectroscopy can therefore study the surfaces of samples in the range of a fraction of a μm to several μm provided intimate and uniform contact is maintained with the ATR element. It is particularly important to maintain uniformity of contact if further data processing such as spectral subtraction is to be made. A torque wrench to press the sample against the ATR crystal element should be used to obtain reproducible contact. The surfaces of the ATR element should be clean and highly polished. For the best results the sample should be pliable and have a smooth surface. In practice, a spectrum is run of the ATR element alone then in contact with the sample. The latter spectrum is then ratioed to the former, since FTIR instruments are single beam instruments, producing the spectrum of the surface of the sample alone. Depth profiling of surfaces is possible by varying the crystal material and the angle of incidence. An interesting comparison of ATR and PAS depth profiling of some biocompatible polymer surfaces showed that PAS was more sensitive to surface impurities and segregation [11]. The outstanding advantage of the FTIR-ATR spectroscopy is the ability to study the solid/aqueous solution interface, previously inaccessible to dispersive IR spectrometers because water is a strong IR absorber. Two examples will suffice to show its capability.

The first example is protein adsorption on foreign surfaces from whole blood. Interactions between flowing blood and a foreign surface are of great practical significance when medical polymers such as those used in heart valves, indwelling catheters, dialysis membranes, vessel grafts, and other artificial organs are implanted in the body. Most typical polymers in contact with blood give rise to a number of complex reactions culminating in a blood clot (thrombus). FTIR-ATR spectroscopy has contributed significantly to our understanding of clot formation [12,13]. Since FTIR spectra can be collected in fractions of a second, the kinetics of the process can be readily studied. Figure 8.4 contains the spectrum of flowing blood, a reference saline blank,

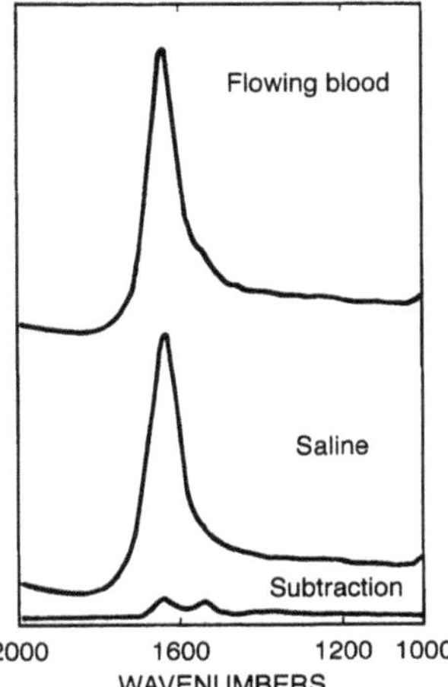

Fig. 8.4. Spectra collected during flowing blood experiment: *top:* representative spectrum collected during run *middle:* spectrum of saline blank *bottom:* subtraction of flowing blood spectrum minus saline spectrum

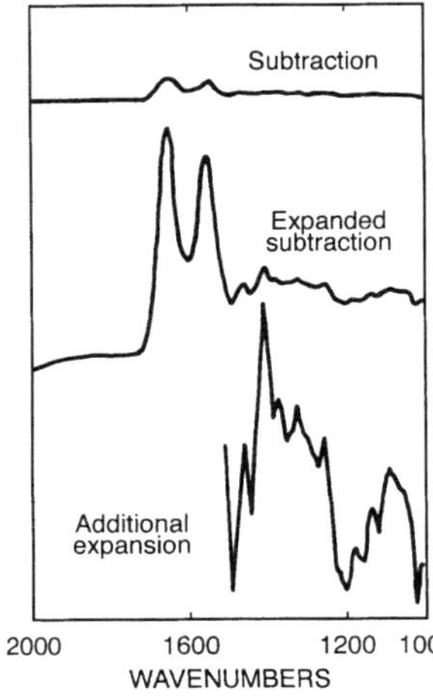

Fig. 8.5. Expansion of difference spectrum Fig. 8.4

and the spectrum obtained by subtracting the latter from the former. Figure 8.5 is a continuation of Fig. 8.4, showing additional expansion of the subtraction spectrum. These spectra were obtained in 0.8 s. The presence of the amide I (1650 cm^{-1}) and amide II (1550 cm^{-1}) peaks due to proteins so close to the strong OH binding vibration of water (1640 cm^{-1}) shows that successful spectral subtraction can be obtained in less than 1 s close to strong interfering peaks. Figure 8.6 shows the total amount of protein adsorbed, as measured by the amide II peak (1550 cm^{-1}), as a function of time. Careful analysis of the spectrum below 1550 cm^{-1} allows a more detailed eludication of the proteins adsorbed. The ATR crystal may be bare, as in this case, or coated with a very thin film of polymer (100–1000Å) for a study of its biocompatibility.

The second application of FTIR–ATR spectroscopy is the adsorption of flotation collectors on mineral surfaces. FTIR–ATR spectroscopy is particularly suitable for studying the in situ adsorption of flotation collectors on mineral surfaces. Previously only ex situ methods were available for studying the process, i.e. the mineral with adsorbed collector was removed from the aqueous environment, dried and the spectrum recorded. However, virtually nothing is known about the effect of removing the water from the sample. FTIR–ATR spectroscopy allows ready subtraction of the water and mineral spectra to leave only the spectrum of the adsorbate in situ [14]. The experimental procedure is as follows. Firstly, the mineral is vacuum evaporated onto an ATR element to produce a thin film ($\sim 0.05\,\mu m$). A word of caution is required here. It is important to be sure that the mineral is unchanged during evaporation. The term "thin film" is a misnomer, as it is known that the "thin film" consists of very small crystallites. Evaporation of tin (IV) oxide produces a variety of oxides. Fortunately in this case the lower oxides

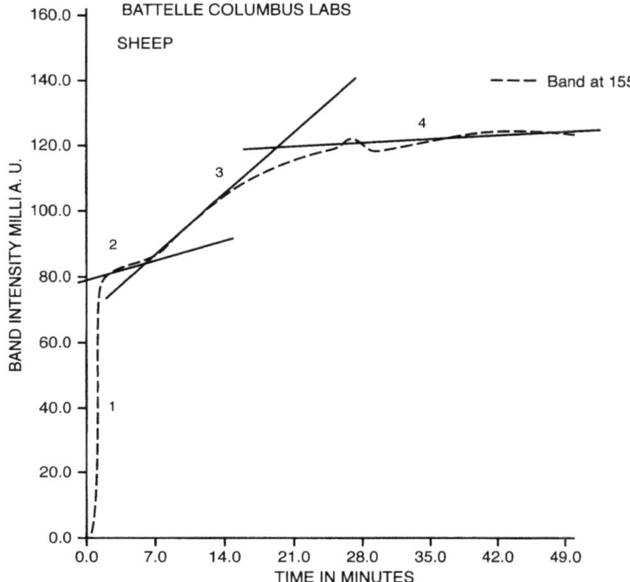

Fig. 8.6. Plot of amide II ($1550\,\text{cm}^{-1}$) band intensity versus time of blood flow showing total amount of protein adsorbed in the time period indicated. Spectra collected every 5 s

may be converted to tin IV) oxide by annealing in oxygen. Fluorite and other minerals have been successfully produced as "thin films". If thin films cannot be obtained it may be possible to obtain spectra on suspensions of the mineral, e.g. Goethite, or use a transmission technique if the mineral is infrared transparent in the appropriate region, for example sphalerite. Secondly, a spectrum is run of the ATR element coated with the mineral film in the presence of water. The experimental set-up is shown in Fig. 8.7. The water is then

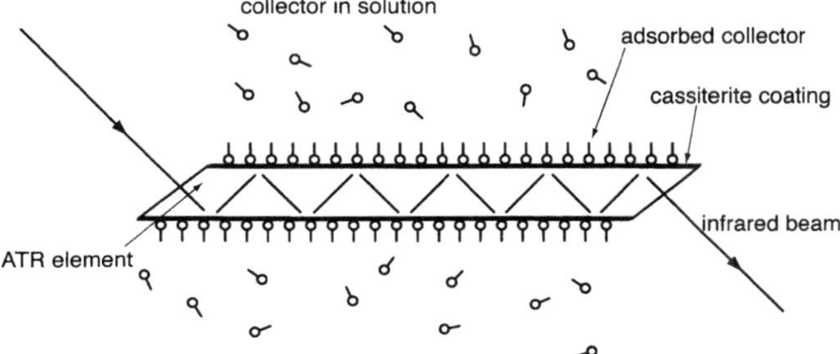

Fig. 8.7. Schematic representation of the use of ATR method to study collector adsorption

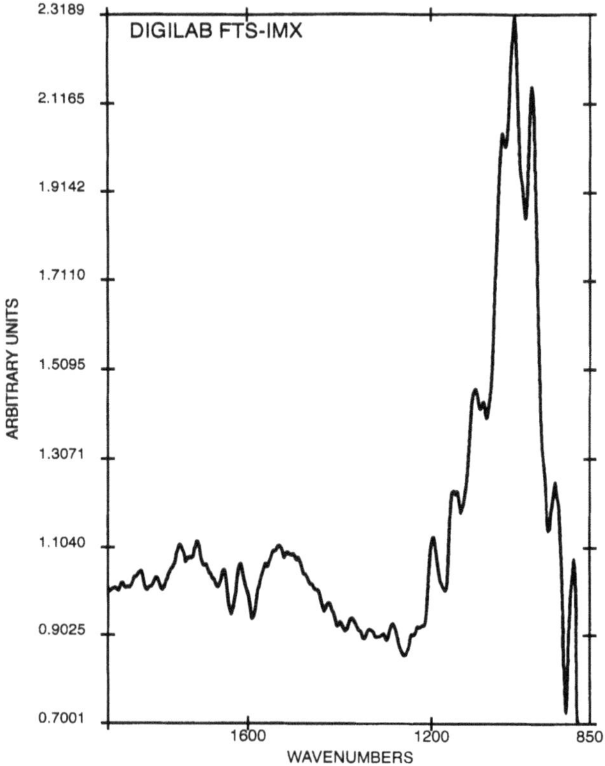

Fig. 8.8. FTIR-ATR spectrum of 300 ppm SPA adsorbed on a thin film (~ 0.05 μm) of tin (IV) oxide at pH 4.5 (100 scans)

pumped out and replaced with the collector solution. Spectra may then be run at appropriate intervals and difference spectra obtained showing the adsorbate on the mineral surface. Figure 8.8 shows the result of the adsorption of styrene phosphoric acid (SPA) on cassiterite. Figure 8.9 is the spectrum of styrene phosphoric acid. Comparison with the corresponding SPA complexes shows a close resemblance between the adsorbate species and the corresponding tin (IV) and titanium (IV) complexes. On adsorption of the collector the bands due to the $\nu(P{=}O)$, $\nu_{as}(P{-}O)$ and $\nu_{s}(P{-}O)$ vibrations of the phosphonic acid group are replaced by a complex broad band due to the resonance stabilized phosphonate group. The kinetics of the adsorption process may be followed in situ along with the effect of such variables as pH, temperature and ionic composition. At the same time spectra of the solution species may be obtained for comparison with the adsorbed species.

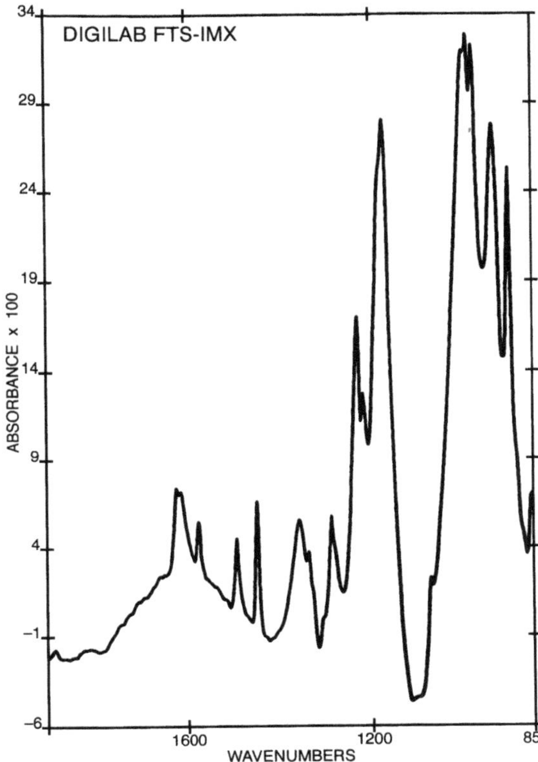

Fig. 8.9. FTIR spectrum of SPA in KBr disc

In the pH range studied the mechanism of adsorption is as follows:

R—P(=O)(O$^-$)(OH) + (HO)$_2$Sn< (surface) —slow→ R—P(=O)(OH)—O—Sn(OH)< + OH^-

monoanion of SPA; cassiterite surface

↓ fast

R—P(=O)(—O—)$_2$Sn< + H_2O

bidentate surface complex of SPA

8.2.3 Photoacoustic Spectroscopy (PAS)

UV-visible photoacoustic spectroscopy has been used for some time to investigate solid samples; however, PAS was not applied to the IR region until the advent of FTIR. Only when photoacoustic detection was combined with an interferometer and the large capacity data system in the FTIR instrument could PAS be used in the IR region. The signal to noise (S/N) ratio in dispersive instruments with PAS is far too low to make it a useful technique.

The solid sample is placed in a sealed cell of small volume containing an infrared transparent gas such as helium to carry the photoacoustic signals. The IR radiation absorbed by the solid sample is converted to heat by a radiationless transfer process. When the heat propagates to the sample surface and is transferred at the sample–gas interface into the surrounding gas, pressure variations of the gas are generated because the intensity of the IR radiation is modulated by the interferometer at an audio frequency region. The Michelson interferometer modulates the IR beam as does a chopper in a dispersive IR instrument. The gas expands and contracts at the modulation frequency resulting in a pressure wave within the sealed cell. The photoacoustic signal is detected by a sensitive microphone and preamplified in the cell. Subsequently the signal is Fourier-transformed to yield a single beam FTIR/PAS spectrum. The single beam spectrum is referenced to that of carbon black since it can absorb all IR radiation and convert it to photoacoustic signals. A schematic represention of the PAS sample cell is shown in Fig. 8.10.

The strength of the signal I_{PAS} given by the relation,

$$I_{\mathrm{PAS}} = \text{number of photons} \times \text{energy per photon} \times \text{thermal diffusivity}$$

The thermal diffusivity of the sample depends on the particular modulated frequency of the interferometer incident on the sample. The thermal diffusion length μ_s, in cm can be calculated from the following equation,

$$\mu_s = (2k/\rho c\omega)^{1/2} , \tag{8.5}$$

where k is the sample's thermal conductivity in cal. cm s^{1-} C^{-1}, ρ is the density of the sample in g cm^{-3}, and ω is the angular modulation frequency in radian s^{-1}. The angular modulation frequency can be calculated from the modulation frequency, f, as follows;

$$\omega = 2\pi f , \quad f = u\bar{\nu} \tag{8.6}$$

where u is the optical velocity of the interferometer in cm s^{-1} and $\bar{\nu}$ is the infrared frequency in wavenumbers (cm^{-1}). For a Michelson interferometer a mirror displacement of x results in an optical path difference of $2x$, i.e. the "optical velocity" is twice the "mirror velocity". For FTIR instruments using a Genzel interferometer the "optical velocity" is four times the "mirror velocity", so they achieve the same resolution for half the mirror travel. For example if the "mirror velocity" is 0.155 cm s^{-1} (equivalent to an "optical velocity" of 0.31 cm s^{-1} with a Michelson interferometer) then the modulation

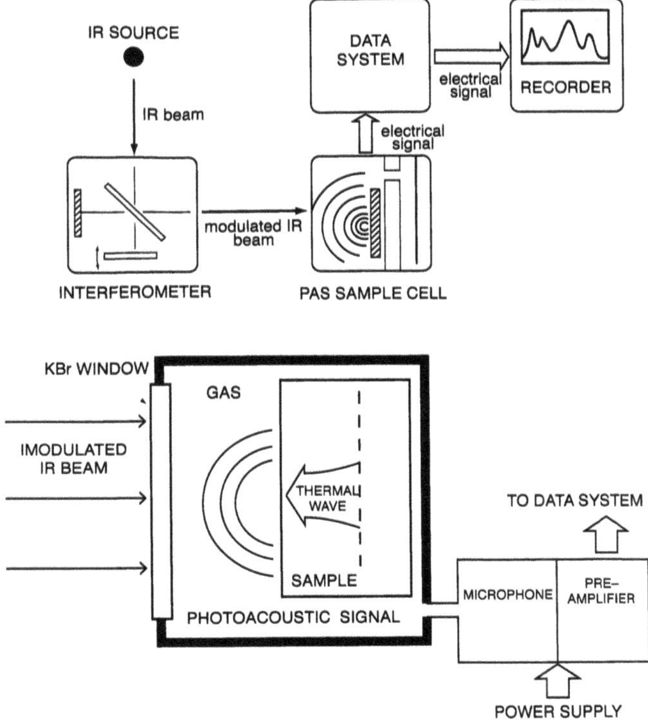

Fig. 8.10. Schematic of FTIR/PAS sample cell

frequencies at $1000\,\mathrm{cm\,s^{-1}}$ and $4000\,\mathrm{cm\,s^{-1}}$ will be 320 and 1280 Hz respectively, i.e. in the audio frequency range. That is, as far as the FTIR instrument is concerned, each frequency component of the infrared source contributes a frequency component to the interferogram. In principle one can obtain the spectra of different layers of the sample by varying the mirror velocity of the Michelson interferometer. The lower the mirror velocity the greater the depth of penetration. However if the radiation is absorbed within a length that is smaller than the thermal diffusion length of the sample, photoacoustic saturation sets in. Therefore one must be extremely careful interpreting the results of PAS experiments on depth profiling particularly at low modulation frequencies as varying photoacoustic saturation effects of different bands in the spectrum may occur.

An interesting application of PAS is the non-destructive depth profiling of cellulose materials and polymers to determine the penetration of chemical additives. In the textile industry, sizing agents are applied to warp yarns to increase their resistance to abrasion during weaving. After weaving the sizing agents are removed through a desizing process which produces desired

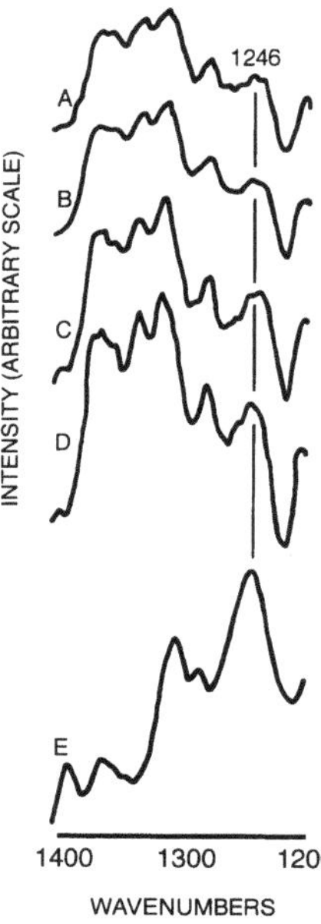

Fig. 8.11. Photoacoustic infrared spectra of cotton yarn sized with polyurethane collected at *(A)* $0.235\,\mathrm{cm\,s^{-1}}$, *(B)* $0.396\,\mathrm{cm\,s^{-1}}$, *(C)* $0.665\,\mathrm{cm\,s^{-1}}$, *(D)* $1.119\,\mathrm{cm\,s^{-1}}$, *(E)* polyurethane sizing agent

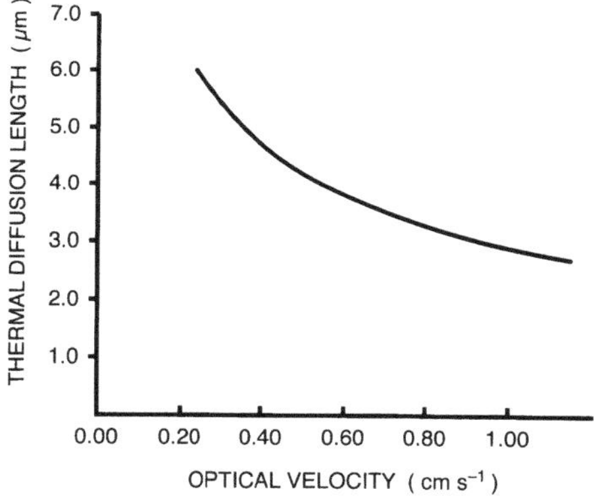

Fig. 8.12. Thermal diffusion length of cellulose at $1730\,\mathrm{cm^{-1}}$ versus the optical velocity of the interferometer

properties. Figure 8.11 shows the PAS-FTIR spectra of cotton yarn sized with a polyurethane at different interferometer velocities [15].

The intensity of the peak at $1246\,\mathrm{cm^{-1}}$, due to the polyurethane sizing agent, increases dramatically as the optical velocity changes from $0.235\,\mathrm{cm\,s^{-1}}$ to $1.119\,\mathrm{cm\,s^{-1}}$, indicating that the agent is concentrated at the surface. Figure 8.12 shows the thermal diffusion length of cellulose at $1730\,\mathrm{cm\,s^{-1}}$ for different optical velocities. Hence it may be concluded that the polyurethane sizing agent is more concentrated in the top 2.8 μm and gradually decreases to a depth of 6.8 μm.

8.2.4 Infrared Emission Spectroscopy (IES)

Normally, when using dispersive infrared spectroscopy, the infrared beam is split into two beams, one the reference beam and the other the measuring beam. The resultant absorbance is the negative log of the ratio of the intensities of the measured to reference intensities. In a Fourier transform instrument the background is measured as a single beam and the measured spectrum ratioed to this background spectrum. In IES, the emission spectrum is actually this single beam reference or background spectrum, modified by the instrument response function. Emission spectra are obtained by ratioing the emission of the sample to that of a reference usually a black body source, which emits a continuum of radiation according to Planck's law. Real samples will always emit less energy than that of the black body. Consequently for any given temperature, the ratio of the energy emitted by the sample to that of the black body at any given wavelength is the emissivity.

FTIR emission spectroscopy may be carried out on a FTIR spectrometer, modified by replacing the IR source with an emission cell. A description of the cell and principles of the emission experiment have been published elsewhere [18]. Approximately 0.2 mg of, for example, a finely powdered mineral is spread as a thin layer on a 6 mm diameter platinum surface and held in an inert atmosphere within a nitrogen purged cell with an additional flow of argon over the sample, during heating. The emission spectra may be collected at intervals of either 5, 25, 50°C or 100°C, over the range 100–1000°C, and then repeated on cooling at 300 and 100°C. The time between scans (while the temperature is raised to the next hold point) is approximately 10 seconds. The spectra are usually acquired by coaddition of 256 scans for temperatures 100–300°C (approximate scanning time 140 seconds) and 64 scans for temperatures 400–1200°C (approximate scanning time 45 seconds), with a nominal resolution of $4\,\mathrm{cm}^{-1}$. The emittance spectrum at a particular temperature is calculated by subtraction of the single beam spectrum of the platinum backplate from that of the platinum + sample, and the result ratioed to the single beam spectrum of an approximate black body (graphite). Emittance values vary from 0 to 1 with a scale equivalent to a transmission spectrum. The data are linearised with respect to concentration where required, by transforming to units of $-\log_{10}[1-\mathrm{emittance}(\nu)]$. Spectral manipulation such as baseline adjustment, smoothing and normalisation may be performed using the applications software package GRAMS (Galactic Industries Corporation, Salem, NH, USA).

Applications. IES has many and varied applications. To illustrate the use of IES, two examples from firstly the synthesis of a molecular sieve (a pillared montmorillonite) and secondly an example of a mineral surface of industrial importance are used. A typical set of IES data is shown in Figure 8.13 which displays thc infrared spectra of an Al_{13} pillared montmorillonite from 200 to 750°C at 50°C intervals. To show the advantages of the technique, the letters

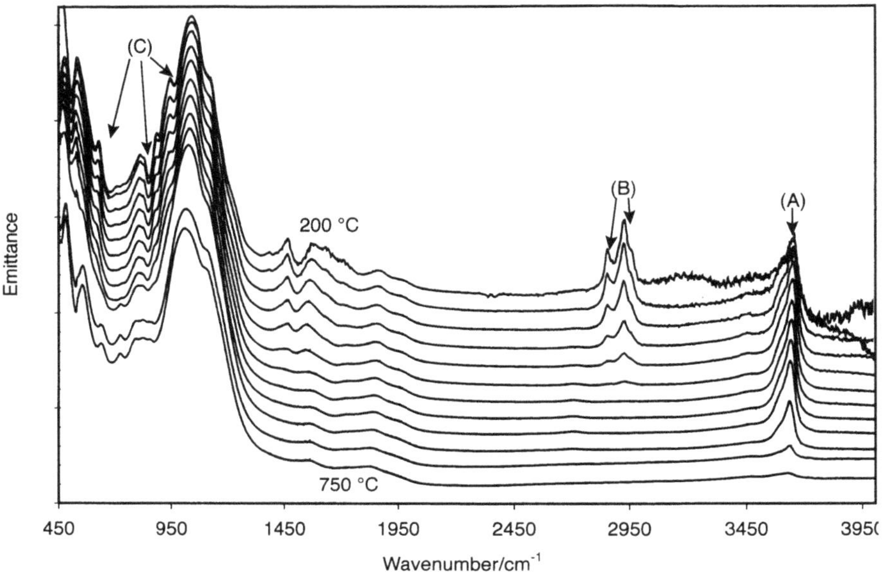

Fig. 8.13. Infrared emission spectroscopy of an aluminium 13 pillared montmorillonite

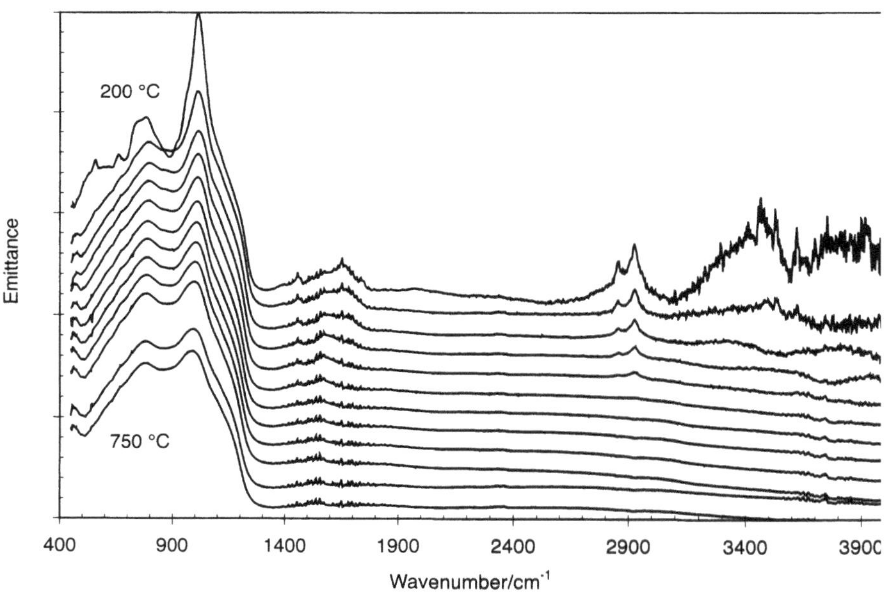

Fig. 8.14. Infrared emission spectra of a natural gibbsite

show the types of changes that are observed upon thermal treatment. Spectral changes at [A] report the dehydroxylation of the montmorillonite, at [B] the combustion of organics in the sample and at [C] the changes in the structure as the clay is pillared. A second example is the thermal treatment of natural gibbsite (Figure 8.14). Clearly the processes of dehydration, dehydroxylation, combustion of organics and the phase change of gibbsite to an aluminaphase may be observed.

8.3 Fourier transform Raman spectroscopy

Vibrational spectroscopy enables the study of surfaces of materials at the molecular level. The techniques of infrared spectroscopy and Raman spectroscopy are complementary: infrared probes the vibrational state transitions involving changes in dipole moment; Raman spectroscopy those transitions involving changes in polarisability. Thus vibrations which involve both a change in dipole moment and polarisability will provide signals in both the infrared and Raman. Vibrations which involve a change in polarisability only will be detected only in the Raman. The advantages of using Raman spectroscopy lie in the fact that it detects primarily symmetric vibrations and results in simpler spectra without overtone and combination bands. Raman allows simpler sampling arrangements so that solids can be studied without the preparation of, for example, a pellet or mull as in IR. The disadvantages of Raman spectroscopy rest with the weakness of the signal, the occurrence of thermal or photodegradation and the occurrence of fluorescence, which may swamp the Raman signal.

Fourier transform Raman spectroscopy (FTRS) is based on a novel technique, in which instead of using visible or UV light as is the case with conventional dispersive Raman spectroscopy, excitation with near IR radiation and detection by a Fourier transform infrared spectrometer is used [20,21]. The approach of using near IR excitation radiation combined with the conventional FTIR spectrometer confers a number of major advantages: fluorescence is greatly reduced, signal to noise is improved by the co-adding of scans and the use of the longer wavelength of the excitation radiation reduces the problems associated with the photo-degradation under the incident laser.

The information obtained from Raman spectral studies is normally not the same as that from infrared spectroscopy as the Raman technique generally measures symmetric vibrations and the physics of the Raman emission is very different from the physics of infrared absorption where the spectral results depend on changes in dipole moments. Raman spectral studies complement infrared studies and provide additional information. Raman spectroscopy can provide information on the low frequency region of the mineral surfaces [22–24]. Such information can be difficult to obtain using mid-range FTIR spectroscopy. FTRS has allowed the study of, for instance, clay mineral surfaces that were previously difficult to measure because of laser induced fluorescence and provided ready access to the extensive data handling facilities that are available with a commercial FT-IR spectrometer.

FTRS spectra may be obtained using several instruments: e.g. the Perkin-Elmer 2000 series Fourier Transform spectrometer fitted with a Raman accessory or the Biorad series 2 FTIR spectrometer. The Perkin-Elmer 2000 series FTIR spectrometer equipped with a Raman accessory comprises a Spectron Laser Systems SL301 Nd-YAG laser operating at a wavelength of 1064 nm, and a Raman sampling compartment incorporating 180° optics. The FT-IR spectrometer contains a quartz beam splitter capable of covering the spectral range 15,000-4000 cm^{-1}. The Raman detector is a highly sensitive indium-gallium-arsenide detector and is operated at room temperature. Under these conditions Raman shifts would be observed in the spectral range 4000–150 cm^{-1}. Spectra are corrected for instrumental function and detector response. Raman spectra may also be obtained using a Biorad series 2 FTIR spectrometer equipped with a Raman accessory comprising a Spectraphysics T10-1064C Nd-YAG diode laser operating at 1064 nm. Measurement times of between 0.5 and 2 hours are used to collect the Raman spectra with a signal to noise ratio of better than 100/1 at a resolution of 2 cm^{-1}. A laser power of 100 mW is used. This power is low enough to prevent damage to most surfaces, but was found to be sufficient to produce quality spectra in a reasonable time. No significant heating, as may be evidenced by the lack of thermoluminescent background, is observed. Raman spectra are collected as single beam spectra and are corrected for instrumental effects. Spectral bands are ratioed to the instrument profile function determined by recording the spectrum of a calibrated grey body, calibrated at the National Standards Laboratory at C.S.I.R.O., Linfield, Sydney, Australia.

Applications. The ability of FT RS to adequately measure the Raman spectra of kaolinite and other clay mineral surfaces is apparently very much dependent on the crystallinity at the surface. The more crystalline the mineral surface, the more easily the FT Raman spectrum of the clay is determined. The spectra of powdered or ground samples are more difficult to obtain. This is not to say that disordered or powdered samples cannot be measured, it simply means that data collection times are longer. The intensity of the FT Raman scattering depends on the Raman cross section, which depends on the long range order of the kaolinite crystals. Consequently the FT Raman spectra of disordered clays are more difficult to obtain. Figures 8.15 and 8.16 illustrate the FT Raman spectra of the hydroxyl stretching and low frequency region of a series of minerals. These minerals are often used for adsorption studies, surface modification and alteration through a range of chemical reactions to prepare mesoporous materials. The differences in the hydroxyl stretching frequencies may be readily observed. Each mineral also has its own characteristic low frequency spectrum [25].

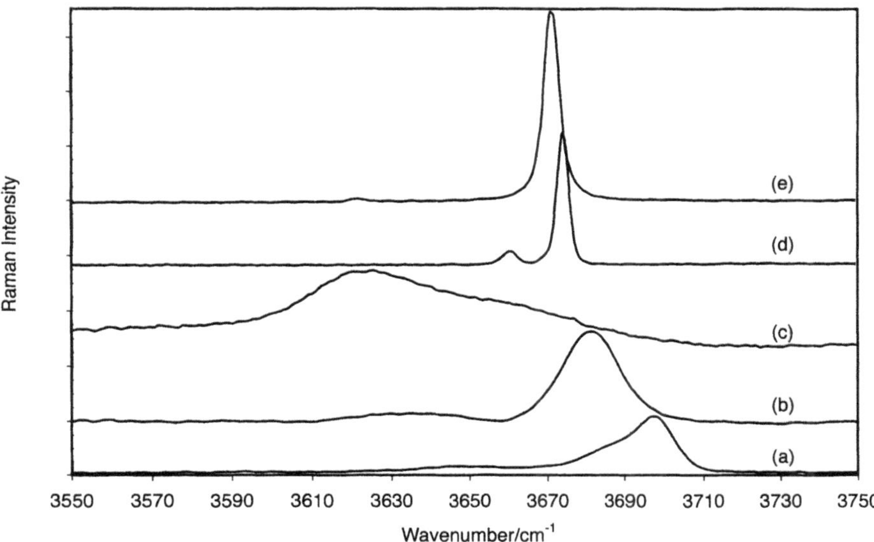

Fig. 8.15. Raman spectra of the hydroxyl stretching region of *(a)* chrysotile, *(b)* montmorillonite, *(c)* vermiculite, *(d)* tremolite, *(e)* pyrophllite

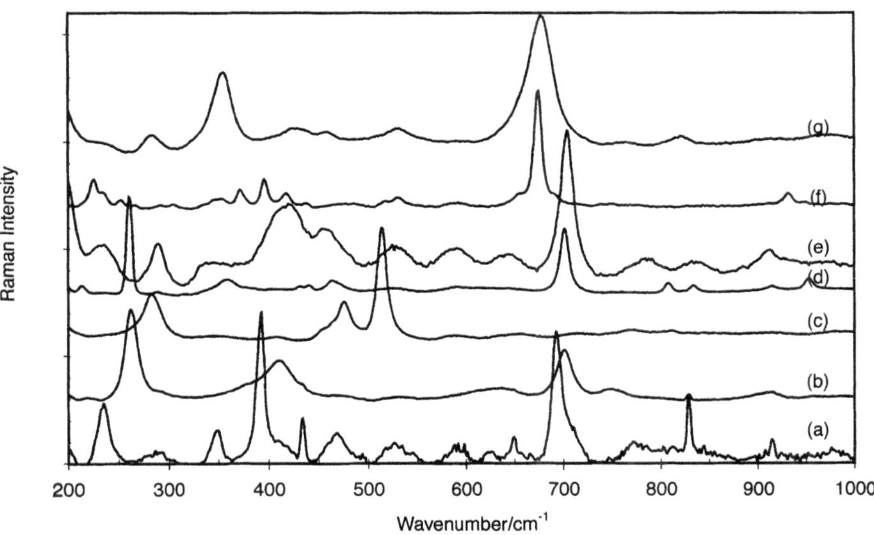

Fig. 8.16. Raman spectra of the low frequency region of *(a)* chrysotile, *(b)* orthoclase, *(c)* pyrophllite, *(d)* montmorillonite, *(e)* muscovite, *(f)* tremolite, *(g)* vermiculite

8.4 Raman Microscopy

Although the technique of Raman microscopy has been in existence for a considerable length of time, the application of the technique to the study of material and mineral structure and surfaces has been limited. The advent of low energy near IR lasers and the use of charge coupled devices (CCD) has meant that the Raman microprobe can be used for the analysis of materials and in particular mineral surfaces. For spectra at 298 K, minerals are placed on a polished stainless steel surface on the stage of an Olympus BHSM microscope, equipped with 5×, 20× and 50× objective lenses. No sample preparation was needed. The microscope is part of a Renishaw 1000 Raman microscope system, which also includes a monochromator, a filter system and a charge coupled device (CCD). Raman spectra are excited by a Spectraphysics model 127 HeNe laser (633 nm), recorded at a resolution of 2 cm^{-1} in sections of 1000 cm^{-1} for 633 nm excitation. Repeated acquisitions using the highest magnification, are accumulated to improve the signal to noise ratio in the spectra. For the 298 K spectra, data are collected at 20 s intervals for 10 minutes at maximum magnification. Spectra are calibrated using the 520.5 cm^{-1} line of a silicon wafer. It should also be noted that the filters in the Renishaw spectrometer start to eliminate the Rayleigh line at about 150 cm^{-1}. This makes the determination of bands below 200 cm^{-1} difficult and unreliable.

Spectra obtained with the microprobe normally consist of Raman bands superimposed on a background, which is a combination of the fluorescence and the instrument function. In the case of the HeNe laser and the measurement of the hydroxyl stretching region, the background is predominantly due to fluorescence and consists of a sloping linear baseline that is easily corrected. For the HeNe laser and the determination of the low frequency region, the background is predominantly a combination of the instrument function that is dependent on the behaviour of the notch filters, the grating-detector responses, and the fluorescent background. This background is induced by elements placed in the optical path between the sample and the entrance slit of the spectrometer. The background, although a complex function, is easily measured using the incandescent white light from the microscope. The measured spectra are then ratioed to this instrumental background to give the spectra as illustrated in the figures in this chapter. Such a method ensures the correct instrumental function is taken into account and that the true spectrum is measured. Different instrumental profiles are obtained for the operation of different lasers.

Normally with conventional dispersive Raman spectroscopy and to a lesser extent, FTRS data collection times can be excessively long, particularly for clay minerals. Often several thousand FT Raman spectra must be coadded to obtain a spectrum of sufficient quality. In the case of Raman microprobe spectroscopy with multiplex detection, data collection times are very short and spectra are easily obtained within 1 to 10 minutes. Typically a collection time of 60 seconds, in which 10 spectra are co-added, is used with the maximum magnification. Power at the sample is in the 0.1 to 1 mW range. This use of such low power is a major advantage particularly over that used for FT RS where typically the power used for clays is 100 mW for a spot size of 200 μm. For Raman microscopy, the laser spot size is 0.8 μm. Such low power means that there will be no significant heating effects and consequential damage to the clay mineral.

One of the disadvantages of Raman microprobe analysis is the fluorescence of mineral samples, which is particularly pronounced when using the HeNe laser at 633 nm and other lasers of similar wavelengths. Each of the laser lines used to excite the Raman spectra of the mineral surface causes fluorescence that subsides with time. Even so, difficulty in measuring the low frequency region with the 633 nm excitation can be experienced and the use of the diode laser operating at 780 nm does not alleviate the problem. Obtaining spectra at 77 K produces an instrumental background shift which may be related to the change in the fluorescence of the sample at liquid nitrogen temperature.

An advantage of Raman spectroscopy is that bands in the 200 to 400 cm^{-1} region are easily measured whereas such bands are not easily observed using FTIR techniques. Dispersive Raman spectroscopy inherently has the mineral sample spectrum superimposed on the intense Rayleigh line. The use of Raman microscopy employing CCD detectors is greatly advantageous because of signal to noise improvement. Detectors in both FT Raman spectrometers and conventional dispersive instruments are photon noiselimited. With a CCD detector this is no longer true: typically the noise from a CCD detector is at least 100 times less than from a semiconductor detector. The disadvantage of using Raman spectroscopy operating at the diode laser frequency of 780 nm is the loss of scattering efficiency. Raman scattering decreases as the fourth power of the wavelength of the scattered radiation, so the intensity of the Raman scattering at 780 nm compared to say 532 nm is ~5 times less. For the 633 nm (HeNe) laser, this factor is 2. Nevertheless, the disadvantage of this loss of efficiency is greatly outweighed by the advantages of the CCD detector. Further the advantage of using lasers operating in the 633 or 780 nm range rests with the reduction of the laser induced fluorescence which is a problem when using green or blue excitation wavelengths.

Applications. Raman microscopy lends itself to the study of aqueous sols and gels and surfaces in contact with water. Water is a weak scatterer and is not readily detected using Raman spectroscopy. Raman spectroscopy also has the advantage that there is no sample preparation and a very small sample is required. To illustrate the applications of Raman spectroscopy, an example of a kaolinite surface modified with dimethyl sulphoxide (DMSO) is used. The preparation involves reacting the kaolinite with an aqueous solution of DMSO. Figure 8.17 shows the changes in the Raman spectra of the hydroxyl stretching region for both a low and high defect kaolinite which has

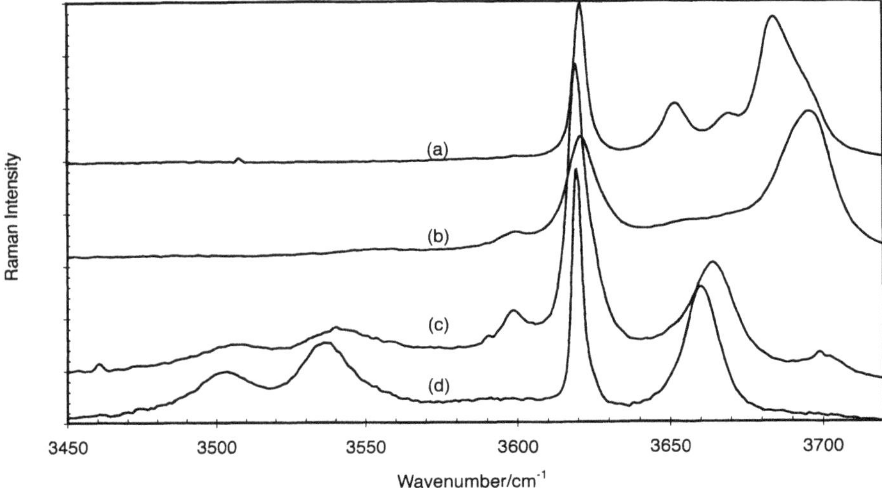

Fig. 8.17. Raman spectra of the hydroxyl stretching region of *(a)* low defect kaolinite, *(b)* high defect kaolinite, *(c)* DMSO intercalated low defect kaolinite, *(d)* DMSO intercalated high defect kaolinite

been intercalated with dimethylsulphoxide. Several features may be observed: (a) the decrease in intensity of the hydroxyl stretching bands attributed to the inner surface hydroxyl groups near $3660\,cm^{-1}$; (b) The appearance of additional bands at $\sim 3660\,cm^{-1}$ in the intercalated kaolinite which are attributed to the formation of the hydrogen bond complex formed between the inner surface hydroxyl groups and the DMSO; (c) The appearance of new bands at 3505, 3540 and $3600\,cm^{-1}$. These bands are attributed to the incorporation of water into the kaolinite surfaces. Figure 8.18 shows the modification of kaolinite surfaces through intercalation with potassium acetate under a range of conditions. Spectrum (a) is the spectrum of the untreated kaolinite; spectrum (b) is that of the kaolinite surfaces modified through hydrogen bonding with potassium acetate at 1 bar and 25 °C, spectrum (c) is that at 1 bar and 100 °C, spectrum (d) at 20 bars and 220 °C and (e) at 2 bars and 120 °C. Intercalation of the kaolinite through hydrogen bonding of the inner surface hydroxyls with potassium acetate causes a dramatic decrease in intensity of the bands at 3693, 3685, 3668 and $3650\,cm^{-1}$. At the same time the appearance of a new band at $3605\,cm^{-1}$ attributed to the inner surface hydroxyl groups hydrogen bonded with the acetate is observed. Intercalation under the conditions of 1 bar and 100 °C caused the kaolinite to show an increase in defect structures. These spectra show the suitability of Raman microscopy to the study of mineral surfaces and the changes in the Raman spectra brought about by the surface phenomena.

One of the major advantages of Raman spectroscopy is that it does not easily detect the presence of water. Thus aqueous systems and wet samples

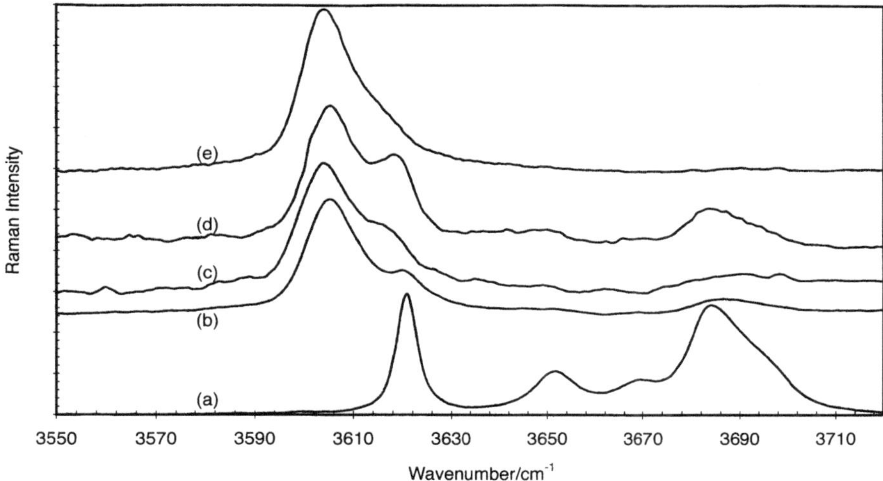

Fig. 8.18. Raman spectra of the hydroxyl stretching region of *(a)* low defect kaolinite (ldk), *(b)* ldk intercalated with KAc at 1 bar and 25°C *(c)* at 1 bar and 100°C *(d)* at 2 bars and 200°C

may be measured. This has particular application to the study of sols and gels. Because water is a strong infrared absorber, infrared spectra from components of aqueous systems are very difficult to obtain. Further silicate glasses have weak Raman spectra and the Raman spectra of aqueous sols may be measured *in situ* in a glass sample container [26,27]. The spatial resolution of a Raman microprobe is about 1 μm. This may be compared with infrared microscopy where because of the wavelength of light used, the spatial resolution at best is 20 μm. The full potential application of Raman spectroscopy to the study of surfaces is yet to be realised.

Acknowledgement. The assistance of Dr K Kuys in obtaining and checking references is gratefully noted.

References

1. T. Theophanides (ed.): *Fourier Transform Infrared Spectroscopy* (Reidl, Dordrecht, 1984)
2. P.M. Fredericks, J.B. Lee, P.R. Osborn, D.A.J. Swinkels: Appl. Spectrosc. **39**, 311 (1985)
3. C.Q. Yang, R.R. Bresee, W.G. Fateley, T.A. Perenich: *The Structure of Cellulose*, ACS Symposium Series Vol. 340 (1987)
4. A. Bewick, S. Pons: *Advances in Infrared and Raman Spectroscopy*, Vol. 12 (Wiley, New York 1985)
5. S.I. Yaniger, D.W. Vidrine: Appl. Spectrosc. **40**, 174 (1986)

6. J.R. Ferraro, L.J. Basile (eds.): *Fourier Transform Infrared Spectroscopy* Vols. 1–3 (Academic, New York 1978–82)
7. P.R. Griffiths, J.A. Haseth: *Fourier Transform Infrared Spectrometry* (Wiley, New York 1986)
8. J.M. Cases, P. de Donato, M. Kongolo, L. Michot: Colloids and Surfaces **36**, 323 (1989)
9. J. Fahrenfort: Spectrochim. Acta **17**, 689 (1961)
10. N.J. Harrick: *Internal Reflection Spectroscopy* (Wiley, New York 1967)
11. J.A. Gardella, G.L. Grobe, W.L. Hopson, E.M. Eyring: Anal. Chem. **56**, 1169 (1984)
12. S. Winters, R.M. Grendreau, R.I. Leininger, R.J. Jacobsen: Appl. Spectrosc. **36**, 404 (1982)
13. R.I. Leininger, T.B. Hutson, R.J. Jacobsen: Annal. New York Acad. Sci. **516**, 173 (1987)
14. K.J. Kuys, N.K. Roberts: Colloids and Surfaces **24**, 1 (1987)
15. C.Q. Yang, R.R. Brisee, W.G. Fateley: Appl. Spectrosc. **41**, 889 (1987)
16. P. Kubelka, F. Munk: Z. Tech. Phys. **12**, 593 (1931)
17. P. Kubelka: J. Opt. Soc. Am. **38**, 448 (1948)
18. A.M. Vassalo, P.A. Cole-Clarke, L.S.K. Pang, A. Palmisano: J. Appl. Spectrosc. **46**, 73 (1992)
19. R.L. Frost and A.M. Vassallo: Clays and Clay Minerals **44**, 635 (1996)
20. D.B. Chase: J. Am Chem Soc. **108**, 7485 (1986)
21. D.J. Cutler: Spectrochimica Acta **46A**, 131 (1990)
22. R.L. Frost, J.R. Bartlett, P.M. Fredericks: Spectrochimica Acta **49A**, 667 (1993)
23. R.L. Frost: Clay and Clay Minerals **43**, 191 (1995)
24. R.L. Frost: Clay Minerals **32**, 73 (1997)
25. R.L. Frost and Llew Rintoul: Applied Clay Science **11**, 171 (1996)
26. Paul A. Venz, John R. Bartlett, Ray L. Frost and James L. Woolfrey: Spectrochimica Acta **53A**, 969 (1997)
27. Venz, JR Bartlett, RL Frost and JL Woolfrey: Raman spectroscopic studies of titania hydrolysates, sols and gels. Sol-Gel Processing of Advanced Materials, Ceramic Transactions Vol. 81, 35–46 (L.C.Klein, E.J. Pope, S. Sakka and J.L. Woolfrey, Eds.) (1998)

9 Rutherford Backscattering Spectrometry and Nuclear Reaction Analysis

S.H. Sie

9.1 Introduction

The RBS (Rutherford Backscattering Spectrometry) and NRA (Nuclear Reaction Analysis) are a subset of what is generally known as ion beam analysis (IBA) methods, performed with energetic (typically a few hundred keV to a few MeV) ion beam from accelerators. The energies involved are such that the interaction between the projectile and the target is insensitive to molecular or atomic shell effects, and the methods are thus thus not suitable for measurements of chemical effects. The methods are non-destructive, and provide the elemental composition and/or structure, namely the depth profiles from the surface region spanning the first few hundred layers of atoms to a few microns depth. Other main advantages are the rapidity of analysis (few minutes), and the direct and simple way the information can be obtained from the data. The methods are amenable to simple calibration procedures to facilitate quantitative, standardless analysis. RBS is based on elastic Coulomb scattering between the projectile and the target nuclei, and is usually applied to obtain data for most if not all elements present in the specimen. In contrast, NRA is based on nuclear reactions which are element specific. The most commonly used beam in RBS is ^4He (alpha particles) with 1–4 MeV energies. Protons are also used for RBS, typically with energies between 100 keV and 2 MeV, but the beam is more suited for NRA applications. The range of depths accessible by these methods is about 1–10 microns, beyond which complications due to straggling effects result in loss of sensitivity and/or of depth resolution. At very low energies, the effect of screening of the atomic electrons must be taken into account and this is covered in another section. The other main IBA method not discussed here is PIXE (Particle Induced X-ray Emission), which yields stoichiometric information over even greater depths (20–30 microns) and high sensitivity (parts-per-million detection limit).The methods are most suitable for characterization of mainly inorganic compounds as thin films or bulk samples, giving composition variation or impurity distribution as a function of depth with resolution of few tens of Angstrom (NRA) to a few hundreds Angstrom (RBS). Applications to organic or biological samples are limited by radiation and thermal damage to the specimen. Analysis of conducting samples is straightforward, whereas for semi- or non-conducting material, a conductive coating has to be applied on the surface to prevent charge build-up, which can affect the result. Typically a thin layer of carbon is applied; Au is often used as well. The choice is determined to minimize interference with the analyzed material. In RBS, the element identification is poor for high Z due to limits in the detector resolution, and the sensitivity for light element detection is also limited, especially in the presence of high Z matrix. NRA is limited to only those elements exhibiting strong nuclear reactions, occurring mainly in light nuclei. Interpretation of data can be ambiguous when the sample is not homogeneous, for instance when precipitation, segregation or clustering of impurities in the matrix occur. The surface texture of the sample must be smooth for accurate measurements. In RBS surface roughness would affect the result, in that in effect the target presents micro-surfaces at various angles to the incident beam, thus smearing the results. This can in fact be used to measure surface roughness. If microbeams are available, surfaces with rough texture can still be analyzed if the undulation is larger than the microbeam size. In NRA, surface roughness will degrade the depth resolution at the surface, and non-flatness can result in error due to angular distribution effects.

The measurements are usually carried out in vacuum. While high vacuum (pressure in the 10^{-5} to 10^{-6} mbar range) is quite adequate for typical measurements, lasting a few minutes, an ultra high vacuum may be required for prolonged measurement on the same spot to prevent carbon build-up on the target due to break-up of residual hydrocarbons in the chamber under beam bombardment. This would affect NRA in shifting the beam energy, and may cause interference in RBS when measurements of light elements (e.g. O) are involved. External beam measurements are possible, where the beam is brought out of the accelerator through a thin window or through differentially pumped aperture. In such case the RBS detector is usually operated in an inert gas atmosphere to minimize background problems. At any rate the resolution of the detector suffers, and this kind of measurements are usually conducted only for special cases, e.g. in-situ measurements of large objects.

Table 9.1 summarizes the salient features of the methods to be used as a rough guide. Special cases can extend the general limits shown. The usefulness of the methods is enhanced with microbeams, which for energetic beams are typically limited to a few microns in diameter for viable beam intensities.

In the following, basic principles of the methods are presented and examples of applications to both simple and complex samples are given as illustrations. The scope is limited to applications in stoichiometric analysis, and thus the channelling method will not be discussed in detail. For more detailed treatment of the subject, and all other IBA methods, the reader can refer to a number of excellent handbooks, the latest being 'Handbook of Modern Ion Beam Materials Analysis' edited by Tesmer and Nastasi [1]. Proceedings of conferences based on the theme of ion beam analysis are also excellent resource material and usually contain the repertoire of the applications in various disciplines. One of the most well known series is the biennial

Table 9.1. Important parameters and capabilities of RBS and NRA

	RBS	NRA
Typical probing beam	H, He^4	H, He^4, ^{15}N, ^{19}F
Beam energies	0.2–4 MeV	0.5–20 MeV
Beam currents	1–100 nA	10–1000 nA
Sample structure: layers	yes	yes
bulk	yes	yes
Depth resolution [μm]	> 0.03	> 0.01
Range [μm]	> 10	< 3
Sensitivity [ppm]	> 500	> 100
Element range: Z	> 1	< 30
Z discrimination: high Z	poor	–
low Z	good	good

International Ion Beam Analysis conference, now approaching its 20th year. The latest proceedings published as an issue of Nuclear Instruments and Methods is given in [2].

9.2 Principles

Both RBS and NRA involve measurements of energy and flux of the radiations resulting from the interaction of the ions and the atoms of the material under analysis. In RBS, they are the beam projectiles themselves after being scattered by the target; and in NRA they could be gamma-rays or fragments of the reacting nuclei. Unless stated otherwise the general nomenclature is as follow:

Z_1, m_2 – ion beam projectile atomic number and mass (in amu)
Z_2, m_2 – target atomic number and mass
E – incident ion beam energy (MeV)
q – charge state of the incident ions
e – electronic charge (1.6×10^{-19} Coulomb)
I – ion beam current (Amperes)
F – ion beam flux $= \int I\, dt/eq$
N – target molecular density (molecules/cm^3)
s – solid angle of the detector (steradians) subtended at the target.

Analysis of the spectrum requires a knowledge of the stopping power and the cross section of the reaction. The stopping power is defined as the rate with which the beam particles lose energy in traversing the target material. The cross-section is defined as the probability of an incident beam particle producing the reaction product of interest for individual atoms in the target, expressed as an effective area presented to the beam by the target atoms. The yield from a layer of the target where the beam of energy E loses δE is given by:

$$dY = N\frac{d\sigma}{d\omega}\delta E \Big/ \frac{dE}{dx} \tag{9.1}$$

where:

$\frac{d\sigma}{d\omega}$ = differential cross section of the reaction product (cm^2)

$\frac{dE}{dx}$ = stopping power of the target matrix for beam at energy E (MeV/cm) .

The count rate, dN/dt, in the detector depends on the flux of the beam particles, on the detector and on absorption effects. The last of these arise from self-absorption processes or additional absorbers used with the detectors.

$$\frac{dN}{dt} = dY\, se_f f_1 f_2 I/eq \tag{9.2}$$

where:

e_f = efficiency of the detector

f_1 = transmission the reaction product through the target matrix

f_2 = transmission of detector absorbers.

To obtain the yield for analysis of the data, this basic formula can be integrated.

9.2.1 Stopping Power

Energetic ions lose energy mainly through electronic excitation of the stopping medium atoms, which does not involve change in direction, referred to as electronic stopping. At low velocities ($< c/137$), nuclear collisions become more frequent, giving rise to an additional energy loss mechanism known as nuclear stopping, usually involving many changes of directions. But this occurs towards the end of the ion trajectory, and thus will not affect or be observed in the data obtained. For instance for a Sn target, the nuclear stopping contribution (estimated theoretically [3]) to the total is only about 1% for 1 MeV proton rising to ~15% at 0.5 MeV. Nuclear stopping correction can be significant and must be included in determining the range R for low energy particles. The corrections to R range from $\sim 10^{-4}$ for protons in Be to ~0.2 for the heaviest element in heavy element medium.

Extensive measurements over the past two decades have established the values for the stopping power of protons and alpha particles in various elemental material. The data are not as extensive for heavy ions, but these can be related to the stopping power of protons by a scaling factor based on the so-called effective charge. The current usage of stopping power is based on an empirical parametrization. The first comprehensive tabulation of stopping power of all elements in selected media was published by Northcliffe and Schilling [4]. Subsequently *Ziegler* and *Andersen* [5] refined the parametrization for protons and He^4, incorporating many of new experimental data that became available in the intervening years, and extended the parameterization to all ions. These values agree with available experimental data to within 3–10%.

Molecular stopping power can be derived from the elemental values according to *Bragg* and *Kleeman* [6], generally known as Bragg's rule. For a general compound $A_{n1}B_{n2}\ldots$ the stopping power is given by:

$$\frac{dE}{dx} = f_A \left(\frac{dE}{dx}\right)_A + f_B \left(\frac{dE}{dx}\right)_B + \cdots \tag{9.3}$$

where

$$f_A = \frac{n_1 m_A}{M}, \quad f_B = \frac{n_2 m_B}{M}, \ldots$$
$$M = n_1 m_A + n_2 m_B + \cdots$$

Deviations from this rule have been observed for gaseous organic compounds as well as oxides and nitrides, but they are generally less than 20%. The rule applies well for higher velocities, especially for protons [7].

9.2.2 Straggling

The statistical nature of the energy loss at low velocities involving many binary collisions gives rise to fluctuations in the energy loss, resulting in the straggling effect. This effect progressively broadens the energy distribution of the incident particle as it loses energy, degrading the depth resolution in RBS and decreasing the sensitivity of NRA. A theoretical estimate of this effect was first given by *Bohr* [8]:

$$\Omega_{\mathrm{B}}^2 = 4\pi Z_2 Z_1^2 e^4 N\, dR \tag{9.4}$$

where Ω_{B} is the width of the energy distribution at one standard deviation (corresponding to FWHM/2.355), and dR is the target thickness. Another theoretical estimate [9] incidated a smaller effect than Bohr's theory, particularly in the low energy region and for heavier elements. It is instructive to use Bohr's estimate to determine whether straggling should be considered when information at depths of a few microns is required. For instance for Sn, at $1\,\mathrm{mg/cm^2}$ depth ($1.74\,\mu\mathrm{m}$) the Bohr estimate gives 7.5 keV FWHM. The total effect on backscattered particles is the sum in quadrature of this and another for the outgoing path, and the detector resolution. For a detector resolution of 16 keV, the resulting effect is a broadening of the resolution to 19 keV.

9.3 Rutherford Backscattering Spectrometry

In a typical RBS analysis the scattered beam particles are usually detected at the "back angles" (> 90° with respect to the beam direction) using a surface barrier detector. The energy distribution and the yield contain respectively the identity and of the concentration of the target nuclei. In this method, f_1 and f_2 in (9.2) are unity, as is the detection efficiency e_f. However the energy of the scattered particle is attenuated by the target matrix. The particles scattered from a layer thus have an energy uniquely determined by the depth of the layer, and the species of the scattering atom. The yield is determined by the Rutherford scattering cross section:

$$\frac{d\sigma}{d\omega} = \left(\frac{Z_1 Z_2 e^2}{4E}\right)^2 \frac{4[\cos\theta + \sqrt{1-(m_1 \sin\theta/m_2)^2}]^2}{\sin^4\theta\sqrt{1-(m_1\sin\theta/m_2)^2}} \tag{9.5}$$

where θ is the scattering angle in the laboratory system.

The strong quadratic dependence on Z_2 can be seen from this formula, and the angular dependence is strongly peaked towards the forward angle, and at the back angles near 180° the cross section is virtually constant.

The scattered beam energy E' is determined from the kinematics of an elastic collision; its ratio to the incident beam energy is the kinematic factor:

$$K = \frac{E'}{E} = \left(\frac{m_1\cos\theta + \sqrt{m_2^2 - m_1^2\sin^2\theta}}{m_1+m_2}\right)^2 . \tag{9.6}$$

The angular dependence of K is strongest near 90° and the contrast for different m_2 is higher for larger m_1, as can be seen in Fig. 9.1, calculated for H and ^{4}He.

Consider now a scattering from a thick target set at an angle θ_1 with respect to the beam, and detection at angle θ_2 (Fig. 9.2). At a depth dt, the incident energy E_0 is attenuated by $dE = (dE/dx)_{E_0}\, dt/\cos\theta_1$, and the scattered energy is $K'E'' = K'(E_0 - dE)$ where K' is the kinematic factor at E''. The scattered particle energy is further attenuated on the way out by $dE' = (dE/dx)_{E''}\, dt/\cos(\theta_2)$. This particle is separated in energy from that scattered from the surface by δE. For infinitesimal dt a quantity S can be defined:

$$\delta E = S\, dt \tag{9.7}$$

where

$$S = K\left(\frac{dE}{dx}\right)_{E_0}\Big/ \cos\theta_1 + \left(\frac{dE}{dx}\right)_{E''}\Big/ \cos\theta_1 \, .$$

The quantity S reflects the total stopping power for incident and scattered particles, and can be used to define the depth scale. This general expression

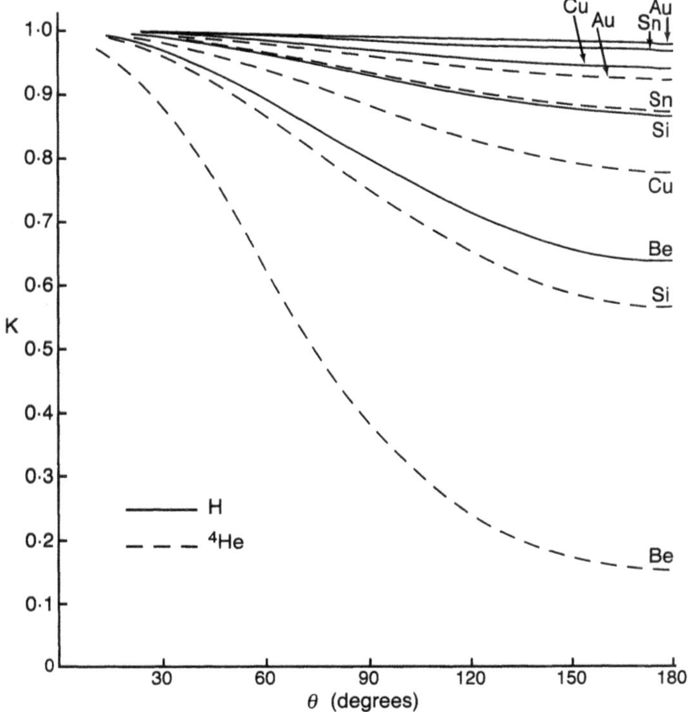

Fig. 9.1. Kinematic factors for H and ^{4}He beams on selected targets as a function of angle

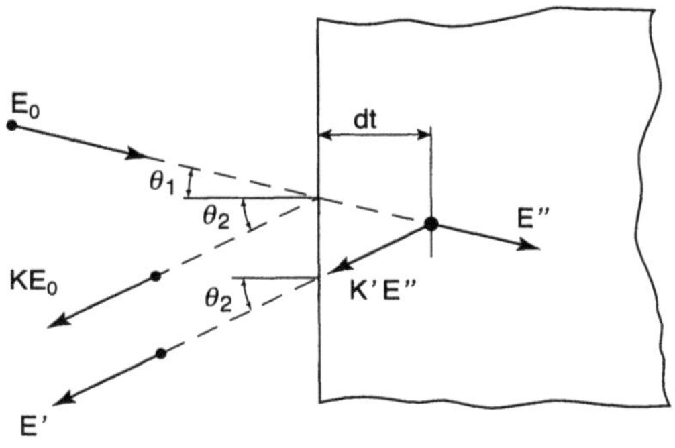

Fig. 9.2. Kinematics of Rutherford scattering from a near surface layer

is the basis for manual analysis of the spectrum, which is practical for thin targets.

For thick and complex targets numerical simulation of spectrum is necessary in the analysis. While many laboratories tend to produce their own RBS simulation programs, some are available from a number of authors, e.g. [10,11]. In calculating the simulated spectrum the response of the surface barrier detector must be taken into account. The response of the detector to monoenergetic particles appears as a Gaussian distribution in an energy dispersive pulse height analyzer with the width dominated mainly by the electronic noise of the system. Incomplete charge collection and multiple scattering effects in the detector produce a continuum extending from the full energy peak to low energies, which could amount to as much as 10% of the total counts.

9.3.1 Experimental Considerations

Most RBS measurements are carried out with H or ^{4}He ions. Heavier ions such as Li or B can be used to gain better mass resolution, but due to poor detector resolution for such ions, this is applied only in special cases. Typical conditions for RBS measurements are as follows:

1. The particles are detected in surface barrier detectors set between 90° and 170°. For maximum mass resolution, the angle must be close to 180°, which can be achieved with annular detectors. The resolution for this detector however is not as good as that for standard detectors (12–16 keV). Larger solid angles can degrade the depth resolution due to kinematic effects. The count rate in the detector is usually maintained below 10 000 counts/s to maintain good resolution and reduce electronic pulse pile-up effects.
2. Targets are usually set normal to the incident beam, except when 90° detection is required, in which case the target is set at some angle sufficient to prevent blocking of the detector by the target frame and to minimize the straggling effect.
3. The beam energy is usually chosen to avoid nuclear reactions, which result in deviations from the Rutherford cross-section. However, these can be corrected by means of relative measurements against suitable standards, since the other quantities governing the yield are unaffected by nuclear effects.
4. Low beam currents (below 1 nA) are desirable to minimize the possibility of damaging the sample, but may be difficult to measure accurately due to secondary electron emissions from the target and leakage currents in the beam integration circuit.

9.3.2 Examples

Figure 9.3 shows a simulation spectrum for a target consisting of a thin layer of Sn on a thick Cu substrate, analyzed using a 2.0 MeV ^{4}He beam.

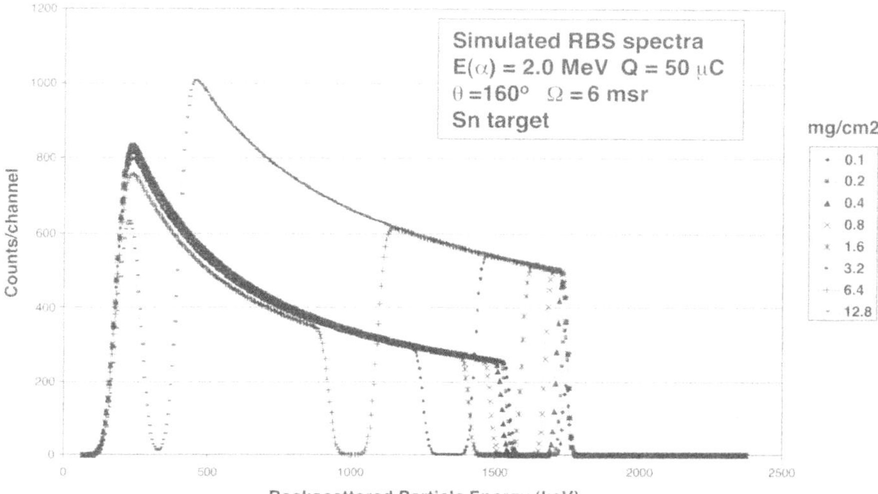

Fig. 9.3. Calculated RBS spectra from a series of targets consisting of a Cu substrate with a thin surface layer of Sn of various thicknesses. The spectra are computed for 2.0 MeV ^{4}He beam and a detector at 160° with respect to the beam direction subtending a solid angle of 6 millisteradians at the target for 50 μCoulomb integrated charge

The detector is a surface barrier detector set at 160°, subtending a 6 msr solid angle and its response simulated by a Gaussian (16 keV FWHM) with a constant continuum tail amounting to 10% of the total counts. The Sn layer appears as a Gaussian with a total area (counts) A, for thin layers, i.e. when the energy loss through it is less than the detector resolution. The thickness of the layer, as areal density, is given by:

$$W(\mathrm{g/cm^2}) = \frac{A}{Fs\frac{d\sigma}{d\omega}} \, .$$

The peak height would increase with target thickness until the maximum height H is reached. Further increase results in the increase of the width of the peak with a trapezoidal shape. The Sn peak is partially separated from the spectrum from Cu through kinematic effect. The Cu spectrum shows a high energy edge corresponding to the interface layer, and extends to the low energy region showing the characteristic shape of the spectrum from a thick target. The depth scale for the Sn peak can be determined from the dispersion of the spectrum. If dE is the energy per channel of the spectrum, then the height of the spectrum H at the edge, corresponding to the surface is given by:

$$H = F\left(\frac{d\sigma}{d\omega}\right)_{E_0} \frac{sN\,dE}{S_0} \, .$$

At greater depth, dE would correspond to a different depth interval corresponding to the appropriate energy at that depth. More conveniently, this depth scale can be established from the simulated spectra calculated for the various target thickness greater than the detector resolution.

Figure 9.4a shows the calculated RBS spectra for a series of targets consisting of a $0.5\,\mathrm{mg/cm^2}$ layer of various elemental material, for the same experimental condition. The width of the spectra generally increases with lower atomic number, although occasionally this does not hold, for instance in the case of Pt and W. The general trend is primarily due to the kinematic factor effect (9.6) and stopping power, obscuring effects of the different atomic density. Figure 9.4b shows the calculated spectra for an assortment of materials on a Cu substrate. The materials range from pure Sn to pure Pb with an assortment of their compounds in between. Each spectrum shows similar features: a trapezoidal peak corresponding to the layer at higher energy and a lower one corresponding to the substrate. The features observed can be related to (9.5, 9.6); the energy term E in (9.5) results in the increase of yield towards lower energy, and the height of the spectrum (the yield per unit energy interval), is proportional to the atomic number, as can be expected from (9.5). The separation between the surface layer and the substrate in the present example is due mainly to kinematic effects (9.6).

These examples illustrate that RBS is an ideal tool for analysis of thin layers of a few microns thickness for stoichiometric information. However, its applicability is limited to analysis of relatively simple structures and compositions. Unless the components are resolved kinematically, compositions of ternary and more complex mixtures are difficult to obtain unambiguously. The method is also limited to depth ranges where the scattered particle can still escape the target. This varies between 5 and $10\,\mu\mathrm{m}$ for most materials for proton and alpha beam energies below the Coulomb barrier, i.e. where nuclear reactions start.

9.3.3 Special Cases

The channelling effect: when the target is a single crystal, the backscatter yield will be reduced when the incident beam is aligned with the major (low index) axes [12]. This phenomenon can be used to probe the site of an impurity atom. For interstitial sites, the backscatter yield will increase again towards the random direction value.

Resonant scattering: for a number of light nuclei, e.g. ^{16}O, ^{12}C, a number of resonances (marked by increase of yield) occur for certain energies. Application of this is similar to that which will be discussed further in the NRA section below.

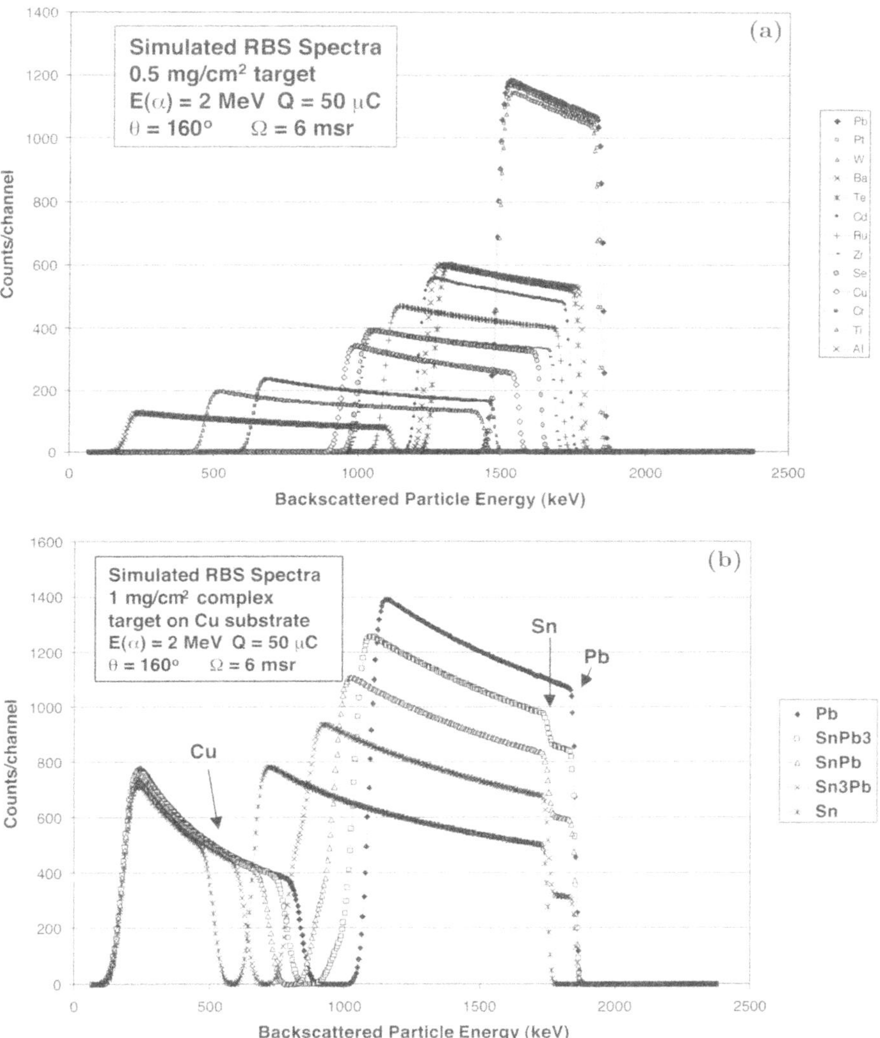

Fig. 9.4. (**a**) Calculated RBS spectra from a series of targets of $0.5\,\mathrm{mg/cm^2}$ layer of various elemental material. The spectra are computed for the same conditions as in Fig. 9.3. The width of the spectra generally increases with lower Z. This is mainly due to kinematic and stopping power effect. Occasionally this trend does not hold, e.g. in the case of Ba. (**b**) Calculated RBS spectra from a series of targets consisting of a Cu substrate with a $1\,\mathrm{mg/cm^2}$ surface layer of Sn, Pb and their various compounds. Note the variation of the height and width of the surface peak as a function of the material

Forward recoil: in this technique, under bombardment with heavy ions, the recoiling nuclei are detected in the forward direction. This provides some depth profile information but it is usually limited to a depth of around 1 μm due to straggling effects [13].

Non-resonant reaction: nuclear reactions can have a much higher cross-section than Rutherford scattering, and can thus be used to enhance the detection of selected elements.

9.4 Nuclear Reaction Analysis

A class of nuclear reactions is known as resonant reactions, defined as those where the reaction cross-section increases dramatically, as much as few hundred fold over the Rutherford cross section, for a narrow range of energy (resonance width). Resonances are common for reactions induced by protons and deuterons, and to a lesser extent by ^{3}He and ^{4}He particles, on light nuclei (up to $Z_2 \sim 30$). These resonances can occur below the Coulomb barrier energy given by $E_\mathrm{c} \sim Z_1 Z_2/(m_1^{1/3} + m_2^{1/3})$ MeV in the centre of mass system.

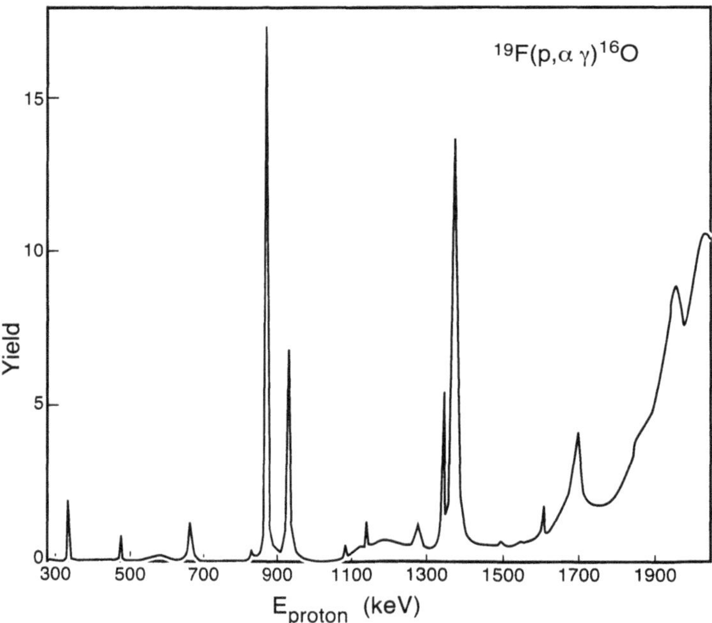

Fig. 9.5. Resonances in the reaction between a proton beam and a fluorine target, appearing as peaks in the reaction yield, e.g. gamma rays, as a function of the incident beam energy

They are known as sub-Coulomb transfer reactions. The most common reactions are capture reactions which show resonances corresponding to discrete high energy states in the product (compound) nucleus. In proton capture reactions, alpha particles are sometimes emitted, and can be used to detect the resonance. These reactions occur at specific energies for specific elements, and this is the feature exploited in NRA. At the resonant energy, the reaction occurs at the surface, sampling the surface region for which the energy loss of the beam corresponds to the resonance width. For higher beam energies the reaction will occur deeper in the target where the resonance energy is reached. Hence this method can be used to determine the concentrations of the particular atom involved in the resonance as a function of depth, giving the depth profile of the specific element. With typical resonance widths in the few keV range depth resolutions of a few nm are obtained.

For example, the resonances in the reaction between a proton beam and a fluorine target can be seen in Fig. 9.5 as peaks in the reaction yield against the incident beam energy. The criteria for selection of the reactions for NRA applications are that the cross section should be large and that the radiation can be detected readily. Gamma rays are commonly detected in this reaction.

9.4.1 Formalism

In the resonant NRA method the yield is obtained only at the resonance energy, and from a layer corresponding to the natural width of the resonance. The correction factor f_1 (9.2) depends on the nature of the radiation; it is 1 for particles as long as they are not stopped within the target. For gamma rays, the usual attenuation must be computed. The intensity of a gamma ray after traversing a medium with a linear absorption coefficient μ' is given by:

$$I = I_0\,\mathrm{e}^{-\mu' x} \tag{9.8}$$

where:

I_0 = initial intensity (no of counts)

μ' = linear attenuation coefficient of the medium for gamma ray of energy E_g, in cm^{-1} unit

x = thickness of the material traversed in cm

A more convenient variable is the areal density ($\mathrm{g/cm^2}$), and the corresponding coefficient is referred to as mass attenuation coefficient $\mu\,(=\mu'/\varrho$, where ϱ is the density). For a compound medium, Bragg's rule of additivity applies. Thus for a compound $A_{n1}B_{n2}\ldots$ the mass attenuation is given by:

$$\mu = \left(m_A^* x^* \mu_A + m_B^* y^* \mu_B + \ldots\right)/mt \tag{9.9}$$

$$mt = m_A^* x + m_B^* y + \ldots$$

Extensive tabulation of μ is available, based on both theoretical calculation and empirical data [16]. Parametrization of the coefficients are given by *Theisen* and *Vollath* [17].

For high energy gamma rays this absorption could be negligible, hence $f_1 \sim 1$. The absolute value of the cross-section can only be determined empirically, but in most applications measurements are carried out against a standard. The "raw" profile, i.e. the yield as a function of incident beam energy are further unfolded to take into account the effect of straggling, to give the true profile. The Bohr estimate can be used to determine whether this effect is significant for the range of energy (depth) involved in the measurement.

9.4.2 Experimental Considerations

In NRA measurements, the following points should be considered:

1. The cross-section in (9.1) must contain the angular distribution effect of the detected radiation. Most resonant reactions show a very strong angular distribution, and care must be taken in deciding the take-up angle of the detector.

2. When the resonant reaction produces discrete, relatively low energy gamma rays (below $\sim 2\,\mathrm{MeV}$) usually emitted in the last stages of the product nucleus deexcitation, Ge detectors are used. The high resolution of this detector (typically $< 2\,\mathrm{keV}$ at $1.33\,\mathrm{MeV}$) results in a good signal to background ratio.

3. In measurements involving high energy gamma rays ($> 2\,\mathrm{MeV}$), NaI detectors are commonly used for better efficiency. The poor energy resolution (typically 6% of the full energy peak) results in poorer signal to background ratio. The main sources of background are gamma rays from $^{40}\mathrm{K}$ ($1.461\,\mathrm{MeV}$) and one of the daughters of $^{232}\mathrm{Th}$ decay ($2.614\,\mathrm{MeV}$ from the decay of $^{208}\mathrm{Tl}$) from building materials, and cosmic rays.

4. To improve the signal to background ratio for the reaction gamma-rays, a close geometry is necessary and can be achieved by appropriate design of the target chamber. Background can be reduced further by absorbers, which are however only effective for the room background and less so for cosmic rays. More active reduction of cosmic rays can be achieved by using anti-coincidence shields.

5. If sharp resonances are used, it is desirable to carry out the measurements in ultra high vacuum conditions to reduce the problem of carbon build-up on the target that will shift the energy of the beam. This can be achieved by means of a cold finger (cooled by liquid nitrogen) around the target.

6. The surface texture is also important when sharp resonances are used. Any surface roughness would broaden the depth resolution due to smearing of the beam energy.

9.4.3 Examples

Figure 9.6 shows the hydrogen profile from a set of samples of hydrogenated amorphous silicon, obtained using the resonance at 6.412 MeV fluorine beam energy, corresponding to the 0.340 MeV resonance (Fig. 9.5) when protons are used on fluorine target [18]. In this particular case, the inverse of the resonant reaction induced by protons is used, where now one uses the ^{19}F as the beam to detect hydrogen. The 6.13 MeV gamma rays emitted by the reaction are detected in NaI detectors. The surface peaks on two of the samples indicate a depth resolution of 44 nm, corresponding to the resonance width of 3 keV. The range of applicability is usually limited to a few μm, because of increasing diffuseness of the beam with depth due to multiple scattering. In the present case, the onset of another resonance corresponding

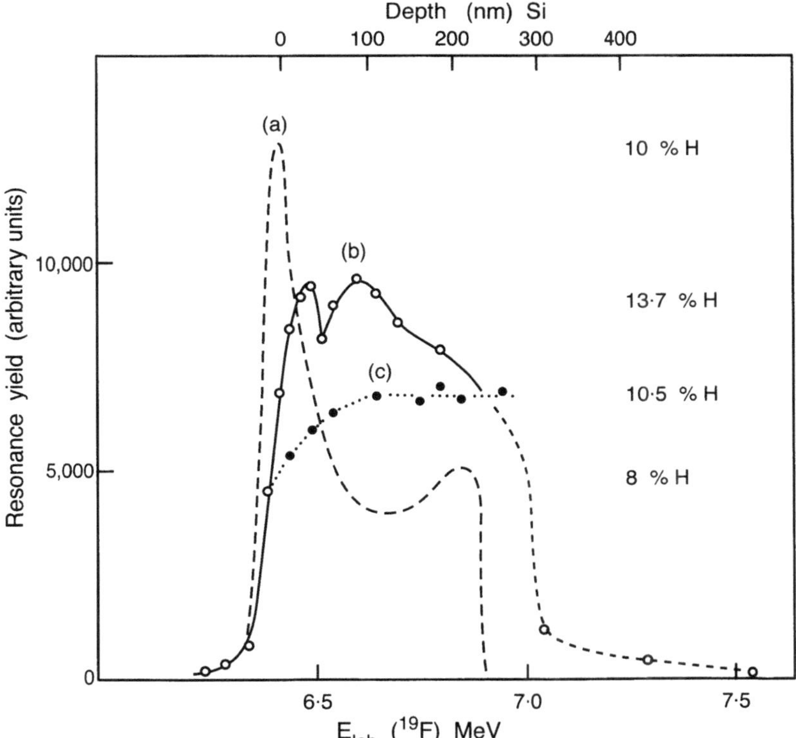

Fig. 9.6. A hydrogen profile of a selection of *a*-Si:H samples, obtained using the resonance at 6.412 MeV fluorine bombarding energy. Two of the samples show surface peaks which may arise from adsorbed moisture. The width of the surface peak indicates the depth resolution of 44 nm obtained with this resonance. Note the nonlinearity of the hydrogen content scale due to change in the stopping power with composition

A

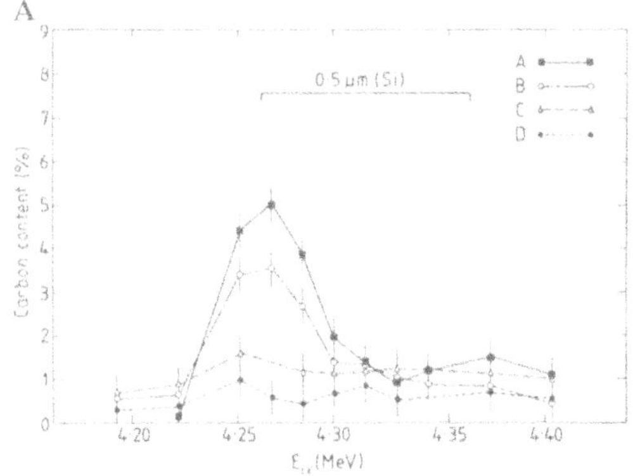

B

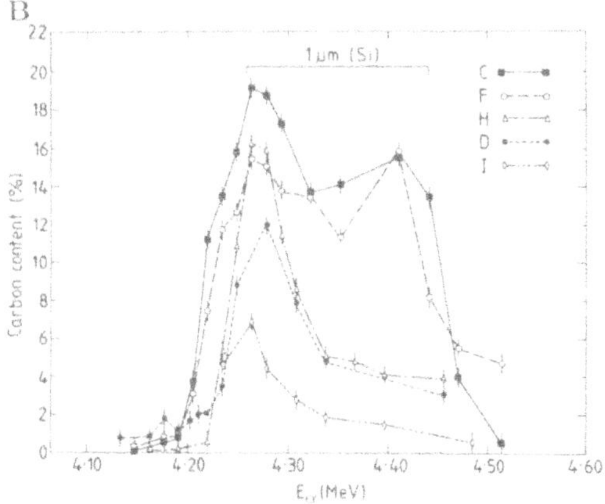

C

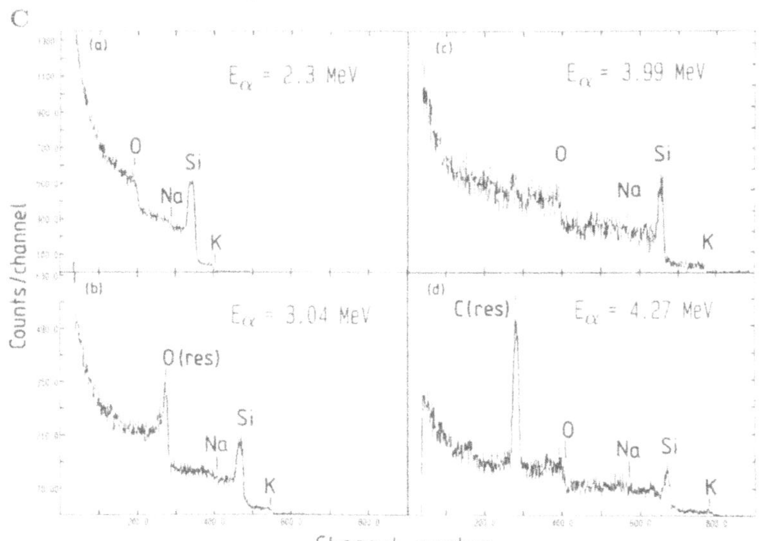

Fig. 9.7. Carbon profiles (**A** and **B**) from a series of samples of amorphous hydrogenated Si-C compound, obtained using the alpha resonance at 4.26 MeV. The corresponding backscattered particle spectra (**C**) show the onset of the resonance as well as that due to oxygen at 3.036 MeV. The oxygen arises from the glass substrate

to the 0.484 MeV in Fig. 9.5, limits the applicability of the reaction above to a depth of only 1 μm. The sensitivity of this method is limited by background to around 100 ppm. This can be improved when anti-coincidence techniques are applied to reduce the cosmic ray background. A detection limit as low as 10 ppm has been reported [19].

The resonance yield can be calibrated against a hydrogen-rich target of known composition. Organic polymers (e.g. kapton, polyethylene, mylar) are convenient but they are easily damaged by the beam mainly due to thermal effects. Care must be taken in the calibration run to minimize such effects by running very low level and diffuse beam on the target. The damage can be observed by monitoring the yield as a function of time or beam dosage, and the true yield can be obtained by extrapolation to zero time or beam dosage. More stable standards can be obtained using implanted material [18] or hydrogenated amorphous silicon.

Sub-Coulomb barrier resonances in scattering of protons and alpha particles on some light nuclei can be utilized in this method, where the detected radiation is the scattered particle. The main advantage are the somewhat longer range of applicability, and higher efficiency of detection afforded by particle detectors [$f_1 = 1$ in (9.2)]. As an example, Fig. 9.7 shows the depth profiles of carbon obtained from several samples of thin layers of hydrogenated amorphous Si-C compound using the 4.26 MeV resonance of alpha scattering on carbon, and the corresponding spectra obtained at various bombarding energies showing the onset of the resonance [21]. The figure also shows the resonance of oxygen from the glass substrate, at 3.036 MeV. These particular resonances are peaked at 180° with respect to the beam direction, hence the ideal measurement geometry is consistent with the standard RBS requirements.

9.5 Summary

Two of the suite of ion beam analysis techniques, namely RBS and NRA, and variations thereof, are versatile and powerful non-destructive methods for characterizing the elemental composition and structure of the surface region, extending from a few hundred atomic layers to several microns. In this respect they complement other surface analytical methods which probe fewer atomic layers of the surface, and their chemical states. The chemical insensitivity of the forces governing the interaction between the probing beam and the sample results in methods which are highly quantitative, being amenable to straightforward calibration procedures.

Acknowledgement. The simulation spectra presented in this article were produced by Leszek Wielunski using the computer program HYPRA.

References

1. J.R, Tesmer, M. Nastasi eds. *Handbook of Modern Ion Beam Materials Analysis* Materials Research Society, Pittsburgh, Pennsylvania, 1995.
2. Nucl. Instr. and Meth. in Phys. Res. **B136–138 (1998)**
3. J. Lindhard, M. Scharff, H.E. Schiott: Kgl. Danske Videnskab. Selskab., Mat. Fys. Medd. **33**, 14 (1963)
4. L.C. Northcliffe, R.F. Schilling: Nuclear Data Tables A**7**, no. 3–4 (1970)
5. H.H. Andersen, J.F. Ziegler: *Hydrogen Stopping Powers and Ranges in All Elements* (Plenum, New York 1977) and related titles in the series "The Stopping and Ranges of Ions in Matter", ed. by J.F. Ziegler (Pergamon, New York 1980)
6. W.H. Bragg, R. Kleeman: Phil. Mag. **10**, 318 (1905)
7. J.F. Ziegler, J.M. Manoyan: Nucl. Instr. Meth. in Phys. Res. B**35**, 215 (1988)
8. N. Bohr: Kgl. Danske Videnskab. Selskab., Mat. Fys. Medd. **18** no. 8 (1948)
9. W.K. Chu: Phys. Rev. A**13**, 2057 (1976); also in [1, page 1]
10. P.A. Saunders, J.F. Ziegler: Nucl. Instr. Meth. in Phys. Res. **218**, 67 (1983)
11. L.R. Doolittle: Nucl. Instr. Meth. B**9**, 344 (1985)
12. B.R. Appleton, G. Foti: In Ref. [9.1] p. 67
13. B.L. Doyle, P.S. Peecy: Appl. Phys. Lett. **34**, 811 (1979)
14. J.A. Davies, J.S. Forster, S.R. Walker, Nucl. Instr. and Meth. in Phys. Res. **B136–138** (1998) 594
15. W. Assmann, J.A. Davies, G. Dollinger, J.S. Forster, H. Huber, Th. Reichelt, R. Siegele, Nucl. Instr. and Meth. in Phys. Res. **B118** (1996) 242
16. J.H. Hubbell: Atomic Data **3** no. 3 (1971)
17. R. Theisen, D. Vollath: *Tables of X-Ray Mass Attenuation Coefficients* (Stahleisen, Düsseldorf 1967)
18. S.H. Sie, D.R. MacKenzie, G.B. Smith, C.G. Ryan: Nucl. Instr. Meth. in Phys. Res. B**15**, 525 (1986)
19. H. Damjantschitsch, M. Weiser, G. Heusser, S. Kalbitzer, H. Mannsperger: Nucl. Instr. Meth. in Phys. Res. **218**, 129 (1983)
20. J.F. Ziegler et al.: Nucl. Instr. Meth. **149**, 19 (1978)
21. S.H. Sie, D.R. MacKenzie, G.B. Smith, C.G. Ryan: Nucl. Instr. Meth. in Phys. Res. B**15**, 632 (1986)

10 Materials Characterization by Scanned Probe Analysis

S. Myhra

10.1 Introduction

The STM (see list of acronyms at end of book) was invented by Binnig et al. in 1982 [1]. The two main protagonists, G. Binnig and H. Röhrer, were subsequently awarded the Nobel Prize for physics. Thus began the age of SPM. Much of the early development and excitement generated by the unequivocal demonstration of spatial resolution on the scale of the single atom and of local spectroscopies have been described in the literature [2–4]. The rationale for including a chapter on SPM in a book on surface analysis can be inferred from Fig. 10.1 (adapted from Röhrer [5]). The impact of SPM in the broad field of surface and interface science and technology can be illustrated by its prominence at a conference in Birmingham in September 1998, which brought together a representative cross-section of the international surface science community through the 14th International Vacuum Congress, 10th International Conference on Solid Surfaces, 5th International Conference on Nanometer-scale Science and Technology and 10th International Coference on Quantitative Surface Analysis. Of some 1350 invited and contributed papers approximately 20% were based substantially on SPM techniques and methodologies, while the corresponding indices for the 'traditional' techniques and 'other' were 34% and 46%, respectively.

The demonstration of the 7×7 reconstruction of the Si (111) surface was arguably the result which put STM on the scientific map. Irrefutable evidence for single-atom real-space resolution, or even more convincingly for single *missing* atoms and atomically resolved I-V spectroscopy, were also significant events [6].

The second member of the SPM family, the AFM, was demonstrated in 1986 [7]. Subsequently, other tip-to-surface interactions (e.g. magnetostatic, electrostatic, thermal radiation, etc.) have given rise to additional members. As well, most of the original techniques have had offsprings by way of variations on the underlying theme (e.g. contact, non-contact, intermittent contact, lateral force and force-vs-distance modes for the AFM).

In hindsight it may be that the invention of the STM was not, in itself, the most significant aspect of the work of the group at IBM. Rather it was the thinking which led to the STM, and its technical implementation, which paved the way for a plethora of scanned probe techniques. For purposes of

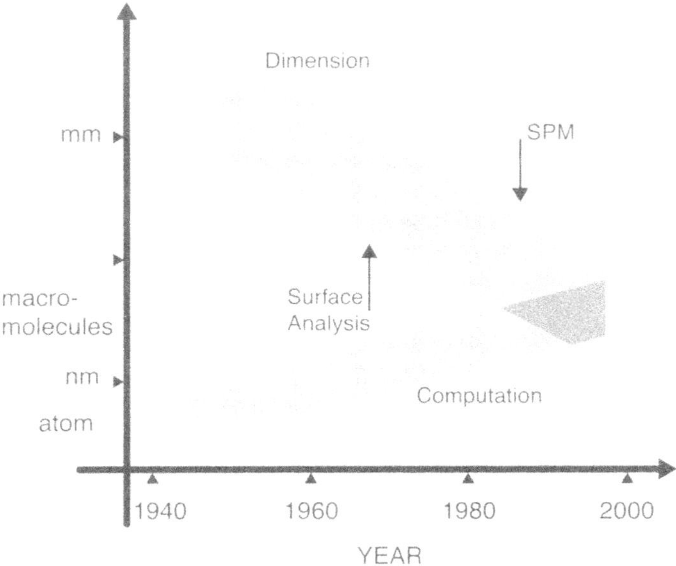

Fig. 10.1. Trends in computationally intensive modelling of multi-atom systems, compared with decrease in dimensions of devices, and juxtaposed with the ability of SPM characterization and manipulation. The arrows show year of first deployment of particular techniques

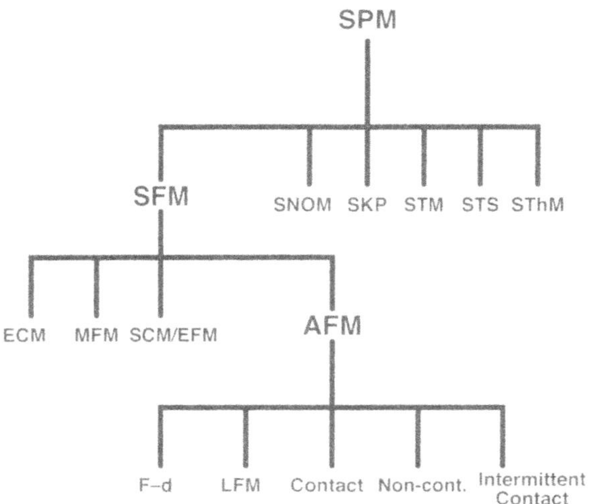

Fig. 10.2. The SPM family tree in an abbreviated form. The acronyms are defined at the end of the chapter

Table 10.1. SPM and traditional techniques – complementary strengths and weaknesses (adapted from [8])

Requirement	SPM	Other Techniques
Surface nanostructure	STM/Contact AFM	LEED (reciprocal space)
	real space	FIM and ISS (real space)
Morphology/topography	STM/AFM	SAM/SEM/ion microscopy
	×1 000 000	×10 000
	good z-resolution	poor z-resolution
Grain structure	STM/AFM	SAM/SEM/Imaging SIMS
Phase structure	—	SAM/Imaging XPS/SIMS
Electronic structure		
Valence/conduction states	STS (local)	UPS/IPES/EELS (non-local)
Core states	—	XPS/AES
Composition	—	XPS/AES/ ...
Mechanical properties	AFM/F–d	Nano-indentation
Frictional properties	LFM/F–d	—
Optical properties	SNOM	Optical microscopy
	×100 000	×1000
	good z-resolution	poor z-resolution
Magnetic structure	MFM ('local')	—
Thermal properties	SThM	Scanning calorimetry
Surface conductivity	SCM (local)	Differential charging (non-local)
'Buried' information	—	Profiling XPS/AES/SIMS/ ...
In situ capabilities	Vac/gas/liquid	Usually UHV
In situ reaction dynamics	Potentially very good	Generally limited

classification, and in the interest of being internally consistent from a definitional point of view, one should note the following: SPM is the name of the family; an abbreviated family tree is shown in Fig. 10.2. As indicated above the acronym is usually taken to refer to both the instrument, 'microscope', and the method, 'microscopy'. The SFM branch has emerged as the largest and most popular, in terms of usage and installed instrumental capacity. The AFM sub-branch includes the most widely used technique(s). Thus it merits further sub-division into its various operational modes:

- Contact – when the interaction is short-range, usually the net force is repulsive
- Non-contact – when the net interaction is longer-range and attractive
- Intermittent contact, sometimes called tapping – when the lever is stimulated to oscillate at an eigen-frequency with an amplitude so that the tip enters the short-range force-field of the surface
- Lateral force – when the system is controlled in the constant short-range force mode, but the mapping parameter is responsive to the lateral, 'frictional', force component
- F–d – when the surface map reflects changes in the force-versus-distance relationship

Table 10.2. SPM Techniques - General figures of merit in context of surface analysis[1] (adapted from [8]).

Technique	Interaction	z-resolution (nm)	xy-resolution (nm)	'Interaction' vol.(m^3)	Probe/tip	Information outcome
STM/STS	electron tunnelling	0.005	0.05	$< 10^{-30}$	sharp tip	structure/topography, electron spectroscopy, 'work' function[2]
AFM contact	inter-atomic forces (short-range)	0.02	0.15	$< 10^{-28}$	cantilever + tip	topography, molecular structure
AFM non-contact	(long-range)	0.2	1	10^{-26}	"	topography
AFM intermittent contact	(short-range)	0.1	0.5	$< 10^{-27}$	"	topography
AFM contact	(lateral)	n.a.	0.3	$< 10^{-28}$	"	tribology, friction
AFM F–d	inter-atomic forces	0.2	1	10^{-26}	"	mechanical properties, adhesion, force constants
MFM non-contact	magnetic force	1	5–50	$< 10^{-25}$	cantilever + magnetised tip	magnetic 'topography', domains, walls, magnetic 'spectroscopy'
EFM/SCM non-contact	electrostatic or capacitive force	5	10–100	$< 10^{-23}$	cantilever + insulating tip	topography, patch charge, capacitance
ECM contact	electrical conductivity	0.05	1	$< 10^{-26}$	cantilever + conducting tip	topography, electrical conductivity, break-down voltage, I–V analysis
SThM non-contact	heat loss	1	100	$< 10^{-21}$	thermocouple tip	topography, thermal conductivity, emissivity
SNOM	photons	2	5	$< 10^{-25}$	small aperture fibre tip	topography, optical properties, spectroscopy (?)

[1] An attempt has been made to provide an 'interaction volume' as is often quoted for surface analytical techniques.
[2] The 'work' function refers to an effective barrier height.

10.2 The Surface Analytical Context

In comparison with the 'traditional' surface analytical techniques, which took off in the 1960s and have now reached stages of relative maturity, SPM is a relative newcomer. Thus it is natural that its merits and demerits should be considered in the context of the established methodologies so that the complementarities are clearly apparent, and in order to have rational criteria for deployment of the technique(s) and method(s) which are most likely to lead to the solution of a problem within the broad area concerned with surfaces and interfaces. Table 10.1 represents an attempt at a comparison of relative strengths and weaknesses. Some of the attributes and figures of merit of SPM microscopies/spectroscopies are summarised in Table 10.2.

10.3 Generic SPM Systems

An SPM system, Fig. 10.3, may be thought of as consisting of: a surface, having a number of structural and physico-chemical characteristics; an interaction possessing a characteristic strength and range; and a tip, which can be located and controlled in the spatial and temporal domains, with its own characteristics. Therefore one can make the elementary observation that, given *sufficient* knowledge about *any* two of the three elements, it is possible in principle to gain insight about the third. This observation demonstrates

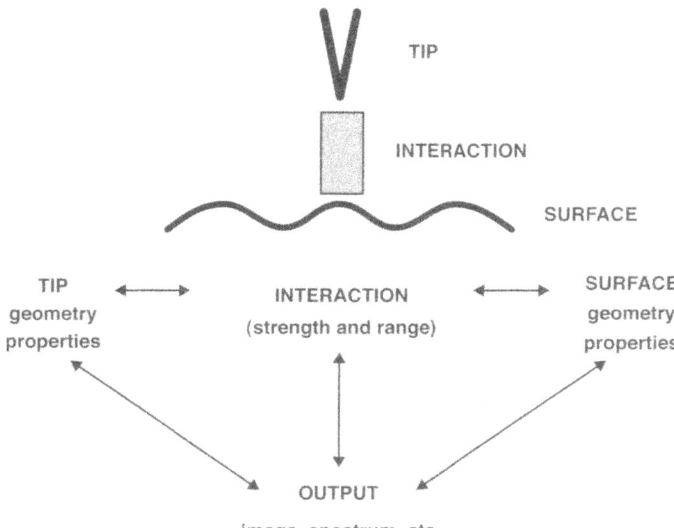

Fig. 10.3. The essential elements of an SPM system, and a representation of how the information obtained must of necessity be a 'convolution' of contributions from the three elements

the intrinsic versatility and inherent richness of the SPM system. In practice, exploitation of the system is predicated on definitions and acceptance of *sufficiency* of knowledge. The consequences of the limitations regarding sufficiency will be explored below. Also, it should be noted that the terminology of 'microscop(e/y)' implies that in the overwhelming majority of instances the 'surface' is taken to be the unknown. Thus most of the discussion of the SPM system will concentrate on the conventional microscop(e/y) aspects/applications, but some attention will be given to other configurations. The schematic description in Fig. 10.3 also makes it abundantly clear that a resultant image must represent a complicated convolution of the above three elements.

The vast majority of present SPM facilities are based on two generic types of instruments.

Type I: The most popular configuration, by a very wide margin, is an air-ambient multi-technique/mode platform, which usually also supports scanning in a fluid ambient. Such instruments are comparatively inexpensive, very flexible, relatively user-friendly and reliable, and are providing excellent service as routine tools for surface and interface characterisation. However, being even more surface-specific than the 'traditional' UHV-based techniques there are fundamental limitations due to inability to control the ambient environment. The temperature range is also limited to near-ambient. Thus STM, and STS in particular, is of relatively little utility for most materials, except at the fluid/solid interface, and thermally activated processes/properties cannot be probed over a sufficiently wide temperature range. Likewise, inevitable adsorbed moisture and contamination largely nullifies the availability of MFM and other true non-contact operational modes. Those limitations are reflected in the fact that most STM analyses are carried out under extreme UHV conditions, while the SFM group of techniques dominate the applications of air-ambient instruments.

Type II: Is based on operation within a UHV envelope, and tends to be single technique, usually STM. Some recent systems do also support AFM in contact and non-contact modes. Some special purpose UHV instruments may incorporate hot/cold stages, transfer systems, specimen preparation chambers, and various combinations of complementary surface/interface analysis techniques. Such instruments are expensive, not user-friendly and relatively inflexible. Their main utility is for applications where atomic or near-atomic resolution is required, and where cleanliness of reactive surfaces is mandatory. They are not suited for rapid turn-around of specimens or for routine analysis.

Type III: The gap between Types I and II is gradually being filled. A typical Type III instrument retains the multi-mode functionality and flexibility of the air-ambient instruments, while at the same time providing a controlled ambient envelope, offering down to 10^{-4} Pa clean turbo-molecularly pumped base vacuum, and having the facility for back-filling with any gaseous species up to atmospheric pressure. Hot (to 800°C) and cold (to 80K) stages may be optional attachments which will permit investigation of thermally activated surface and interface processes.

10.4 Physical Principles

The following will constitute a brief, and necessarily qualitative, description of the physical principles which account for the information content of the SPM system. The discussion will be limited to STM/STS and to AFM and its most commonly used variants.

10.4.1 STM/STS

Electron tunnelling is the elementary process which accounts for the operation of STM/STS. The approach adopted by Tersoff and Hamann [9] remain an intuitive and useful tool for interpretation of data. Additional material can be found in the review literature [5,10,11]. The energy level scheme of a tunnel barrier is shown in Fig. 10.4. The respective wavefunctions outside and inside the barrier are sinusoidal and exponential; the latter is particularly relevant to the problem at hand. The tunnel current is given by a summation over elastic, hence $(E_2 - E_1)$, tunnelling channels linking occupied states

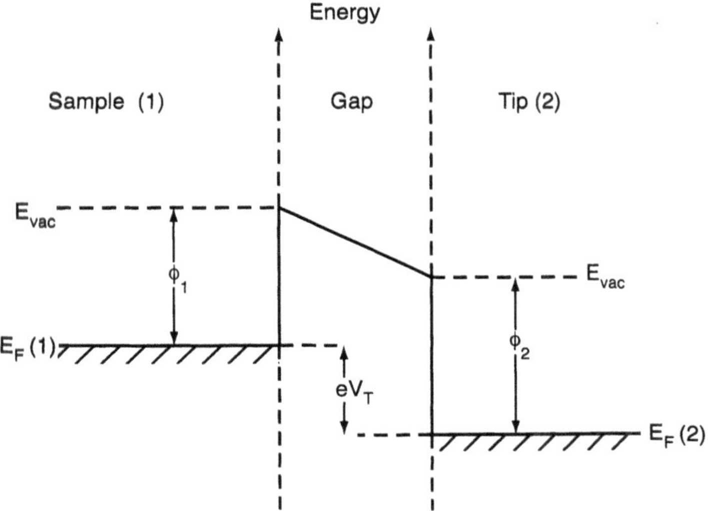

Fig. 10.4. The energy level scheme of a tunnel barrier. Sample and tip are denoted (1) and (2), respectively. Work function is ϕ, and tunnel voltage is V_T

(as described by the Fermi-Dirac function f) on one side of the barrier with unoccupied states $(1-f)$ on the other side

$$I = \frac{2\pi e}{h} \sum_{1,2} f(E_2)\left[1 - f(E_1 + eV_{\mathrm{T}})\right] |M_{1,2}|^2 \, \delta(E_2 - E_1) \tag{10.1}$$

The two distributions of states are displaced by eV_{T}, where V_{T} is the tunnel voltage. Figure 10.4 illustrates the case of tunnelling being two metals from occupied states below E_{F} to unoccupied states above E_{F}. The matrix element $M_{1,2}$ is the transfer-Hamiltonian of Bardeen; and is a weakly varying function.

A simple one-dimensional case of two identical free-electron metals with identical work functions, ϕ, will provide useful insight. Applying boundary conditions, the wavefunctions in the gap may be written as

$$\Psi_1 = \Psi_1^{\circ}\, e^{-kz} \quad \text{and} \quad \Psi_2 = \Psi_2^{\circ}\, e^{-k(s-z)} \tag{10.2}$$

where the width of the tunnel barrier is s; k is a real number (typically $10\,\mathrm{nm}^{-1}$), and is given by $k = (2m\phi/h^2)^{1/2}$. It is straightforward to show that

$$I \propto \sum_{1,2} \left|\Psi_1^{\circ}\right|^2 \left|\Psi_2^{\circ}\right| 2e^{-2ks} \tag{10.3}$$

The expression gives qualitative insight into the tunnelling process, irrespective of tip shape, if the wavefunction anchored in the tip can be approximated by an s-wave. Furthermore, if the overlaps of the respective wavefunctions within the barrier are small, and if the tails of the two functions are of similar shape, then

$$I \propto \sum_{1} \left|\Psi_1\right|^2 \delta(E_1 - E_{\mathrm{F}}) \tag{10.4}$$

The local density of states in the sample surface at the position of the tip can be written as

$$\rho(E_{\mathrm{F}}, \boldsymbol{r}) = \sum_{1} \left|\Psi_1(\boldsymbol{r})\right|^2 \delta(E_1 - E_{\mathrm{F}}) \tag{10.5}$$

The description above allows one to make a number of qualitative statements about the STM/STS system.

- The exponential dependence in (10.3) of the tunnel current I on the width of the tunnel gap accounts for the extreme spatial resolution in the z-direction; a change in s of 0.1 nm will change I by a factor of e^2.
- Equations (10.4, 10.5) show that an STM image, formed by mapping constant tunnel current, represents a map of constant density of states. Thus, unless there is a spatial one-to-one correlation between the positions of atoms, i.e. nuclei, and maxima/minima in the density of states map, then it is the electronic structure that is being revealed. Conversely, the symmetries in the image must reflect the surface structural symmetries, if there is a one-to-one correspondence.

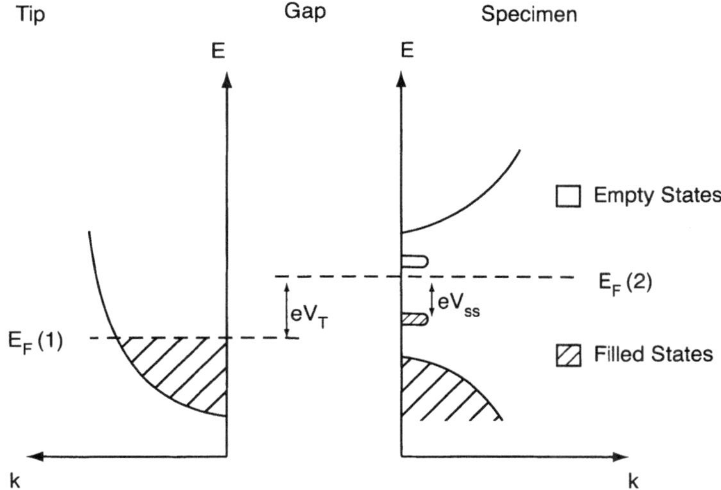

Fig. 10.5. STS illustrated schematically. Electron states in the tip are being swept past bulk and surface states in the specimen surface. The probability of elastic tunnelling at a particular tunnel voltage depends on availability of occupied and empty states and manifests itself as a change in tunnel current

- The lateral resolution in the x–y plane is due, in part, to the exponential dependence of I on the barrier width. As well, the lateral decay of the s-state located in the apex atom of the tip will confine the interaction laterally to a radius of *ca.* 0.1 nm. Thus there will be single atom resolution due to the lateral spatial decay of the density of states function. Of equal significance is the fact that the elastic tunnel process picks out particular states in the sample and tip, as a consequence of energy conservation. Thus the tunnel voltage, V_{T}, sets an energy window, with a width of a few kT. Consequently one may choose a tunnel voltage for maximum resolution, i.e., an energy for which the corresponding electron states have the greatest spatial variation. These observations account for the relative ease with which 'atomic' resolution can be obtained for covalent materials (e.g. Si); this is a consequence of the strong spatial dependence of the density of states across the unit cell at the extrema of bands. Conversely, similar resolution is much more difficult to obtain for metals where the spatial variations are more gentle.
- The ability to 'tune' the energy window of the tunnelling process by changing eV_{T} is the basis for STS. Successive maps of the surface at different values of V_{T} will reveal the real-space locations of the corresponding equi-energy contours of the density of states. Local spectroscopy can be carried out by fixing the lateral and vertical position of the apex of the tip with respect to the surface and recording an I_{T}–V_{T} curve. The principle of the method is illustrated in Fig. 10.5, adapted from [12]. The I_{T}–V_{T}

data will not be interpretable as a simple plot of the density of states function, however, since there will be an exponential dependence of the tunnel probability with V_T from the matrix element M. This and other effects can be eliminated, in part, by measuring $\frac{dI}{dV}/\frac{I}{V}$ which is a useful dimensionless quantity related to the local band structure. As well as being local, STS has the additional merits of extreme surface specificity, literally the first monolayer, and of being able to probe both valence and conduction band states by the simple expediency of reversing the polarity of V_T, (and thus reversing the tunnel process). Investigations of localised surface states are particularly useful and rewarding with STS. In the context of the 'traditional' surface spectroscopies, one should note that STS can take the place of the combination UPS/IPES. The additional merits and convenience of STS *vis-à-vis* UPS/IPES make this an attractive proposition [6,13].

- From the description of the tunnel process it can be seen that the tunnel current is exponentially dependent on k, as well as on barrier width. The decay constant, k, is a function of the effective barrier height, and is therefore related to the work function of the surface being probed. Thus it is possible to extract the relative spatial variation of the work function from the STM map.
- It will also be apparent that STM/STS can be applied only to materials which are tolerably good conductors. The criterion for 'goodness' is that the effective sample resistance, R_s, must be such that $I_T R_s \ll V_T$. Hence the technique will work well for clean surfaces of metals, alloys, semimetals, and doped semiconductors. The presence of thin (1 nm) insulating surface barriers (e.g. oxide layers) can be accommodated by allowing them to act as tunnel barriers. Because of the extreme surface specificity of STM/STS, a UHV environment is necessary when the surface is reactive. One non-UHV area of application in which the STM has much to offer is that of a fluid environment. In particular, electrochemical STM, where the tip can be a local electrode as well as a local probe, has revealed a great deal of detailed information [14].

10.4.2 Scanning Force Microscopy (SFM)

An SFM probes a surface by sensing a force, or its gradient, between the surface and a tip; therein is the underlying principle of operation. The attention in this section will be principally with the AFM and its various operational modes.

The principles of the SFM cannot be described with the same degree of generality and neatness as for STM/STS. Neither can the image formation process be understood at the same level of detail and physical insight. Continuum theories will provide only a rough phenomenological description of the tip-to-surface system, which may be adequate for most purposes, but the full power of the SFM will only be realised once detailed microscopic theories can be deployed on a routine basis.

Atoms respond to short-range repulsive exchange interactions and longer range attractive ones. The tip-to-surface system may be modelled crudely by pair-wise interactions between atoms in the tip and in the surface. The interactions combine to produce a potential well with a location of lowest energy, the binding energy E_b, and a point at which there is zero net force. The shape of the well defines a force gradient; the magnitude of the force gradient is relevant for defining the operational mode and for the design of the force sensing element of an SPM instrument. The contact mode has its operating point in the repulsive force regime, where the force gradient is high, and the repulsion is due to the overlaps of the electron clouds of atoms in close proximity. The simplest description of the interaction in the contact mode is in terms of the Lennard–Jones potential function; it turns out, however, that a proper description of the contact mode requires a full molecular dynamics simulation [15]. Intuitive and qualitative insight may be gained from describing the system by two coupled springs. At equilibrium, when the tip is tracing out equal force contours, then

$$k_{\text{eff}} z_{\text{int}} = k_{\text{N}} z_{\text{L}} \tag{10.6}$$

where k_{eff} and k_{N} are the effective force constants for the tip-to-surface interaction and the normal spring constant of the lever, respectively. Excursions from the bottom of the potential well and from the point at which the lever is undeflected are z_{eff} and z_{L}, respectively. Since in general $k_{\text{eff}} \gg k_{\text{N}}$, then $z_{\text{eff}} \ll z_{\text{L}}$, and a contact mode image will be a very good approximation to the actual surface topography. The simple description also underscores the need to image a 'soft' material with a probe with a low spring constant.

The non-contact mode, on the other hand, has its operating point in the attractive force regime, where the force gradient is low and the attraction is due to dispersion forces of the van der Waals type and other interactions. Modelling of forces acting on extended bodies in close proximity is a non-trivial exercise, and is beyond the scope of the present Chapter. Most of the current attempts to reconcile theory with practice are based on a seminal monograph by Israelachvili [16]. An AC resonance technique is generally used to control the feed-back loop in the non-contact mode. The sampling of the effective potential well is then via a driven and damped oscillator where both amplitude and phase will depend on the average force gradient. The sensitivity is greatest when the operating point is at the steepest part of the resonance curve for the probe. Thus a non-contact image will represent a map of constant force-gradient.

10.4.3 Intermittent Contact Mode

Aspects of the modes described under Sections 10.4.1 and 10.4.2 can be brought together. The lever is driven at resonance at or close to one of its eigenmodes, but the amplitude is increased until the tip makes intermittent excursions during each oscillatory cycle into the part of the potential well dominated by the short-range repulsive interactions (hence the alternative designation of tapping mode). The feed-back loop will now lock in either on a constant decrement in amplitude or on a constant change in phase, where the amplitude decrement/phase change is substantially dependent on effects of the short-range interactions.

10.4.4 F–d Analysis

The AFM analogue of I–V spectroscopy in STM is F–d analysis; the force sensing lever is the dispersive element of the system. A quasi-static F–d analysis can be undertaken by holding the tip at a particular x–y location far away from the surface. The tip is then driven towards the surface at a rate which is slow in comparison with the mechanical response of the system. The net force is sensed during the approach, contact and retraction parts of the cycle. Two idealized response curves are shown below in Fig. 10.6, with stage travel and lever deflection plotted on the horizontal and vertical axes, respectively. The vertical units can be converted into force sensed/applied by the lever by the simple expediency of multiplying the deflection with k_N. The curve in Fig. 10.6a represents the case of a tip and a surface, both of which are incompressible, and with the surface being covered with a thin aqueous film. The various segments represent: Approach half-cycle: AB – tip and surface are well-separated, no interaction; BC – the tip senses the attractive interaction from the meniscus layer, the force constant of interaction exceeds k_N, and the tip snaps into contact with the 'hard' surface; CD – tip and surface are incompressible and the stage travel distance must be equal to the lever deflection. Retract half-cycle: DE – the system retraces itself since all deformations/deflections are elastic; EF – the meniscus interaction has increased due to capillary action, and there will be a greater lift-off instability/discontinuity (than for the snap-on); FA – the return to large separation and no interaction. This kind of curve is representative of events for a Type I instrument due to adsorbed moisture. The meniscus interaction is generally a nuisance feature in that it will mask other surface mechanical effects. With a Type I instrument the meniscus can be eliminated by doing F–d analysis under water (or some other fluid ambient); a Type II instrument will of course pump away the meniscus. The 'hard' surface F–d curve is used to calibrate the detection system so that a measurable detector response can be related accurately to the lever deflection.

The schematic curve in Fig. 10.6b shows lever deflection as a function of tip indentation of the specimen surface or separation from the surface, the difference between stage travel and lever deflection, and thus illustrates other surface mechanical aspects of the system which are accessible to F–d analysis [17]. The shape of the curve assumes that at no stage is the force

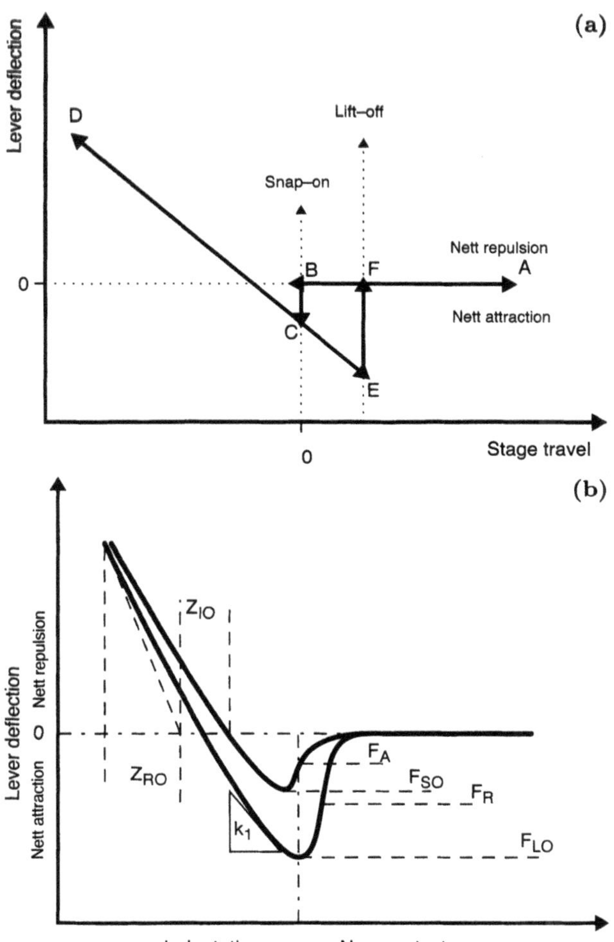

Fig. 10.6. Schematic F–d curves. Lever deflection is plotted along the vertical axis. **(a)** Interaction between a 'hard' surface and an incompressible tip is shown, with stage travel as the independent variable. The effect of the meniscus aqueous layer is apparent, and gives rise to instabilities at the points of snap-on and lift-off. **(b)** An incompressible tip interacting with a compressible surface and being subjected to adhesive interactions at the interface. The horizontal axis now shows explicitly the indentation of the surface by the tip

constant of interaction greater than k_N (hence no instabilities of the snap-on/lift-off varieties). The information content of the generic F–d curve in Fig. 10.6b may be summarized as follows:

- The force constant of interaction, k_i, is simply the slope of the curve, $k_N z_L/|z_d - z_L|$, where z_d refers to stage travel.
- The forces at 'contact' on approach and retract, F_A and F_R, may be defined at the inflection points where the force constants of interaction are the greatest.
- The snap-on and lift-off forces, F_{SO} and F_{LO}, are measured at the points of greatest net attraction. The latter is generally taken to be the force of adhesion.
- The distances z_{RO} and z_{IO} are measures of the elastic recovery and the plastic indentation, both at zero lever loading.
- The extent of hysteresis in the system is given by the area enclosed by the curves.

The parameters defined above cannot readily be related to the familiar macroscopic definitions of mechanical properties, e.g. hardness, adhesion, Young's modulus, tensile strength, flexural strength, etc, unless the system can be specified further (e.g. tip shape, contact area, surface free energy of tip, etc) and unless additional assumptions are made (homogeneity, isotropy, surface topography, etc).

A consequence of the discussion above is the need to match the force constant of the lever to that of the interaction being investigated in order to extract maximum information. If there is mis-match, then either the lever is the only compliant element, and no information is obtained about the surface, or the surface is the only compliant element, and the deflection of the lever is not measurable.

10.4.5 Lateral Force Microscopy (LFM)

The essential feature of the LFM is that the feed-back loop maintains control as in the conventional contact mode, but that torsional deformations of the lever are monitored by sensors orthogonal to those that generate the signal for the AFM control loop. The difference signal from the orthogonal lateral sensors as a function of position is then used to generate a friction force image. LFM is gaining in popularity as a method capable of producing image contrast from lateral differentiation; it has also extended tribology into the nano-range, and can offer insights into mechanisms of friction and wear on mesoscopic and atomic scales. The LFM arrangement is shown schematically in Fig. 10.7.

The tip-to-surface interaction, assuming that the tip is moving in a direction perpendicular to the long axis of the lever, can be described (in the

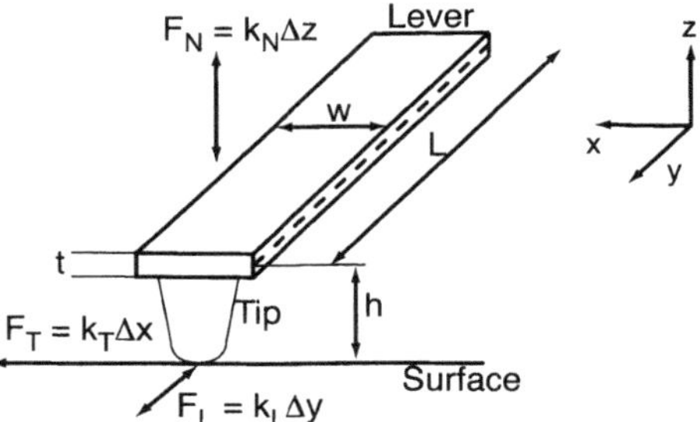

Fig. 10.7. The LFM system. The lever applies and senses tip-to-surface forces with components $F_{\mathrm{N}} = k_{\mathrm{N}} z$, $F_{\mathrm{T}} = k_{\mathrm{T}} x$ and $F_{\mathrm{L}} = k_{\mathrm{L}} y$

meso/macro-scopic regimes), by

$$F_{\mathrm{T}} = \mu(F_{\mathrm{N}} + F_{\mathrm{N,adh}}) \tag{10.7}$$

or

$$F_{\mathrm{T}} = \mu F_{\mathrm{N}} + F_{\mathrm{N,adh}} \tag{10.8}$$

where F_{T} is the torsional friction force, F_{N} is the normal force imposed, $F_{\mathrm{N,adh}}$ is the additional normal force due to attractive interactions between tip and surface, and μ is the coefficient of friction. The choice of expression depends on the adhesive interaction being static or dynamic; both expressions assume multi-asperity contact. A non-linear dependence will arise from single-asperity Hertzian contact, i.e.

$$F_{\mathrm{T}} = AF_{\mathrm{N}}^{n} + BF_{\mathrm{N}} \tag{10.9}$$

where A and B are 'constants'. The $n \neq 1$ exponent arises from the change in contact area with increasing load [18]. In the regime most closely resembling that explored by traditional methods we have that $F_{\mathrm{N}} = k_{\mathrm{N}} z$. Since the z-resolution is typically 0.01 nm, the minimum detectable normal force is 10^{-4} nN; assuming a maximum deflection of the lever of 10^{3} nm, the maximum normal force transmitted through the tip will be 10^{5} nN. The friction force is $F_{\mathrm{T}} = k_{\mathrm{T}} x$ where k_{T} is related to the torsional resistance of the lever and the height of the tip, h. The deflection of the tip from its original undeflected position is x, for the chosen geometry. The angle of rotation of the lever at the position of the tip will be x/h. Given a resolution of 10^{-7} rad, and taking the tip height to be 10 µm, this converts to a minimum detectable x being approximately 10^{-3} nm, in principle, which is degraded to 5×10^{-2} nm by the scanning stage. The conventional coefficient of friction is then

$$\mu = \frac{k_{\mathrm{T}}}{k_{\mathrm{N}}} \frac{x}{z} \tag{10.10}$$

with a resolution limit of $\mu \approx 10^{-3}$ for the current generation of instruments and commercially available probes. The adhesive interaction is determined by F–d analysis, while wear can be induced at high loads, and quantified by conventional AFM imaging.

10.5 Procedures for 'Best Practice'

In comparison with the more mature electron-optical microscopies and surface analytical techniques 'best practice' and 'quality assurance' have yet to be established for SFM at the present stage of the learning curve. A great deal of useful material on methods and procedures can be found in the literature, although widely scattered, and sometimes not in a readily accessible format. The present state of affairs has been reviewed [19] and some of the most salient points are summarized below. It will be convenient to consider an SFM system in terms of the relevant elements and their functions, as in Table 10.3 below.

Table 10.3. Calibratable parameters/variables of elements/functions of an SFM system

Element/Function		Requirement*
Scanning stage:	X–Y plane	X–Y calibration
		X–Y orthogonality
		X–Y linearity
	Z-motion	Z calibration
		Z orthogonality
		Z linearity
	X–Y–Z	Thermal Drift
Sensor stage:	Deflection (c_N)	Linearity
		Conversion factor (nA(V) to nm)
	Lateral/torsional (c_T)	Linearity
		Conversion factor (nA(V) to radian)
Lever:	k_N (deflection)	Relative/absolute value
	k_T (lateral/torsion)	Relative/absolute value
	k_L (lateral/longitudinal)	Relative/absolute value
	Resonance	Frequency and envelope
Tip:	Aspect ratio (A_r)	Value
	Apex shape (R_{Tip}, etc)	Value and description
	Tilt	Value and description
	Height (h)	Value
Non-contact:	Amplitude	Value
	Effective height	Value
	Effective force	Value and description

* In most cases the 'requirement' refers to determination of *actual*, as opposed to generic, values for parameters or variables which affect the interpretation and reproducibility of results.

10.5.1 Spatial Characteristics of Scanners

An SFM system may have useful lateral resolution down to 0.1 nm, while the field of view may be as large as $100 \times 100\,\mu m^2$; thus the linear dynamic range may be a factor of 10^6. In practice the range will be covered by two or more stages, e.g. a 1 μm stage for high resolution scanning, a 10–25 μm general purpose stage, and a 100 μm one when a large field of view is required. Stages will generally be supplied fully calibrated with a suitable set of parameters recognisable by the relevant control software package. In many cases the as-received calibration will suffice. However, non-ideal characteristics such as nonlinearity, hysteresis, aging and creep affect the accuracy, resolution and reproducibility of spatial mapping.

In general x–y plane calibration procedures require the use of standard specimens. Standard specimens are usually supplied by the manufacturer, e.g. HOPG (highly oriented pyrolytic graphite) and mica, useful for x–y calibration of the highest resolution stage, and micro-machined semiconductor grids for calibration over larger fields of view. As well, one can prepare well-ordered specimens from monodisperse latex spheres of known diameter [20–22]. The spheres are available in the size range from 50 nm to several μm, and can be obtained with certification traceable to primary size standards. Perfect hexagonally close-packed mono- and multi-layer structures can readily be prepared, thus allowing calibration of all relevant x–y plane parameters (linearity, orthogonality and magnification). Point defects will be found occasionally, and can serve as useful markers for determination of resettability and drift. The features of a typical specimen are shown in Fig. 10.8; it will be apparent that contour lines taken along particular directions and 2-D fast Fourier transform techniques, if available, are convenient tools.

Two of the most important attributes of SFM are, firstly, that the depth of field is limited only by the dynamic range of the stage in the z-direction, and, secondly, that the spatial resolution in the z-direction may be better than 0.1 nm (absolutely) and 1 part in 10^4 (relatively).

The precision in the z-motion will be conditioned by the performance of the piezoelectric actuator. Once again it will be convenient to resort to use of standard specimens. Single d-spacing features on cleaved surfaces (e.g. MgO, Si(111) and NaCl) will be useful in some cases, but it is more common to use VLSI standards for calibration along the z-direction. A flat surface with a known tilt angle, will give rise to a z-excursion over the field of view comparable to the z dynamic range, and can be used to explore linearity and hysteresis [23]. Alternatively, the known slope of a pyramidal tip may be used for the same purpose [24]. A direct and convenient laser interferometric method has been described [25,26]. As in the case of the x–y plane calibration, layers of monodisperse latex spheres are excellent for z-calibration [22,27]. Monodisperse Au particles have also been used [28].

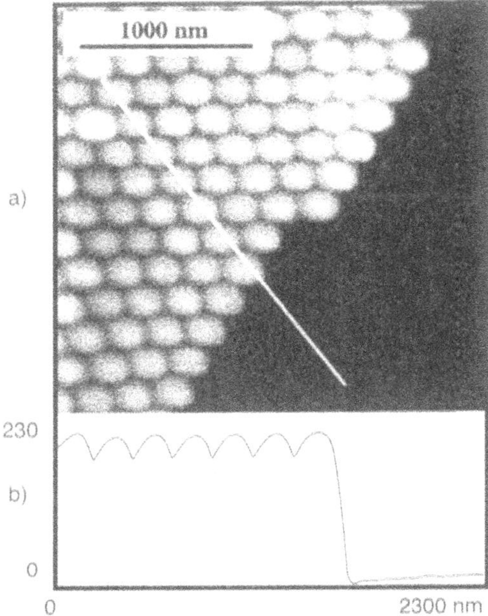

Fig. 10.8. **(a)** Contact mode image of a polystyrene bead layer (200 nm spheres) at an edge. **(b)** A z-direction line profile along a close-packed axis, extending to the edge, shows that the scanning stage is calibrated accurately along x-, y- and z-axes [22]

10.5.2 Determination of c_N and c_T

It is common practice (e.g. [29]) to calibrate the sensitivity of the optical lever arrangement, and thus obtain c_N, in units of nm deflection per unit of detector signal, from the linear part of an F–d curve when the tip is pressed against a 'hard' surface; the travel of the sample stage in the z-direction is then equal to that of the lever at the tip.

The position of the PSD is generally adjusted with lead screws along axes orthogonal to the incident laser beam [30]; the facility allows adjustment of zero difference signal when the lever is unloaded. Thus given the length of the optical path from the lever to the detector, and of the pitch of the adjustment screws, angular deflections can readily be measured independently as functions of PSD signal by translating the PSD known distances perpendicular to the beam. The angular sensitivity factors can readily be related to c_T and c_N, and the results are independent of beam shape and spot location. Moreover, the linearity can readily be checked.

10.5.3 Determination of Spring Constants

Both beam and V–shaped levers are in common usage. The supplier will generally provide a generic value for k_N; the actual value for a particular lever may differ by a factor of two, however. At the present time suppliers do not specify k_T, much less k_L. The expressions for k_N and k_T for a 'long' beam-shaped lever are given by [31].

$$k_N = \frac{Et^3 w}{4L^3} \tag{10.11}$$

$$k_T = \frac{Et^3 w}{6(1+\nu)Lh^2} \; . \tag{10.12}$$

where E is Young's Modulus, t is the thickness of the lever, ν is Poisson's ratio, and other dimensions are defined in Fig. 10.7. For completeness, the expression for k_L is

$$k_L = \frac{Et^3 w}{6Lh^2} \tag{10.13}$$

The 'difficult' quantities are t and E, the former being difficult to determine even with SEM imaging and the latter being dependent on crystallographic orientation, dopant concentration and stoichiometry. However, given that k_N can be determined reliably, the two other spring constants can be calculated relatively accurately from

$$k_T = k_N \frac{2L^2}{3(1+\nu)h^2} \tag{10.14}$$

$$k_L = k_N \frac{L^2}{3h^2} \tag{10.15}$$

In the case of LFM it can be shown that the limits of detectability for μ is given by

$$\mu = \frac{F_T}{F_N} = \frac{2L^2}{3(1+\nu)h^2} \frac{x}{z} \tag{10.16}$$

For a typical lever $\mu \approx 4 \times 10^2 x/z$. For the current generation of instruments x_{min} and z_{max} are 0.05 nm and 20 µm, respectively. Consequently there is a limit of detectability of $\mu \approx 10^{-3}$. Also it can be shown that $k_T \approx k_L \approx$ 400 kN.

Modelling of the torsional, longitudinal and normal spring constants of V-shaped levers is rather more difficult [32] due to the more complicated geometry and multiple bending modes; closed-form expressions have been obtained, which allow in-principle evaluation of the constants.

Several procedures for determining k_N to a precision of 10–30% are described in the literature, [31] and references therein. These can be divided into two broad categories, being based either on comparison against external standards or on the use of analytical or numerical expressions. A convenient version of the former variety has been reported [29] which requires once-off determination of k_N for a particular Si lever, using SEM analysis to measure the geometrical dimensions and taking E as the accepted value for Si of the appropriate crystallographic orientation. Other levers can then be calibrated *in situ* by F–d analysis using the standard lever as the 'specimen'. Equations (10.14, 10.15) may then be used to calculate k_T and k_L. A more direct method using slopes of a facetted $SrTiO_3$ surface for determining k_L has been described [33].

10.5.4 Determination of Actual Tip Parameters in the Mesoscopic Regime

The outcome of an AFM experiment will depend on the shape of the tip. The as-received shape will vary from tip to tip. Moreover, the shape is in general a dynamic property likely to alter during the conduct of an experiment, by exchange of material between tip and surface. Thus it is desirable to check tip shape before and after, or even during, an experiment. A convenient method for tip characterisation is 'reverse' imaging [34]. The essential aspect of the method consists of scanning an 'unknown' tip over a surface feature having a shape approximating that of a delta-function spike. The result is an image of the tip. The outcome is due to the spike having much higher aspect ratio, A_r = (height of tip)/(width at the base), and much smaller tip radius, R_{Tip}, than those of the 'unknown' tip. The supplier's nominal value for A_r will be reliable, in general, while the nominal value of R_{Tip} is less reliable. Imaging other objects with known geometries (e.g. polystyrene spheres, gold particles, TMV, etc.) can produce values for subsets of those parameters [35–38]. While reverse imaging can provide accurate information about tip shape in the mesoscopic domain (> 5 nm), information at the atomic level of detail cannot be obtained by any routine methodology.

A commercially available sharpened tip is often an adequate approximation to a delta-function in comparison with a standard pyramidal tip [19,31]. An up-turned tip can be mounted as a standard specimen, or even be incorporated into the actual experiment. The tip height can be determined from a line profile with a precision of about 5%. Fig. 10.9 shows reverse images of tips at high resolution. An apex angle of 70° and $R_{Tip} = 35$ nm can be deduced from Fig. 10.9a; the results are in good agreement with the nominal values of 70.5° and 40 nm, respectively.

The practical importance of tip analysis is apparent from Fig. 10.9b. Reverse imaging over two $BaCO_3$ spikes show that the actual 'tip' consisted of two blunt apexes ($R_{Tip} \approx 125$ nm), offset by some 500 nm laterally and 65 nm vertically. The use of non-ideal tips will result in anomalous outcomes, even though the spring constants and other relevant parameters may be identical and/or well-characterised.

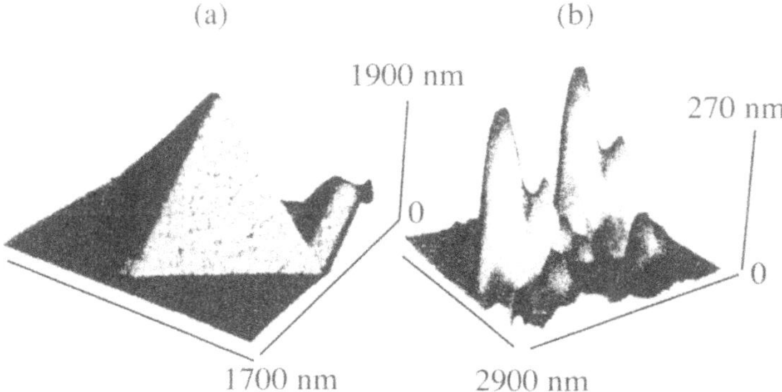

Fig. 10.9. Reverse images of AFM tips. (a) Tip shape can be determined from a line profile. (b) Shape of an altered tip

10.6 Illustrative Case Studies

10.6.1 Surface and Defect Structures of WTe_2 Investigated by UHV-STM

The layered transition-metal dichalcogenides have found practical application as solid lubricants and are under investigation for applications in catalysis. Being layered and relatively unreactive in air they are also attractive for STM/AFM analysis. The formula unit building blocks are joined along the c-axis by van der Waals interactions, so that atomically flat surfaces of macroscopic dimensions can be prepared by cleavage between adjacent chalcogenide layers. The electronic properties of the compounds range from semi-metallic (e.g. WTe_2) to small band-gap insulators (e.g. WSe_2). Several members of the family exhibit hexagonal symmetry (e.g. $MoTe_2$ and MoS_2). Thus it is difficult to correlate unambiguously the structures in atomically resolved STM maps with the underlying crystal structure and the proposed electronic structure. WTe_2, on the other hand, has an orthorhombic unit cell, Fig. 10.10, with two non-equivalent sites for both the W and Te species. Accordingly it should be easier to relate the STM map to the proposed contributions of orbitals located in the top Te layer and the W layer below. The summary outcomes below are drawn from two successive studies [39,40].

Typical atomically resolved images are shown in Fig. 10.11. The images also show two distinctly different point defect configurations which were assigned tentatively to vacancies occupying the two non-equivalent Te sites. Defect dynamics was apparent; high-speed repeat scanning over a field of view showed creation and annihilation of defects in real time, from which it was inferred that surface migration was taking place. Since WTe_2 is a semimetal with a pseudo-bandgap, conditions for optimum image contrast were obtained at tunnel voltages of a few meV. Taking the simple Hamann-Tersoff model as a guide it was shown that the tunnel gap was then less than 0.3 nm,

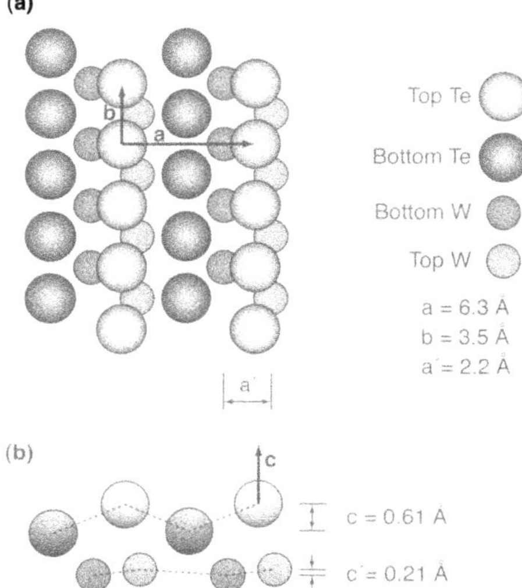

Fig. 10.10. Schematic of the WTe_2 structure, (**a**) horizontal plane, (**b**) vertical plane. The 'top' and 'bottom' Te atoms are indicated by large light and dark spheres, respectively, while the 'top' and 'bottom' W atoms are drawn similarly as small spheres. The unit cell vectors are drawn, as are some of the most relevant dimensions and displacements

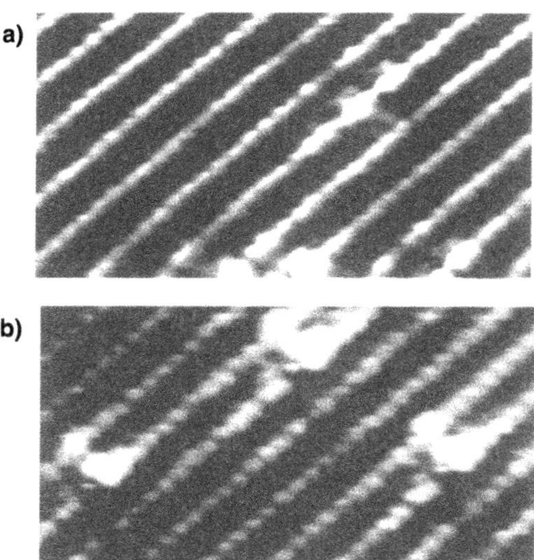

Fig. 10.11. Atomically resolved images of WTe_2, obtained at negative sample bias, showing the general surface structure and point defects. The spacing between the close-packed rows is 0.63 nm

which is a distance comparable to the nearest neighbour distances (ranging from 0.27 to 0.37 nm) in the W–Te block. The 'independent electrode' assumption must therefore be of limited validity in the present case, and a better description of the system should be in terms of a travelling bound state where the orbital located in the apex atom of the W-tip is mixed with the orbitals located at the surface atom sites. The expected break-down of the simple model was confirmed by images which showed that the character of the map altered reversibly (repeat scans over the same field of view showed sudden, frequent and reversible alterations in surface structure as judged from the grey-scale map). The presence of point defects were used as spatial markers as well as to show that the state of the tip affected the STM map of the electronic structure of the surface. It is hypothesized that mobile surface species are intermittently transferred to and from the tip, thus affecting the character of the travelling bound state.

10.6.2 Organic Thin Film and Surface Mechanical Characterization

Thin organic films are the essential elements for many emerging and traditional industries, e.g. [41], based on products and processes such as adhesives, composites, semiconductor planar lithography, paints, coatings in the food industry, magnetic information mass storage media and advanced lubricants [42–47].

While 'traditional' surface and interface analysis of thin films has much to offer, AFM methodologies and F–d mode, in particular, has relevance in the present context [48]. The description below is based on accounts in the literature [49,50].

Deployment of F–d analysis on thin films will result in a system consisting of three interacting compressible elements (lever, film and substrate) which can be modelled by three linear springs and two interface interactions. An idealised generic F–d curve is shown in Fig. 10.12. For simplicity it may be assumed that the tip has a rectangular shape and that the compressed surface will adapt to the shape of the tip. It can then be shown [49] that the spring constant of the compressible substrate is defined for large force loadings, when the thin film is fully compressed, by

$$k_{\mathrm{sub}} = \frac{k_{\mathrm{N}} S_{\mathrm{sub}}}{1 - S_{\mathrm{sub}}} \tag{10.17}$$

where $S_{\mathrm{sub}} = z_{\mathrm{L}}/z_d$ is the slope of the F–d curve defining lever deflection versus stage displacement, both referenced to the point of first contact. For smaller total compressions, when the film is compliant, it can be shown that the film thickness and the spring constant of the film are given by

$$T = \frac{z_d(1)}{S_{\mathrm{sub}} - S_{\mathrm{layer}}} \tag{10.18}$$

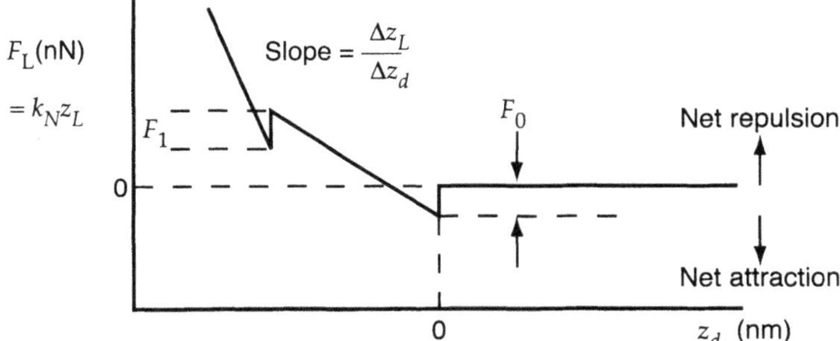

Fig. 10.12. Generic F–d curve for a system with three compressible elements in series – substrate (k_{sub}), thin layer/film (k_{layer}) and lever (k_{N}). Two attractive interactions are sensed by the tip, when it touches the outside surface of the layer/film (F_0), and when the layer is compressed to negligible thickness and the tip senses the layer-to-substrate interface (F_1)

$$k_{\text{layer}} = k_{\text{N}} \frac{S_{\text{layer}} S_{\text{sub}}}{S_{\text{sub}} - S_{\text{layer}}} \tag{10.19}$$

where S_{layer} is the slope of the F–d curve in the region where the film is compliant, and $z_d(1)$ is the tip position at the point where the tip 'senses' the film-to-substrate interface.

(a) Application to Treated Wool and Magnetic Tapes. Long-chain fatty acids are attached by ester linkages to surfaces of wool fibre. The surface layer thus consists of 18-methyleicosanoic acid chains, which are oriented away from the interface and form a hydrophobic surface [51,52]. The layer has a critical role in determining surface properties of the wool fibre, which in turn are important in many aspects of wool processing.

The wool fibre surfaces exhibited longitudinal striations, as in Fig. 10.13. The structure is shown in the line profile drawn perpendicularly to the direction of texturing. The spacing of the features was some 500 nm while the vertical amplitude of corrugation was *ca.* 20 nm. The AFM analysis shows that the striations remained prominent after exposure to water.

Magnetic video tapes are multilayer structures consisting of a polymer backing, a magnetic film, and one or more chemically protective layers which are also intended to lower friction and adhesion *vis à vis* the read/write head. Tribological investigations of video tapes and head sliders have been carried out by Bhushan et al. [53,54]. Two types of tape were analysed: VH8–PET polymer backing/mag. layer/lubricant film; and DVC–PET polymer backing/mag. layer/DLC/lubricant film, where DLC is diamond-like carbon. The lubricant film consists of polar long-chain molecules (modified perfluoropolyether). The tape surfaces were relatively featureless; typical surface

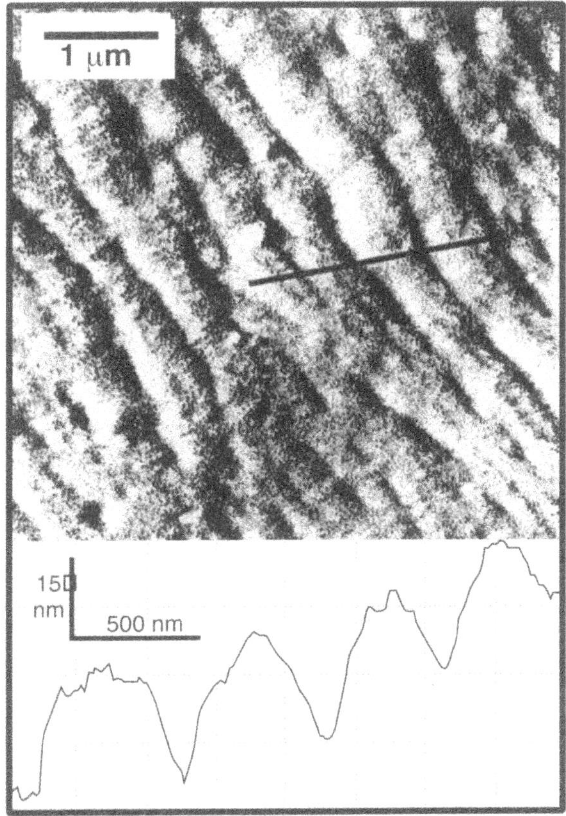

Fig. 10.13. AFM image show details of the surface texturing for wool. The spacings and heights of surface texturing are shown by the line profile

roughness was 11.4 ± 1.5 and 8.1 ± 1.0 nm for the VH8 and DVC tapes, respectively.

The outcome of F–d analyses on scoured wool is illustrated in Fig. 10.14; similar results were obtained from the video tapes. There is an excursion during approach due to net repulsive interaction which is followed by an adhesive 'snap-on', and then by another excursion into the repulsive force regime. In the context of the present measurements the thin film can be thought of as a linearly compressible element of finite thickness. The three-element system will be compressed so that the slope of the curve will reflect the effective stiffness of the combined system. When the lipid layer is near its full compression, the tip senses the adhesive interaction at the film-to-substrate interface and is pulled into contact with the substrate. From that point onwards the film is passive, having effectively been compressed to a

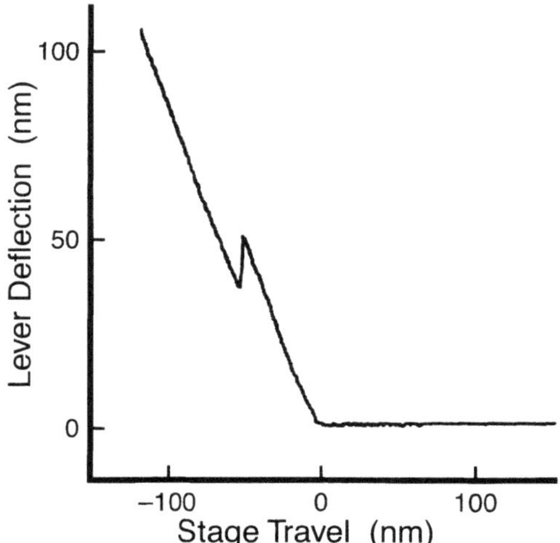

Fig. 10.14. F–d data obtained under water on scoured wool. Note the similarity with the generic curve in Fig. 10.12

Table 10.4. Outcomes of AFM thin film analysis

Specimen	k_{layer} (N/m)	k_{sub} (N/m)	E_{layer} (MPa)	$F_{\text{ahs}}^{\text{int}}$ (nN)	T (nm)
Wool	0.13 ± 0.02	0.12 ± 0.01	8–15	0.3 ± 0.1	2.5–7
VH8	0.018 ± 0.003	0.08 ± 0.01	2.8 ± 1.6	2.7 ± 0.8	7.3 ± 2.0
DVC	0.028 ± 0.005	> 3*	2.4 ± 1.9	0.5 ± 0.15	8.2 ± 2.5

*The lower bound is determined by the stiffness of the lever.

rigid layer of negligible thickness, and the system reverts to having only two compressible elements.

The inferred spring constants of the substrate and the film, and the thickness of the film are given in Table 10.4. The spread in values for T is due to surface roughness, uncertainty in tip shape, and uncertainty in choosing the point of initial contact. It has been shown, e.g. [55], that when a parabolically shaped tip indents a compliant surface, then Young's Modulus, E, can be inferred from

$$F_{\text{load}} = \frac{4(R_{\text{para}})^{1/2}}{3} \frac{E}{1-\nu^2} z^{1.5} \tag{10.20}$$

The load is the sum of the restoring force of the compressed layer and the adhesive interaction, $F_{\text{load}} = k_{\text{layer}}T + F_0$, $z = T$ is the extent of indentation/compression, and $R_{\text{para}} = R_{\text{Tip}} = 50\,\text{nm}$. The results are given in Table 10.4; the spread in values for E is a direct reflection of uncertainties in determination of T. The adhesive 'snap-on' force at the interface was also determined.

10.6.3 AFM Analysis of 'Soft' Biological Materials

(a) Tobacco Mosaic Virus (TMV). TMV is a well-characterized and convenient biological object; it has a cylindrical shape with a diameter of some 18 nm and a typical length of 300 nm. It is composed of identical protein subunits and a single strand of RNA, with the subunits having a helical arrangement with a pitch of 2.3 nm and 49/3 units per turn. The structure has a central channel of 4 nm diameter.

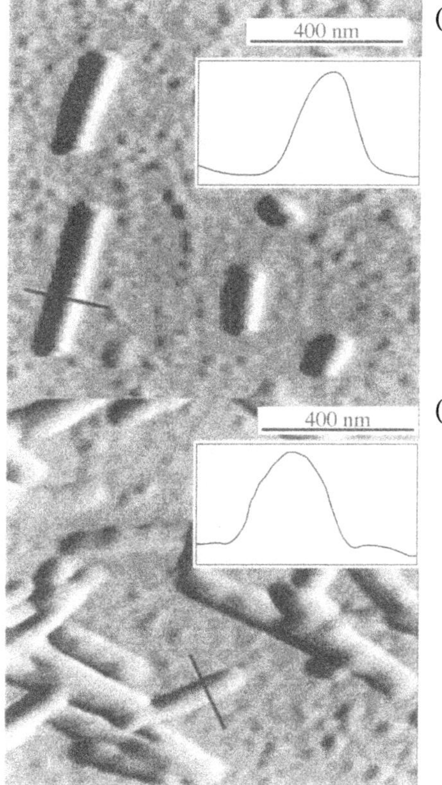

Fig. 10.15. TMV attached to mica substrates and imaged in **(a)** contact and **(b)** intermittent contact modes. The contour lines in the insets show the measured height of the TMV and the tip-induced lateral broadening

Table 10.5. Apparent dimensions of TMV

Imaging mode	N*	Height (nm)	Width (nm)
Contact	50	19.3 ± 1.2	82.2 ± 10.1
Intermittent contact	25	10.7 ± 1.0	143.3 ± 7.9
Lateral force	25	n.a.	79.1 ± 6.5

*Number of measurements

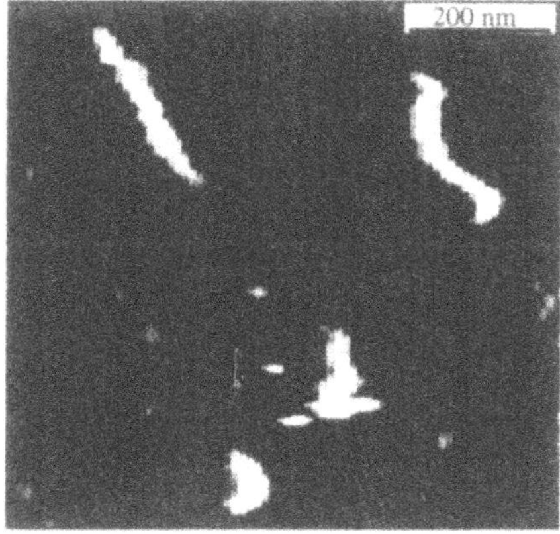

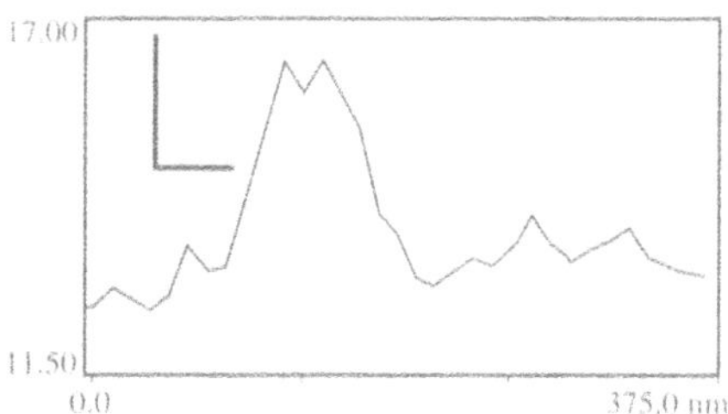

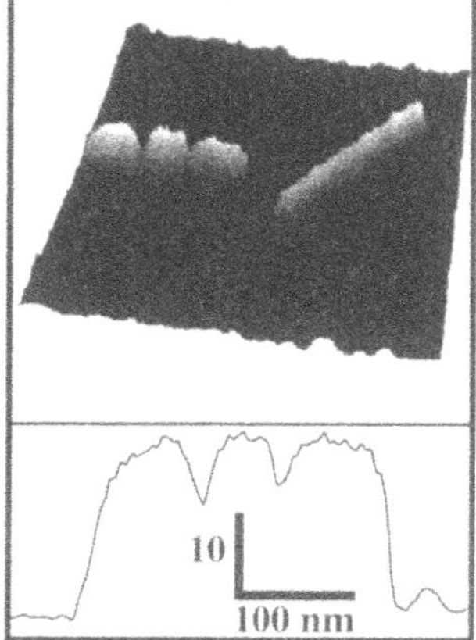

Fig. 10.16. Nano-dissection of TMV. *(Top)* Selective removal of material reveals the central channel. *(Bottom)* Notches can be cut in the TMV structure by repetitive contact mode scanning along a line

The outcomes of contact and intermittent contact mode imaging are shown in Fig. 10.15 [56]. Dimensional analysis of numerous images was carried out, and the results are summarized in Table 10.5. It is apparent that the height in the contact mode is in good agreement with the known diameter of TMV, while the widths exceed the diameter due to convolution of the shape of the tip with that of the object.

Repeated scanning at higher force loadings in the contact mode resulted in controlled dissassembly of a TMV. The top panel in Fig. 10.16 show that the central channel in the TMV can now be revealed and resolved, and small sections of the TMV can be removed selectively [56].

(b) *In Vitro* Analysis of Cells by AFM. The principal merits of the AFM for investigation of cellular processes are due to its functionality in a fluid environment, and to the interaction being relatively unintrusive. Thus live cells can be imaged *in situ* in a biocompatible fluid. Due to the extreme

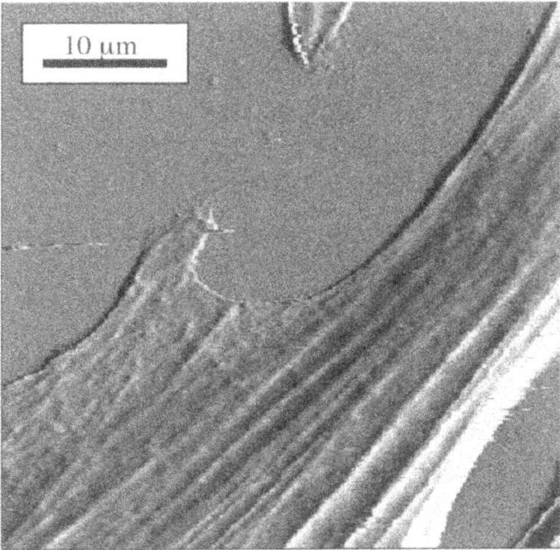

Fig. 10.17. Contact mode image obtained *in vitro* of a fibroblast

deformability of the plasma membrane, cytoskeletal and other intra-cellular structural elements can be mapped by the tip. The AFM can also track continuously the temporal evolution of cellular processes [57–59]. The results below have been reported in full elsewhere [60].

Primary human skin fibroblasts were grown on sterile untreated glass coverslips and sterile tissue culture dishes. The latter substrates are pre-treated by the manufacturer so as to present a negatively charged and hydrophilic surface, thus ensuring better cell adherence.

On relatively rare occasions good imaging conditions, in the contact mode, were obtained for cells on coverslip substrates. The resolution was similar to that obtained in previous studies on live glial cells [61,62], and can be accounted for with a membrane deformation model. On most occasions, however, imaging of cells grown on coverslips revealed only a very general outline of the cell, with complete absence of fine structure.

Imaging of cells grown in treated culture dishes was dramatically different in comparison with those described above. Irrespective of lever force loading (up to *ca.* 20 nN), size of the field of view, or scan speed, well-resolved images were now the rule. A typical image of a live human fibroblast is shown in Fig. 10.17. Continuous scanning was carried out over single cells for periods of several hours without any significant deterioration in image quality.

F–d analysis was carried out, Fig. 10.18; only approach curves are shown. The upper curves demonstrate the low effective stiffness (shallow slope) of the weakly supported plasma membrane for a live cell in comparison with a relatively stiffer region (steeper slope) for the same cell and with the 'hard' substrate (straight line with the steepest slope). The data were plotted as $\ln F_N$ *versus* $\ln z$ (force applied by the lever *versus* indentation).

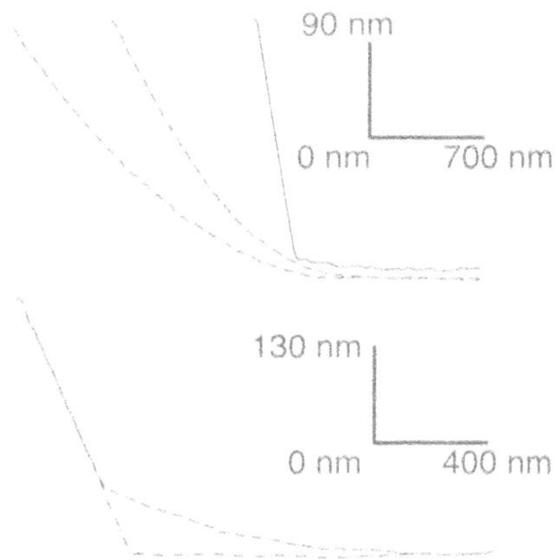

Fig. 10.18. F–d analysis: *(Top)* Calibration curve (straight line with steepest slope on right) and results of indenting a soft *(left curve)* and a stiffer region *(middle curve)* of the plasma membrane; *(Bottom)* the hard substrate will provide support for the cellular structures being compressed

The outcomes were straight lines having slopes of 2, thus justifying the assumption of conical tip shape for $z \gg R$, in accord with the expression $\Delta z_{\text{cone}} = \left[F\left(\frac{\pi}{2}\right) \tan \alpha (1 - \nu^2)/E\right]^{1/2}$, where α is the opening half-angle of the tip. The effective Young's Modulus, E, was found to be 4.6 and 17 kPa, respectively. The former represented the effective modulus for a weakly supported membrane section, while the latter higher value was obtained from a location supported by a stiffer cytoskeletal substructure. The lower graph in Fig. 10.18 demonstrates additional stiffening at high force loading, when cytoskeletal elements become trapped between the tip and substrate.

The improved imaging conditions due to use of an activated substrate described above suggest that strong adhesive tip-to-surface interactions, resulting in shear stress, are principally responsible for variable imaging conditions of live cells. The treatment of the surface of the tissue culture dish used in these experiments is proprietary, but the trade literature claims negative charging and hydrophilicity for the surface. AFM Si_3N_4 tips should in principle be oxidised and be hydrophilic, however, in practice they will be hydrophobic due to hydrocarbon contamination. Adhesive interactions between functionalised tips and surfaces have been studied by Frisbie et al. [63], who found that the strengths of adhesive lift-off forces ranged from 8 (COOH/COOH) to 3 (CH_3/CH_3) to < 1 nN (COOH/CH_3). A phospholipid monolayer may be expected to have hydrophobic patches, and one would therefore expect strong adhesive interactions at both bilayer-to-substrate and tip-to-bilayer interfaces. Hence the tip will bond to sites in the phospholipid surface and the resulting shear forces will disassemble the membrane. Large segments of membrane material and other debris will be carried along transiently by the tip, and will be the cause of loss of spatial resolution and instability of imaging conditions. The results suggest that the activated substrate acts as a selective scavenger thus preventing build-up of debris on the tip.

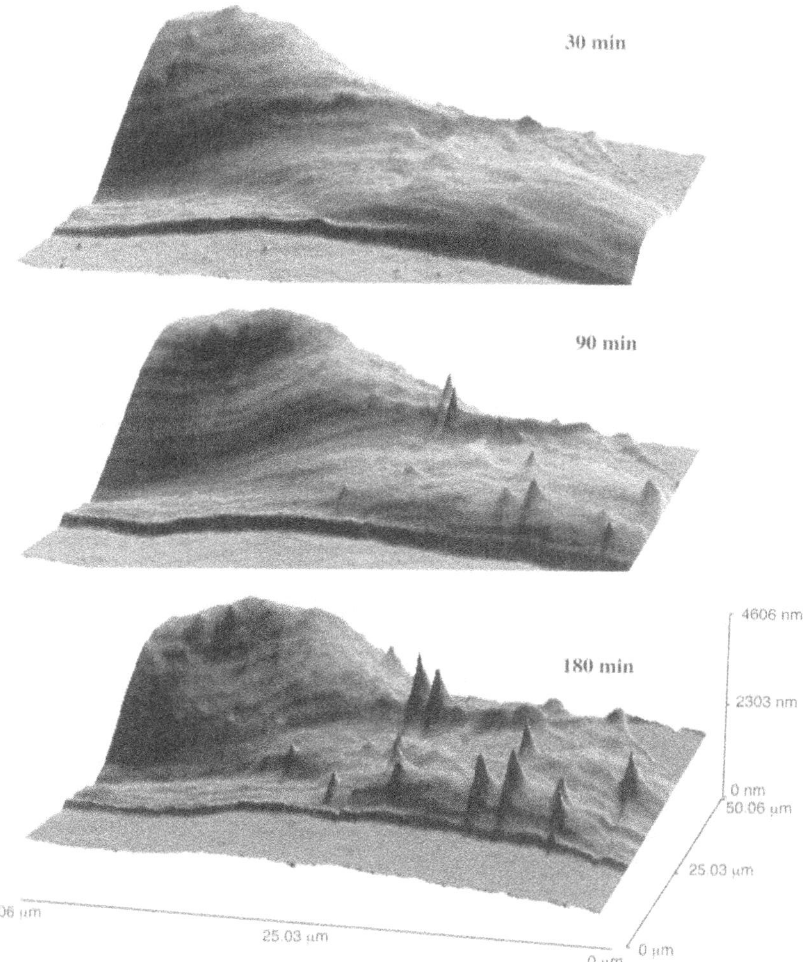

Fig. 10.19. A sequence of *in vitro* images showing intra-cellular nucleation and growth of formazan crystals

Fibroblasts were treated with cytochalasin and imaged *in situ* during the evolution of the system. A sequence of images [60] suggests that the integrity of the cytoskeletal structure is degraded inwards from the plasma membrane. The observation is in accord with a model ascribing the dynamics of actin filaments to addition (loss) of actin monomers near to (away from) the lamellipodial plasma membrane. Thus if the addition process is inhibited, then the filaments will retract from the membrane. The sequence of images in Fig. 10.19, obtained *in vitro*, shows intracellular growth of formazan crystals during the application of the MTT test, which is commonly applied to a cell population in order to determine the viable fraction.

10.6.4 Nanotribology of Solid Lubricants

(a) Diamond-Like Carbon (DLC) Films. DLC films have many attractive properties which can be tailored for a range of actual or potential applications [64–67]. In particular, the tribological properties of DLC films are beginning to attract interest. Since many applications are concerned with macroscopic surfaces in sliding contact, the traditional macroscopic methods (e.g. pin-on-disk or ball-on-disk) are often the most relevant indicators of performance. However, the trend is toward smaller devices and point contacts. Accordingly functionality must also be validated on the (sub)μm-scale, requiring measurements which are spatially resolved on the mesoscale and indeed into the nano-regime. A study was undertaken in order to measure, on the mesoscopic scale, friction and surface adhesion for DLC coatings from two fabrication routes, and to explore the influence of the two most likely service environments, air and vacuum [68].

Materials from two fabrication routes, IBAD (ion-beam assisted deposition) and RF-CVD (radio-frequency assisted chemical vapour deposition), were considered. The primary LFM data were obtained from the 'friction loop', an example of which is shown in Fig. 10.20. The average lateral force along the track is given by half the average peak-to-peak signal of the loop converted into lateral tip deflection multiplied by the torsional spring constant, i.e. $\langle F_{\mathrm{L}} \rangle = c_{\mathrm{T}} \langle V_{\mathrm{L}}/2 \rangle k_{\mathrm{T}}$. A subset of the processed experimental results obtained with a *blunt*, $R_{\mathrm{Tip}} > 100$ nm, tip are shown in Fig. 10.21. The results can be described by a multi-asperity model, which predicts a linear

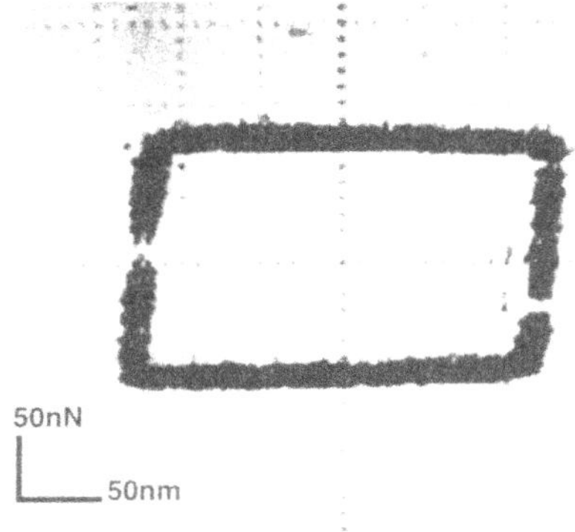

Fig. 10.20. Typical friction loop. The track length plotted along the horizontal axis was *ca.* 200 nm. The lateral force on the tip is plotted along the vertical axis. The slopes of the near-vertical sections represent static friction when the stage reverses its direction of travel

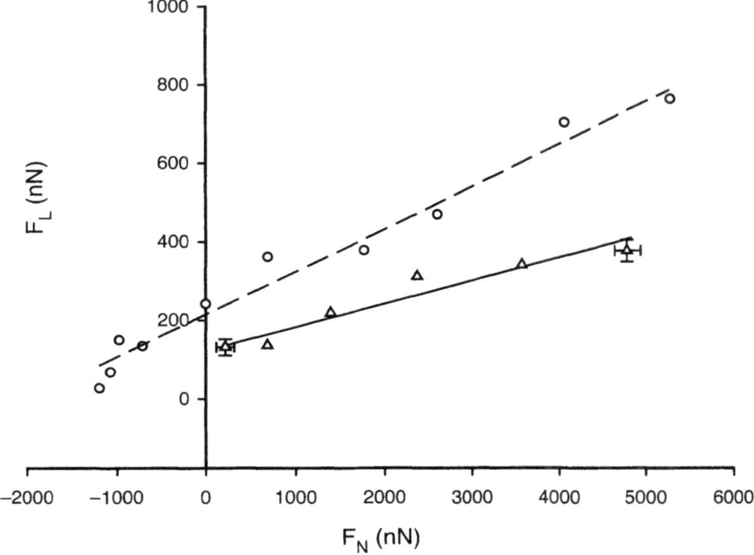

Fig. 10.21. Friction data for an IBAD specimen showing F_{L} as a function of F_{N}

dependence irrespective of the gross shape of the tip. The sets of data were therefore fitted by straight lines from which coefficients of friction were calculated. The results are listed below in Table 10.6.

The results from analysing a CVD-route film with a *sharp* tip, $R_{\mathrm{Tip}} = 10$–20 nm, are shown in Fig. 10.22. A non-linear dependence is apparent. In the single-asperity Hertzian regime it is predicted for low loads that $F_{\mathrm{L}} = cF_{\mathrm{N}}^{2/3}$, with c being a constant. A curve was fitted to the data and exhibited good agreement with the 2/3-dependence.

Table 10.6. Tribological properties for DLC films inferred from data base

Material/designation	Condition	μ
IB-assisted deposition		
A1(1)	air/vac	0.08 ± 0.03
A1(2)	air/vac	0.075 ± 0.03
A1(3)	air/vac	0.09 ± 0.03
A2	air	0.09 ± 0.03
A2	vac	0.045 ± 0.02
RF-CVD deposition		
B1(DLC4)	air	0.06 ± 0.02
B2(DLC10)	air	0.06 ± 0.02
B2(DLC10)	vac	0.125 ± 0.04

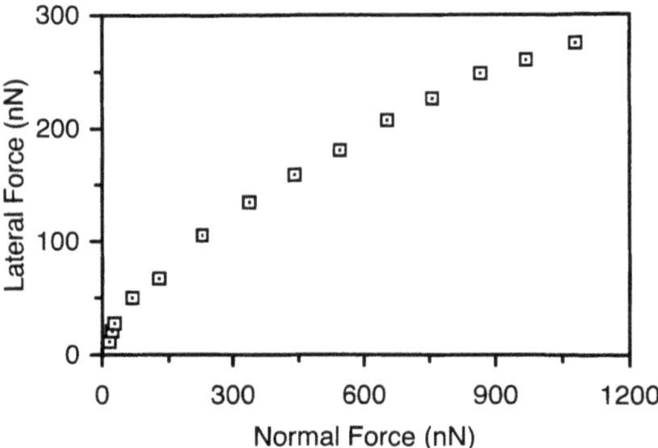

Fig. 10.22. Friction data obtained with a *sharp* tip on the CVD-type surface

All linear curves fitted to the friction data for DLC specimens intersect the F_L axis at positive values (the zero load lateral force, F_{LO}) and extrapolate to intercept at negative values on the F_N axis (the *dynamic* lift-off force required to obtain zero lateral force, F_{NO}). The observation shows that there is an adhesive interaction between tip and surface. The values of F_{NO} for DLC specimens ranged from 500 to 2200 nN. These values are far too high to be explained by simple Van der Waals interactions, or by other chemical effects. The alternative possibilities are concerned with meniscus forces and/or tribo-generated electrostatic interactions. The former should lead to a correlation with *static* F–d lift-off forces, and a meniscus force should be substantially dependent on the ambient (i.e., air versus vacuum). No such correlation or dependence was observed. Detailed consideration of the data base showed that the meniscus makes only a minor contribution to the adhesive interaction for DLC surfaces. The electrostatic interaction, on the other hand, should be largely absent in the F–d mode since the measurement is static, as was observed. Also, one would expect the insulating RF–CVD specimens to exhibit a greater electrostatically generated adhesive interaction than the conducting IBAD specimens. The results were in accord with that expectation.

(b) An Ultra-Low Friction Ceramic. Improved methods of synthesis and characterisation have identified a layered high-temperature ceramic, Ti_3SiC_2, [69–71] with an unusual combination of properties. Electrical and thermal conductivities exceed those of many metals; there is excellent resistance to oxidation up to 1400°C and it has tolerance to thermal shock; it has good compressive strength, a high Young's modulus, relatively low hardness, and there is evidence for ductility; and the material is readily machinable with standard tools. The mechanical properties suggest that the material could possess unusual fracture mechanisms and low friction.

Fig. 10.23. FeSEM images showing layer structure and preferential cleavage planes *(top)*, and platey wear and fracture debris *(bottom)*

Cleavage faces were prepared in air by fracture, abrasion or scraping, characterised by FeSEM, and then analysed by lateral force microscopy (LFM) in air or vacuum. The images in Fig. 10.23 show multiple parallel cleavage of a trapped grain (FeSEM), and platey wear and fracture debris. The surface preparation methods were chosen in order to mimic realistic tribological service conditions. The surfaces were also investigated following exposure to air for durations ranging from minutes to several months. Measurements were obtained in vacuum (2×10^{-3} Pa) or in air for all surfaces immediately after preparation under ambient laboratory conditions.

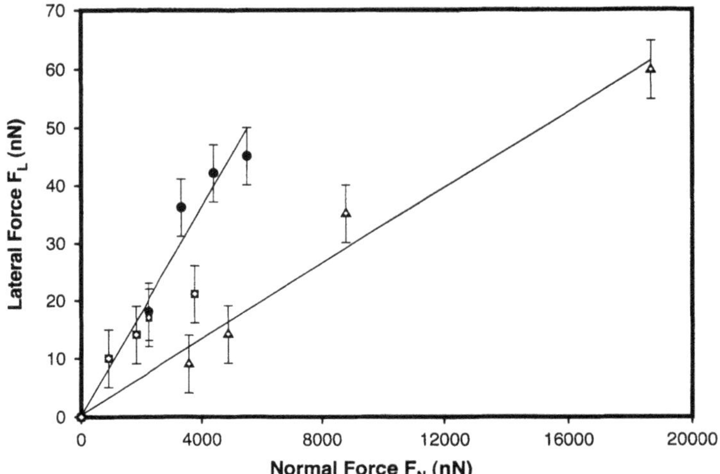

Fig. 10.24. Representative data for measurements of F_{L} versus F_{N}. The data points refer to: (△) an abraded face analysed in vacuum; (•) a scraped face exposed to air for 45 mins, then analysed in vacuum; and (□) a scraped face analysed in vacuum

Table 10.7. Coefficients of friction for various surface conditions

Surface preparation and type*	Atmospheric exposure	μ
FF/FR	< 30 mins	$(4 \pm 2) \times 10^{-3}$
FF/RR	< 30 mins	0.1–0.25
AF/FR	< 15 mins	$(3 \pm 1) \times 10^{-3}$
SF/FR	< 15 mins	$(5 \pm 2) \times 10^{-3}$
SF/FR	45 mins	$(1 \pm 0.2) \times 10^{-2}$
AF/FR	1.5 hr	$(5 \pm 2) \times 10^{-3}$
AF/RR	6 hr	0.12–0.16
FF/RR	2 days	0.15–0.21
FF/FR	6 month	$(4 \pm 2) \times 10^{-3}$

*Code: FF=fracture face, SF=scraped face, AF=abraded face, FR=flat region, RR=rough region.

Typical results for F_{L} as a function of F_{N} are shown in Fig. 10.24. Other outcomes are summarized in Table 10.7. Tribological characterisation was also carried out by a standard macroscopic method on a polished disc-shaped specimen some 1.8 mm in diameter, against a lightly peened stainless steel sheet, at loads in the range 0.15–0.9 N. The measurements were taken in air and the resultant value of $\mu = 0.11$–0.12 was in good agreement with that obtained by LFM on 'rough' surfaces. The relatively high value is most likely due to the multi-asperity contact area extending over many grains in the surface, so that the low-μ portions will consequently be 'shorted' out.

Acknowledgements. The insights and outcomes reported above represent a collective effort over several years by many industrious and creative people from several institutions. I am especially indebted to co-workers in the SPM Group at Griffith University and in the Materials Characterization Service at AEA Technology (formerly UKAEA-Harwell). Funding and other support has been forthcoming from many sources, including the Australian Research Council, Griffith University, CSIRO and the AEA Corporate Research Programme. The loan of an SPM instrument from JEOL facilitated progress on several of the projects described above.

References

1. G. Binnig, H. Röhrer: Helv. Phys. Acta **55**, 726 (1982)
2. See the two special issues, July and September, of IBM J. Res. Develop. **30** (1986)
3. G. Binnig, H. Röhrer: Sci. Amer. **253**, 50 (1985)
4. G. Binnig, H. Röhrer: Surf. Sci. **126**, 236 (1983)
5. H. Röhrer: Ultramicroscopy **42–44**, 1 (1992)
6. P. Avouris: J. Phys. Chem. **94**, 2246 (1990)
7. G. Binnig, C.F. Quate, C. Gerber: Phys. Rev. Lett. **56**, 930 (1986)
8. S. Myhra, in *Handbook of Surface and Interface Analysis*, Eds. J.C. Rivière and S. Myhra (Marcel Dekker, NY, 1998) p. 401.
9. J. Tersoff, D.R. Hamann: Phys. Rev., B **31**, 805 (1985)
10. L.E.C. van Leemput, H. van Kempen: Rep. Prog. Phys. **55**, 1165 (1992)
11. J. Shen, R.G. Pritchard, R.E. Thurstans: Contemporary Physics **32**, 11 (1991)
12. W.J. Kaiser, R.C. Jaklevic: Surf. Sci. **181**, 55 (1987)
13. W.J. Kaiser, R.C. Jaklevic: IBM J. Res. Develop. **30**, 411 (1986)
14. T.R.I. Cataldi, I.G. Blackham, G.A.D. Briggs, J.B. Pethica, H.A.O. Hill: J. Electroanal. Chem. **290**, 1 (1990)
15. D. Sarid: *Scanning force microscopy* (Oxford University Press, New York, 1991)
16. J.N. Israelachvili: *Intermolecular and surface forces*, 2nd ed. (Academic Press, San Diego, 1992)
17. N.A. Burnham, A.J. Kulik, G. Gremaud: in *Procedures in scanning probe microscopy*, R.J. Colton, ed., (Wiley, 1996)
18. U.D. Schwarz, P. Köster, R. Wiesendanger: Rev. Sci. Instrum. **67**, 2560 (1996)
19. C.T. Gibson, G.S. Watson, S. Myhra: Scanning **19**, 564 (1997)
20. Y. Li, S.M. Lindsay: Rev. Sci. Instrum. **62**, 2630 (1991)
21. R. Wurster, B. Ocker: Scanning **15**, 130 (1993)
22. M. van Cleef, S.A. Holt, G.S. Watson, S. Myhra: J. Microsc. **181**, 2 (1996)
23. J. Fu: Rev. Sci. Instrum. **66**, 3785 (1995)
24. F. Jensen: Rev. Sci. Instrum. **64**, 2595 (1993)
25. M. Jaschke, H.-J. Butt: Rev. Sci. Instrum. **66**, 1258 (1995)
26. D.D. Koleske, G.U. Lee, B.I. Gans, K.P. Lee, D.P. DiLella, K.J. Wahl, W.R. Barger, L.J. Whitman, R.J. Colton: Rev. Sci. Instrum. **66**, 4566 (1995)
27. C. Odin, J.P. Aimé, Z. El Kaakour, T. Bouhacina: Surf. Sci. **317**, 321 (1994)
28. S. Xu, M.F. Arnsdorf: J. Microsc. **173**, 199 (1994)

29. C.T. Gibson, G.S. Watson, S. Myhra: Nanotechnology **7**, 259 (1996)
30. N.P. D'Costa, J.H. Hoh: Rev. Sci. Instrum. **66**, 5096 (1995)
31. C.T. Gibson, G.S. Watson, S. Myhra: Wear **213**, 72 (1997)
32. J.M. Neumeister, W.A. Ducker: Rev. Sci. Instrum. **65**, 2527 (1994)
33. D.F. Ogletree, R.W. Carpick, M. Salmeron: Rev. Sci. Instrum. **67**, 3298 (1996)
34. L. Hellemans, K. Waeyaert, F. Hennau: J. Vac. Sci. Technol. B, **9**, 1309 (1991)
35. P. Markiewicz, M.C. Goh: Langmuir **10**, 5 (1994)
36. J. Vesenka, S. Manne, R. Giberson, T. March, E. Henderson: Biophys. J. **65**, 992 (1993)
37. J.E. Griffith, D.A. Grigg, M.J. Vasile, P.E. Russell, E.A. Fitzgerald: J. Vac. Sci. Technol. B **9**, 3586 (1991)
38. P. Markiewicz, M.C. Goh: Rev. Sci. Instrum. **66**, 3186 (1995)
39. A. Crossley, S. Myhra, C.J. Sofield: Surf. Sci. **318**, 39 (1994)
40. J.A.A. Crossley, C.J. Sofield, S. Myhra: Surf. Sci. **380**, 568 (1997)
41. *Materials and Processes for Surface and Interface Engineering*, Y. Pauleau, ed., NATO ASI Series E, Vol. 290 (Kluwer Acad. Publ., Dordrecht, 1995)
42. *Organic Coatings I*, A. Wilson, H. Prosser and J.W. Nicholson, eds., (Appl. Sci. Publ., 1987)
43. *Interfaces in Polymer, Ceramic and Metal Matrix Composites*, H. Ishida, ed., (Elsevier, 1988)
44. *Organic Coatings*, AIP Conf. Proc. 354, P.-C. Lacaze, ed., (AIP Press, 1995)
45. *Corrosion Protection by Organic Coatings*, M.W. Kendig and H. Leidheiser, Jr., eds., (Electrochem. Soc., 1987)
46. J. Seto, N. Asai, I. Fujiwara, T, Ishibashi, T. Kamei, S. Tamura: Thin Solid Films **273**, 97 (1996)
47. *Fundamentals of Friction: Macroscopic and Microscopic Processes*, I.L. Singer and H.M. Pollock, eds., (Kluwer, Dordrecht, 1991)
48. *Forces in Scanning Probe Methods*, H.-J. Güntherodt, D. Anselmetti and E. Meyer, eds., NATO ASI, Series E: Applied Sciences-Vol. 286, (Kluwer, Dordrecht, 1995)
49. J.A.A. Crossley, C.T. Gibson, L.D. Mapledoram, M.G. Huson, S. Myhra, D.K. Pham, C.J. Sofield, P.S. Turner, G.S. Watson: Micron **31**, 659 (2000)
50. C.T. Gibson, G.S. Watson, L.D. Mapledoram, H. Kondo, S. Myhra: Appl. Surf. Sci. **144–145**, 618 (1999)
51. P.W. Wertz, D.T. Downing: Lipids **23**, 878 (1988)
52. A.P. Negri, H.J. Cornell, D.E. Rivett: Aust. J. Agric. Res. **42**, 1285 (1991)
53. B. Bhushan, V.N. Koinkar: Wear **180**, 9 (1995)
54. V.N. Koinkar, B. Bhushan: Wear **202**, 110 (1996)
55. A.L. Weisenhorn, M. Khorsandi, S. Kasas, V. Gotzos, H.-J. Butt: Nanotechnology **4**, 106 (1993)
56. G.R. Bushell, G.S. Watson, S.A. Holt, S. Myhra: J. Microsc. **180**, 174 (1995)
57. Z. Shao, J. Mou, D.M. Czajkowsky, J. Yang, J.-Y. Yuan: Adv. in Physics **45**, 1 (1996)
58. S. Kasas, N.H. Thomson, B.L. Smith, P.K. Hansma, J. Miclossy, H.G. Hansma: Int. J. Imaging Syst. Technol. **8**, 151 (1997)
59. M.F. Arnsdorf, S. Xu: J. Cardiovasc. Electrophysiol., **7**, 639 (1996)
60. G.R. Bushell, C. Cahill, F.M. Clarke, C.T. Gibson, S. Myhra, G.S. Watson: **36**, 254 (1999)
61. E. Henderson, P.G. Haydon, D.S. Sakaguchi: Science **257**, 1944 (1992)

62. V. Parpura, P.G. Haydon, E. Henderson: J. Cell Sci. **104**, 427 (1993)
63. C.D. Frisbie, L.F. Rozsnyai, A. Noy, M.S. Wrighton, C.M. Lieber: Science **265**, 2071 (1994)
64. R.F. Davis: Physica B. **185**, 1 (1993)
65. H. Tsai, D.B. Body: J. Vac. Sci. Technol. A **5**, 3287 (1987)
66. F.M. Kimock, B.J. Knapp: Surf. Coat. Technol. **56**, 273 (1993)
67. A. Matthews, S.S. Eskildsen: Diamond Relat. Mater. **3** 902 (1994)
68. A. Crossley, C. Johnston, G.S. Watson, S. Myhra: J. Phys. D: Appl. Phys. **31**, 1 (1998)
69. M.W. Barsoum, T. El-Raghy: J. Am. Ceram. Soc. **79**, 1953 (1996)
70. S. Myhra, J.W.B. Summers, E.H. Kisi: **39**, 6 (1999)
71. A. Crossley, E.H. Kisi, J.W.B. Summers, S. Myhra: J. Phys. D: Appl. Phys. **32**, 632 (1999)

11 Low Energy Ion Scattering

D. J. O'Connor

Low energy ion scattering (LEIS) is the study of the structure and composition of a surface by the detection of low energy (100 eV–10 keV) ions (and atoms) elastically scattered off the surface. This technique is a subset of ion scattering spectrometry which involves the use of incident ions with energies ranging from 100 eV to over 1 MeV. The range of measurements possible over such a large range of energies extends from purely atomic layer resolution at the low energy end to analysis to depths of the order of microns at the high energy end. Some of the high energy effects (> 250 keV) are covered in Chap. 9, while the intermediate energy range (medium energy ion scattering) has been successfully developed as a near surface structure probe mentioned briefly in Chap. 1. The use of low energy ions to measure the surface structure of solids was established by *Smith* [1] in 1968. In that study the basic elements of LEIS were established and these have been built on over the past 20 years to develop into a powerful surface atomic layer structure and composition probe [2–6]. It has been successfully applied to a wide range of practical surface problems which include the surface composition analysis of:

- binary alloys
- catalysts
- cathode surfaces
- polymers
- surface segregation
- adsorbates
- surface structure
- adsorbate site identification

It can answer the following questions to a precision which is target dependent;

- What is on the surface?
- How much is on the surface?
- Where is it located (relative to other atoms)?

11.1 Qualitative Surface Analysis

LEIS involves the bombardment of the surface with either inert gas or alkali ions and measuring the energy distribution of the ions (or less commonly the neutrals) scattered off the surface. As the energy loss to electronic processes (termed inelastic energy loss) is relatively small at low energies it is possible to identify the mass of the target atoms from the energy of the scattered projectiles. By applying the principles of the conservation of energy and momentum the scattered projectile energy, E_1, is given by

$$\frac{E_1}{E_0} = \left(\frac{\cos\theta + \sqrt{\mu^2 - \sin^2\theta}}{1+\mu} \right)^2 , \tag{11.1}$$

where E_0 is the projectile energy, the scattering angle is θ, and the ratio of the target mass to the projectile mass is represented by μ. From this simple analysis it is possible to determine the mass (and usually the identity) of atoms situated on the outermost layer of the solid. It is exactly this approach which allows the determination of the target masses from the ion scattering spectrum in Fig. 11.1 where (11.1) predicts the E/E_0 values observed for Cu and Au as the surface components. In the absence of recoil peaks (not observed for scattering angles greater than 90°) the heavier the target the higher will be E/E_0.

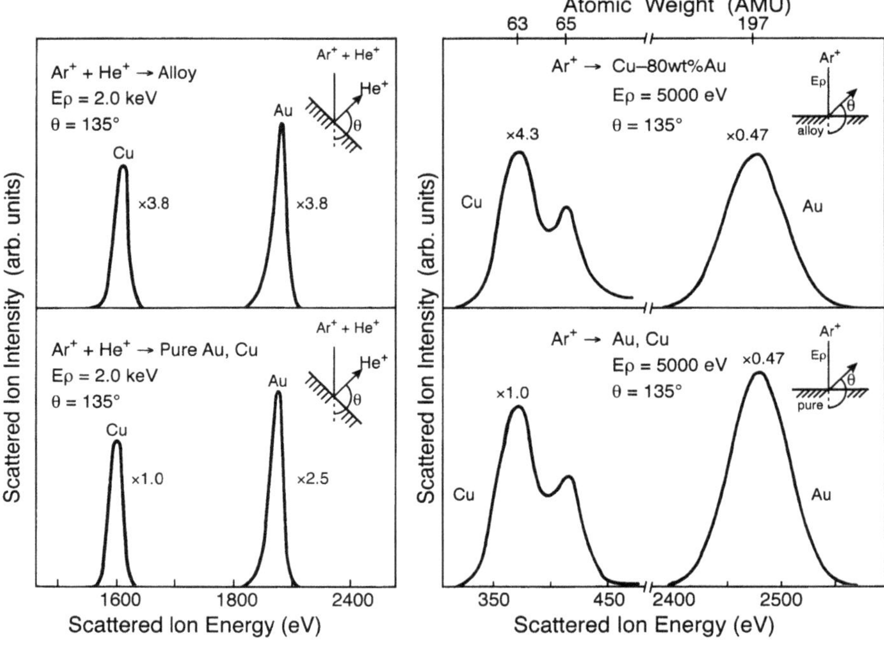

Fig. 11.1. Comparison of the analysis of a Au–Cu(43%) alloy with He and Ar projectiles demonstrating the increased mass resolution attainable with heavier projectiles. From [7]

The ability to resolve different masses on a surface is a function of the ratio of the projectile mass to the target mass. It is easy for He to resolve C ($\mu = 3$) and O ($\mu = 4$) under normal conditions but it would be impossible for He to resolve W ($\mu = 46$) from Au ($\mu = 49.25$). To improve the mass resolution for heavier target masses it is necessary to increase the projectile mass and/or increase the scattering angle. This is more clearly illustrated in Fig. 11.1 where the use of He and Ar as projectiles is compared and it is evident that with He it is possible to resolve Cu and Au, while for Ar it is also possible to resolve the isotopes of Cu. However if the projectile used is heavier than the target atom there is a limiting angle θ_1 beyond which no single scattering is observed. θ_1 is given by

$$\theta_1 = \arcsin(m_2/m_1) . \tag{11.2}$$

Scattered projectiles can be observed at all scattering angles whenever the projectile mass is less than the target mass. In some cases the recoiling target atom is captured and analyzed, in which case the energy of the recoiling atom is given by

$$\frac{E_2}{E_0} = \frac{4\mu \cos^2 \phi}{(1+\mu)^2} \tag{11.3}$$

where E_2 is the energy of the particle recoiling at angle of ϕ to the incident direction. The angle of recoil is limited to 90° so in many systems the detection angle is set at greater than 90° in give the dual benefit of improving the mass resolution and removing the potentially complicating features introduced by recoiling projectiles.

In some cases the identification of the existence of an element on a surface is sufficient to allow some conclusions to be reached. The simplest example of this form of analysis was performed by *Smith* [1] who analyzed a clean Ni surface onto which CO had been adsorbed. In the spectrum scattered ions from Ni and from O were observed, but no ion yield was measured from the C atoms. It was concluded from this that the CO molecule was bonded perpendicularly to the Ni surface with the C atom forming the bond hence the C atoms were shadowed from the incident ions by the O atoms and no scattered ion yield was observed off C.

11.2 Advantage of Recoil Detection

The existence of recoils can be exploited in some applications for the specialized tasks of detecting light elements and for the very sensitive detection of electronegative adsorbates. The detection of light elements is extremely difficult directly with ion scattering however the recoils are easily identified and measured [4–12] and has been used to identify hydrogen on surfaces. While it is possible to observe recoil peaks in a positive ion spectrum it is often difficult to separate this contribution from other ion scattering processes

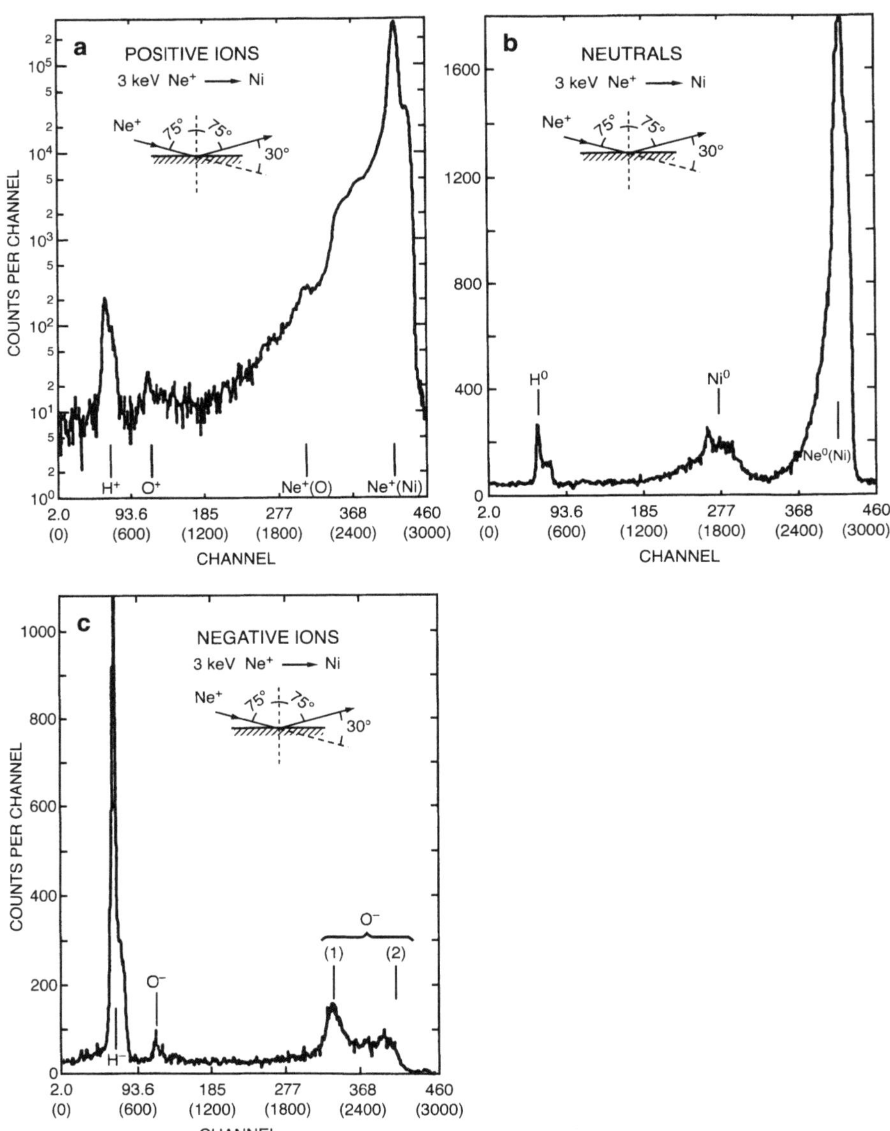

Fig. 11.2. The measured scattered and recoiled spectra obtained for different charge states when a Ni surface is bombarded by 3 keV Ne^+. From [8]

with inert gas ion yields at the same energy. To overcome this limitation it is possible to take advantage of the fact that some elements will escape the surface in positive, negative and neutral charge states while in general the inert gas projectiles only escape as positive ions or neutrals (Fig. 11.2). In the case of electronegative elements (O, Cl, etc.) the negative charge fraction can be as large or larger than the positive so by measuring the negative ion spectrum the contribution from the electronegative elements stands out without interferences from the projectiles and as a result it has been estimated that as little as 10^{-5} of a monolayer can be measured [12].

11.3 Quantitative Analysis

11.3.1 Scattered Ion Yield

In many analyses it is sufficient to identify the presence of an element on a surface to allow some conclusion to be arrived at, however in some applications the scattered ion yield is used to determine the composition of the surface layer of a sample. The appropriate relationship between these quantities is

$$Y_a^+ = N_0 N_a (d\sigma_a/d\Omega)\Delta\Omega P^+ T(E) \; . \qquad (11.4)$$

Y_a^+ is the measured ion yield for element a, N_0 in the number of projectiles, N_a is the number of atoms of element a per unit area on the surface, $\Delta\Omega$ is the collection angle the analyzer presents to the target and $T(E)$ is the transmission function of the analyzer and detector. All these terms can be determined to a high degree of precision. The principal uncertainties in the analysis arise from the differential scattering cross section $(d\sigma_a/d\Omega)$ and the charge exchange factor (P^+). These two aspects will be dealt with in the following sections.

11.3.2 Differential Scattering Cross Section

The differential scattering cross section represents the cross-sectional area that each atom presents to the beam for a particular scattering event and it can be determined by integration if the scattering potential is known. In LEIS the projectile energy is sufficiently large that only the repulsive part of the interatomic potential needs to be considered in determining the cross section and a screened coulomb approximation to the repulsive potential of the form given in (11.5) is normally used to represent the interaction,

$$V(r) = \frac{Z_1 Z_2 e^2}{4\pi\varepsilon_0 r}\phi(r/a) \; . \qquad (11.5)$$

Here Z_1 and Z_2 are the atomic numbers of the interacting particles, r is the interatomic separation and a is the screening length. While the Moliere approximation [13] (11.6) has been used extensively and successfully for many years it has recently been surpassed in accuracy by the ZBL potential [14] (11.8) which is based on the wavefunctions of a large range of elements.

$$\phi_{\mathrm{M}}(x) = 0.35\exp(-0.3x) + 0.55\exp(-1.2x) + 0.10\exp(-6x)\ . \tag{11.6}$$

The screening length a is related to the Bohr radius, a_0, by

$$a = 0.88534a_0/\left(\sqrt{Z_1}+\sqrt{Z_2}\right)^{3/2}\ , \tag{11.7}$$

$$\phi_Z(x) = 0.1818\exp(-3.2x) + 0.5099\exp(-0.9423x) + 0.2802\exp(-0.4029x) + 0.02817\exp(-0.2016x)\ . \tag{11.8}$$

One of the reasons for the success of the ZBL potential is the different expression used for the screening length, a, given by

$$a = \frac{a_0}{Z_1^{0.23}+Z_2^{0.23}}\ . \tag{11.9}$$

This potential has been shown [15] to be the best fit to the range of currently available measurements of the interatomic potential and in absolute terms it may be accurate to a precision of the order of 10%, however the ratio of the cross sections of two atoms will be of higher precision and may be better than 1%.

A more detailed description of the numerical determination of the scattering cross section has been given by *Torrens* [16], however as a general guide the cross section increases with

- increasing projectile and target atomic number
- decreasing projectile energy
- decreasing scattering angle.

11.3.3 Charge Exchange

When an inert gas ion is in the vicinity of a surface there is a significant probability that it will be neutralized as the ground state of the inert gas atom is a lower energy state than the conduction electrons of the surface (see Fig. 11.3). Thus a proportion of scattered inert gas projectiles will suffer neutralization and that proportion will be a function of the time spent in the vicinity of the surface. Typically only 0.5–5% of the projectiles scattered off the surface layer are ions while the ion fraction for projectiles scattered off subsurface layers is typically less than 0.1%.

When alkali ions are used as projectiles the escaping ion fraction is significantly different. As the is comparable to the work function of most materials, the projectiles come to a charge equilibrium with the surface and the ion fraction is both large and independent of the time spent near the surface at these energies. In this case the ion fraction for projectiles scattered from subsurface layers is little different from those scattered from the surface layer.

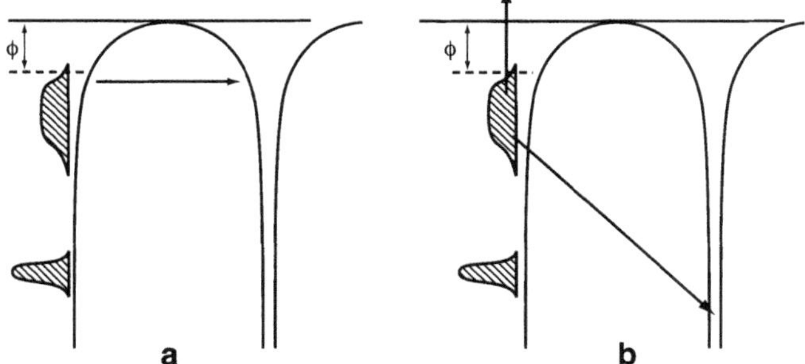

Fig. 11.3. The principal charge exchange processes responsible for the neutralization of low energy ions near surfaces. Part (**a**) depicts the resonance process in which there is a match between the energy level of an excited state in the projectile and a filled state in the band structure of the solid. The Auger process (**b**) involves an electron filling the ground state of the projectile and a second electron in the band acquiring the excess energy

The transfer of an electron from the solid to the projectile can occur by two principal processes. The simplest is the resonance process in which the electron transfers directly from a level in the solid to a vacant level in the projectile at the same energy (Fig. 11.3a). The second major process is the Auger process (discussed in Chap. 6) in which an electron from the solid transfers to a lower energy state in the projectile and the excess energy is transferred to another electron of the solid (Fig. 11.3b). A complete description of the charge exchange process is much more complicated with the inclusion of the possibility of surface electronic states, the changing energy levels of the projectile due to the interaction with its image charge and with the electrons of the solid. As well there are currently two unresolved views of the charge exchange process which favour either an interaction of the projectile with the distributed electron distribution of the solid or a description which favours a discrete interaction with each atom of the solid.

While a completely predictive theory for charge exchange is not yet available recent intense effort, both experimental and theoretical, has established a basis for the description of charge exchange which is most appropriate to inert gas ions. *Hagstrum* [17,18] developed a model for very low energy ions (1–10 eV), and while this is for much lower energies than encountered in LEIS and some approximations were not valid at these energies, it proved to successfully describe the transition rates. Recent, more rigorous, theoretical studies have yielded the same basic equations given by the Hagstrum description. The Hagstrum model considers a transition rate which varies exponentially with distance from the surface and can be integrated over the trajectory of the projectile. The resulting expression for the ion fraction P^+ is given by

$$P^+ = \mathrm{e}^{-v_c/v_\perp} \tag{11.10}$$

where v_c is a characteristic velocity (composed of a transition rate and a screening length) which should be dependent on the identity of the projectile and the target, while the second term, $v_\perp$, is the perpendicular component of velocity of the projectile and is a measure of the time the projectile spends in the vicinity of the surface. While in the simplest model v_c is expected to be energy independent it is found experimentally to depend on the projectile velocity [19–21]. While it is not possible to yet predict the value of v_c from the properties of the solid and the target, the most recent experimental results reveal that the detailed electronic structure is not significant and that the most important parameter may be the free electron density [21]. This implies that the quantification of LEIS is reliable as all scattered projectiles off a surface will experience the same transition rates hence all have a predictable probability of neutralization.

An alternative approach to charge exchange treats the transition rates as discrete interactions (11.11) with individual atoms and the net ion yield is

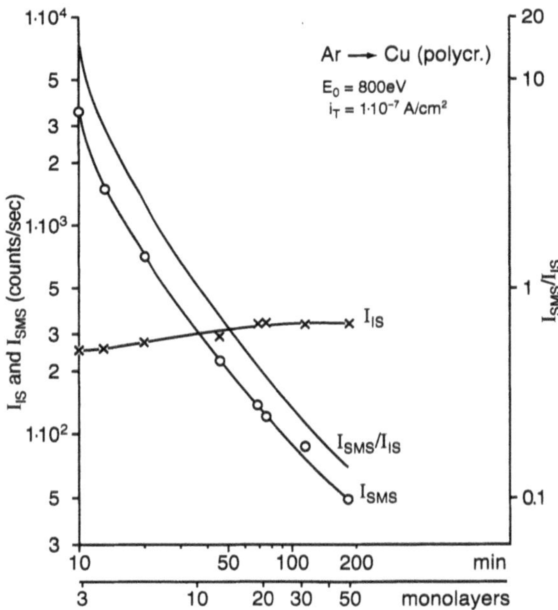

Fig. 11.4. Demonstration of the relative sensitivity of LEIS and SIMS to the existence of contamination on a Cu surface. While the Cu ion yield in SIMS decays by two orders of magnitude as the surface concentration of Cu increases during sputter cleaning, the ISS Cu yield increases only 10%. From [22]

then the integration of these independent interactions over the trajectory of the projectile,

$$P^{+} = \mathrm{e}^{-ar} \quad (11.11)$$

where a is the screening distance and r is the distance between the projectile and the target atom. This model would suggest that the ion fraction depends on the path followed in a mixed surface and that it would be difficult to estimate the charge exchange probability in practical applications. Further work is in progress to establish which of the two models of charge exchange best describe all observations.

Despite this uncertainty it is possible to demonstrate that the role of charge exchange in LEIS is less significant than in SIMS where similar processes occur and it is well known that the ion fraction of secondary ions is very sensitive to low levels of contamination on the surface. The relative sensitivities of LEIS and SIMS to surface contamination were demonstrated by *Grundner* et al. [22] as a Cu surface was cleaned by ion bombardment and the SIMS signal for Cu decreased by a factor of 100 while the LEIS signal for Ar scattered off Cu increased by approximately 10% in keeping with the increased concentration of Cu on the surface as contaminants were removed (see Fig. 11.4). Thus LEIS, while not as sensitive to low levels of contamination as SIMS, will more accurately describe the changing composition of a surface.

While the scattering cross section and the neutralization probability are major uncertainties in the quantification of LEIS, it is nevertheless possible to perform quantitative analysis despite them. The form of these measurements fall into the following categories:

1. Relative measurements
2. Standards

11.3.4 Relative Measurements

A number of early studies concentrated on the linearity of the LEIS yield as a function of surface concentration. The principal problem is the lack of a reliable set of calibrated surface composition standards, so comparison was made to the Auger electron spectrometry (AES)signal (or in one case the Rutherford backscattering spectrometry[23], RBS, yield) to establish whether a linear relationship existed between the LEIS yields and sub-monolayer concentrations when an impurity was adsorbed onto a clean surface. Early [24–28] studies certainly identified a direct relationship between the LEIS and the AES yields under conditions of up to one monolayer of adsorbate (Fig. 11.5). Implicit in this result is the assumption that the ion fraction of the scattered projectiles is independent of the composition of the surface of the solid.

This is not a conclusive test as the AES signal has an escape depth which is larger than the one or two atomic layers that LEIS probes so it is proof only

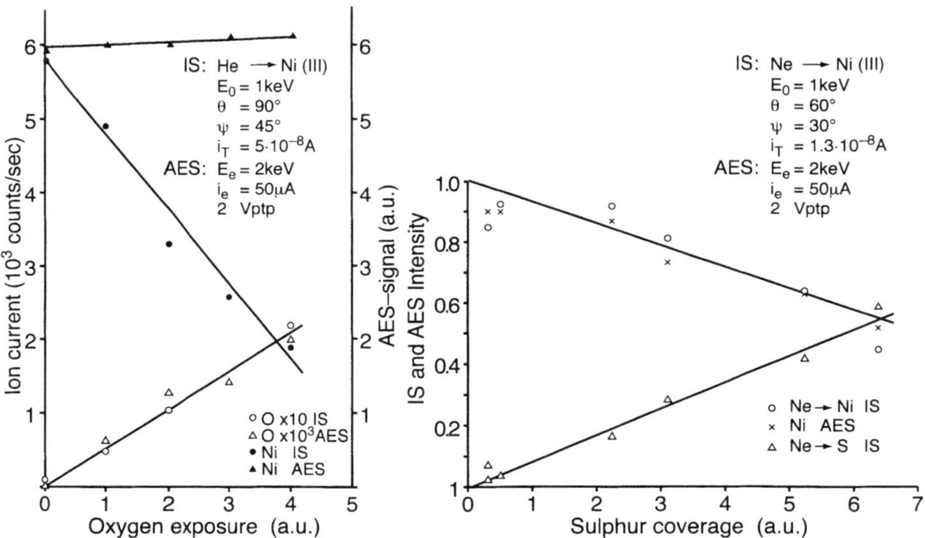

Fig. 11.5. Demonstration of the linearity of LEIS yield response with adsorbed material for O and S on a clean Ni surface. The enhanced surface sensitivity of LEIS is evident in the case of the O adsorption as the LEIS yield from Ni decreases with increasing dose while the AES yield from Ni increases marginally. From [25]

if it is assumed that all the adsorbate remains on the surface. Examples can be found where a linear relationship between the LEIS and AES yields breaks down [29], in which case the interpretation of this difference may be that the adsorbate moves below the surface so that it continues to be detected by AES and not by LEIS, or that there is a change to the electronic properties of the surface which affects the neutralization rate of the ions used in LEIS.

In other studies LEIS has been applied to compare the surface composition of clean and contaminated catalysts to identify the contaminate. In such studies only the existence of an unexpected element is sufficient to allow a conclusion without the need for quantitative compositional analysis. In the following list of studies of surfaces it was not necessary to establish an absolute scale of concentration but instead the identification of a particular element on the surface or a measurement of the relative change in concentration of an element as some change or profiling is undertaken was sufficient.

- Pb impurity in AgBr [30]
- Ni catalysts on alumina and silica [31]
- $BaTiO_3$ and $Gd_2(MoO_4)_3$ [32]
- Ba surface enrichment in Tungsten cathodes [33,34]
- Be and Sn segregation in Cu [35]
- $BaO{:}CaO{:}Al_2O_3$ [36]

- Ni-Mo-Al_2O_3 catalysts [37]
- Iron based glasses [39]
- Surface composition of $Pt_{10}Ni_{90}(111)$ alloy [40]
- Preferential sputtering of TiC by Ar^+ [41]
- Segregation of Pb on Yttrium-Iron-Garnet [60]

In these cases the assumption that the ion fraction is constant was sufficient to establish a relative concentration scale.

11.3.5 Standards

The analysis of surfaces using standards has been performed with both elemental and molecular standards. In the case of alloy analysis the most convenient reference materials are targets of the pure elements, thus to analyze a compound or alloy [42] of two components A and B which have surface concentrations of N_a and N_b (in atoms cm^{-2}) respectively, it is first necessary to establish the measured ion yield I_a^0 and I_b^0 off pure surfaces of A and B which have surface concentrations of N_a^0 and N_b^0 respectively. From the measured ion yields for the alloy surface, I_a and I_b respectively, the surface concentrations can be determined.

$$N_a = \left(I_a/I_a^0\right) N_a^0 \tag{11.12}$$

and

$$N_b = \left(I_b/I_b^0\right) N_b^0 \ . \tag{11.13}$$

If only A and B are present, then the relative concentrations can be determined from the total number of surface sites N_0 $(= N_a + N_b)$ as

$$N_a/N_0 = \frac{S I_{a<}}{S I_a + I_b} \tag{11.14}$$

where

$$S = \frac{I_b^0}{I_a^0}\frac{N_a^0}{N_b^0} = \frac{I_b^0}{I_a^0}\left(\frac{d_b}{d_a}\right)^2 \ . \tag{11.15}$$

The sensitivity factor S has been measured [43] for Cr and Fe off clean and oxide surfaces and to within experimental accuracy the same value (1.5 ± 0.1) was obtained. In the same study it was established that the relative sensitivity factors for O/Fe was 0.11 and for O/Cr was 0.08, highlighting the difficulty in detecting O with LEIS. There have been variations on this approach [44,45] with similar findings which support the assumption that the ion fraction is little affected by the nature of the surface or the environment of the scattering elements.

The use of the standards approach is complicated for O (and similar adsorbates) as a pure O target is not available. One approach to overcome this

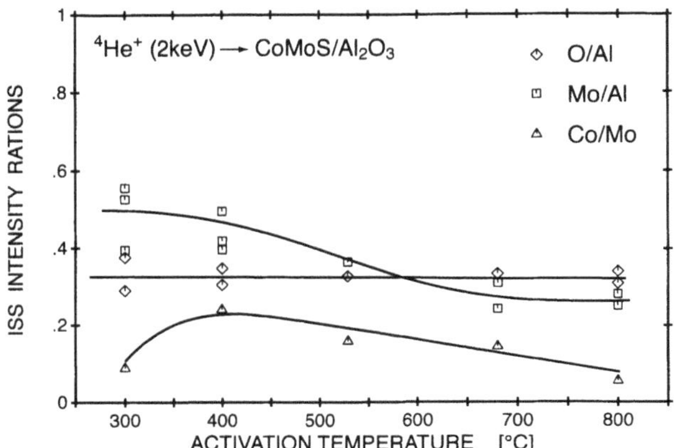

Fig. 11.6. The evolution of the relative concentrations of principal elements in the surface of a $CoMoS/\gamma\text{-}Al_2O_3$ catalyst measured by LEIS as a function of temperature. From [47]

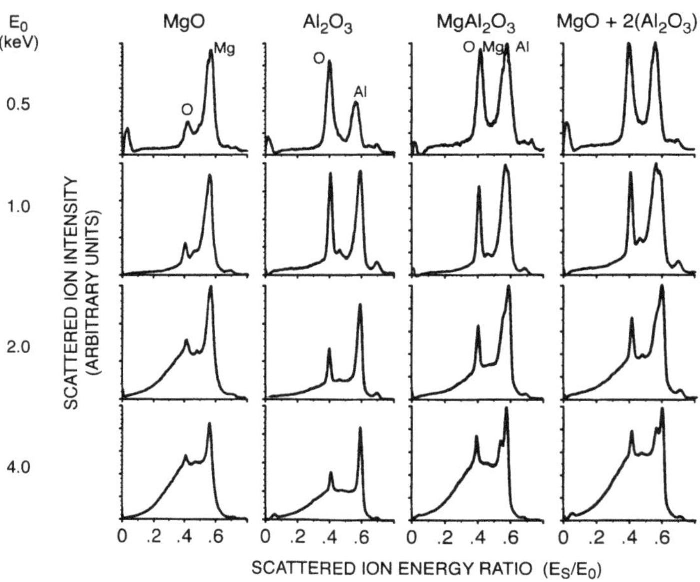

Fig. 11.7. A demonstration of surface analysis of insulators by spectrum synthesis using standards. The columns demonstrate the effect of projectile energy on the background observed and the relative peak amplitudes from different elements. The first two columns are measurements of standards and the third is for a sample of $MgAl_2O_4$ which is compared in the fourth column with the best fit synthesis of the standards as $MgO + 2(Al_2O_3)$. The nonstoichiometric result may reflect the true state at the surface or may result from systematic variations in the specimen surface, crystal orientation effects or density differences between the component materials and the compound. From [61]

difficulty has been employed in analyzing catalytic surfaces involving oxides. To analysis a $BiPO_4$-MoO_3 surface samples of $BiPO_4$ and MoO_3 as pure samples were used as standards [46]. In a later study [47] of $CoMoS/Al_2O_3$ catalytic surfaces spectra from Co_9S_8, MoS_2 and Al_2O_3 were used as standards and the relative concentrations of the principal elements were monitored as a function of activation temperature (Fig. 11.6).

In an analysis of magnesium aluminate and magnesium silicate surfaces *McCune* [48] used surfaces of MgO, Al_2O_3 and SiO_2 as the standards, and by spectrum synthesis derived a surface composition of the surfaces as $MgO(Al_2O_3)_2$ and $(MgO)_{1.5}SiO_2$ rather than the expected bulk stochiometries of $MgOAl_2O_3$ and $(MgO)_2SiO_2$ respectively. In this analysis the projectile energy was varied over the range 0.5–4 keV with good agreement at all energies between the synthesized data and the real surface data (Fig. 11.7). The departures from bulk stoichiometry were attributed to systematic variations in the specimen surfaces, crystal orientation effects and differences between the densities of the component materials and the compounds. Without an independent analysis of the surface composition of these surfaces to the same depth resolution it is not possible to attribute the different compositions to a misleading LEIS analysis or to a true surface compositional variation.

11.4 Surface Structural Analysis

The use of LEIS in surface structural analysis most commonly involves one of two basic principles – multiple scattering, or shadowing and blocking of the ions. These methods have been successfully used to determined atomic positions to an accuracy of ± 0.1 Å.

11.4.1 Multiple Scattering

As the projectile energy and the scattering angle decrease the probability of multiple scattering increases and it is often significant in LEIS. An ion can be scattered through an angle of θ by a single scattering event as described above or by two weaker collisions (double scattering) whose total scattering angle is also $\theta(\theta - \theta_1 + \theta_2)$, see Fig. 11.8. Other multiple scattering processes are possible with the next most common the 'zig-zag' sequence [49] which is a double scattering process where the two target atoms do not lie in the same plane as the incident and exit trajectories. The projectile energy after two weak collisions is greater than that observed for a single scattering event which results in a clearly defined multiple scattering peak. The relative magnitude of the double scattering peak is inversely dependent on the interatomic spacing of the target atoms and this can be used in a straightforward manner to determine some basic structural information about a surface.

In Fig. 11.9 the ion yield from single (large peaks) and multiple scattering peaks (small peaks) is shown as a function of the azimuthal angle (relative to

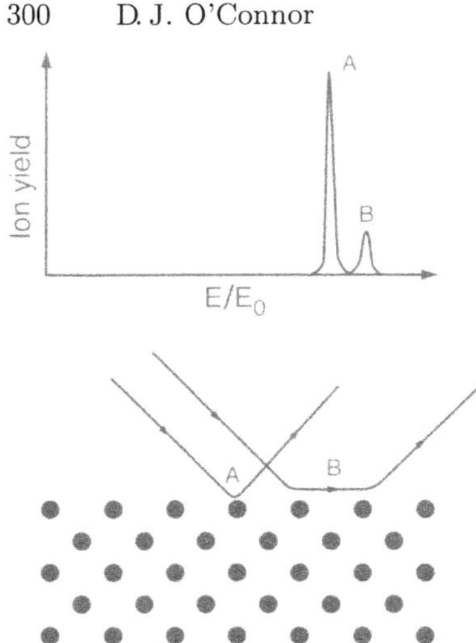

Fig. 11.8. At low energies both single and double scattering sequences as illustrated here can be observed

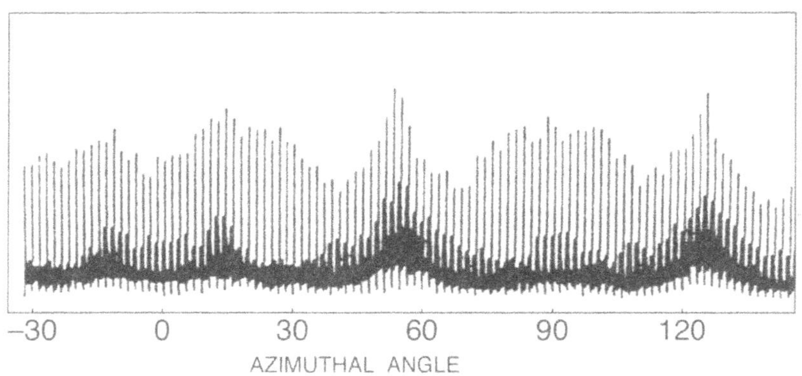

Fig. 11.9. The ion yield from single (large peaks) and multiple scattering peaks (small peaks) is shown as a function of the azimuthal angle (relative to the [100] direction) of the target for 6 keV Kr incident at 25° to a W(110) surface. The peaks in the multiple scattering at 55° and 125° are evidence of the [111] directions, while the influence of multiple scattering at 0°-[100], the 90°-[110] and ±25°-[113] can be observed

the [100] direction) of a W(110) surface. The peaks in the multiple scattering at 55° and 125° represent the influence of the [111] directions on multiple scattering yields, while the multiple scattering from the [100] (0°), the [110] (90°) and [113] (±25°) surface directions are less pronounced as expected by their interatomic spacing.

11.4.2 Impact Collision Ion Surface Scattering (ICISS)

Perhaps the most direct use of shadowing and the most easily interpreted is the ICISS developed by *Aono* [50]. To explain its use it is first necessary to understand the concept of a shadow cone in ion scattering (Fig. 11.10). If a projectile is incident upon a target atom (which we will assume has a greater mass than the projectile) and it strikes 'head on', i.e. with zero impact parameter, then it will be scattered straight back along its incident trajectory. (The impact parameter is the perpendicular distance between the initial undeviated trajectory and the initial target position.) If instead it is incident with a small impact parameter it will be scattered through a large scattering angle, and as the impact parameter becomes larger the scattering angle becomes smaller. Behind the target atom there is a shadowed region, or excluded zone into which no projectiles can penetrate. The shape of this shadow cone can be predicted [51] with a suitable knowledge of the interatomic potential and the general features are that the shadow cone radius will increase with increasing atomic number of the collision partners and with decreasing projectile energy. The flux at the edge of the shadow cone is enhance over the incident ion flux as most projectiles pass close to the edge of the cone. By suitable choice of geometry this shadow cone can be used to locate atoms above, in, and below the surface layer. This technique has been applied to scattered inert gas ions, scattered alkali ions and scattered neutrals with equal success in different applications [52–57].

If an ion beam is incident at a shallow angle to a surface (see [50]) and the detector is at a large scattering angle (which means the projectile must have a near zero impact parameter collision to be detected) then no scattering will be observed as each atom in the surface lies in the shadow of the preceding atom. As the angle of incidence of the ion beam is increased the shadow cone rotates about each scattering centre and at some critical angle of incidence the edge of

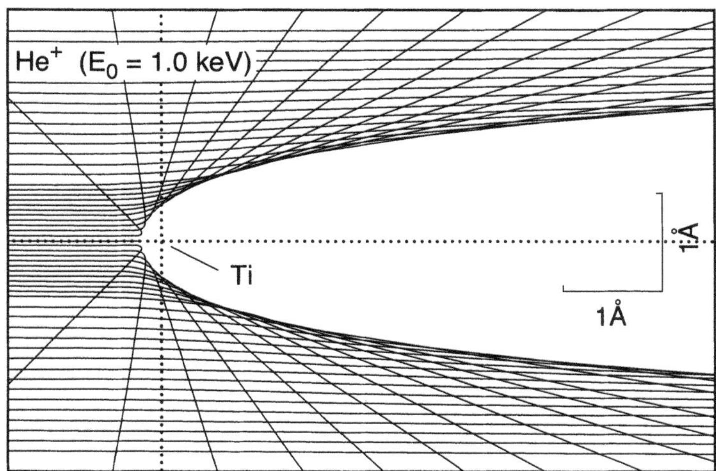

Fig. 11.10. The shape of a typical shadow cone for 1 keV He incident upon a Ti atom. The clear area behind the atom is the excluded region or shadow cone. From [50]

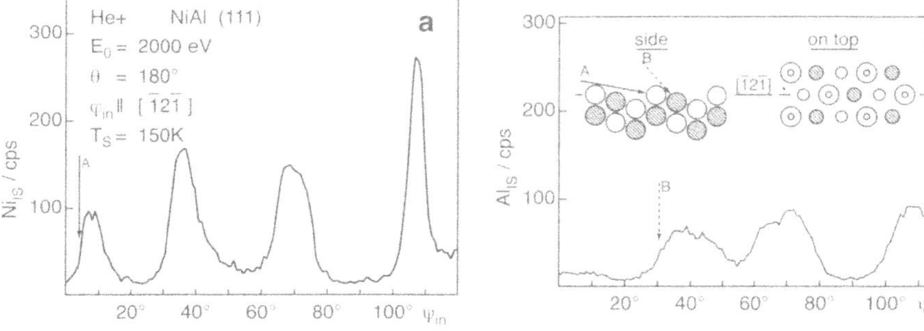

Fig. 11.11. Results of He scattering yield from Ni (**a**) and Al (**b**) on a NiAl(111) surface both demonstrating that there is no Al terminated surface layer and allowing the determination of the relative positions of the surface atoms. From [59]

the shadow cone will intersect the neighbouring atom. The scattered ion yield will be zero below the critical angle and beyond this angle it will peak (from the flux enhancement) then fall to an intermediate value. The measurement of the interatomic spacing comes from the critical angle determination and a knowledge of the shape of the shadow cone. An early application of this technique was the location of C under the surface layer of Ti in a TiC(111) surface [58]. In this application the C sits asymmetrically between the surface Ti atoms and hence there are two critical angles depending on from which direction the ion beam is incident. The position of the C is 87 ± 8 pm below the Ti surface layer.

In a more recent demonstration of the power of this form of analysis [59] a variation of this technique was applied to the NiAl(111) (Fig. 11.11) surface in conjunction with LEED and STM to determine the termination structure of this binary ordered alloy. In Fig. 11.11a the yield from the Ni scattered projectiles yields information on the relative positions of surface Ni atoms while that in Fig. 11.11b reveals that the Al is confined to the second layer.

11.5 Experimental Apparatus

The apparatus used in ion scattering is common to other surface analysis techniques. It is first essential to have a mass analyzed monoenergetic ion beam of inert gas ions or in some applications alkali ions. While it is possible to achieve fine beam spots with ion beams, space charge limitations ensure that the current density decreases with energy and compounded with that the scattered ion yield is sufficiently small that beam spot sizes less than 100 μm are impractical. In most common applications 0.5–5.0 keV He or Ne ions are used for analysis and the beam size and current depends on the available ion source or sample damage considerations but it is most common to use a 0.1–0.5 μA ion beam with a diameter of approximately 1 mm.

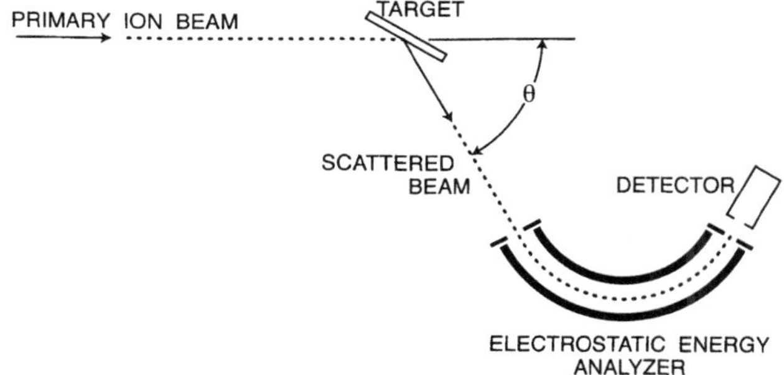

Fig. 11.12. Schematic representation of the layout of a basic LEIS experimental system

LEIS is an extremely surface sensitive technique so the sample must be analyzed in a UHV environment otherwise contamination will build up before practical analysis can be performed. The normal layout of the experimental apparatus is shown in Fig. 11.12. The electrostatic energy analyzer is the same construction as the analyzers used in the electron spectrometries but with the opposite polarities applied to all lens and analyzer elements. The natural line widths in LEIS are principally made up of the kinematic broadening and thermal vibration broadening. The kinematic broadening results from the energy analyzer accepting particles which have suffered a range of scattering angles as a consequence of its finite acceptance angle and this term becomes increasingly significant as the mass ratio (μ) decreases. The magnitude can be estimated from (11.16)

$$\Delta E_1 = E_0 \frac{dK}{d\theta} \Delta\theta \,. \tag{11.16}$$

The second contribution arises because the target atom is not stationary and thus the projectile collides with a moving target. As a result the energy width of the scattered ions is broadening with a width given by

$$\Delta E_2 = \sqrt{E_{\text{th}} E_0} \tag{11.17}$$

where E_{th} is the mean thermal vibration energy. At room temperature E_{th} is 0.025 eV so the thermal energy contribution for a 1000 eV projectile is 5 eV or 0.5%. As the natural widths of LEIS peaks are greater than those encountered in electron spectrometries the energy resolution ($\Delta E/E$) need be no better than 0.5% in most practical circumstances. In some instruments the solid angle is increased to increase sensitivity but this results in a greater kinematic broadening and diminished mass resolution. The kinematic broadening increases with increasing projectile mass and with increasing collection solid angle.

References

1. D.P. Smith: J. Appl. Phys. **38**, 340 (1967)
2. H. Niehus, W. Heiland, E. Taglauer: Surface Science Reports **17**, 213 (1993)
3. H.H. Brongersma, P.A.C. Groenen and J.-P. Jacobs: "Application of low energy ion scattering to oxidic surfaces", in: Science of Ceramic Interfaces 2, Ed. J. Nowotny, Elsevier (1994) pp. 113–182.
4. J.W. Rabalais: Surf. Sci. **300**, 219 (1994)
5. M.M Sung, V. Bukov, A. Albayati, C. Kim, S.S Todorov, J.W Rabalais: Scanning Microscopy **9**, 321 (1995)
6. C. Kim, C. Hoefner, V. Bykov, J.W. Rabalais: Nucl Instr Meths in Phys Res B **125**, 315 (1997)
7. H.J. Kang, R. Shimizu, T. Okutan: Surf. Sci. **116**, L173 (1982)
8. P.J. Schneidner, W. Eckstein, H. Verbeek: Nucl. Instrum. Meth. **218**, 713 (1983)
9. B.J.J. Koeleman, S.T. de Zwart, A.L. Boers, B. Poelsema, L.K. Verhey: Nucl. Instrum. Meth. **218**, 225 (1983)
10. J.A. Schultz, R. Kumar, J.W. Rabalais: Chem. Phys. Lett. **100**, 214 (1983)
11. B.J.J. Koeleman, S.T. de Zwart, A.L. Boers, B. Poelsema, L.K. Verhey: Phys. Rev. Lett. **56**, 1152 (1986)
12. D.J. O'Connor: Surf. Sci. **173**, 593 (1986)
13. G. Moliere: Z. Naturforsch. **2A**, 133 (1947)
14. J.P. Biersack, J.F. Ziegler: In *Ion Implantation Techniques*, ed. by H. Ryssel, H. Glawischnig, Springer Ser. Electrophys. Vol. 10 (Springer, Berlin, Heidelberg 1982) pp. 122–156
15. D.J. O'Connor, J.P. Biersack: Nucl. Instrum. Meth. B**15**, 14 (1986)
16. I.M. Torrens: *Interatomic Potentials* (Academic, New York 1972)
17. H.D. Hagstrum: Phys. Rev. **96**, 336 (1954)
18. H.D. Hagstrum: In *Inelastic Ion-Surface Collisions*, ed. by N.H. Tolk, J.C. Tully, W. Heiland, C.W. White (Academic, New York 1977)
19. R.J. MacDonald, D.J. O'Connor: Surf. Sci. **124**, 423 (1983)
20. R.J. MacDonald, D.J. O'Connor, P.R. Higginbottom: Nucl. Instrum. Meth. Phys. Res. B**2**, 418 (1984)
21. D.J. O'Connor, Y.G. Shen, J.M. Wilson, R.J. MacDonald: Surf. Sci. **197**, 277 (1988)
22. M. Grundner, W. Heiland, E. Taglauer: Appl. Phys. **4**, 243 (1974)
23. H. Verbeek: In *Materials Characterisation Using Ion Beams*, ed. by J.P. Thomas, A. Cachard (Plenum, New York 1978)
24. E. Taglauer, W. Heiland: Appl. Phys. Lett. **24**, 437 (1974)
25. E. Taglauer, W. Heiland: Surf. Sci. **47**, 234 (1975)
26. P.J. Martin, C.M. Loxton, R.F. Garrett, R.J. MacDonald, W.O. Hofer: Nucl. Instrum. Meth. **191**, 275 (1981)
27. A. Sagara, K. Akaishi, K. Kamada, A. Miyahara: J. Nucl. Mater. **93/94**, 847 (1980)
28. H.H. Brongersma, G.C.J. Van der Ligt, G. Rouweler: Philips J. Res. **36**, 1 (1981)
29. H. Niehus, E. Bauer: Surf. Sci. **47**, 222 (1975)
30. Y.T. Tan: Surf. Sci. **61**, 1 (1976)
31. M. Wu, D.M. Hercules: J. Phys. Chem. **83**, 2003 (1979)

32. L.L. Tongson, A.S. Bhaila, I.E. Cross, B.E. Knox: Appl. Surf. Sci. **4**, 263 (1980)
33. W.L. Baun: Appl. Surf. Sci. **4**, 374 (1980)
34. C.R.K. Marrian, A. Shih, G.A. Haas: Appl. Surf. Sci. **24**, 372 (1985)
35. C. Creemers, H. Van Hove, A. Neyeens: Appl. Surf. Sci. **7**, 402 (1981)
36. W.V. Lampert, W.L. Baun, B.C. Lamartine, T.W. Haas: Appl. Surf. Sci. **9**, 165 (1981)
37. H. Jeziorowski, H. Knozinger, E. Taglauer, C. Vogdt: J. Catal. **80**, 286 (1983)
38. J.P. Jacobs, S. Reijne, R.J.M. Elfrink, S.N. Mikhailov, H.H. Brongersma: J. Vac. Sci. Technol. **A12**, 2308 (1994)
39. Th. Berghaus, H. Neddermeyer, W. Radlik, V. Rogge: Phsica Scripta T**4**, 194 (1983)
40. J.C. Bertolini, J. Massarddier, P. Delichere, B. Tardy, B. Imojik, Y. Jugnet, Tran Minh Duc, L. Temmerman, C. Creemers, H. van Hove, A. Neyens: Surf. Sci. **119**, 95 (1982)
41. H.J. Kang, Y. Matsuda, R. Shimizu: Surf. Sci. **134**, L500 (1983)
42. T.A. Flaim: Research Publication GMR-1942, 1975, Research Laboratories, General Motors, Warren, Michigan
43. R.P. Frankenthal, D.L. Malm: J. Electrochem. Soc. **123**, 186 (1976)
44. D.G. Swartzfager: Anal. Chem. **56**, 55 (1984)
45. M.A. Wheeler: Anal. Chem. **47**, 146 (1975)
46. P. Bertrand, J.-M. Beuken, M. Delvaux: Nucl. Instrum. Meth. **218**, 249 (1983)
47. J.-M. Beuken, P. Bertrand: Surf. Sci. **162**, 329 (1985)
48. R.C. McCune: Anal. Chem. **51**, 1249 (1979)
49. D.J. O'Connor, R.J. MacDonald: Radiat. Eff. **45**, 205 (1980)
50. M. Aono: Nucl. Instrum. Meth. B**2**, 374 (1984)
51. O. Oen: Surf. Sci. **131**, L407 (1983)
52. M. Aono, Y. Hou, R. Souda, C. Oshima, S. Otani, Y. Ishizawa: Phys. Rev. Lett. **50**, 1293 (1983)
53. J. Moller, H. Niehus, W. Heiland: Surf. Sci. Lett. **166**, L111 (1986)
54. H. Niehus: Surf. Sci. Lett. **166**, L107 (1986)
55. J.A. Yarmoff, R.S. Williams: Surf. Sci. Lett. **165**, L73 (1986)
56. M. Aono, Y. Hou, C. Oshima, Y. Ishizawa: Phys. Rev. Lett. **49**, 567 (1982)
57. R. Souda, M. Aono, C. Oshima, S. Otani, Y. Ishizawa: Surf. Sci. Lett. **128**, L236 (1983)
58. M. Aono, C. Oshima, S. Zaima, S. Otani, Y. Ishizawa: Japan. J. Appl. Phys. **20**, L829 (1981)
59. H. Niehus, W. Raunau, K. Besocke, R. Spitzl, G. Comsa: Surf. Sci. Lett. **225**, L8 (1990)
60. S. Priggemeyer, A. Brockmeyer, H. Dotsch, H. Koschmeider, D.J. O'Connor, W. Heiland: Appl. Surf. Sci. **44**, 255 (1990)
61. R.C. McCune: Anal. Chem. **51**, 1249 (1979)

12 Reflection High Energy Electron Diffraction

G.L. Price

Reflection high energy electron diffraction (RHEED) was first used in the study of a cleaved calcite crystal by *Nishikawa* and *Kikuchi* in 1928 [1]. They observed diffraction spots, and lines attributed to diffuse scattering. Other early workers included *Germer* (1936) [2] who took diffraction patterns from galena and *Miyake* (1936) [3] who examined oxide surfaces. *Uyeda* et al. [4] used RHEED with metal films in 1940 and with adsorbed organic molecules in 1950. Commercial RHEED equipment was developed through the 1950s but this operated at 10^{-4} to 10^{-6} Torr and hence only on dirty surfaces. However its application was as an alternative to X-ray diffraction: RHEED's forward scattering nature and comparatively high scattering cross section (10^8:1) made it competitive. As ultra high vacuum equipment became common in the 1960s, systems were equipped with RHEED guns, but good LEED was then possible, and LEED largely displaced RHEED as a diffraction technique for clean single crystal surfaces. The main reasons for this neglect was that RHEED gives quantitative results only on extremely flat surfaces which were not easily prepared and offered no theoretical advantages over LEED in the central surface science question of the surface atomic structure. Also the LEED apparatus was easily constructed and much cheaper than the tra-

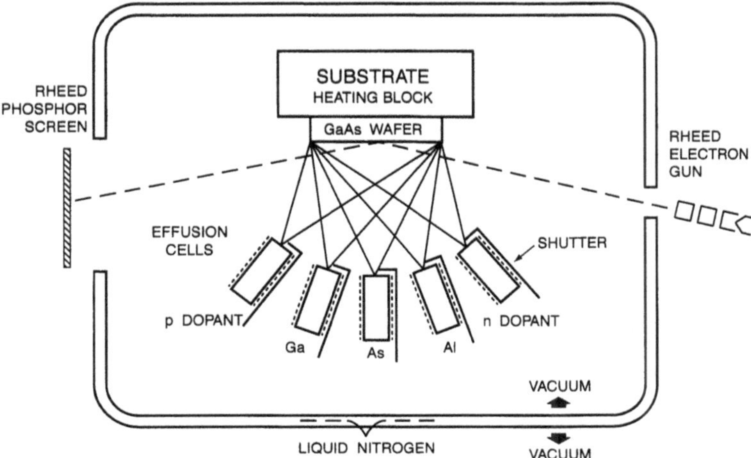

Fig. 12.1. Schematic diagram of an MBE system showing the RHEED geometry

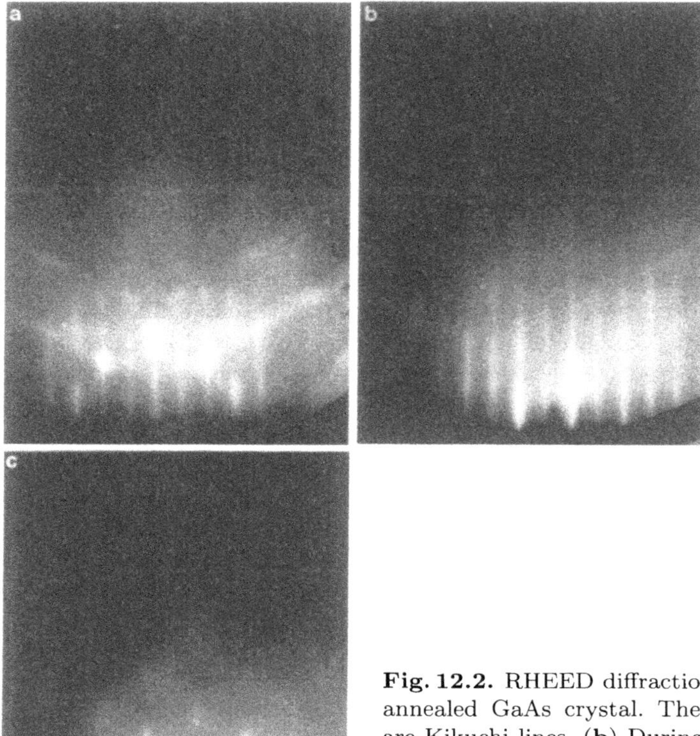

Fig. 12.2. RHEED diffraction patterns for (**a**) annealed GaAs crystal. The diagonal streaks are Kikuchi lines. (**b**) During growth of GaAs: the rougher surface causes Laue zone and other fine detail to be lost. (**c**) Three dimensional growth of InGaAs on GaAs. Much of the background variation is due to adsorbed arsenic on the phosphor screen

ditional magnetically focussed RHEED guns. A good review of RHEED and LEED before 1970 is given by *Bauer* [5]. The growth technique of molecular beam epitaxy (MBE) brought about a renaissance in RHEED in the 1980s. For this reason the applications of RHEED discussed in this chapter will be drawn from MBE.

The RHEED geometry is shown in Fig. 12.1. An electron beam, of energy 5 to 30 keV, is directed at a glancing angle of $\sim 1°$ at a single crystal held in the centre of an ultra-high vacuum chamber. Diffracted electrons fall on a phosphor screen on the other side of the chamber giving a typically streaked pattern shown in Fig. 12.2. The simplicity and robustness of the technique is immediately apparent. For flat surfaces and energies $\leqslant 10$ keV, only a simple TV electron gun and a phosphor screen are necessary. No accelerating voltages need be applied to the screen and grids are not required. Precautions against stray magnetic and electric fields are minimal compared with

LEED because of the high electron energies. The glancing nature keeps the apparatus well clear of other techniques in the chamber.

MBE and its variants such as metalorganic and gas source MBE, have wide application throughout material science for the growth of single crystals [6]. It is used to grow metals, semiconductors and ceramics and is a very general method of crystal growth with particular application where extreme purity and precision are required. MBE is simultaneously a mass production and an advanced research technique. It is a unique combination of surface science and production technology. As RHEED is the essential surface science tool for routine MBE, it has been raised to new prominence and very detailed studies have been done in an endeavour to exploit its potential. In this chapter, the examples will be confined to III–V semiconductors as these were the prototype materials for MBE. However the technique applies generally to all other MBE growths.

In an MBE apparatus as shown in Fig. 12.1, the single crystal substrate faces an array of eight or more ovens, each containing elements such as gallium, indium, silicon, aluminium and arsenic. If as an example, a layer of $n-$type GaAs is required, a single crystal GaAs substrate is held at $\sim 580°$ C; the Ga, As and Si dopant shutters are opened and the growth commences. The Ga and Si has unit sticking probability but the As simply supplies an overpressure and is incorporated as needed. The crystal alloy layers are grown epitaxially at about one atomic layer per second. These materials are subsequently transformed into electronic and optoelectronic devices. Generally the ovens are $\leqslant 1200°$ C, the substrate is $\leqslant 700°$ C and the chamber may contain 10^{-5} Torr of highly corrosive arsenic vapour. The quality of the semiconductor is critically dependent on the exact surface reconstruction. RHEED is uniquely fitted to this role because of its robustness and geometry. It is one of the few techniques of any kind that can monitor crystal growth *in situ*.

12.1 Theory

The relation between the surface reciprocal lattice and the diffraction pattern is shown in Fig. 12.3a. The RHEED Ewald sphere takes a section through the surface rods and streaks are observed on the phosphor screen rather than the spots of the LEED case. To obtain the full reciprocal lattice map which is given directly by LEED, the crystal substrate must be rotated about its normal; the Ewald sphere then sections each plane in turn. The LEED pattern shown in Fig. 12.3b is a complex gallium rich reconstruction referred to as the centred 8×2. The three main azimuthes which should be observed in RHEED are also drawn. The fine detail of the eighth order is often not observed by RHEED and the pattern is referred to as a 4×2. Practically this is of little consequence as the growth conditions are established to avoid this reconstruction by decreasing the metal/arsenic ratio and obtaining the arsenic rich 2×4 (LEED $c(2 \times 8)$): the fine detail is not required. When the transition

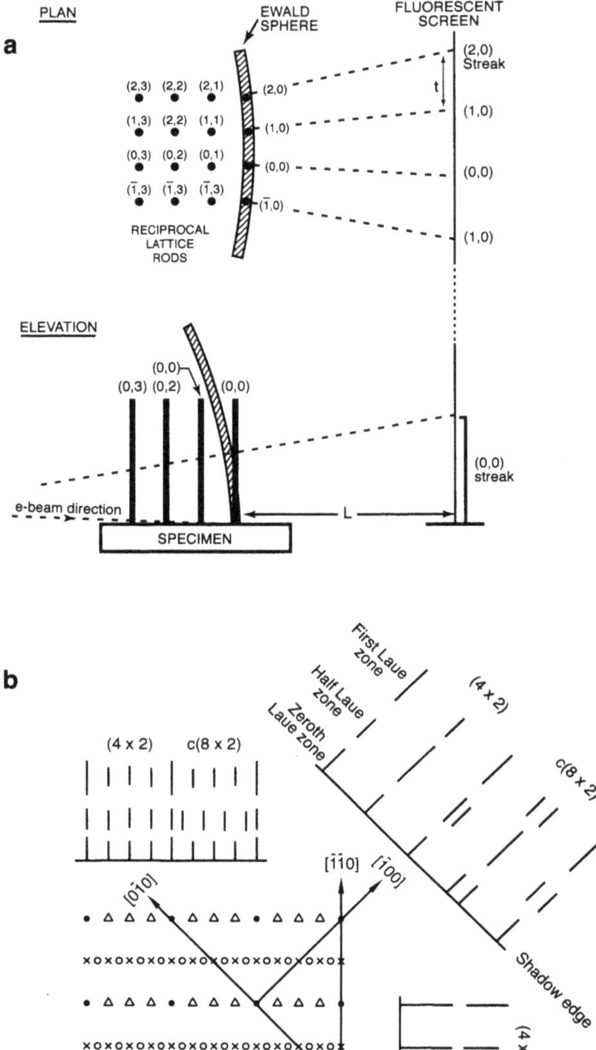

Fig. 12.3. (**a**) Sectioning of the two-dimensional reciprocal lattice rods give the observed RHEED streaks (after [6]). (**b**) Comparison of the LEED and RHEED patterns for the GaAs $c(8 \times 2)$ reconstruction (after [7])

from the metal to arsenic rich reconstruction occurs, the four streaks of the [110] change to two. If the surface roughens, then transmission rather than reflection patterns are obtained as shown in Fig. 12.2c. The beam, skimming over the surface, penetrates peaks and ridges. The streaks are replaced with points since the surface reciprocal lattice rods are now replaced by the reciprocal lattice itself. Roughness of order of an atomic layer can be detected by RHEED. This is extremely useful when monitoring crystal growth and it cannot be easily done by LEED.

To estimate the depth sensitivity of RHEED refer to the mean free path (λ) curve of Fig. 1.8. At 10 keV, the mean free path normal to the surface is $\lambda \sin\gamma$ where $\gamma \sim 1°$ is the glancing angle. The result is about one to two atomic layers. The electron wavelength is ~ 0.1 Å which is an order of magnitude smaller than an atomic layer. Thus the diffraction is sensitive to the surface and the electrons are easily scattered by surface steps and terraces. The diffraction, like LEED, is dominated by multiple scattering and simple kinematic arguments cannot be applied to the streak intensities.

Both LEED and RHEED are surface sensitive, but they differ in their properties parallel to the surface. There are three aspects: the area which can be examined, the coherence length of the diffraction and the ratio of ordered to diffuse scattering. LEED is usually used at normal incidence with a beam size of $\sim 1\,\mathrm{mm}^2$. The glancing nature of RHEED means that it is imprecise in the direction of the beam, of order of a mm with a simple electrostatic gun, while accurately positioned perpendicular to the beam. More sophisticated high energy and magnetically focussed guns can improve the accuracy by an order of magnitude or more. The coherence length is a measure of the sensitivity of the diffraction to the long range order of the surface. It is the maximum distance between reflected electrons which are able to interfere. By the uncertainty principle, $\delta x = 2\pi/\delta k$. The electron momentum uncertainty has two contributions, δk_α due to the beam convergence half angle α, and dk_E due to the finite energy spread δE. Using $E = h^2k^2/2m$ and assuming α very small, it can be shown that [8]

$$\delta k_\alpha = 2k\alpha \sin\gamma \tag{12.1a}$$

and

$$\delta k_E = (k\delta E/2E)\cos(\gamma + \alpha) \tag{12.1b}$$

where γ is the glancing angle. In LEED, $\alpha \sim 10^{-2}$ rad, $\delta E \sim 0.5$ eV, $E \sim 100$ eV, $\gamma = \pi/2$, and $\delta x \sim 100$ Å. In RHEED, $\alpha \sim 10^{-4}$ rad, $\delta E \sim 0.5$ eV, $E \sim 10^4$ eV, $\gamma \sim 10^{-2}$ rad, and $\delta x \sim 1000$ Å. Thus RHEED is more sensitive to long range order. Variations in RHEED streaks have been attributed to monatomic steps 2000 Å apart on a GaAs crystal in an MBE system [9]. A large coherence length does not necessarily make the diffraction pattern more susceptible to disorder. Disorder gives a diffuse background spread evenly over a large solid angle. Large increases in this background have little affect on the sharp, high intensity diffraction features. Thus the diffraction process tends to select out the ordered parts of the surface. For this reason and because of the beam spread across the crystal, and the glancing angle geometry, RHEED can often give a better diffraction pattern then LEED off a patchy or dirty surface.

12.2 Applications

The main traditional application of RHEED is the monitoring of surface reconstructions and surface roughness which has been mentioned with reference to Fig. 12.2. The importance of this knowledge lies in the primary nature of RHEED. For example, in an MBE growth of an III–V alloy such as $Al_xGa_{1-x}As$ on GaAs, the quality of the resulting material depends on the substrate temperature, the Al and Ga fluxes, the arsenic flux and minute (sometimes in the case of oxygen leaks or source contamination, immeasurable) background contamination. The coupling between these parameters and what is actually measured – the oven thermocouple readings, the pyrometer measurement of substrate temperature, daily ion-gauge flux calibrations, is somewhat loose and can alter day to day. It is often difficult to distinguish between the effects of altering the III/V ratio, changing the substrate temperature or lowering the total growth rate. Over a range of parameter values which provide epitaxial growth there are a number of different reconstructions, but only one of which is optimum. Even with the simple growth of GaAs on GaAs, the chief reconstructions with either increasing temperature or increasing III/V ratio are $c(4 \times 4)$, 2×4, 3×1 and 4×2 with optimum growth in the 2×4. In more complex alloys such as AlGaAs on GaAs, slight contamination which causes high trap densities in the resulting epilayer, can be immediately identified by the behaviour of the reconstructions. Optimum growth areas are found within a reconstruction and these can be identified by the oscillation techniques described below.

Before going into the detail which can be extracted during crystal growth with RHEED, an account of the different growth processes will be useful. Growths which give RHEED streaks, not spots, have a two dimensional nucleation and growth behaviour. This means that the adatoms remain in the sample plane. At the initial states of growth on an atomically flat surface, the adatoms migrate over distances measured by a migration length $l = l_0 \exp(-E_l/kT)$ where E_l is the migration energy. The adatoms are scavenged by a number of processes including: nucleation, where they form an island with other adatoms; step propagation, binding to a step which can be the side of an island or a substrate terrace; and desorption. Three dimensional growth involves adatoms jumping up on top of islands and forming three dimensional crystals which give rise to RHEED spots.

RHEED has recently been found able to calibrate these two-dimensional growths and to give detailed information on surface migration kinetics. As growth proceeds, the specular beam oscillates in intensity, the period of the

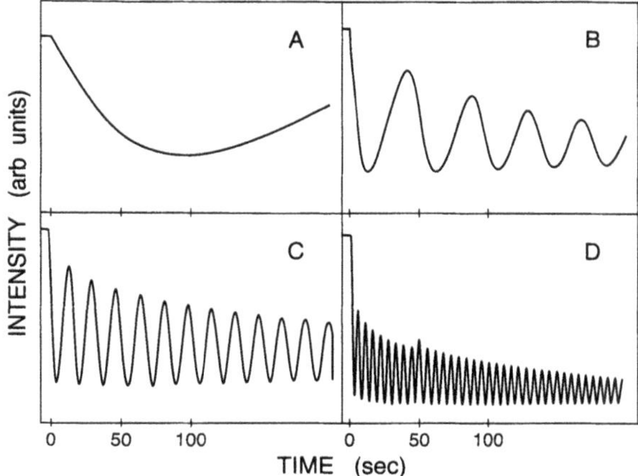

Fig. 12.4. Observed oscillations on a GaAs crystal for increasing flux rates. In *A* the flux is so slow that the substrate steps can scavenge the adatoms before nucleation. The vicinal angle is 1.0 mrad (after [10])

oscillations being equal to the time taken to grow a single monolayer of material. Typical oscillations are shown in Fig. 12.4 and a schematic explanation is given in Fig. 12.5. The first few layers are essentially complete before another begins. Since the layer thickness is much larger than the de Broglie wavelength of the electrons (for GaAs $2.83\,\text{Å} \gg 0.1\,\text{Å}$), the electrons are easily scattered out of the specular beam by the step edges. The step edge concentration is a minimum for a completed layer and a maximum for half a monolayer coverage – hence the oscillations. This simple explanation must be modified to explain the damping of the oscillations as the growth continues. If adatoms fall in a 'ravine' between two steps of width less than the migration length, they will be gettered by the steps edges and nucleation and creation of more step edges will not be possible. As the coverage increases this becomes more likely but so does nucleation of a second layer on top of the first. The result is an increase in the concentration of step edges with growth, tending to a limiting equilibrium value. In RHEED this is seen by the damping of the oscillations and their disappearance, corresponding to the steps gettering all arriving adatoms. A good illustration of this process is shown in Fig. 12.6. Al has a smaller migration length than Ga. GaAs is grown until there are no oscillations. When the aluminium is added oscillations are seen again as the smaller migration length initiates nucleation. When the Al is removed, nothing happens as the growth continues to be by step propagation. An estimate of the migration length can be found by growing at a fixed flux but for differing temperatures on a vicinal surface. On such a surface, cut at a slight angle to a (100) plane, there are terraces of known length (Fig. 12.9). The result is shown in Fig. 12.7. The migration length increases with temperature, and at a certain temperature, growth is only by step propagation.

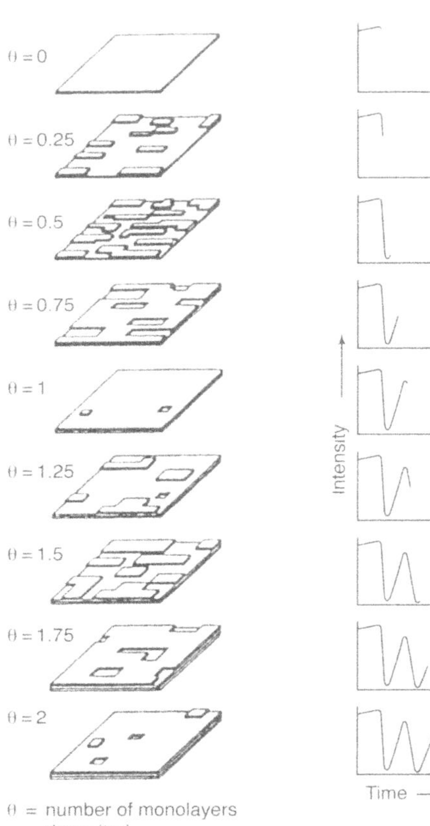

Fig. 12.5. A first order growth model of the intensity oscillations (after [11])

From this, values of l_0 and E_l can be derived. One estimate for Ga atoms on a GaAs (100) surfaceis $l_0 = 4\,\text{Å}$ and $E_l = 0.3\,\text{eV}$.

It is expected that the best quality growths will be obtained if step propagation is the main growth mode as the randomness of island nucleation and dendritic growth is reduced. One method is to adjust the growth conditions for minimum damping, endeavouring to keep the rate of step nucleation to a minimum. Claims have also been made that vicinal surfaces produce better quality materials. An extreme application of this principle is migration enhanced epitaxy (MEE). If growing GaAs by this technique, first a layer of gallium is grown, then separately a layer of arsenic and so on. The method assumes that the migration length of Ga on a bound As layer or As on a bound Ga layer to be much greater than if free Ga and As coexist in the same plane. This is borne out by the RHEED data of Fig. 12.8. Here the MEE oscillations are caused by specular reflection changes between Ga and As terminated surfaces, not scattering by step edges. There are smooth transitions from As to Ga terminated surfaces and the oscillations do not damp

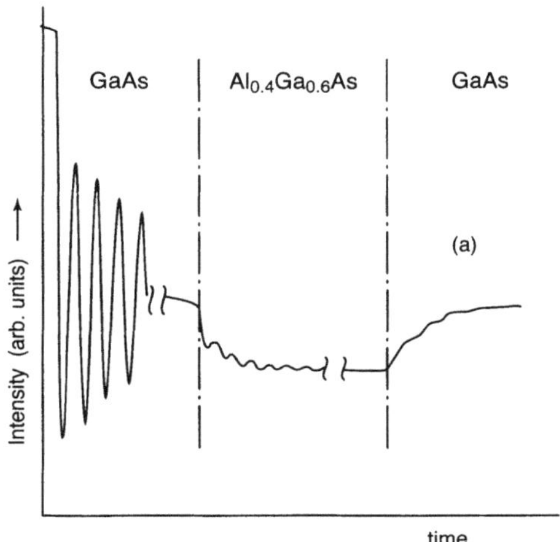

Fig. 12.6. Intensity oscillations across GaAs/AlGaAs/GaAs interfaces. The short migration length of the Al restarts the oscillations (after [12])

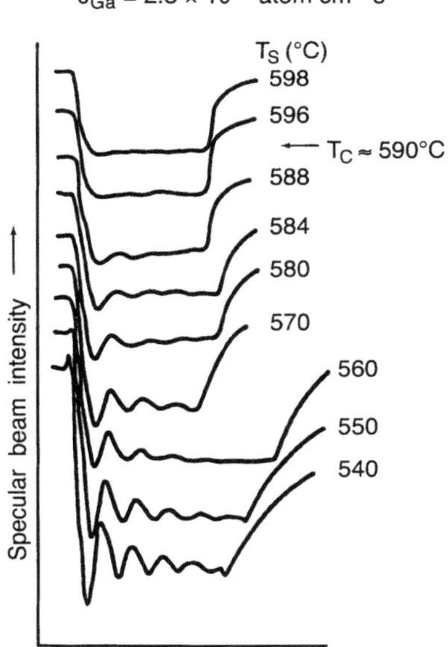

Fig. 12.7. Growth of GaAs on a vicinal surface as a function of temperature. At T_c, the migration length is of order the step separation (after [13]

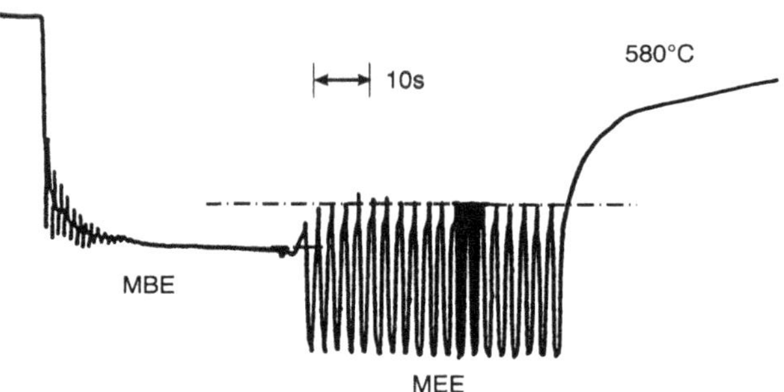

Fig. 12.8. RHEED oscillations by MBE followed by MEE. The same 2×4 reconstruction remained through the growth (after [14])

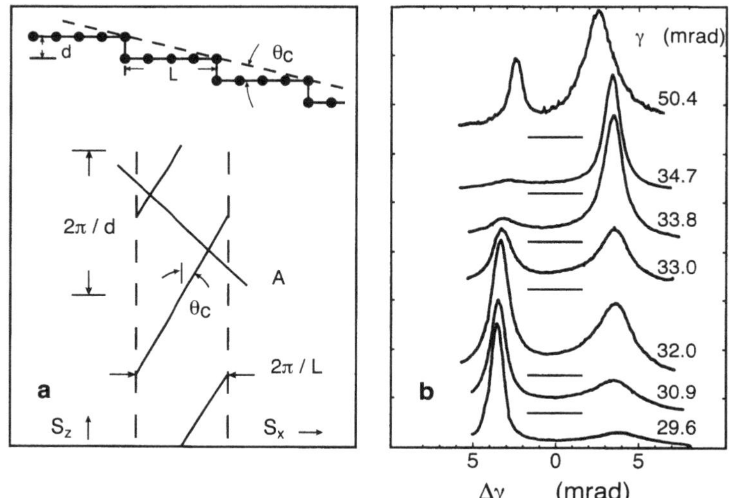

Fig. 12.9. (**a**) Schematic of a regular staircase and its reciprocal lattice. Curve A is part of the Ewald sphere for a beam directed down the staircase with a RHEED streak split into two. (**b**) Intensity profiles along a (00) streak as a function of glancing angle γ (after [9])

out with growth. With this technique, high quality materials can be grown at much lower temperatures than normally used, 300° C instead of 600° C.

As previously mentioned, RHEED is highly dynamic; multiple scattering dominates the diffraction and simple kinematic models can not in general be used. One exception is the application of RHEED to measure the terrace widths on vicinal surfaces. The direct and reciprocal lattice for a terraced vicinal surface (not to scale) is shown in Fig. 12.9a. The diffracted intensity is the product of the diffraction due to a terrace of atoms times the diffraction due to a grating of step edges. The rod shown by the dashed lines has the reciprocal width of the terrace; the slashes have the reciprocal length of

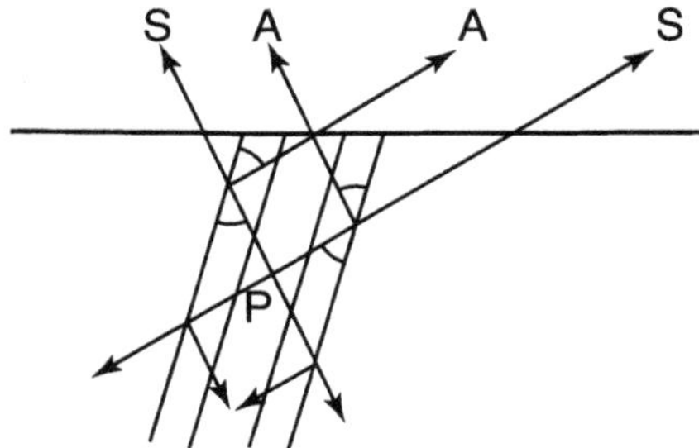

Fig. 12.10. The formation of Kikuchi lines. If diffuse scattering is modeled by emission from a point P in the bulk, then diffracted electrons A and S add and subtract from the diffuse background. Dynamical effects which includes unequal absorption for differing paths, gives A and S different intensities: cancellation does not occur and cones of high and low intensity electrons are emitted from the crystal

the risers and the angle of the vicinal surface. Depending on the angle of incidence, the Ewald sphere will intersect one or more of the slashes. Figure 12.9 shows intersection with two slashes, producing splitting in the observed RHEED streak. The vicinal angle can be straightforwardly estimated from this data using

$$\beta = (2\pi/kd)\theta_c \cos\phi / \left(\theta_c \cos\phi + \gamma\right) \tag{12.2}$$

where β is the measured splitting angle, θ_c is the angle between the vicinal surface and the low index bulk plane, $\gamma = \pi/2 - \theta$ is the glancing angle and ϕ is the azimuthal angle with $\phi = 0$ down the staircase.

So far we have ignored multiple scattering, but one very important multiple scattering phenomenon which is observed in RHEED and LEED is Kikuchi lines. They appear as sharp dark or light lines crossing the phosphor screen (Fig. 12.2a) and they move rigidly with the crystal. Because of this they are used to check the crystal orientation and, with experience, their sharpness provides a qualitative guide to surface cleanliness. Their origin is shown in Fig. 12.10. Diffusely scattered electrons are diffracted inside the bulk crystal. These diffracted electrons lie in cones which intersect the phosphor screen as arcs of large radii. Whether the line is light or dark depends on whether the diffraction is in or out of the diffuse background. This diffraction is from the bulk; the Kikuchi electrons carry underlying information of the solid and not the surface. Their effect is often noticed in oscillation growth experiments. If a Kikuchi line intersects that part of the specular streak which is being monitored, a π out of phase component is observed in the intensity oscillations. This is caused by electrons, which have been scattered from the specular beam by surface steps, contributing to the population of diffuse electrons and being diffracted by the Kikuchi process back into the specular streak. At half a monolayer coverage when the specular beam intensity is a minimum, the scattering in from the Kikuchi line is a maximum. An example

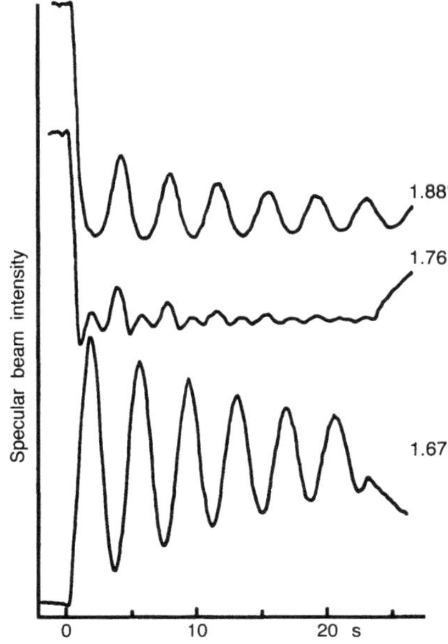

Fig. 12.11. The appearance of harmonics in the oscillations of the specular beam as the glancing angle is varied. At 1.67° the main contribution is from a Kikuchi line. There is a π phase shift compared with the surface pattern at 1.88° (after [13])

is shown in Fig. 12.11. Care must be taken in oscillation experiments to filter this harmonic out.

References

1. S. Nishikawa, S. Kikuchi: Nature **122**, 726 (1928)
2. L.H. Germer: Phys. Rev. **50**, 659 (1936)
3. S. Miyake: Nature **139**, 457 (1936)
4. R. Uyeda: Proc. Phys. Math. Soc. Japan **22**, 1023 (1940); *ibid* **24**, 809 (1942); Y. Kainuma, R. Uyeda: J. Phys. Soc. Japan **5**, 199 (1950)
5. E. Bauer: In *Techniques of Metals Research*, Vol. 2, ed. by R.F. Bunshah (Interscience, New York 1969) p. 501
6. E.H.C. Parker (ed.): *The Technology and Physics of Molecular Beam Epitaxy* (Plenum, New York 1985)
7. J.H. Neave, B.A. Joyce: J. Cryst. Growth **44**, 387 (1978)
8. D.P. Woodruff: In *The Chemical Physics of Solid Surfaces and Heterogeneous Catalysis*, Vol. 1, ed. by D.A. King, D.P. Woodruff (Elsevier, Amsterdam 1981) p. 108
9. P.R. Pukite, J.M. Van Hove, P.I. Cohen: J. Vac. Sci. Technol. B**2**, 243 (1984)
10. J.M. Van Hove, P.I. Cohen: J. Cryst. Growth **81**, 13 (1987)
11. B.A. Joyce, P.J. Dobson, J.H. Neave, K.Woodbridge, J. Zhang, P.K. Larsen, B. Bolger: Surf. Sci. **168**, 423 (1986)
12. B.A. Joyce, J. Zhang, J.H. Neave, P.J. Dobson: Appl. Phys. A**45**, 255 (1988)
13. P.J. Dobson, B.A. Joyce, J.H. Neave, J. Zhang: J. Cryst. Growth **81**, 1 (1987)
14. Y. Horikoshi, M. Kawashima, H. Yamaguchi: Jap. J. Appl. Phys. **27**, 169 (1988)

13 Low Energy Electron Diffraction

P. J. Jennings and C.Q. Sun

One of the most powerful techniques available for surface structural analysis is low energy electron diffraction (LEED). It is widely used in materials science research to study surface structure and bonding and the effects of structure on surface processes. However because it usually requires single crystals and ultrahigh vacuum conditions it has limited value for applied surface analysis, which is often concerned with polycrystalline or amorphous materials. LEED has many similarities to X-ray and neutron diffraction but it is preferred for surface studies because of the short mean free path of low energy electrons in solids.

13.1 The Development of LEED

LEED was discovered accidentally in 1924 by *Davisson* and *Kunsman* in the course of their studies of the secondary emission of electrons from a nickel crystal when it was bombarded by a beam of monoenergetic electrons [1]. In 1926 *Elsasser* suggested that the observed anisotropy in the elastic scattering of electrons from nickel surfaces was due to diffraction [1]. He used the de Broglie relationship to derive an electron wavelength from the momentum p and then related the electron's wavelength to the accelerating voltage V in the electron gun

$$\lambda = \frac{h}{p} \approx \sqrt{\frac{150}{V}}\,\text{Å} \tag{13.1}$$

where V is in electron volts and $V \lessapprox 1\,\text{keV}$. Thus for an electron with a kinetic energy of 150 eV the de Broglie wavelength is $\approx 1\,\text{Å}$, which is similar to the spacing between rows of atoms in a crystal. In 1927 *Davisson* and *Germer* [2] carried out a systematic study of the scattering of electrons from Ni(111) and found that the maximum in the reflected intensity of the elastically scattered electrons at any angle α satisfied the plane grating formula

$$n\lambda = a \sin\theta \tag{13.2}$$

where a is the spacing between adjacent rows of atoms, λ is given by de Broglie relationship in (13.1) and n is an integer. At about the same time *Thomson* found diffraction rings when high energy electrons were transmitted through thin metal films. He related the diameter of these rings to de

Broglie's wavelength and therefore helped to demonstrate the wave nature of the electron. Davisson and Thomson shared the 1931 Nobel Prize for Physics for the discovery of matter waves. Thus LEED had a momentous birth in the confirmation of de Broglie's hypothesis about the wave nature of matter. *Germer* realized that LEED had great potential for surface studies when he wrote in 1928 [4] that, "information of the nature of that which we have obtained may turn out to be of great scientific importance. The whole problem of chemical catalysis concerns itself with what occurs on the surface of a solid body. We believe that this electron diffraction offers the only means which has been suggested for a direct study of questions of this nature".

In the succeeding years high energy electron diffraction, based on Thomson's work, developed rapidly into the field of electron microscopy. However LEED stagnated for many years because researchers found the technique too difficult and too irreproducible. The problem was that it was tedious and time consuming to collect the data and the vacuum technology of the day was inadequate to prepare and maintain clean metal surfaces. During the 1930s most researchers abandoned LEED because of these problems. However in 1934 *Ehrenburg* [5] developed a fluorescent screen which enabled him to display a complete LEED diffraction pattern instantaneously. He showed that a typical LEED pattern is a two dimensional array of spots which arise from diffraction by a two dimensional layer or mesh of scattering centres. The LEED pattern therefore reflects the symmetry and structure of the two dimensional reciprocal lattice of the crystal surface. However, despite this advance the LEED experiments were still plagued by a lack of reproducibility due to contaminated surfaces. *Farnsworth* [6] in the USA continued to work with LEED through the period from 1930 to 1960 and gradually he improved the apparatus and developed a technique for data collection and analysis. By 1960 ultrahigh vacuum technology was available and LEED had come of age. During the 1960s the field developed rapidly and great advances were made in technique and theory. By 1975 the field was mature with a well established technique and a reliable theory for simulating and analyzing the experimental data. LEED researchers have achieved many substantial successes in the years since then. Some of the most successful applications are described in Sect. 13.6 of this chapter. These days LEED is recognized as the major technique for surface structural studies. These surfaces may be ordered or partially disordered. The LEED technique is also suitable for studying the surfaces of alloys and compounds as well as simple metals and semiconductors.

13.2 The LEED Experiment

In principle the LEED experiment is very simple [7]. A narrow beam of monoenergetic electrons with an energy between 0 and 500 eV is directed onto a planar single-crystal surface at a given angle as shown in Fig. 13.1. A

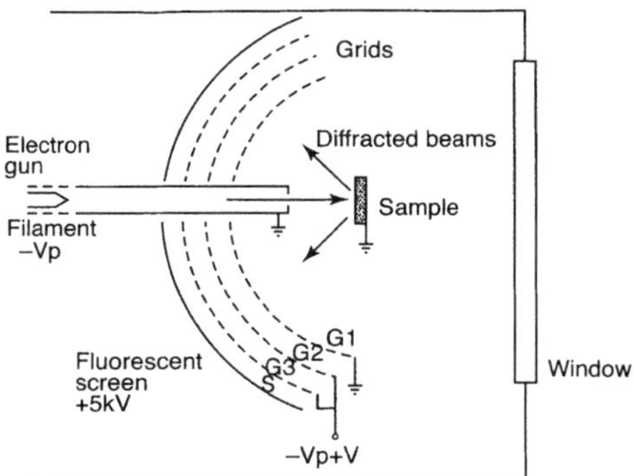

Fig. 13.1. Typical LEED system showing the fluorescent screen S and the hemispherical grids $G1$, $G2$ and $G3$

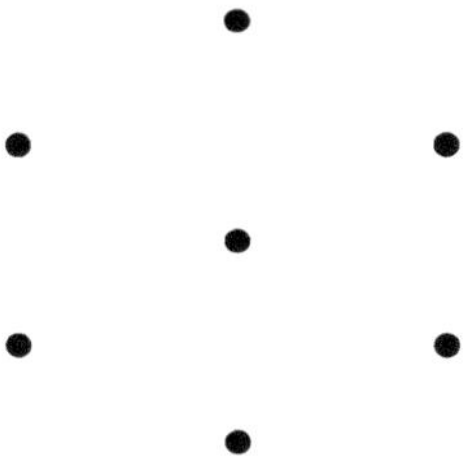

Fig. 13.2. LEED spot pattern for normal incidence on the (111) surface of Cu with a primary beam energy of 80 eV

number of diffracted beams of electrons with the same energy as the incident beam are produced in the backward direction. The spatial distribution of these beams and their intensities as a function of angle and energy of the incident beam provides information which can be used to analyze surface structure. Thus, in contrast to X-ray diffraction in which the wavelength is usually held fixed, in LEED there is this extra degree of freedom of changing wavelength via changing energy.

It is customary in LEED to photograph spot patterns produced by the surface layers as shown in Fig. 13.2 and to measure the intensity of one or more beams as a function of the incident beam energy for a fixed angle and azimuth of incidence as shown in Fig. 13.3. The LEED spot intensity also varies as a result of other influences such as annealing and adsorption. The analysis of LEED spot profiles at constant energy and incidence angles can provide information about surface roughness, critical points for phase transitions and surface structural changes during chemisorption. [15]

In practice the LEED experiment, like all surface experiments, is difficult to carry out and difficult to analyze for the following reasons:

- The sample surface must be well-oriented, planar and either clean or contaminated in a reproducible manner with some adsorbate.

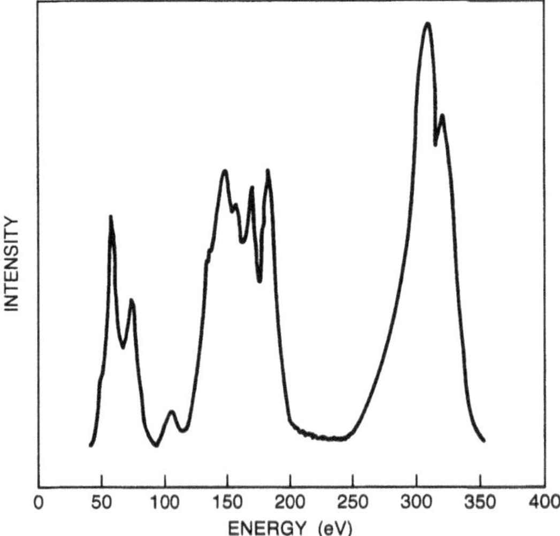

Fig. 13.3. A LEED intensity versus energy plot for Cu(001) at an angle of incidence of 42° along the ⟨11⟩ azimuth

- The experiment must be carried out in ultra high vacuum to minimize surface contamination by surrounding gases. The background pressure in the LEED chamber should be less than 10^{-10} Torr (10^{-8} Pa).
- Sample manipulation requires sophisticated devices.
- Balancing out of stray magnetic fields is essential.
- The accurate measurement of diffraction angles is difficult, [8].
- The collection of intensity data is time consuming although computer based data logging does facilitate this, [9].

The LEED apparatus contains four essential components, as are shown in Fig. 13.1:

1. The *electron gun.* This need not be very sophisticated. The electrons are generated either by on-axis, indirectly heated cathodes or by off-axis tungsten filaments, with energies in the range 0–1000 eV. The electron current is usually a monotonically increasing function of the gun voltage reaching 1–2 μA above about 150 V. The effective diameter of the electron beam is about 1 mm and the energy spread is typically about 0.5°. This implies a coherence width of the incident beam at the sample of 200-550 Å. Thus LEED is only sensitive to small regions of the surface of the order of 200–500 Å over which the periodicity of the lattice is maintained. Steps, kinks and surface imperfections are generally not visible but manifest themselves as background noise.
2. The *goniometer* which holds and orientates the sample. This is usually complicated and costly. Most goniometers allow rotation about an axis in the plane of the surface and about an axis perpendicular to the surface. The sample holder must also allow for temperature control as the samples will need to be heated and cooled (often to liquid nitrogen temperatures).

3. The *detector* to monitor the diffracted beams. The most common is the post-diffraction accelerator of *Ehrenburg* [5]. Electrons accelerated through a potential difference V impinge on the surface and are diffracted. Less than 5% of the backscattered electrons are elastically scattered and the remainder are inelastically scattered. These travel through the field free region to grid $G1$. A small negative voltage between $G1$ and $G2$ serves to reject most of the inelastically scattered electrons while the elastically scattered electrons reach $G3$ and a large positive voltage between $G3$ and the fluorescent screen accelerates them to F where they produce fluorescence on impact. The observer looking through the window will see bright spots at the main points of impact. This spot pattern is called a LEED pattern. The brightness of the spots is proportional to the intensities of the corresponding diffracted beams which can therefore be monitored with a spot photometer – this function is usually digitized and stored on a magnetic disk for subsequent plotting and processing.
   The screen is hemispherical in profile and centred on the position of the sample.
4. The whole apparatus is enclosed in an UHV *chamber* capable of reaching and maintaining pressures as low as 10^{-11} Torr. At these low pressures a surface will be covered by a monolayer of gas in about 10 h (if the sticking probability is one).

13.2.1 Sample Preparation

Preparation of samples for surface studies proceeds in two stages:

1. *Outside the vacuum chamber:* this includes selection and orientation of the sample, usually an ultrapure single crystal ingot; followed by cutting, lapping and polishing of the sample to produce a flat, smooth surface oriented within 0.5°.
2. *Inside the vacuum chamber* at pressures $\approx 10^{-10}$ Torr (10^{-8} Pa) a series of cleaning procedures aimed at eliminating all foreign atoms from the surface. Common heat treatments are O_2 (to remove carbon) or H_2 (to remove oxygen) followed by scattering or ion bombardment in cycles with annealing (e.g. Ni, Cu). Some materials must be heated to high temperatures to evaporate surface impurities (e.g. W, Mo). Cleavage along favourable crystallographic planes is useful for Si, Ge, Be and Zn.

13.2.2 Data Collection

For purposes of surface structure determination we must photograph the LEED pattern and collect the intensities of a number of diffracted beams as a function of incident beam energy. The angle of incidence θ and the azimuthal angle ϕ must also be specified. The angle is defined relative to the surface normal and the azimuth is defined relative to a specified direction in the surface – (generally one of the lattice directions). The intensities are normally collected by means of a spot photometer or by means of a microprocessor-controlled video camera. The data must be processed before it can be used for structure analysis. This involves:

a) normalisation to crystal incident current (since gun current varies with V)
b) contact potential difference correction (between sample surface and crystal surface)
c) background subtraction
d) corrections for grid transparency as the spot moves across the surface.

13.3 Diffraction from a Surface

If an electron wave with wavelength λ falls upon a two-dimensional net of scattering centres a number of scattered waves are produced. The plane grating condition applies for normal incidence and defines the diffraction maxima or spot pattern

$$n\lambda = a \sin \theta . \tag{13.3}$$

The intensity of the spots varies with energy because the atomic scattering factors are functions of energy and the scattering also occurs from successive layers in the crystal. Thus the Bragg condition also applies for scattering from subsequent layers.

The spot pattern contracts as the energy increases according to (13.3). For the two dimensional scattering the appropriate diffraction conditions are

$$\boldsymbol{k}_{\|} - \boldsymbol{k}'_{\|} = \boldsymbol{g} \tag{13.4}$$

where $\boldsymbol{g}$ is a reciprocal lattice vector

$$\boldsymbol{g} = l\boldsymbol{b}_1 + m\boldsymbol{b}_2 \tag{13.5}$$

and $\boldsymbol{k}_{\|}$ and $\boldsymbol{k}'_{\|}$ are the surface components of the incident and diffracted wave vectors.

Thus the diffracted beams can be labelled by vectors of the reciprocal net $\boldsymbol{g} = (l, m)$.

The Ewald sphere construction provides a useful graphical representation of these two-dimensional diffraction conditions. The construction proceeds as follows:

a) each of the points of the 2-D reciprocal lattice is the origin of a rod,
b) the incident $\boldsymbol{k}$ vector is oriented so that its origin is at the 00 point and its length is $1/\lambda = k/2\pi$,
c) a sphere is drawn with the end point A of the incident vector as origin,
d) the possible diffraction conditions are given by (13.4) or by the intersection of sphere with the reciprocal lattice rods. The direction of the beams is also given by this construction as shown in Fig. 13.4.

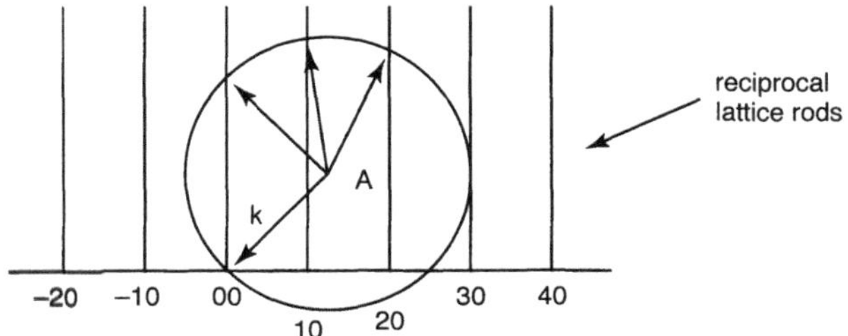

Fig. 13.4. Ewald sphere construction for LEED

Thus the LEED pattern which consists of the intersections of all of the diffracted beams $\boldsymbol{k}'$ with the Ewald sphere is an image of the reciprocal net when it is viewed along the normal direction to the surface at infinite distance.

As the energy of the incident electrons increases the radius of the sphere will increase and the spot pattern will shrink. Actually this form of the Ewald sphere construction is only valid for scattering by a single layer. For 3-D scattering, as in X-ray diffraction, the rods become points in the 3-D reciprocal lattice. LEED also has some 3-D features as the electrons do penetrate some distance into the crystal and the rods can therefore be thought of as having 3-D lattice points embedded in them. When the Ewald sphere intersects the rod near a 3-D lattice point there will be a maximum in the intensity of that spot. Such maxima are called Bragg peaks.

13.3.1 Bragg Peaks in LEED Spectra

Consider an incoming beam with wavevector $\boldsymbol{k}_0$ being scattered by a succession of identical, equally spaced layers of the crystal as in Fig. 13.5. The amplitude of the scattered wave is given by

$$\begin{aligned}\psi_{\mathrm{sc}} = &\left(\rho_{0\nu} + \tau_{00}\omega_0\tau_{\nu\nu} + \tau_{00}\omega_0\tau_{00}\omega_0\rho_{0\nu}\omega_\nu\tau_{\nu\nu}\omega_\nu\tau_{\nu\nu} + \cdots\right)\mathrm{e}^{\mathrm{i}\boldsymbol{k}_\nu\cdot\boldsymbol{r}} \\ &+ \text{higher order terms}\end{aligned} \tag{13.6}$$

where ρ is the reflection coefficient for a layer, τ the transmission coefficient and ω_ν is the phase factor $\exp(\mathrm{i}\boldsymbol{k}_\nu^\perp d)$, where $\boldsymbol{k}_\nu$ is the wave vector of the diffracted beam.

These contributions may be summed to infinity to give

$$\psi_{\mathrm{sc}} \approx \frac{\rho_{0\nu}}{1 - \tau_{00}\tau_{00}\omega_0\omega_\nu}\mathrm{e}^{\mathrm{i}\boldsymbol{k}_\nu\cdot\boldsymbol{r}} \,. \tag{13.7}$$

The maximum of this function with respect to energy can be determined by assuming that to a first approximation $\tau_{00} \approx \tau_{\nu\nu} \leq 1$. Thus the maximum occurs for $\omega_0\omega_\nu \approx 1$, i.e. for

$$\left(\boldsymbol{k}_0^\perp + \boldsymbol{k}_\nu^\perp\right) d = 2\pi n \,. \tag{13.8}$$

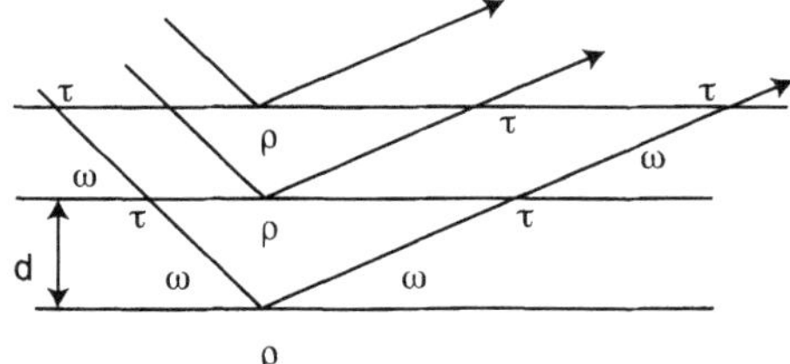

Fig. 13.5. Formation of Bragg peaks in LEED

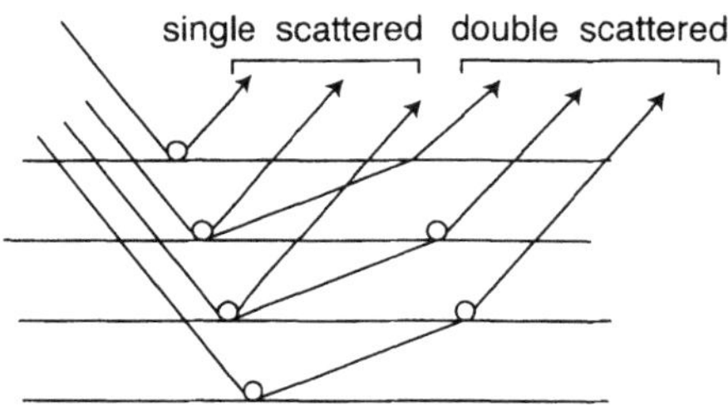

Fig. 13.6. Mechanism of production of secondary Bragg features in LEED

This is the condition for a Bragg peak in the diffracted beam ν. For the 00 or specular beam it becomes

$$\left(\boldsymbol{k}_0^{\perp} + \boldsymbol{k}_{\nu}^{\perp}\right) d = 2\pi n \, . \tag{13.9}$$

Actual LEED intensity spectra also exhibit a number of secondary Bragg features which are due to multiple scattering.

The mechanism of production of secondary Bragg features is illustrated in Fig. 13.6. A non-specular beam at the Bragg angle may be scattered back into the specular beam as it leaves the crystal thus echoing its Bragg feature in the specular beam.

The presence of these secondary Bragg features in LEED intensity curves demonstrates that multiple scattering is important. The reason for this situation is that LEED occurs via coulombic scattering rather than via induced electric dipoles as in the case of X-rays. Slow electrons are strongly scattered by a crystal surface and thus penetrate only a short distance. This is useful for surface work but it also implies that a multiple scattering or dynamical theory is required to simulate the observed intensity curves.

13.4 LEED Intensity Analysis

The Bragg peak locations are dependent on the surface structure of the crystal and thus the aim of LEED intensity analysis is to deduce surface structure from the measured intensity curves [10,11]. The usual approach is to postulate a structural model of the surface and to use this as the basis for a simulation of the LEED intensity spectra. The simulated and measured spectra are compared visually or qualitatively and the model is refined to obtain the best possible fit. The simulations are complex and the analysis is difficult and indirect. The procedure involved is basically:

a) the potential in the solid is calculated from the superposition of Hartree Fock atomic potentials. It is spherically averaged using the muffin-tin approximation to obtain the potential of an average atom in the solid;
b) the coulomb scattering from this muffin-tin atom is calculated using the method of partial waves to give a set of scattering phase shifts (as a function of electron energy);
c) the reflection and transmission coefficients for each layer are calculated from the atomic data and the assumed two-dimensional structure using a multiple scattering theory such as the Green's function or KKR method;
d) the single layers are combined to form a multilayer crystal using the transfer matrix which takes account of the scattering of each incident beam into the other beams at each layer.

This produces a set of intensities for each of the diffracted beams as a function of the energy and diffraction angles and for the assumed surface structure. The analysis proceeds using simulation techniques in which the crystal model is optimized to obtain the best fit of the computed and measured LEED spectra as shown in Fig. 13.7. The process involved in the calculation is

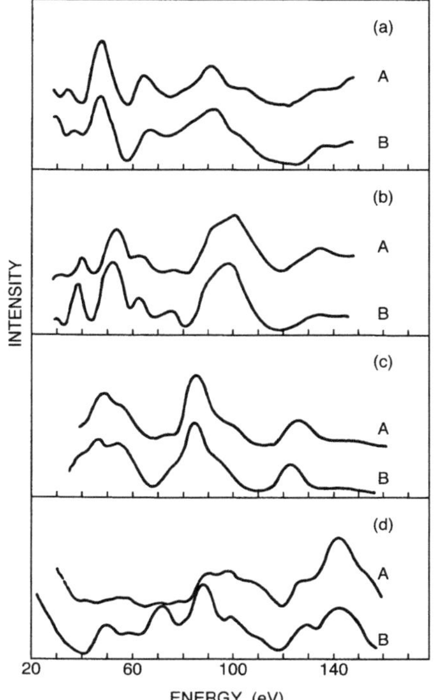

Fig. 13.7. Illustration of the simulation of LEED intensity curves for various surfaces (A: experimental, B: theory

similar to that used in band structure calculations for solids. In fact the LEED intensity curves are dependent on the band structure for the solid along a particular direction through the Brillouin zone corresponding to the direction of the primary beam $\boldsymbol{k}_0$. The Bragg peaks occur at gaps in the band structure (well above the Fermi level) while the secondary Bragg peaks occur at partial band gaps. If an electron is incident on the crystal surface at an energy where a band gap is present there are no propagating states in the solid for it to enter and thus it is strongly reflected. Thus Bragg peaks correspond to absolute gaps in the band structure. Secondary Bragg peaks occur where the density of propagating states in the solid is low.

13.5 LEED Fine Structure

The LEED intensity curves also show a number of narrow, complex features called LEED fine structure at very low primary energies. These features arise from the scattering of diffracted beams by the surface barrier. They are found only at low energies and they usually form a series of peaks converging on the threshold energy for a new diffracted beam as shown in Fig. 13.8 [12].

The mechanism of production of these features is shown in Fig. 13.9. It involves a new beam which has begun to propagate in the crystal but still does not have sufficient energy to surmount the surface barrier. This beam is totally internally reflected by the surface barrier and after subsequent diffraction by the crystal it may be diffracted back into the incident beam where it

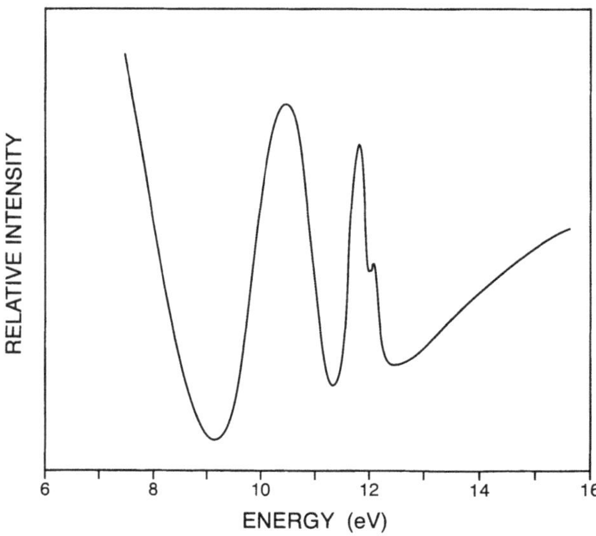

Fig. 13.8. LEED fine structure for Cu(001) for an angle of incidence of 60° along the ⟨11⟩ azimuth

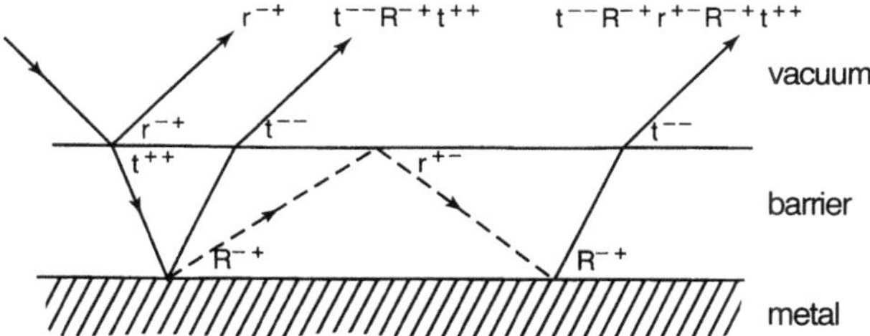

Fig. 13.9. Mechanism for the production of LEED fine structure

can produce interference effects in that beam. The spacing of the interference fringes is an indication of the structure of the surface barrier. The analysis of the LEED fine structure enables us to deduce the structure of the barrier and to map the valence band structure of metals and chemisorbed surfaces [16–18]. This is an important component in the theory of many types of electron emission phenomena from solids such as photoemission, thermionic emission and field emission.

13.6 Applications of LEED

LEED is the basic structural technique for solid surface analysis for crystalline materials. It has been applied to many different types of surface but usually it is best suited to structural studies on clean, well-ordered surfaces of simple metals and semiconductors.

13.6.1 Determination of the Symmetry and Size of the Unit Mesh

The LEED pattern provides a map of the reciprocal lattice of the surface. The (00) spot or specular beam can be identified as the spot towards which all of the others gravitate as the energy of the electron beam is increased. At normal incidence we can determine the rotational symmetry of the pattern by inspection. We can also determine the size of the unit mesh from the pattern. To do this we photograph the LEED pattern and measure b_1 and b_2 on the photograph. From a knowledge of the geometry of the LEED chamber we can calculate the diffraction angle θ given by

$$n\lambda = a \sin\theta \,. \tag{13.10}$$

Hence from a knowledge of θ and λ we can calculate the direct mesh lengths a_1 and a_2, and compare them with the predicted values. Thus we can determine whether the surface and substrate nets are identical. This is usually the case for most simple transition metal surfaces. However most semiconductor surfaces show reconstructed unit meshes which are formed because of the covalent bonds which are broken during the fracture of the surface.

13.6.2 Unit Meshes for Chemisorbed Systems

If a small amount of gas is allowed to adsorb on a metal surface it often forms an ordered structure. Such structures are usually different from those of the clean surface as shown in Fig. 13.10. Studies of LEED patterns make it possible to determine the symmetry and relative size of the unit mesh of the adsorbed layer. If the pattern shows new spots which do not belong to the clean surface net then we must conclude that the surface net is different from the substrate net.

Fig. 13.10. LEED patterns for normal incidence on clean Ni(001) and for $c(2 \times 2)$ O/Ni(001) with a primary beam energy of 80 eV

This pattern analysis is very important and LEED is now established as a major tool for analyzing the structure of chemisorbed overlayers and reconstructed clean surfaces.

Such pattern analysis forms the basis for models of the chemisorption process and is vital to an understanding of catalysis and corrosion. However pattern analysis only tells us the relative size, symmetry and orientation of the overlayer. It tells us nothing about the location of the overlayer relative to the substrate. This is the goal of LEED intensity analysis.

13.6.3 LEED Intensity Analysis

The analysis of LEED intensities generally proceeds from the measured intensity versus energy curves for the specular and several non-specular beams. Because of the strong multiple scattering in LEED a dynamical theory is required. The simple kinematical analysis used in X-ray diffraction is unsuccessful in LEED. Dynamical calculations involve heavy computational effort and sophisticated simulation packages are widely available [19]. Studies have extended from the simplest metals to some complex structures using techniques such as tensor-LEED [20] and diffuse-LEED [21]. A referential database for ordered surface structures is also avalable on-line [22]. For example the position of adsorbed selenium atoms on the $\sqrt{2} \times \sqrt{2}\,R45°$ Se/Ag(001) has been studied by several authors. They all find that the adsorbed selenium atoms are located at alternate fourfold silver sites [13]. Recently it has been shown that LEED spectra can be inverted to produce the three-dimensional coordinates of atoms neighbouring a reference atom without prior knowledge of what types of atoms are present [23].

Similar studies were done for O on Ni but at first they disagreed about the Ni-O bond length, [12]. Finally after considerable dispute it was found that the O layer is located about 0.9 Å above the topmost nickel layer [14]. This result has recently been disputed by researchers working with ion scattering spectroscopy. In general the field is still in an immature state. LEED intensity analysis is the most widely accepted surface structural technique and it has been extended to the study of complex surfaces, although the analysis requires considerably more computational effort and sophisticated analytical skills [24]. Generally the adsorption site can be found readily and clearly but the accuracy of the bond lengths and surface layer spacings are unsatisfactory. Very little progress was made with the analysis of the structure of surfaces with large unit meshes or surfaces with adsorbed molecules until recently when a computer code for multi-atom and multiple-diffraction was developed [25]. In conjunction with scanning tunneling microscopy and ultraviolet photoelectron spectroscopy, the band structure, bonding and the non-uniform surface potential barrier for the O–Cu(001) surface have been investigated [26,27]. On the basis of work done to date we can make some useful observations about surface structure:

a) the most closely packed surfaces fcc(111), bcc(110) of clean transition metals exhibit very little or no contraction of the top layer;
b) surfaces with intermediate packing density [e.g. fcc(001), bcc(001)] have a contracted first interlayer spacing. The contractions are typically 2% to 10% depending on the material;
c) the more open surfaces such as fcc(110) and bcc(111) are strongly contracted (of the order of 10–15%). The contraction of surface layers provides a justification for the bond-order – bond-length – bond-strength correlation mechanism that may have important impact on the size dependency of nanosturctured materials [28];
d) clean semiconductor surfaces are mostly reconstructed – although some such as Si(111) can be stabilized with minute amounts of impurity. Such reconstruction is usually obvious in the patterns [e.g. Si(001) 2×1, Si(111) 7×7]. The structures are often very complicated and may involve four or five layers;
e) the study of adsorbate structures has shown that surface bond lengths are similar to gas phase covalent bond lengths and the surface is often distorted in the process of adsorption.
f) for surfaces with chemical adsorbates, the first interlayer spacing often expands by 10%–25%, and the second layer spacing contracts by 5%–10%. However, the exact vertical positions of the adsorbates are still under debate.

13.6.4 Surface Barrier Analysis

The analysis of LEED fine structure at low energies provides information about the structure of the surface barrier (Fig. 13.11). From classical theory we expect that the image force law will apply to an electron at large distances from the surface. This predicts that the potential energy $V(z)$ of an electron at a distance z from the surface is given by

$$V(z) = \frac{-q}{16\pi\varepsilon_0(z - z_0)} . \tag{13.11}$$

Quantum theory predicts a saturation of this potential at short distances leading to a constant average inner potential U_0 which is approximately equal to $\phi + E_F$ at low energies (where ϕ is the work function of the surface, and E_F is the Fermi energy). In addition it is customary to include an imaginary component of this potential to simulate the effects of inelastic scattering in the surface region. Various semi-emperical barrier models have been used to analyse LEED fine structure [16,18]. The real and imaginary parts of the one-dimensional surface potential barrier take the following forms [16,18]:

$$\mathrm{Re}\, V(z) = \begin{cases} \dfrac{-V_0}{1 + A\exp[-B(z - z_0)]} & \text{if } z \geq z_0 \text{ (a pseudo-Fermi-}z\text{ function)} \\ \dfrac{1 - \exp[\lambda(z - z_0)]}{4(z - z_0)} & \text{if } z < z_0 \text{ (the classic image potential)} \end{cases} \tag{13.12}$$

$$\begin{aligned} \mathrm{Im}\, V(z) &= \mathrm{Im}\big[V(z) \times V(E)\big] \\ &= \gamma \times \rho(z) \times \exp\left[\frac{E - \phi(E)}{\delta}\right] \\ &= \frac{\gamma \times \exp\left[\dfrac{E - \phi(E)}{\delta}\right]}{1 + \exp\left[-\dfrac{z - z_1}{\alpha}\right]} \end{aligned} \tag{13.13}$$

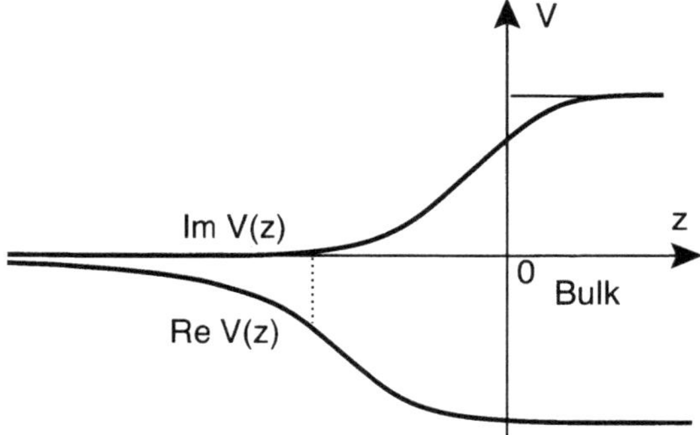

Fig. 13.11. One-dimensional model of the surface potential energy barrier showing the real and imaginary parts as functions of the distance z from the surface

where A, B, γ and δ are constants. V_0 is the muffin-tin inner potential constant of the crystal, α and λ describe the degree of saturation, z_0 is the origin of the image plane inside which electron occupies and $f(E)$ is the local work function that is energy dependent. The z-axis is directed into the crystal. LEED fine structure analysis provides the best available experimental test of barrier models and also enables us to determine the barrier origin z_0. Studies of this type have so far been carried out on Ni, Cu and W surfaces [12]. A high resolution electron spectrometer is required to resolve the fine structure [8]. The potential barrier of the O–Cu(001) surface has also been investigated [27], and it shows strongly localised features.

Recently there have been encouraging advances in this area and it has been possible to extend LEED studies to lower energies ($E < 50\,\mathrm{eV}$) where the analysis is simpler but the experimental demands are greater. Unlike the normal LEED that is dominated by the scattering of a stack of layers of ion cores, the LEED at low energis is dominated by the interaction of low energy electrons with the surface potential barrier and the valence electrons of a very few atomic layers. It is possible with VLEED to determine simultaneously the potential barrier, bonding and the local density of states in the valence band.

13.7 Conclusion

To a large extent LEED has fulfilled the expectations of its early advocates such as *Germer* [4]. However it is now clear that it has certain strengths and weaknesses. Its strengths include:

a) the ability to provide direct and accurate information about surface contamination;
b) its suitability for studies of processes on ordered surfaces including chemisorption, phase transitions and epitaxy;
c) its sensitivity to the surface barrier structure.

LEED has now become a standard technique for investigating structure in surface science experiments and it has provided a wealth of valuable data about structure and bonding on metal and semiconductor surfaces.

However, it also has some well known limitations which include:

a) it is only suited to studies in ultra-high vacuum on surfaces of ordered structures This excludes most industrially and technologically important surfaces. LEED is best suited to fundamental studies of surface processes such as bonding and reconstruction;
b) the analysis of LEED patterns does not give an unambiguous result for complex surface structures. To choose between alternative structural models a LEED intensity analysis is required;

c) LEED intensity analysis requires sophisticated simulation procedures and the computations involved are very time-consuming even for the largest computers for complex adsorption systems. However, there are many interesting problems which involve simple surfaces and adsorbates and these are the current focus of attention;
d) due to the correlation among the structural parameters, the number of numerical solutions may not be unique and physical constraints should be applied. In simulating the LEED data, it would be necessary to replace the traditional approach of individual displacement of the hard spheres with variation of bond geometry and atomic valencies.

For these reasons LEED is one of the most popular research techniques in surface science. However because of its limitations it is advantageous to use it in conjunction with other compatible techniques such as scanning tunnelling microscopy, ultraviolet photoelectron spectroscopy, thermal desorption spectroscopy, x-ray photoemission spectroscopy, Auger electron spectroscopy and work function measurements. Together these techniques are capable of providing a comprehensive understanding of the atomic and electronic processes on ordered surfaces.

References

1. C.J. Calbick: The Physics Teacher, May 1963, pp. 1–8
2. C.J. Davisson, L.H. Germer: Phys. Rev. **30**, 705 (1927)
3. G.P. Thomson: Engineering **126**, 79 (1928)
4. L.H. Germer: J. Chem. Ed. **5**, 1041 (1928)
5. W. Ehrenberg: Phil. Mag. **18**, 878 (1934)
6. H.E. Farnsworth: *Surface Chemistry of Metals and Semiconductors* (Wiley, New York 1959)
7. F. Jona, J.A. Strozier, Jr., W.S. Young: Rep. Prog. Phys. **45**, 527 (1982)
8. G. Hitchen, S.M. Thurgate: Surf. Sci. **24**, 202 (1985)
9. S. Thurgate, G. Hitchen: Appl. Surf. Sci. **17**, 1 (1983)
10. J.B. Pendry: *Low Energy Electron Diffraction; The Theory and Its Application to the Determination of Surface Structure* (Academic, London 1974)
11. M.A. Van Hove, S.Y. Tong: *Surface Crystallography by LEED*, Springer Ser. Chem. Phys. Vol. 2 (Springer, Berlin, Heidelberg 1979); M.A. Van Hove, W.H. Weinberg, C.-M. Chan: *Low Energy Electron Diffraction*, Springer Ser. Surf. Sci. Vol. 6 (Springer, Berlin, Heidelberg 1986)
12. R.O. Jones, P.J. Jennings: Surf. Sci. Rep. **9**, 165 (1988)
13. A. Ignetiev, F. Jona, D.W. Jepsen, P.M. Marcus: Surf. Sci. **40**, 439 (1973)
14. J.E. Demuth, D.W. Jepsen, P.M. Marcus: Phys. Rev. Lett. **31**, 540 (1973)
15. S. J. Murray, D. A. Brooks, F. M. Leibsle, R. D. Diehl, R. McGrath: Surf. Sci. **314**, 307 (1994)
16. R. O. Jones, P. J. Jennings, O. Jepsen: Phys. Rev. B **29**, 6474 (1984)
17. V. N. Strocov, R. Claessen, G. Nicolay, S. Hüfner, A. Kimura, A. Harasawa, S. Shin, A. Kakizaki, P. O. Nilsson, H. I. Starnberg: P. Blaha. Phys. Rev. B **63**, 205108 (2001)

18. C. Q. Sun and C. L. Bai, J. Phys.: Condens. Matt. **9**, 5823 (1997)
19. M. A. Van Hove: LEED program (free) http://electron.lbl.gov/leedpack/leedpack.html
20. V. Blum, K. Heinz: Comp. Phys. Commun. **134**, 392 (2001); K Heinz: Rep. Prog. Phys. **58**, 637 (1995)
21. K. Heinz, D. K. Saldin, J. B. Pendry: Phys. Rev. Lett. **55**, 2312 (1985); U. Starke, J. B. Pendry, K. Heinz: Prog. Surf. Sci. **52**, 53 (1996)
22. M. A. Van Hove: Surface Structure Database http://electron.lbl.gov/ssd/ssd.html
23. S. Y. Tong: Adv. Phys. **48**, 135 (1999)
24. M. A. Van Hove: Prog. Surf. Sci. **64**, 157 (2000)
25. S. M. Thurgate, C. Sun: Phys. Rev. B **51**, 2410 (1995)
26. G. Hitchen, S. M. Thurgate, P. Jennings: Aust. J. Phys. **43**, 519 (1990)
27. C. Q. Sun: Surf. Rev. Lett. **8**, 367 (2001); and Surf. Rev. Lett. **8**, 703 (2001)
28. C. Q. Sun, T. P. Chen, B. K. Tay, S. Li, H. Huang, Y. B. Zhang, L. K. Pan, S. P. Lau, X. W. Sun: J. Phys. D. **34**, 3470 (2001)

14 Ultraviolet Photoelectron Spectroscopy of Solids

R. Leckey

Compared to X-ray photoelectron spectroscopy (XPS) or Auger electron spectroscopy (AES), ultraviolet photoelectron spectroscopy (UPS) is not generally considered to be an analytic technique for the surface characterisation of materials. It is, however, an extremely surface sensitive technique where even a monolayer coverage of an adsorbate or contaminant is sufficient to grossly alter the signal from a given surface. As we shall see, its main strength lies in its unique ability to explore the electronic structure in the conduction/valence band region of a wide variety of solids. As a technique, it can readily be added to other surface science instrumentation and is indeed often offered as an option by manufacturers of XPS/AES equipment. This chapter is consequently included as a brief introduction to the capabilities of UPS, to alert practitioners of other surface science techniques to the information contained in UPS spectra.

UPS relies on the photoelectric effect whereby a beam of monochromatic photons is used to eject electrons from the valence/conduction band region of a material. Traditionally, the photon source used has been a hollow cathode discharge lamp [1] running in an inert gas, the most commonly used resonance lines being He 1 at 21.21 eV and Ne 1 at 16.86 eV. Recently, significantly more intense UV sources utilising magnetic confinement of the plasma, have become available [2]. Since the bandwidth of the conduction/valence band of materials is in the range 5–10 eV, these photon energies are sufficient to probe the entire band structure region of most materials (Fig. 14.1). The photo-emitted electrons consequently have energies typically less than 17 eV using He 1 and consist of two main groups.

The first group of electrons are those that, having been excited within the uppermost few atomic layers, escape into vacuum having suffered no inelastic collisions. The energy, and to some extent the direction of emission of such electrons, may readily be related to their original binding energy within the solid and to the wave vector associated with their original state. The binding energy E_b of each such electron is simply related to the observed kinetic energy E_k via the Einstein formula $E_\mathrm{k} + E_\mathrm{b} = h\nu$ as a first (and usually adequate) approximation, as indicated in Fig. 14.1.

The second group of electrons observed in a UPS spectrum consist of those electrons which have either made one or more inelastic collisions most probably with other bound valence band electrons) or are the secondary electrons which have gained sufficient energy to escape from the material from such a

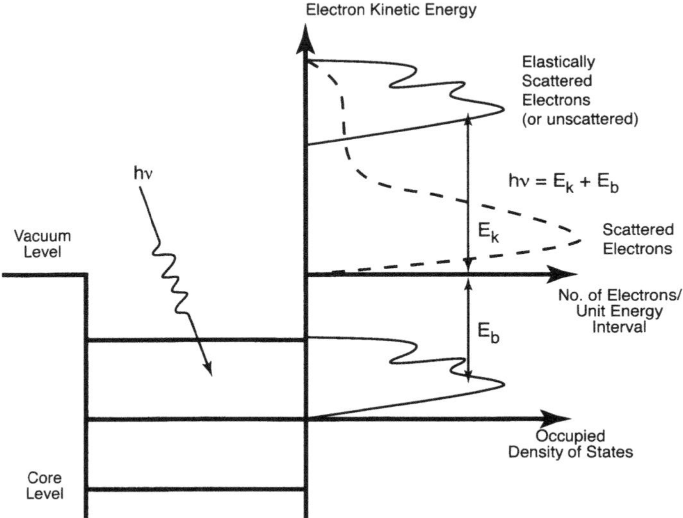

Fig. 14.1. Illustrating the photoexcitation of electrons from a valence band due to monochromatic photons of energy $h\nu$. An idealised energy distribution is also shown with primary emission distinguished from secondary electrons

collision. This group of electrons constitutes a largely featureless low energy peak in all UPS spectra with a shape that resembles somewhat a Maxwellian distribution. There is, regrettably, no established procedure for subtracting this inelastic background from a UPS spectrum at the present time; the effect of the background is clearly most severe for low photon energies. For this reason, among others, the helium ion resonance line at 40.81 eV is sometimes used in laboratory UPS experiments or recourse is made to a synchrotron radiation source.

The main aim of many UPS experiments is to gain information about the distribution of electrons in the outermost valence or conduction band region of the material. It is these electrons which are responsible for the chemical, magnetic, optical and mechanical properties of each material. With sufficiently detailed information about the electronic band structure, we may begin to tailor-make materials with specific properties – such band structure engineering is already occurring in the microelectronics industry. The type of information we need to accomplish such material modification is extremely detailed but can be examined progressively. This may be illustrated by firstly examining information available from the least sophisticated UPS experiment. We may then refine our knowledge of particular materials by examining the interpretation of more exotic experiments involving variable photon energy, angle resolved energy distributions and the effects of the polarisation of both incident photons and emitted electrons. In the present chapter, we can clearly only hope to get a brief flavour of the full complexity of modern UPS.

14.1 Experimental Considerations

The minimum requirements for performing a simple UPS experiment are (a) a resonance radiation lamp, (b) an electron energy analyser capable of operation with 5–150 eV electrons with an energy resolution of $\sim 0.1\,\text{eV}$, (c) an atomically clean polycrystalline sample in a UHV environment. The analyser need not be capable of angular resolution but the energy resolution requirement implies magnetic shielding in the $5 \times 10^{-7}\,\text{T}$ region. Sample surface cleanliness must be maintained on an atomic scale since electrons of $\sim 20\,\text{eV}$ energy have an inelastic scattering mean-free path of $\sim 0.5\,\text{nm}$; this requirement in turn leads to the need for a UHV environment (typically 10^{-10} Torr).

With such equipment, the conduction/valence bandwidth may be readily determined for many materials and the distribution of electrons within the band measured (occupied density of electron states). To illustrate this point, UPS spectra from a range of Ag/Pd alloys are shown in Fig. 14.2 [3].

The probability of a UPS event depends jointly on the number of electrons in the initial state $|\,i\,\rangle$ – the occupied density of states – and on the number

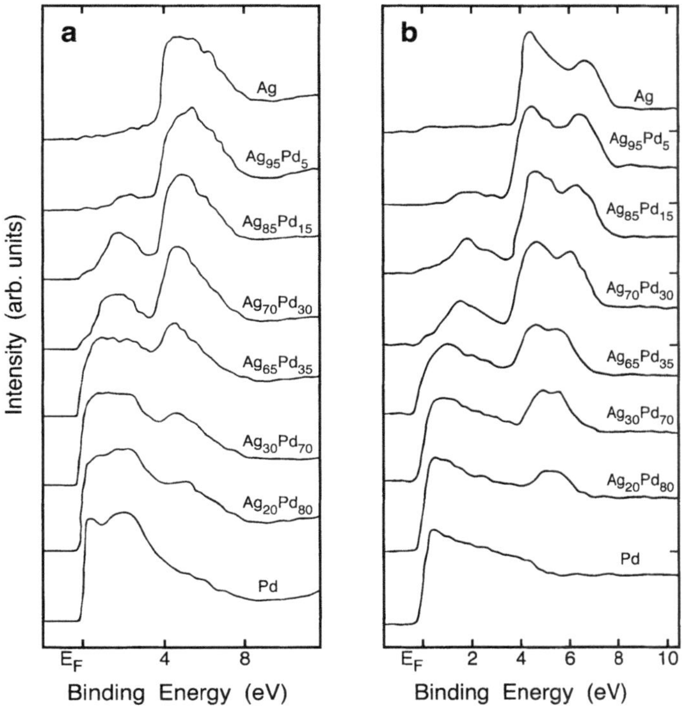

Fig. 14.2. UPS spectra of a series of Ag/Pd alloys taken with **(a)** 21.21 eV and **(b)** 40.81 eV photons under angle integrated conditions. From [3]

of empty states available at the photo-excited energy $\langle f \,|$ – the unoccupied density of states – and on a (dipole) matrix element:

$$\left|\langle f \,|\, r \,|\, i \rangle\right|^2 . \tag{14.1}$$

As the photon energy increases we may expect to find less structure in the unoccupied density of states. At XPS energies, the photo-excited (final) density of states is assumed to be featureless and this assumption holds quite well also at 40.81 eV. Thus both XPS and He II spectra may be expected to mirror variations in the initial (occupied) density of states reasonably accurately whereas a He I spectrum will represent a mixture of these initial and final state functions which is in general difficult to interpret by inspection. The strength of a transition depends on the orbital nature of both initial and final states and experience tells us that emission from d band states is much stronger relative to emission from s or p derived states at 40.81 eV than at 21.21 eV.

Armed with the above information, we can now assert that the He II Ag/Pd alloy spectra of Fig. 14.2 are likely to be a good representation of the width and of the density of states of the conduction band. We can also predict that the most prominent structure comes from band structure states of predominately 'd' nature. As a trivial example of the type of information that follows from these results, we note the manner in which the onset of strong 'd'-band emission moves relative to the Fermi edge as the alloy composition is varied. This in turn controls the optical absorption of these alloys and readily explains the alteration in the colour of these materials with composition.

14.2 Angle Resolved UPS

To proceed beyond the type of experiment outlined above, it is necessary to work with single crystal samples and to acquire spectra with good angular resolution ($\pm 1°$). Under these conditions, we can expect to be able to determine in detail the way crystalline energy levels disperse with wave vector (momentum) in the $E(\boldsymbol{k})$ band structure of the material. The geometry of a typical experiment is shown in Fig. 14.3

Monochromatic photons are shown incident on a clean single crystal surface in UHV at some angle Ψ to the sample normal. The crystal is orientated so that electrons emitted in a selected plane (AZ in Fig. 14.3) containing a symmetry direction of interest can enter an analyser capable of detecting the polar angle of emission θ and of measuring the kinetic energy of each electron. In the past this has been accomplished by stepwise moving an energy analyser possessing a narrow acceptance angle $d\theta$ in small increments of θ, but more recently, display type analysers have been developed which simplify the process and reduce the data collection time required. A review of such analysers appears in [4]. As an example of one such device, our toroidal electron energy analyser is shown in cross-section in Fig. 14.4.

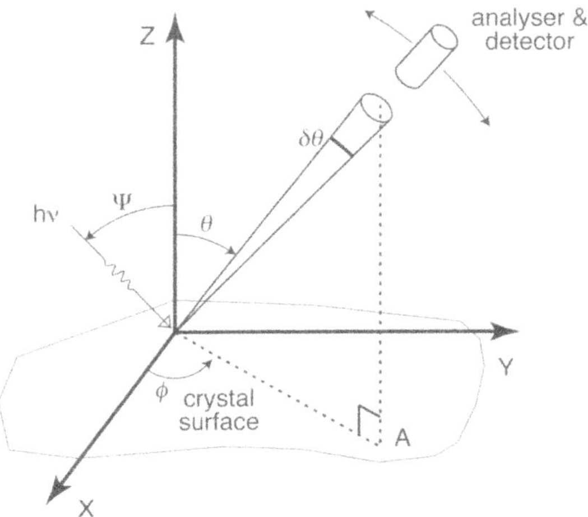

Fig. 14.3. Geometrical arrangement for an angle resolved UPS experiment

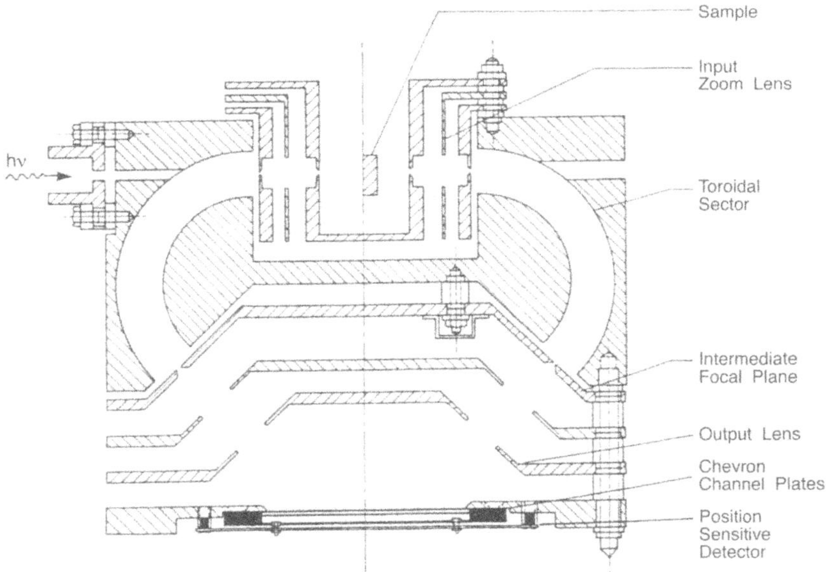

Fig. 14.4. Cross-section of a toroidal electron energy analyser specifically designed for angle resolved UPS. From [5]

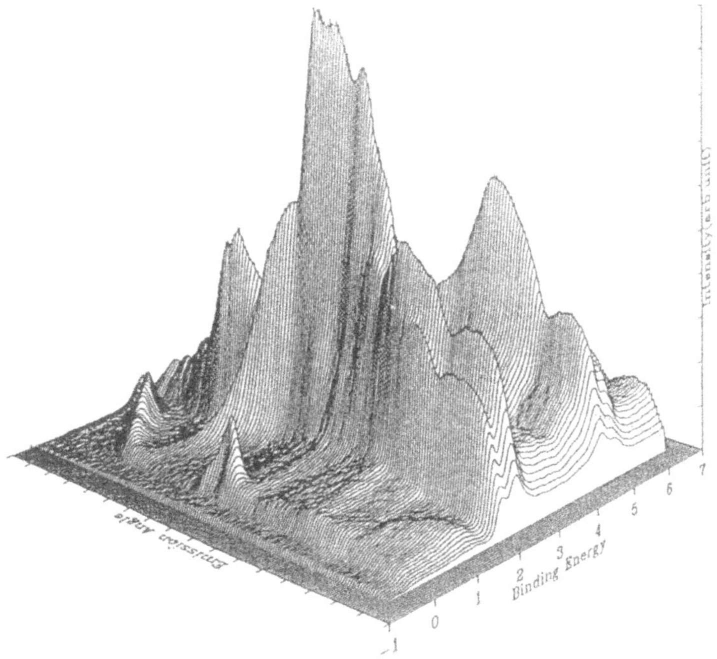

Fig. 14.5. Angle resolved UPS data from Cu (110) in the [112] azimuth. Structures due to *d*- and *s*, *p*- derived bands are clearly visible together with a surface state at Γ as described in the text

In this analyser, [5] electrons emitted in a plane (horizontal in Fig. 14.4 are brought to a ring focus at a position sensitive detector if their energy corresponds with the voltages applied to the various electrodes. The arrival position on the detector is directly related to the polar angle of emission (θ); the kinetic energy of detection may be varied by altering the electrode potentials. The device consequently records $N(E, \theta)$ as shown in Fig. 14.5. Typically, the energy resolution is set to 0.05 eV whereas the angular resolution is close to $\pm 1°$ for the analyser described in [5].

A number of features are immediately evident from Fig. 14.5. Perhaps the most striking is the wealth of detail shown here compared to an angle-integrated spectrum such as one of those shown in Fig. 14.2. The structure visible in Fig. 14.5 is dominated by the 3*d* electrons of the conduction band of copper. They are somewhat atomic like in the manner of core levels. i.e. these electrons are rather closely associated with individual ion cores and are consequently partially localised in space, unlike *s*-*p* type electrons closer to the Fermi energy. Such *s*-*p* electrons are almost completely delocalised band electrons and feel the full effects of the periodic potential of the crystal lattice. Emission from these states can readily be identified as the strongly dispersing feature which breaks away from the relatively flat *d* bands at $\theta = -49°$.

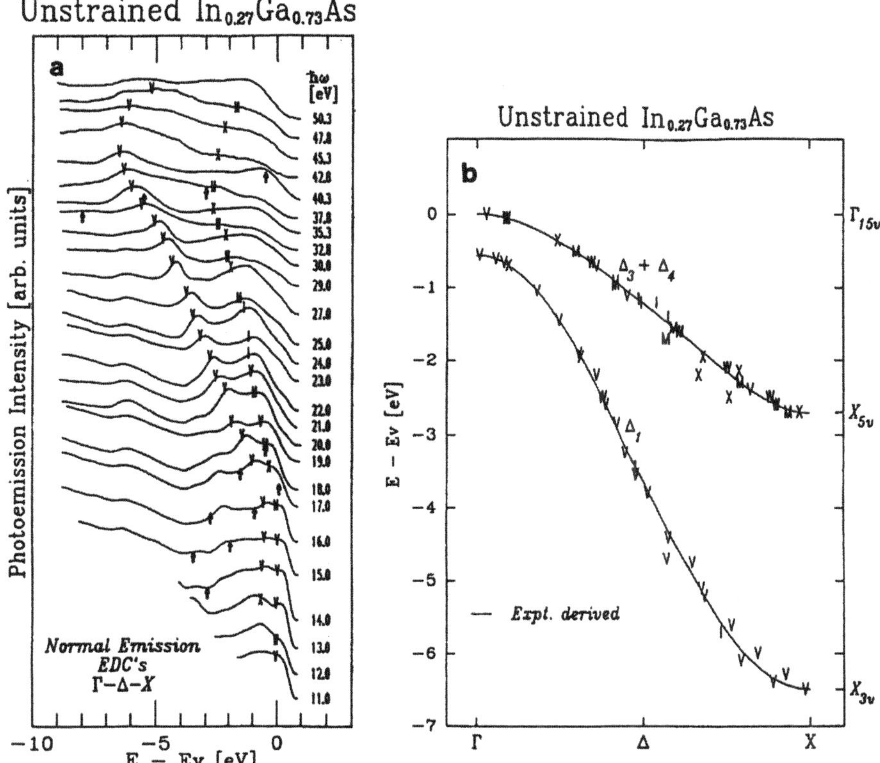

Fig. 14.6. **(a)** Normal emission energy distributions curves from InGaAs (100) for a range of photon energies. **(b)** Valence band dispersion curves derived from this data

The small feature which shows some dispersion and is seen close to the Fermi energy around $\theta = 0°$ is due to a surface state. Such states are not a feature of the bulk band structure but exist as separate solutions of the Schrödinger equation specifically associated with the surface itself. Surface states have an important role to play in all surface sensitive situations e.g. catalysis, adsorption and in the energy alignment of metal/semiconductor interfaces.

The major features of the $E(\boldsymbol{k})$ band structure of a material can most readily be experimentally determined by utilising a synchrotron radiation facility. An example of a set of energy distribution curves obtained from InGaAs over a wide range of photon energies at normal emission is shown in Fig. 14.6, together with the experimental valence band structure derived from this data. Details of the analysis needed to convert the raw data to $E(\boldsymbol{k})$ data may be found in [6].

14.3 Fermi Surface Studies

Knowledge of the topology of the Fermi surface of metals and alloys is of particular importance in materials science and has traditionally been acquired using de Haas van Alphen methods [7] which, however, involve strict requirements regarding crystal perfection and measurement at low (liquid He) temperatures. By acquiring a full hemisphere intensity map of all electrons photo-emitted from the Fermi energy i.e. by measuring $I(\theta, \phi)$ with E_{f} as the initial state, the ARUPS technique can provide an almost direct map of the Fermi surface. The k–space resolution of the photo-emission method is not

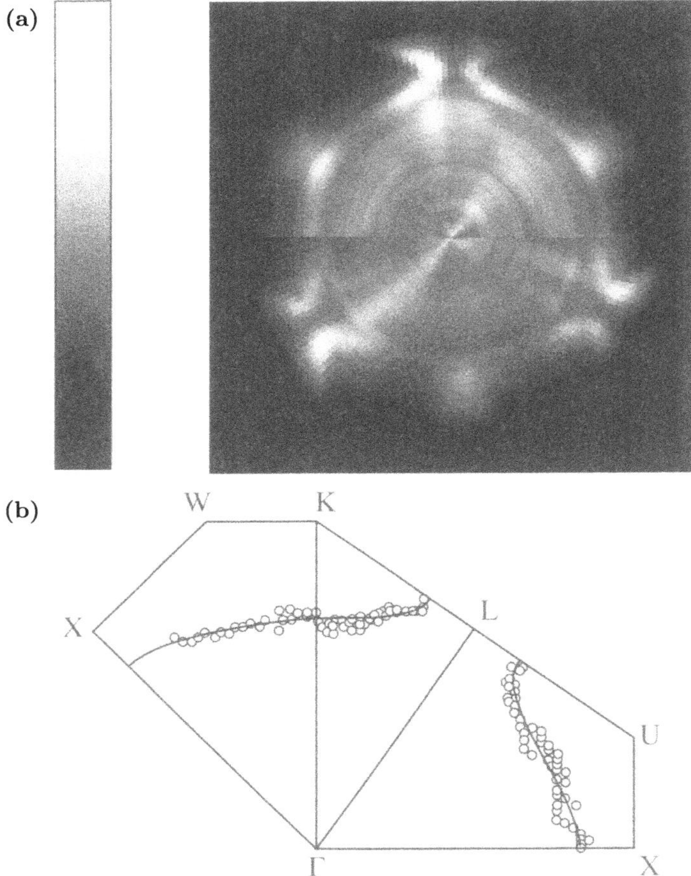

Fig. 14.7. (a) Full hemisphere intensity pattern showing emission at the Fermi energy from Cu (111) using 22 eV p-polarised photons. The presentation is shown on a linear wave vector scale for direct comparison with the Fermi surface topology. **(b)** The Fermi surface of Copper as determined by angle resolved photoemission methods shown on the irreducible sector of the Brillouin Zone

as good as de Haas van Alphen, but the ARUPS method can be performed on standard crystals at elevated temperatures thereby opening up the study of phase transitions. Similarly, the ARUPS method may be applied to substitutionally disordered alloys. An example from our laboratory is included as Fig 14.7. Recently, the ARUPS method has been successfully applied to the measurement of high T_c superconductors [9], to an investigation of the order–disorder transition in Cu_3Au [10] and to the ferro- / para-magnetic transition in Ni [11].

In this brief overview, there has not been space to consider UPS experiments on adsorbates nor from ow dimensional samples, heavy fermion systems etc. Similarly, it has not been possible to consider advances in understanding the magnetic properties of materials using polarised photon sources and polarised electron detectors. More complete descriptions of the capabilities of ARUPS may be found in the review articles [12,13].

References

1. R.T. Poole, J. Liesegang, R.C.G. Leckey, J.G. Jenkin: J. Electron. Spectrosc. Relat. Phenom. **5**, 773 (1974)
2. See, for example, http://www.gammadata.se/
3. A.D. McLachlan, J.G. Jenkin, R.C.G. Leckey, J. Liesegang: J. Phys. F **5**, 2415 (1975)
4. R.C.G. Leckey: J. Electron. Spectrosc. Relat. Phenom. **43**, 183 (1987)
5. R.C.G. Leckey, J.D. Riley: Appl. Surf. Sci. **22/23**, 196 (1985)
6. A. Stampfl, G. Kemister, R.C.G. Leckey, J.D. Riley, F.U. Hillebrecht, J. Fraxedas, L. Ley, P.J. Orders: J. Vac. Sci. Technol. **A7**, 2525 (1989)
7. W.E. Pickett, H. Krakauer, R.E. Cohen, D.J. Singh: *Fermi Surfaces, Fermi Liquids and High Temperature Superconductors* Science **255**, 46 (1992) and references therein
8. A.P.J. Stampfl, J.A. Con Foo, R.C.G. Leckey, J.D. Riley, R. Denecke: Surf. Sci. **331-333**, 1272 (1995). P. Aebi, J. Osterwalder, R. Fasel, D. Naumovic, L. Schlapbach. Surf. Sci. **309**, 917 (1994)
9. P. Schwaller, P. Aebi, H. Berger, J. Osterwalder, L. Schlapbach: J. Electron. Spectrosc. Relat. Phenom. **76**, 127 (1995)
10. J.A. ConFoo, S. Tkatchenko, A.J.P. Stampfl, A. Ziegler, B. Mattern, M. Hollering, L. Ley, J.D. Riley, R.C.G. Leckey: Surf. Interface Anal. **24**, 535 (1996)
11. P. Aebi, T.J. Kreutz, J. Osterwalder, R. Fasel, P. Schwaller, L. Schlapbach: Phys. Rev. Lett. **76**, 1150 (1996)
12. F.J. Himpsel: Adv. Phys. **32**, 1 (1983)
13. R. Courths, S. Hufner: Phys. Rep. **112**, 53 (1984)

15 EXAFS

R.F. Garrett and G.J. Foran

15.1 Introduction

X-ray Absorption Fine Structure (XAFS) spectroscopy is an important technique for determining the local structure around an absorbing element in a sample. It is normally divided into two techniques: Extended XAFS (EXAFS) describing the fine structure more than about 50 eV above an absorption edge, and near edge structure or XANES. In an X-ray absorption spectrum, a series of oscillations in the measured absorption coefficient can be observed on the high-energy side of an absorption edge. These oscillations are due to the fact that the final state wavefunction of the emitted photoelectron consists of an outgoing part and a part that is scattered from neighbouring atoms. The EXAFS oscillations, as they are known, correspond to an interferogram of the spatial distribution of nearby atoms and can be analysed to provide structural information about the local environment of the absorbing atom such as bond length, the number and type of neighbouring atoms, bond angles and a measure of order/disorder and/or chemical lability. EXAFS typically gives very accurate interatomic distances, in particular for nearest neighbours, of the order of ± 0.02 Å or better. Achieving precision in the determination of coordination number is more difficult and errors as large as 20% in the fitted result are common. Careful choice of suitable standard compounds with known structure followed by EXAFS analysis, however, can greatly improve this determination. Similarly, the determination of Debye-Waller factors from EXAFS data is complicated by the fact that this parameter is strongly correlated with other structural parameters used in the refinement and significant uncertainty can arise as a result. Because the X-ray absorption edges of the elements occur at unique energies, XAFS by its nature probes the environment of a selected element only. The selectivity and sensitivity that derive from this feature are often critical in the decision to employ XAFS in a structural analysis.

The purpose of this chapter is to provide an introduction to EXAFS. It is not intended to be an exhaustive theoretical or experimental treatment of the subject; such a task is far beyond the scope of a single chapter, but rather a brief overview of the field with some emphasis on surface sensitive EXAFS techniques. For newcomers to EXAFS, Sect. 15.4 provides a step-by-step description of the data analysis process, usually the most challenging aspect of the experiment. For a full review of EXAFS, a number of excellent books are available [1–3]. The reader is also directed to the proceedings of the XAFS series of conferences [4], and to resources on the World Wide Web [5].

The EXAFS technique, unique amongst X-ray techniques, provides short range structural information without the requirement for long range order in the sample. This is due to the limited mean free path of the emitted photoelectron in most materials. The mean free path is relatively large at low photoelectron kinetic energies (below about 50 eV), reaches a minimum of a few nano-metres in most materials in the 50–100 eV kinetic energy range, and only slowly increases at higher energies. Thus the EXAFS structure in the absorption spectrum is dominated by short range single scattering events above about 50 eV above the absorption edge, as was first pointed out by Sayers, Stern and Lytle [6]. Close to the absorption edge the structure of the spectrum is dominated by transitions to unfilled molecular orbitals or solid state valence bands, and multiple scattering. This structure is known as XANES (X-ray Absorption Near Edge Spectroscopy) or NEXAFS (Near Edge XAFS) [2], and has only recently found adequate theoretical treatment. This region of the absorption spectrum is, however, very useful in many applications as will be described briefly in a later section.

The popularity of EXAFS is, to a large extent, due to the fact that long range order is not required in the sample, i.e. it is not limited to crystalline materials but is applicable to amorphous solids, species in solution and even molecular gases. This has resulted in its use in the study of a wide range of "real world" samples such as biological complexes in solution, homogeneous and heterogeneous catalysts, organic thin films, crystalline and amorphous semiconductors etc.

15.2 Experimental Details

EXAFS is conceptually a very simple experiment, often deceptively so for first time users of the technique. The basic experiment involves measuring the absorption by the sample of a monochromatic X-ray beam as the beam energy is scanned from the absorption edge of interest to a point approximately 1 keV above the edge. Any quantity can be measured from which the absorption of the X-ray beam can be derived. The most common are the transmitted beam intensity and the sample fluorescence yield, but other quantities such as electron emission and even ion desorption have been used in special cases.

A typical EXAFS experimental setup is shown schematically in Fig. 15.1, and the EXAFS experimental station at the Australian National Beamline Facility at the Photon Factory, Tsukuba, Japan, is pictured in Fig. 15.2. While the experiment is very simple in concept, it must be performed with great precision, primarily because of the unfavourable signal to noise conditions. The EXAFS signal is a small oscillation of the absorption probability

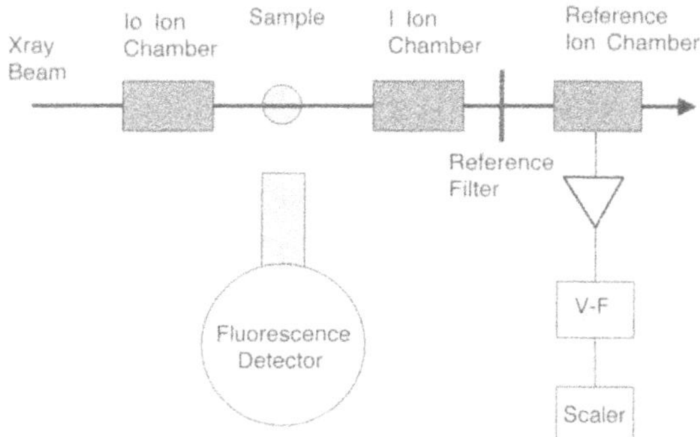

Fig. 15.1. Schematic XAFS experiment showing transmission detection using ion chambers and a fluorescence detector. A typical ion chamber detector counter chain is shown for the reference ion chamber

Fig. 15.2. The XAFS experimental station at the Australian National Beamline Facility at the Photon Factory, Japan. The I_0 ion chamber is visible at the left, followed by a liquid He cryostat for low temperature XAFS. A 10 element high purity Ge SSD is mounted at 90 degrees to the X-ray beam

of the target element, which itself is superimposed on an equal or larger background signal, particularly in the case of transmission mode EXAFS.

The emergence of synchrotron radiation facilities as extremely bright and intense tuneable X-ray sources has revolutionised most branches of X-ray science [7]. The first synchrotron EXAFS experiment was performed by Kincaid and Eisenberger in 1975 [8]. Unlike most X-ray techniques e.g. crystallography, small angle scattering, EXAFS is almost totally a synchrotron based technique. A few laboratory based systems have been developed [9] depending on the Bremsstrahlung output of a rotating anode tube for a continuum source, but are generally restricted to samples with a high concentration of the absorbing element. From its inception, EXAFS for all intents and purposes has been a purely synchrotron based technique, and is one of the techniques which is made possible, as distinct from being greatly enhanced, by the relatively recent availability of synchrotron radiation.

15.2.1 Synchrotron Radiation

Synchrotron radiation is emitted when charged particles moving at relativistic velocities are accelerated. In the usual case of an electron storage ring, the highly relativistic electron bunches are accelerated by the bending magnets which maintain the electron beam on its closed orbit in the ring. The dipole radiation emission pattern is folded into a narrow cone in the forward direction by the Lorentz transformation, resulting in a fan beam emission from each bending magnet. The angular extent of the emission cone is $1/\gamma$ radians, where γ is the ratio of the electron energy to its rest mass. For a 2.5 GeV storage ring (such as the Photon Factory) $1/\gamma$ is of the order of 0.2 milliradians, so that the bending magnet emission is highly collimated in the vertical plane. Figure 15.3 illustrates the angular and energy extent of the synchrotron radiation produced by bending magnets.

The energy spectrum of the emitted radiation is characterised by a smooth continuum with a rapid fall off at high energy. The high energy cutoff depends on the electron energy and the magnetic field strength, and is usually characterised by the critical energy, E_c, which is the median power point of the spectrum and is given, in convenient units, by

$$E_c\,(\mathrm{keV}) = .665E^2\,(\mathrm{GeV})\;B\,(\mathrm{Tesla}) \tag{15.1}$$

The critical energy of a nominal 2.5 GeV storage ring with a 1 Tesla bending magnet field (a typical second generation facility) is thus around 4 keV. As a rule of thumb, useful X-ray intensity can be obtained to approximately five times the critical energy, or to 20–25 keV from the above 2.5 GeV machine.

Synchrotron radiation has a number of advantages for EXAFS over a conventional source:

1. The X-ray intensity is orders of magnitude higher
2. The energy spectrum is continuous from the infra-red to X-ray energies, which is essential for absorption spectroscopy.
3. The radiation is highly collimated, allowing high intensity monochromatic beams to be generated easily
4. The energy spectrum and intensity are very stable
5. The radiation is linearly polarised in the plane of the electron orbit.

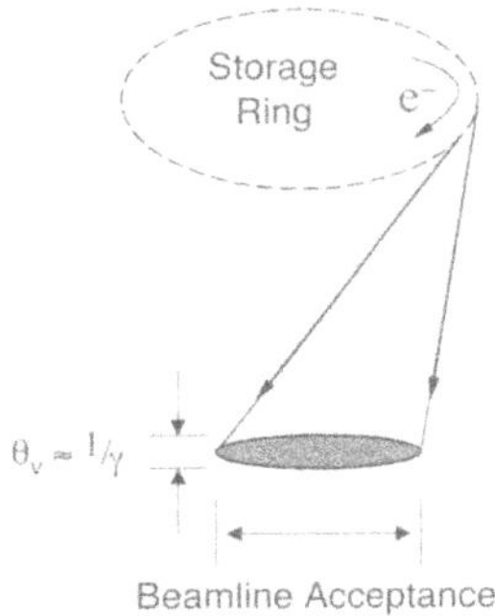

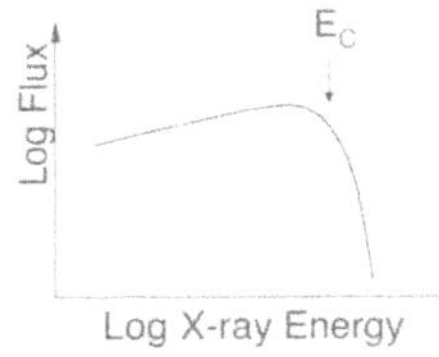

Fig. 15.3. Schematic illustration of the synchrotron radiation produced by a bending magnet source in a storage ring

In addition to the bending magnet synchrotron radiation source described above, periodic magnetic structures called insertion devices can be inserted into straight sections of a storage ring to generate even higher X-ray intensities. These devices cause the electron beam to oscillate as it traverses the straight section, with each oscillation emitting synchrotron radiation. The devices are divided into two classes:

- Wigglers have high fields or long magnetic periods so that the amplitude of the oscillations exceeds the transverse size of the electron bunch. The synchrotron emission from each oscillation adds and the resulting spectrum is that of a bending magnet of the same field multiplied by the number of poles in the wiggler.
- Undulators have smaller magnetic periods, so that the amplitude of the electron beam oscillation is less than the beam size. The emission from each oscillation adds coherently resulting in a energy spectrum composed of a series of harmonic peaks and an X-ray intensity proportional to the square (in the ideal case) of the number of poles.

EXAFS, with its reliance on a smooth stable energy spectrum, is nearly always performed on bending magnet or wiggler beamlines. The majority of EXAFS experiments are not limited by the X-ray beam brightness and to date have been performed on bending magnet and wiggler beamlines. This picture is changing somewhat with a number of undulator based XAFS beamlines now operational at third generation synchrotron facilities.

15.2.2 Synchrotron Beamlines for EXAFS

The detailed design and functioning of a synchrotron beamline is not usually of great interest to the experimenter. However some features of the beamline optics directly impact the experiment and a basic understanding will lead to better results being obtained.

The beamline performs several functions for an EXAFS experiment. Most importantly, it must produce a monochromatic beam with stable energy and intensity from the continuum spectrum emitted by the storage ring. It may also have mirror optics to focus the beam and remove unwanted high energies. It also performs the mundane function of removing the experimental station to a convenient distance from the storage ring, and the very necessary function of containing the radiation and providing personnel safety systems. A schematic beamline is shown in Fig. 15.4.

Fig. 15.4. Schematic synchrotron XAFS beamline

Monochromatic beams are produced by utilising diffraction from perfect crystals. As shown in Fig. 15.4, a double crystal monochromator is almost always used, as the monochromatised X-ray beam is returned to the horizontal plane, and its vertical position is almost independent of energy at the sample position (particularly over the limited energy range of an EXAFS scan). In the most common energy range of 5–20 keV, silicon crystals are nearly universally employed, as highly perfect crystals can be obtained relatively cheaply. Silicon has the added advantage of being very radiation resistant and a good thermal conductor. The energy width of the diffracted beam from silicon is very narrow, of the order of $\Delta E/E = 2 \times 10^{-4}$ (the actual resolution is usually dominated by the vertical angular acceptance of the monochromator) allowing narrow edge features to be resolved. At energies below about 3 keV very large d spacing crystals such as InSb and YB_{66} [10] are used. However there are a difficulties in operating in these energy ranges: high quality monochromator crystals are difficult to obtain and they contain absorption edges which may interfere with the measurement being attempted; and the experiment often must be performed in vacuum to avoid absorption by windows. The result of this is that there are only a hand-full of EXAFS beamlines operating in the 1–3 keV range compared with dozens in the hard X-ray region. Grating monochromators are required below 1–1.5 keV, the so called "soft X-ray" region of the spectrum. While these beamlines are used extensively for NEXAFS, they find limited application for EXAFS.

One consequence of using crystal monochromators is that high order reflections are always present. For the most commonly used Si (111) crystal, the second order reflection is forbidden so that the most intense high order harmonic is the (333) reflection at three times the energy. As EXAFS requires the ratioing of two detector signals to obtain I/I_0, and because the (111) and (333) beams will clearly be absorbed to very different degrees by the detectors and sample, the presence of high order harmonics in the beam is a serious problem. Fortunately the harmonic can be removed by one or both of the following techniques:

1. The high order reflection is intrinsically narrower, and very slightly offset in angle from the fundamental. Thus a small angular misalignment of the two monochromator crystals, called detuning, results in a small decrease in the intensity of the fundamental and a much greater decrease in the harmonic intensity. Reductions of over 100 times in the harmonic fraction of the beam can be easily achieved with this technique.
2. A beamline mirror acts as a low pass energy filter. An ideal mirror has unit reflectivity for X-rays below the critical angle for total external reflection, and the reflectivity falls rapidly above this angle. The critical angle varies with X-ray energy, and with the mirror coating, so that the mirror high energy cutoff can be tuned to reject the unwanted harmonic energy.

In addition, if the absorption edge to be measured is close to the high energy fall-off in the synchrotron radiation spectrum, see Fig. 15.3, then the harmonic contamination will be greatly reduced (because of the rapid drop in intensity at higher energies) and further harmonic rejection may not be necessary.

Synchrotron beamlines for conventional EXAFS in the hard X-ray energy range are all broadly similar to that outlined above (Fig. 15.4). The main points of difference are in the details of monochromator mechanisms and crystals, and in whether focusing optics are employed. However two other configurations which deserve mention are beamlines for time resolved EXAFS. These beamlines fall into two classes: energy dispersive EXAFS and quick EXAFS (QEXAFS).

In energy dispersive EXAFS a single bent crystal is used in Bragg (reflective) or Laue (transmission) geometry [11,12] to focus the synchrotron fan beam onto the sample in the horizontal plane. The bent crystal presents a range of Bragg angles to the beam, and consequently the energy of the monochromatised beam varies as a function of the angle onto the sample. The crystal is bent sufficiently that the entire energy range of the EXAFS spectrum is incident onto the sample. Measurements are made in transmission mode; the beam diverges beyond the sample and is measured by a one dimensional strip detector. In this way the entire EXAFS spectrum is measured simultaneously, allowing for time resolution of the order of tens of milliseconds. Good quality data is very difficult to achieve with this technique. Very uniform samples are required for example, and many samples are not suited to transmission measurements. Additionally, small-angle scattering from the sample superimposed on the energy "fan" of the transmitted beam can seriously degrade the energy resolution and render the data unusable. This problem can be addressed by inserting a scanning slit after the sample at the expense of some time resolution (the slit can also allow fluorescence detection to be used). Despite these potential problems, structural changes can often be tracked quite successfully from a known starting point.

QEXAFS is in essence fast scanning conventional EXAFS [13,14]. Instead of the step-count-step approach of normal EXAFS, the monochromator is continuously scanned and data recorded "on the fly". QEXAFS can equally easily be performed in transmission or fluorescence modes. The achievable time resolution depends on the available X-ray flux and on the stability of the monochromator mechanism which will determine how fast it can be driven and still produce a stable X-ray beam. Time resolutions of seconds to tens of seconds are typical.

15.2.3 Detectors

A typical hard X-ray (i.e. photon energy above $\approx 4\,\mathrm{keV}$) EXAFS experimental setup is shown schematically in Fig. 15.1. The simplest EXAFS technique is the transmission configuration, where the incident and transmitted X-ray intensity is measured by gas ionisation chambers before and after the sample. The X-ray absorption in the gas filled detector produces electron-ion pairs, which are swept apart by a voltage applied between two parallel plates. If the voltage is above the saturation level where recombination of the electron ion pairs is prevented, and below the avalanche region, then the output current is proportional to the X-ray intensity. The applied voltage is usually in the 500–1000 V range for a 1 cm electrode separation. As can be seen in Fig. 15.1, the output current is amplified and digitised by a voltage to frequency converter. For optimum EXAFS data collection, the I_0 detector should absorb about 25% of the X-ray beam, and the I detector 100% of the beam transmitted by the sample. This can be achieved over a wide range of X-ray energies by changing the gas mixture in the detector. Inert gases (He, Ne, Ar, Kr and N_2) and their mixtures are normally used. In some cases a third ion chamber, labelled the reference ion chamber in Fig. 15.1, is used in combination with a reference foil to correct for any drift in the energy scale due to instability of the monochromator, or other beamline elements.

Transmission EXAFS using ion chamber detectors is the method of choice due to its superior signal-to-noise ratio but is limited in application to samples containing a relatively high weight fraction of the absorbing element. Typically if a sample contains a weight fraction of absorber in excess of about 5%, it can be examined by the transmission method. For more dilute elements (and in some cases for surface species on highly absorbing samples) fluorescence detection is normally used. This is almost always the case with solution samples and surface species. A fluorescence detector is generally positioned at 90° to the X-ray beam as shown in Figs. 15.1 and 15.2, to take advantage of the minimum which occurs in the elastic scattered X-rays in the polarisation direction of the beam. For surface EXAFS the detector can be mounted above the sample (looking down) to maximise the solid angle accepted. The first fluorescence detector to find widespread use was the ion chamber detector developed by Lytle et al. [15]. This detector utilises a set of filters to enhance the selectivity of the ion chamber for the fluorescence emission of the element of choice. While the use of fluorescence ion chambers has largely been replaced by solid state detectors (SSDs), they are still used when the concentration of the emitting element results in too high a count rate for a SSD.

Most fluorescence EXAFS is currently performed with intrinsic germanium or lithium drifted silicon SSDs. These detectors have the advantage of relatively high energy resolution (typically around 150 eV FWHM at low count rates) which allows the fluorescence peak of a very dilute element to be separated from emission from other elements and from scattering of the direct beam. The limited count rate capability of these detectors has largely been overcome with the advent of high performance electronics and arrays of 10 or more elements in a single detector [16]. The 10 element germanium detector in use at the Australian beamline at the Photon Factory can be seen in Fig. 15.2.

One final detection system which deserves mention is electron detection [17]. The total electron yield from a sample is proportional to the X-ray absorption and therefore also contains the EXAFS signal, and the electron yield is intrinsically surface sensitive. Electron yield measurement requires the sample to be in vacuum, which limits its use to cases where surface sensitivity is required, or to the soft X-ray region where other techniques are difficult or impossible, and where the experiment is usually performed in UHV anyway. In the soft X-ray region I_0 is normally measured from the photo-current from a grid positioned before the sample, and the I signal can be measured either using the sample drain current if the signal is sufficient, or using an electron multiplier based detector.

15.2.4 The Sample

In transmission EXAFS the best data, in terms of optimum signal to noise and minimum sensitivity to any remaining harmonic contamination, is obtained when the absorption of the sample (μt, where t is the sample thickness and μ is the mass absorption coefficient) is in the range 1-2. Typically this corresponds to a weight fraction of the absorbing element in the sample of about 5% or more. In some cases the sample cannot be modified, but usually the thickness can be adjusted to achieve this optimum absorbance. In cases where the weight fraction of the absorber in a uniform sample is too high to allow the passage of a significant proportion of the X-rays, the sample may be dispersed in an inert matrix with high X-ray transparency such as boron nitride or cellulose. It should be noted that the optimal concentration of absorber will be different for each sample depending on the nature of the other elements in the sample and the type (K, LI, LII etc) and energy of the edge being measured. For this reason, X-ray attenuation coefficients for samples should be calculated prior to preparation to determine the optimal sample concentration to be used.

Dilution of the sample in an inert matrix can also assist in achieving sample homogeneity. Sample uniformity is important as distorted spectra can result from inhomogeneous samples with defects such as variable sample thickness or pin-holes.

For dilute samples, fluorescence EXAFS can provide a superior signal to noise ratio compared to transmission measurements due to the discrimination of the detector. The main sample consideration for fluorescence measurements is concentration. The advent of EXAFS beamlines delivering higher and higher fluxes has progressively pushed down the detection limit to the point that fluorescence EXAFS measurements are now routinely made on solutions in the sub-millimolar concentration range and on solids containing absorber in the parts per million range.

In addition to the count-rate limit of solid state detectors, there is also a maximum concentration beyond which fluorescence detection can not be used to measure EXAFS. This limit derives from the effect of self-absorption. While this limit will vary depending on sample composition, it should be remembered that there is a concentration level in fluorescence measurements beyond which a significant proportion of the fluorescent yield is re-absorbed by the sample resulting in serious distortion of the measured EXAFS amplitudes.

The sample environment is an important factor in EXAFS, in particular the sample temperature. Thermal vibrations degrade the EXAFS signal by broadening the interatomic distance distribution, normally represented by a Debye-Waller factor in EXAFS analysis (see Sect. 15.3.1). It therefore follows that data collected at low temperature is normally superior to room temperature data, and most synchrotron EXAFS beamlines have a low temperature sample stage available. Figure 15.2 shows the closed cycle He cryostat in use at the Australian National Beamline Facility. High temperature measurements are also increasingly popular, to measure *in situ* catalyst systems, ceramics, melting and molten phases, geo-chemical systems etc. High temperature cells/ovens operating over 2000 K have been reported [18]. In addition to the sample temperature, a wide variety of other environmental cells have and can be devised. These include chemical reaction cells [19], electrochemical cells [20,21], pressure cells [22] etc. Catalyst research in particular often requires a combination of high temperature, gas flow and possibly high pressure conditions [23].

15.2.5 Acquiring EXAFS Data

In a typical XAFS experiment the spectral range is conveniently divided into three regions. The first region is the pre-edge that is needed in the data analysis phase to provide an estimate of the trend in the underlying absorption for background subtraction purposes. For this reason, the pre-edge is scanned quickly at low resolution beginning approximately 200 eV below the onset of the absorption edge. Typically an energy step of 10 eV and a count time of one second per point is sufficient.

The second region spans the absorption edge including any near edge spectroscopic features. As the absorption edge is the first-order effect, it is much stronger than the EXAFS signal and as a result can generally be scanned with relatively short count times per point (e.g. 1 second). The edge region is, however, rich in spectroscopic information and so is usually scanned with high resolution. The energy step size that is used will vary depending on the energy of the edge, the monochromator used and even the source properties. For a typical 2nd generation source using a Si(111) monochromator in the range corresponding to the K absorption edges of the first row transition elements (Ti–Zn), a step size of 0.25–0.5 eV would be usual. The edge region covers an energy range of around 30–80 eV.

The third region of a typical XAFS scan is that above the absorption edge where the extended fine structure (EXAFS) is observed. As the absorption data are converted to k-space (see Sect. 15.4 and (15.3)) before analysis, it is convenient and even desirable to scan this region with constant steps in k: a step size of 0.05 Å^{-1} is typical. The EXAFS signal is weak compared to edge features and decays away with increasing k. For this reason, counting times per point in the EXAFS region are usually considerably longer than for the edge region and it is common to ramp the count time per point with k either linearly or with k-squared so that counting statistics improve as the EXAFS signal gets weaker. Normally, the EXAFS region is scanned up to a k value of 14–16 Å^{-1} (16 Å^{-1} equates to about 1000 eV) though very well prepared samples of materials containing strong back scattering atoms can give useful EXAFS out to a k value of 20 or beyond, as can be seen in Figs. 15.6–15.8.

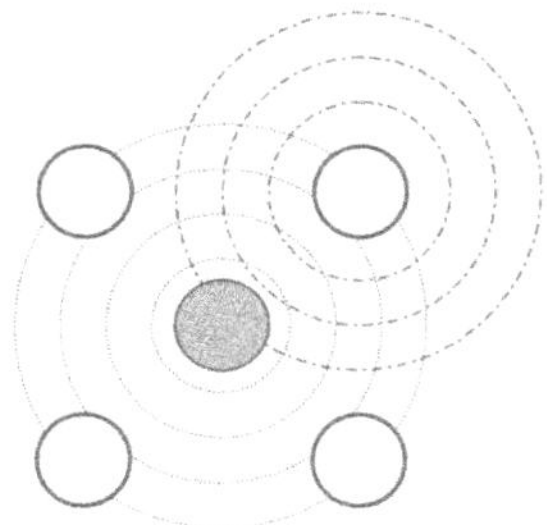

Fig. 15.5. Qualitative picture of EXAFS. The shaded atom is the absorber, and the circles emanating from it represent the outgoing photo-electron. The dot-dashed circles are the backscattered electron wave from a neighbouring atom

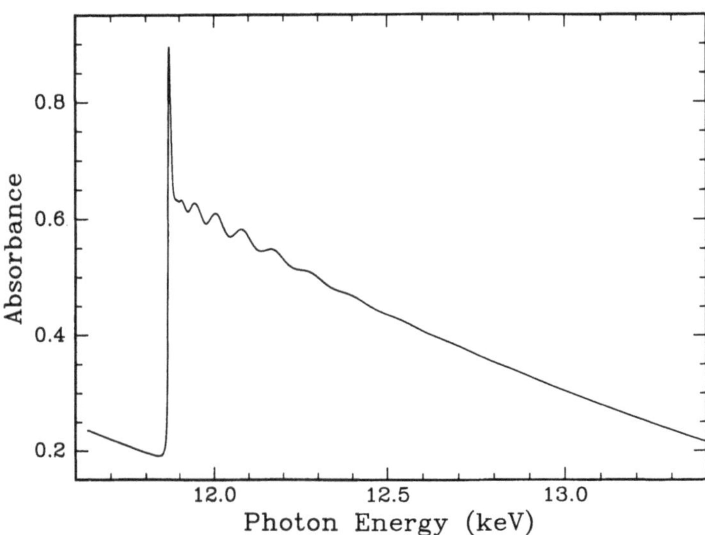

Fig. 15.6. Raw transmission XAFS spectrum of amorphous GaAs, taken at the arsenic k edge [52]

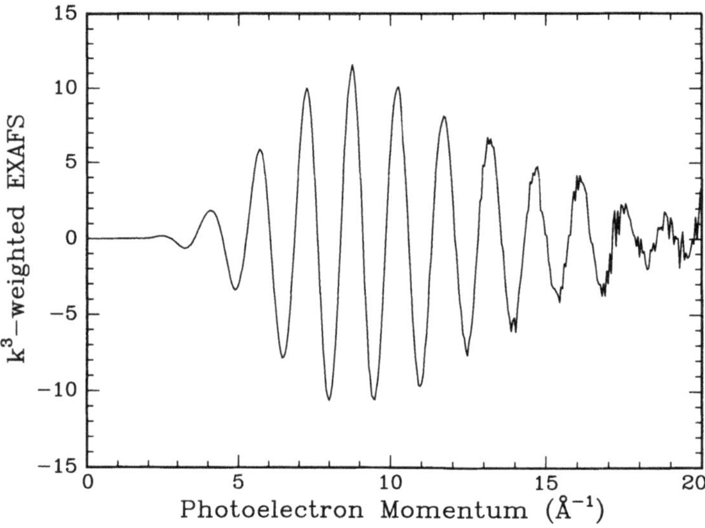

Fig. 15.7. k-weighted XAFS Extracted from the raw spectrum in Fig. 15.6. [52]

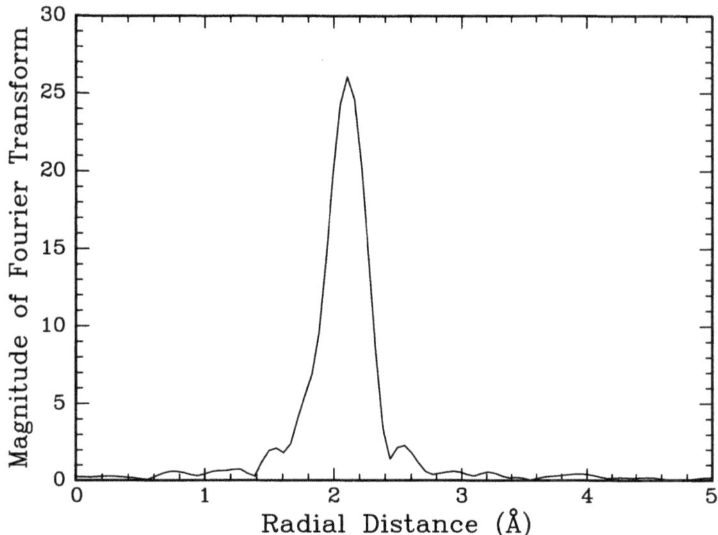

Fig. 15.8. Magnitude of the Fourier transform of the k weighted XAFS shown in Fig. 15.7. Only the first shell has significant intensity due to the amorphous nature of the sample [52]

15.3 Theory of X-ray Absorption

15.3.1 EXAFS

Theoretical attempts to explain EXAFS began soon after the first observations of X-ray absorption spectra [24]. EXAFS theory is now very mature, and excellent and detailed reviews are available [1,3,25,26].

The usual qualitative explanation of EXAFS is illustrated in Fig. 15.5. An incident X-ray photon is absorbed by an atom which emits a photo-electron. The outgoing waves are backscattered by the neighbouring atoms and interfere at the emitting atom, resulting in a small oscillation superimposed on the X-ray absorption spectrum. The oscillations derive from the interference being progressively constructive and destructive as the photo-electron energy increases away from the absorption edge. EXAFS is seen for all but isolated atoms (although recent work has identified fine structure due to scattering of the photoelectron from the absorber itself i.e. atomic EXAFS [27,28]), and was first observed in the 1920s [24]. The scattered electron wave undergoes phase shifts at the scatterer and absorber atoms; if this were not the case a simple Fourier transform would peak at or near the interatomic distances of the neighbouring shells of atoms. Much of the task of EXAFS analysis deals with determining the phase and amplitude functions of the scattering processes.

The EXAFS oscillations therefore derive from interference effects from neighbouring shells of atoms, where a shell is composed of identical atoms at a particular interatomic distance R. The oscillations are usually expressed as a sum of sinusoidal contributions from many shells in the semi-empirical expression [29]:

$$\chi(k) = \sum_i \frac{N_i\, A_i(k)}{k\, R_i^2} \exp\left(\frac{-2R_i}{\lambda}\right) \exp(-2\sigma_i^2 k^2) \sin\left[2kR_i + \phi_i(k)\right] \quad (15.2)$$

Here N is the number of atoms in the ith shell, A is the scattering amplitude function, k is the photoelectron wave vector, λ is the magnitude of the photoelectron mean free path, σ is the Debye Waller factor and ϕ is the phase function. As can be seen from (15.2), the X-ray energy is converted to k in all EXAFS data analysis, via:

$$k(\text{Å}^{-1}) = \hbar^{-1}[15.2m_e(E - E_0)]^{1/2} = 16.2[E - E_0(\text{keV})]^{1/2} \quad (15.3)$$

Where m_e is the electron rest mass, E is the X-ray energy and E_0 is the absorption edge or threshold energy. Equation (15.2) consists of a sinusoidal term including phase shifts, and exponential terms reducing the oscillation amplitude due to the mean free path of the photoelectron, the thermal and static disorder (represented by the Debye-Waller factor), and with k^2. EXAFS data analysis uses (15.2) as a fitting equation, with experimentally or theoretically determined phases and amplitudes, to fit a model structure to the measured EXAFS.

Note that the use of a Debye-Waller factor to represent disorder assumes that the disorder can be approximated by a Gaussian distribution, i.e. it is symmetric. While this assumption is adequate in the majority of situations, a more accurate expression which can incorporate asymmetric distributions is sometimes required. This is clearly the case for surface atoms for example, whose bonding environment is likely to be significantly asymmetric. In these cases a commonly used technique is the cumulant expansion method [26,30,31], where higher order terms are summed to include asymmetry in the disorder distribution.

Polarisation effects are not included in the above treatment. XAFS is in fact polarisation sensitive [32], with a cosine squared dependence on the angle between the electric field vector of the X-rays and the absorber-scatterer axis. This can normally be ignored, as most EXAFS samples are randomly oriented e.g. polycrystalline powders or solutions. With single crystal samples or surface species, however, polarisation must be included in the analysis. In this case EXAFS can give direct information on bond angles and molecular orientation.

Equation (15.2) is essentially a single scattering plane wave approximation, which is usually sufficient to characterise the first shell of neighbouring atoms. Beyond the first shell, multiple scattering is often significant. Fortunately multiple scattering can now be routinely included in EXAFS analysis due to developments in a number of computer codes, principally FEFF5 and higher [33–35] which calculate curved wave multiple scattering EXAFS for model systems.

15.3.2 XANES

Structure close to the absorption edge, to about 50 eV above the edge, is referred to as XANES (X-ray Absorption Near Edge Structure) or NEXAFS (Near Edge XAFS) [2].

Only a brief description of this technique will be given here. In this region of the absorption spectrum the photoelectron energy is low and the mean free path long, so multiple scattering effects are much more dominant than in the EXAFS region. In addition there are normally strong features due to transitions to empty bonding and anti-bonding orbitals in molecular systems, or to atomic-like or unoccupied density of states in solid state systems. The combination of these influences means that XANES is sensitive to the local electronic structure of the absorbing species, its oxidation state (speciation), the number and nature of ligands and the coordination geometry. However the complexity of the processes involved has meant that XANES spectra have proven very difficult to calculate theoretically. Only in very recent times has adequate theory been developed that allows quantitative analysis of the electronic information contained in this region of the absorption spectrum [35]. To date, XANES has most often been used in a fingerprint fashion, with spectra compared to standards to determine the quantity of interest, e.g. the oxidation state of the absorbing element. Both the energy of the absorption edge and the shape of the XANES spectrum changes appreciably with oxidation state for most elements. XANES is proving particularly powerful in environmental science [36], especially when combined with imaging techniques, because samples can be measured in their natural state (e.g. wet) and the speciation of trace elements is key factor in their mobility and toxicity in the environment.

XANES spectra often exhibit strong peaks, both below (so called pre-edge peaks) and above the edge. These peaks are due to dipole transitions of the core electron to an unoccupied state. The localising effect of the core hole produced by the absorption event often produces quasi-atomic energy levels, particularly in non-metallic systems where the core hole is not screened effectively, resulting in very intense peaks. Such a peak can be seen in Fig. 15.6 at the As K edge of GaAs. Dipole forbidden (quadrupole) transitions are also observed with lower intensity. The fact that the transitions producing these strong peaks follow dipole selection rules means that they are polarisation sensitive [32]. This feature has found most use in the study of absorbed hydrocarbon species, whose carbon K edge NEXAFS spectra are usually distinctive [2]. By measuring the polarisation dependence (remembering that synchrotron radiation is plane polarised) of the NEXAFS spectrum by changing the incidence angle of the X-ray beam, the orientation of the adsorbed molecule on the surface can be easily determined.

15.4 EXAFS Analysis

A number of software packages employing a variety of methodologies are available for performing EXAFS analysis (see for example [37] and references therein). The International XAFS society web site lists over 35 analysis packages [5]. They range from schemes in which the EXAFS is calculated based entirely on experimentally determined parameters derived from known standard compounds through to complete ab-initio theoretical treatments. All, however, incorporate the following treatment which represents the standard data analysis protocol. Standards and criteria for EXAFS data analysis have been set down by the International XAFS Working Group [38].

An example of a raw EXAFS spectrum is shown in Fig. 15.6. Some manipulation of the raw EXAFS data is required before analysis can commence. Specta are summed in the case of multiple scans or an array detector being used, and normalised to constant incident beam intensity, i.e. I/I_0 is determined. Actual data correction is usually confined to elimination of monochromator "glitches", which are sudden dips in intensity due to X-rays being diffracted into other crystallographic reflections, and are often seen even after the I/I_0 normalisation. These are typically very narrow, much less than the frequency of EXAFS oscillations, and can usually be safely replaced by an interpolation of the surrounding intensity. Severe glitches indicate problems in normalisation due to sample in-homogeneity (e.g. a pinhole) or the presence of higher order harmonics in the beam. Other problems which may need correction are drift in the X-ray energy from the monochromator and diffraction from the sample if it is crystalline.

EXAFS data analysis may be conveniently divided into the following discrete steps.

1. Data reduction incorporating normalisation of the edge step to unity
2. Conversion to k-space
3. Background subtraction
4. Fourier transformation to r-space
5. Fourier filtering and back transformation
6. Modelling and least squares fitting to the EXAFS equation (15.2)

15.4.1 Data Reduction

Even in the most meticulously planned and prepared EXAFS experiment, it is likely that the concentration of absorber and sample thickness are not precisely known and will probably vary between samples. Similarly, window materials and the sample matrix contribute to the overall absorption and their effect should be corrected for. To compensate for uncertainties in sample concentration and thickness and to isolate the sample absorption from that of other materials in the beam, a normalisation procedure is first carried out in which the data are divided by the size of the edge step. The size of the edge step is generally measures by fitting low order polynomials to the data above and below the edge and extrapolation back to the edge position. It is important to ensure that the fitted regions are remote from the XANES signal which is sensitive to the chemical nature of the sample and which may contain features (e.g. "white lines" – See Fig. 15.6) thus making it unreliable as a measure of overall edge step size.

15.4.2 Conversion to k-space

It is generally accepted that it is desirable to convert the data from energy to k-space prior to background subtraction. This helps ensure that the background fit does not follow the data too closely at high k values where the EXAFS oscillations have large periods on the energy scale.

As indicated by (15.3), conversion to k-space requires that the threshold energy, E_0, be provided. Fortunately it is not necessary that this value be precisely known as small errors in the estimation of E_0 have a negligible effect on structural information derived from the EXAFS region (as opposed to the near-edge region) and in any event, the value of E_0 is refinable in the later stages of EXAFS analysis. As it is relative changes in E_0 between samples and standards that is important, consistency in the definition of E_0 is what is required. For this reason, an edge feature often serves as a useful point to reproducibly define E_0 for the purposes of conversion to k-space.

At this point the EXAFS is generally k-weighted, usually by k^3, to compensate for the damping of the oscillations due to the $1/k$ term and Debye-Waller factor in (15.2). High atomic number back-scatterers have increasing EXAFS amplitude functions at higher k while the backscattering intensity from low Z elements peaks at lower k values. Depending on the chemical composition of the sample and other factors k^2 weighting can be used.

15.4.3 Background Subtraction

The next step in the EXAFS analysis process is the extraction of the EXAFS oscillations from the raw absorption edge spectrum. In principle, this involves the removal from the observed data of the atomic (isolated atom) contribution to absorption. In practice, however even if the atomic contribution were known, the situation is complicated by the contributions to background absorption or scatter from other elements in the beam path. For this reason background subtraction is most often achieved via curve fitting means rather than by theoretical calculation of the atomic contribution.

The background absorption can be removed by first fitting the pre-edge region with a polynomial, extrapolating this function through the full spectral range and subtracting this function to remove the underlying background.

The smooth part of the absorption due to the element being investigated can then be similarly removed by fitting a smooth function to the post-edge part of the absorption spectrum followed by subtraction leaving the EXAFS oscillations as shown in Fig. 15.7. A number of fitting schemes can be used for this purpose but the most commonly used approach is to fit a series of cubic splines to the data constrained to have their first and second derivatives equal at the spline nodes. Some software packages will carry out this fitting process automatically while others allow a more manual approach. In either case, care must be taken with the splining, and with the placement of the nodes separating spline segments to ensure that the spline does not inadvertently fit and therefore remove part of the EXAFS oscillations themselves. Criteria were proposed by Cook and Sayers [39] to aid in producing an acceptable spline fit. The Fourier transform (FT) of the EXAFS signal remaining after spline fitting and subtraction is useful in assessing the quality of the spline. The amplitude of the FT at non-physical short separations, less than 1 Å, should be minimised. In general, FT components of real EXAFS signal are relatively insensitive to the spline parameters while artefacts are more greatly affected and can be minimised by careful choice of fitting parameters.

15.4.4 Fourier Transformation

The EXAFS signal, extracted and k-weighted, consists at this point of sums of damped sine waves corresponding to the different shells of atoms represented in (15.2). The next step in the analysis is therefore to Fourier transform the oscillations, producing a pseudo radial distribution function as shown in Fig. 15.8, with peaks whose position represents the interatomic distance of each shell of atoms, and whose amplitude is roughly proportional to the coordination number. Were it not for the phase shifts in the electron scattering process, the Fourier transform would yield a true radial distribution function. In addition to peaks corresponding to actual shells, extra peaks can be present in the Fourier transform due to the truncation of the EXAFS oscillations and multiple scattering contributions, particularly at larger interatomic distances.

15.4.5 Fourier Filtering and Back Transformation

In standard EXAFS analysis, it is the EXAFS oscillations (Fig. 15.7) that are modelled by theory and fitted to the observed data. However, it is often the case that the EXAFS signal is a complex convolution of contributions from a number of shells. This can make the data difficult to reliably fit due to the large number of parameters. To simplify the situation and reduce the number of fitting parameters, the data can be decomposed into components corresponding the contribution from individual shells by the process of Fourier filtering. Fourier filtering also permits decomposition of the EXAFS oscillations into amplitude and phase components.

In Fourier filtering, a window function is applied to the FT of the observed EXAFS in such a way so as to isolate the peak in the FT corresponding to a single shell. All Fourier components outside the window function are set to zero. As a result, when the windowed FT components are back transformed, only the EXAFS deriving from that shell of atoms remains and this much simplified signal can be more easily handled.

Consecutive Fourier filtering, back transformation and refinement of individual shells greatly simplifies the structural refinement and assists in avoiding false minima during refinement which can often result when too large a number of parameters are simultaneously refined.

15.4.6 Modelling and Least Squares Fitting to the EXAFS Equation

To obtain real structural information, principally the interatomic distances and coordination numbers, curve fitting is used to match the experimental EXAFS to theoretical EXAFS calculated from a model of the local structure of the absorber atom. The refined parameters in the fit are generally the shell radius R, the coordination N, the Debye Waller factor σ, the threshold energy E_0 and a scale factor to account for many-body electronic interactions in the absorption process.

The scattering amplitude and phase (A and ϕ in (15.2)) depend on the identity of the absorbing and scattering atoms, and may be derived either empirically or calculated *ab-initio*. In empirical calculations, the amplitude and phase parameters are derived from analysis of the EXAFS of known model compounds of similar structure to the unknown sample. This approach can yield very accurate results if a good model compound is available, however this is often not the case. The more usual approach in recent years is to calculate the phase and amplitude ab-initio using one of several available codes (FEFF, GNXAS, EXCURVE) [33–35,40,41]. Single scattering calculations using these codes produce very accurate values which are easily obtained without a full model of the structure in question. Multiple scattering can also be incorporated, but the relative positioning of the scattering atoms is obviously important, so accurate results depend on having a good model of the local structure of the absorber.

The curve fitting process is usually used to first refine R and σ, followed by E_0 after which the best fit with an integer value of N can be sought. If the sample is unknown, the identity, number and distance of the scattering atoms in the first shell must be guessed: the distance for example can be approximated by the peaks in the phase corrected Fourier transform. The theoretical and experimental EXAFS are then fit in either real or k space.

Multiple scattering processes are sometimes important, usually beyond the first shell and especially when the molecular geometry around the absorber includes bond angles of around 180 degrees. In these cases, the single scattering method described above is insufficient to reproduce the observed data and a detailed three dimensional model of the environment of the scattering atom must be refined with code capable of calculating the contributions of multiple scattering pathways [33–35].

15.5 Case Studies

15.5.1 Surface EXAFS of Titanium Nanostructure Thin Films

X-ray absorption is not an inherently surface sensitive technique. Both the primary X-ray beam and fluorescence emission are relatively penetrating, sampling depths of at least many microns at hard X-ray energies even in highly absorbing samples. Despite this, EXAFS can be surface sensitive, and indeed the power of EXAFS in determining surface structure has been recognised for many years [3,17,42,43]. Surface sensitive EXAFS is often termed SEXAFS.

Surface sensitivity can be achieved through one of three techniques:

1. A surface or thin film sample. If the absorbing element exists only at the surface then EXAFS will be surface sensitive. This is the case for adsorbates, thin films, many catalyst systems, electrochemical systems etc. The adsorbed layer or film may not contain sufficient absorbing atoms for good quality data in which case one of the following techniques will also have to be employed.
2. Use a surface sensitive detection technique. The predominant method is electron detection. Emitted photo- and secondary electrons have an escape depth of several tens of nano-metres, greatly enhancing the surface sensitivity of the measured EXAFS. It is also possible to measure the yield of ions desorbed by the X-ray absorption process [17], although successful examples are rare.
3. Confine the exciting X-ray beam to the surface region. Surface sensitivity can be enhanced by reducing the angle of incidence of the primary X-ray beam. The sample or probe depth of the beam will reduce with the incidence angle. However if the sample is optically smooth and the incidence angle is less than the critical angle for total external reflection, typically a few milliradians, then the penetration of the X-ray beam is dramatically reduced to the order of a few nano-metres.

The example of surface sensitive EXAFS to be presented here is a study of thin film structure, and uses the glancing incidence geometry of point 3 to enhance the surface sensitivity of the measurement. Glancing angle or "total reflection" EXAFS, where the X-ray beam incidence angle is below the

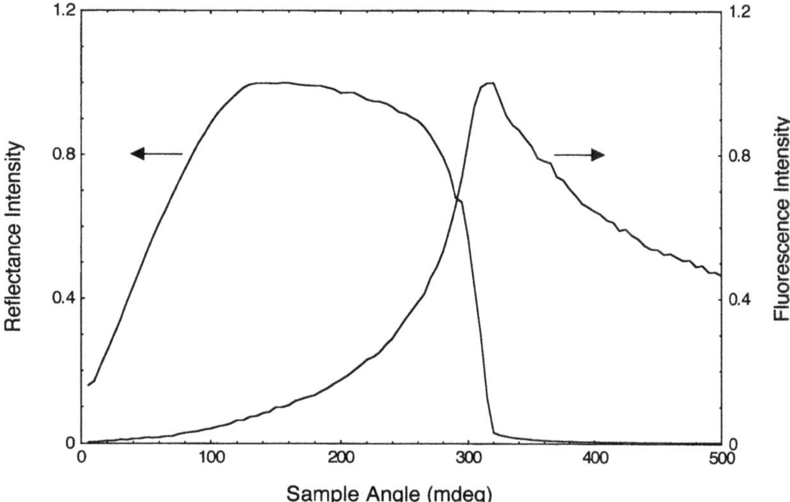

Fig. 15.9. Titania sol gel film X-ray reflectivity and Ti K α fluorescence yield variation with incidence angle at an X-ray energy of 5.2 keV

critical angle, has been used for some time to study surfaces and interfaces [44–48]. EXAFS is present in both the reflected beam and fluorescence emission, although fluorescence detection is generally superior [40,41]. The main limitation of the technique is that flat optically smooth samples or substrates are required for optimum surface sensitivity, and relatively large samples are needed because of the large beam footprint at the grazing incidence angles, which are generally below 0.5 degrees.

Glancing angle EXAFS has been used by Hanley etal. [49,50] to study nanocrystalline titania thin films, which have applications in photovoltaic cells, batteries, electrochromic devices and sensors. There are also possible quantum size effects in these films where the particle size is small. The films were prepared by dipping glass substrates into colloidal anatase sols with a particle size distribution in the tens of nano-metres. The resulting films were 50 nm thick, and were fired at various temperatures. EXAFS spectra were measured at the Ti K-edge at grazing incidence using a 100 micron high beam, and Ti fluorescence was measured with a Ge SSD. The calculated penetration depth of the X-ray beam at the critical angle of 0.36 degrees was about 10 nm. Figure 15.9 shows the X-ray reflectivity and Ti fluorescence emission versus the angle of incidence and is a good illustration of the grazing incidence technique. The sharp fall in reflectivity at the critical angle can clearly be seen, which coincides with a peak in the fluorescence yield as the X-ray beam begins to fully penetrate the film. At higher angles the fluorescence yield drops as the X-ray beam penetrates further into the substrate and less is

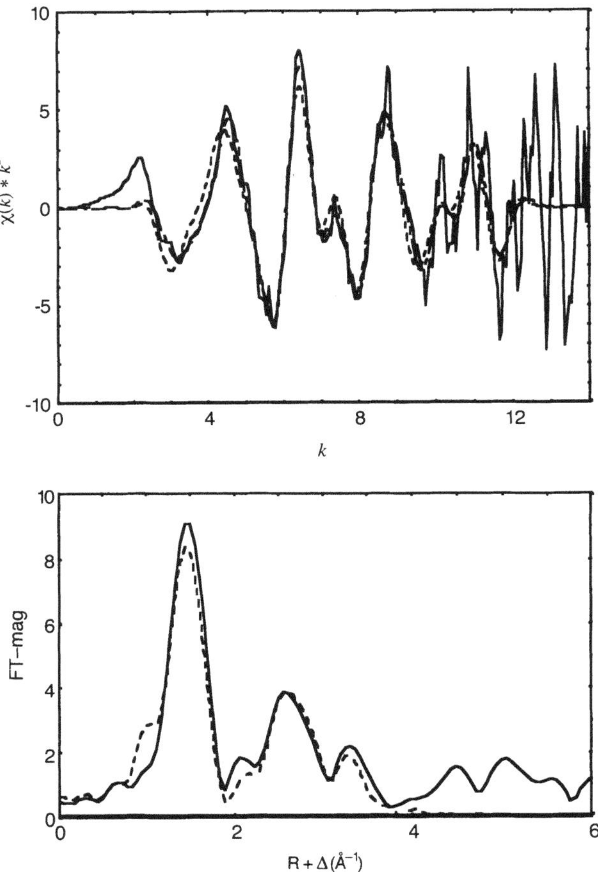

Fig. 15.10. Measured *(solid)* and theoretical *(dashed)* EXAFS from an air dried titania film. The top panel is the k-weighted EXAFS and the bottom is the Fourier Transform. This and Fig. 15.11 illustrate how the measured EXAFS can be fitted in either k-space or real space

absorbed in the thin film. To obtain maximum signal to noise, the EXAFS was measured at the peak of the fluorescence yield, at about the critical angle.

Figures 15.10 and 15.11 compare Fourier filtered observed and calculated data for films dried in air and fired at 800 C respectively. Anatase, the titanium mineral in the sol, has four coordination shells (O-Ti-Ti-O) at distances of 2–4 Å from the absorbing Ti atom. The EXAFS at room temperature is consistent with a distorted anatase structure, with high degrees of disorder in all but the first shell. Increased crystallinity is apparent in films fired at higher temperatures, as shown in Fig. 15.11 for the film fired at 800°C where the decrease in Debye-Waller factors for this higher shells produces much higher

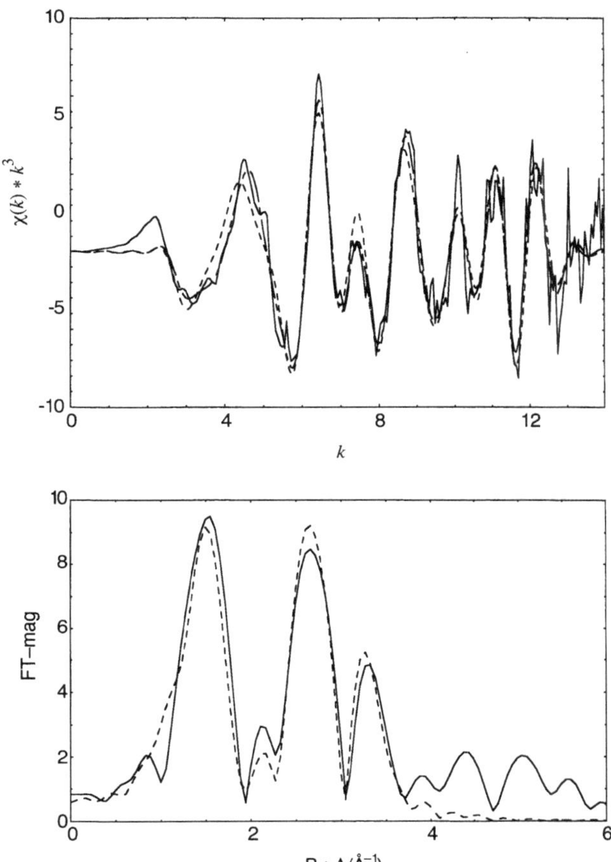

Fig. 15.11. Measured *(solid)* and calculated *(dashed)* grazing incidence Ti K-edge EXAFS of titania film fired at 800°C. *Top:* k-weighted EXAFS; *Bottom:* Fourier transform

intensity in the Fourier transform beyond 2 Å. This work is a good example of the use of the grazing incidence geometry to obtain high quality EXAFS from a thin surface film, a system which would be far too dilute to yield measurable data in normal transmission or fluorescence mode. EXAFS was recorded beyond $k = 12\,\text{Å}^{-1}$, which is usually sufficient for a high confidence solution. The work also illustrates the value of combining many techniques to attack a problem. The nano-crystalline films were also analysed with powder diffraction, electron microscopy, micro Ramen and UV-visible spectroscopy.

15.5.2 Ion-Implantation Induced Amorphisation of Germanium

The study of the crystalline-to-amorphous transition in germanium induced by ion-implantation carried out by Glover et al. [51] is an excellent example of how XAFS can be used not only to study local order but also disorder. A general description of the application of XAFS to the study of disordered systems can be found in [31].

In this case study, crystalline Ge was amorphised by implantation of Ge ions in such a way as to produce a uniform disordered layer approximately 2 microns thick. A range of ion doses was used and the effect of each dose on the local structure studied using EXAFS. The investigators chose a novel, but not widely applicable, technique to enhance the surface sensitivity of EXAFS in addition to uniformly sampling the amorphised surface layer. The amorphised layers were removed from the substrate either by selective etching or polishing and the thin films crushed and dispersed to make transmission EXAFS samples.

Figure 15.12 shows a series of Fourier transforms (not phase corrected) derived from the EXAFS of a range of samples prepared as described above. The important features to note are the steady decrease in intensity of the FT components due to the 2nd and 3rd nearest neighbour shells and the concomitant broadening of the nearest neighbour FT peak. Even without further analysis, the ability of EXAFS to follow and visualise such an increase in local disorder is apparent. The usefulness of EXAFS in studying disordered systems is not limited, however, to such semi-quantitative analysis. Real structural information on the disordered phase can be extracted also.

In systems incorporating a large degree of disorder, the standard harmonic (Gaussian) approximation of the distribution of disorder breaks down and extra terms are required in the data analysis to avoid serious errors. The

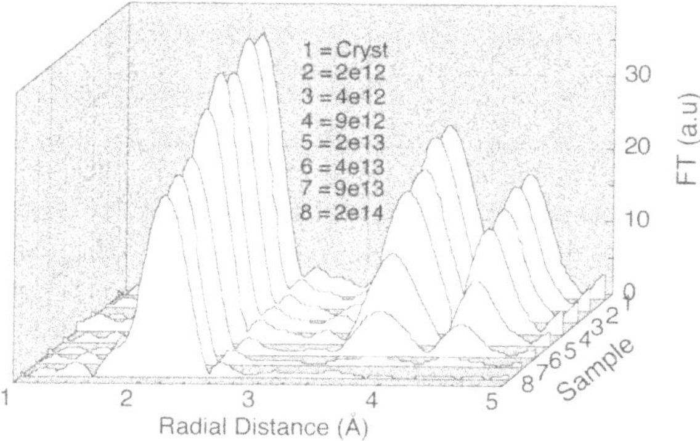

Fig. 15.12. Fourier transforms (non phase corrected) of XAFS data showing the transition from crystalline to amorphous Ge induced by ion implantation. The ion dose ranges from zero (crystalline Ge) to 2×10^{14} ions/cm^2

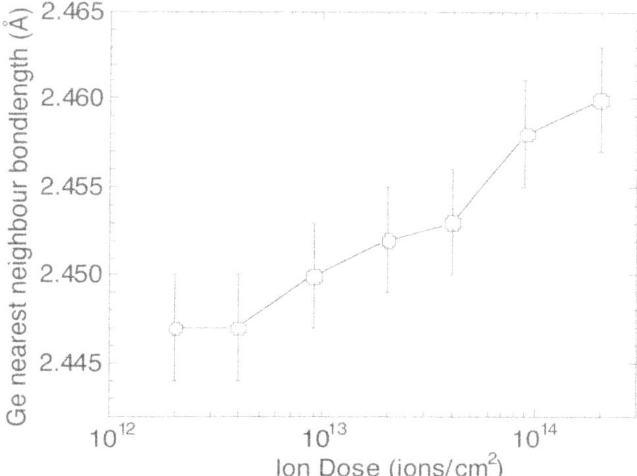

Fig. 15.13. Change in nearest-neighbour bond length in amorphous Ge with implanted ion dose

so called Cumulant Method is often utilised in such cases. The information that can be extracted includes not only the usual EXAFS parameters such as bond length and coordination number but also the degree of anharmonicity in the distribution of the disorder in the system [26,30,31].

In this study the cumulant method to was used to characterise the structure of amorphous Ge and an ion-dose-dependent variation in this structure was identified, even beyond the amorphous threshold. Figure 15.13 shows the change in nearest neighbour bond length that was observed in amorphous Ge as the total ion-implanted dose was increased over the indicated range. In addition to the change in bond length, the cumulant method revealed an increasing anharmonicity in the distribution of disorder as implantation dose was increased. These two important observations were interpreted in terms of increasing defects in the amorphous phase – specifically, higher occurrence rates of three and five-fold coordination around Ge atomic centres.

References

1. "X-Ray Absorption: Principles, Applications, Techniques of EXAFS, SEXAFS and XANES", Eds. D.C. Koningsberger and R. Prins, Wiley, New York, 1988
2. J. Stohr, "NEXAFS Spectroscopy", Springer, Berlin, 1992
3. "X-Ray Absorption Fine Structure for Catalysts and Surfaces", Ed. Y. Iwasawa, World Scientific, Singapore, 1996
4. The proceedings, of XAFS X held in 1998, appears as volume 6 part 3 of the Journal of Synchrotron Radiation, 1999. The XAFS XI proceedings will be volume 8 of J. Synchrotron Rad.

5. See for example the web site of the International XAFS Society at http://ixs.csrri.iit.edu/IXS/index.html
6. D.E. Sayers, E.A. Stern, F.W. Lytle: Phys. Rev. Lett. **27**, 1204 (1971)
7. see for example "Applications of Synchrotron Radiation", eds. H. Winick, D. Xian, M. Ye, T. Huang, Gordon and Breach, New York, 1989
8. B.M. Kincaid, P. Eisenberger: Phys. Rev. Lett. **34**, 1361 (1975)
9. Y. Udagawa, [3, p. 130]
10. J. Wong, Z.U. Rek, M. Rowen, T. Tanaka, F. Schafers, B. Muller, G.N. George, I.J. Pickering, G. Via, B. DeVries, G.E. Brown, M. Froba: Physica **B208/209**, 220 (1995)
11. T. Matsushita, R.P. Phizackerley: J. Appl. Phys. **20**, 2223 (1981)
12. M. Hagelstein, C. Ferrero, M. Sanchez del Rio, U. Hatje, T. Ressler, W. Metz: Physica **B208/209**, 223 (1995)
13. R. Frahm: Nucl. Instrum. Methods **A270** (1988) 578; R. Frahm: Synchrotron Radiation News **8** 38 (1995)
14. L.M. Murphy, B.R. Dobson, M. Neu, C.A. Ramsdale, R.W. Strange, S.S. Hasnain: J. Synchrotron Rad. **2**, 64 (1995)
15. F.W. Lytle, R.B. Greegor, D.R. Sandstrom, E.C. Marques, J. Wong, C.L. Shapiro, G.P. Huffman, F.E. Huggins: Nucl. Instrum. Methods **226**, 542 (1984)
16. S.P. Cramer, O. Tench, M. Yocum, G.N. George: Nucl. Instrum. Methods **A266**, 586 (1988)
17. J. Stohr, R. Jaeger, S. Brennan: Surf. Sci. **117**, 502 (1982)
18. F. Farges, J.-P. Itie, G. Fiquet, D. Andrault: Nucl. Instrum. Methods **B101**, 493 (1995)
19. D.J. Thiel, P. Livins, E.A. Stern, A. Lewis: Nature **362**, 40 (1993)
20. F.A. Schultz, B.J. Feldman, S. Gheller, W.E. Newton, B. Hedman, P. Frank, K.O. Hodgson: "Redox Mechanisms and Interfacial Properties of Molecules of Biological Importance V", Proc. Vol. 93-11, F.A. Schultz and I. Taniguchi, eds. Pennington N.J.: The Electrochemical Society, Inc. pg. 108, 1993
21. I. Ascone, A. Cognigni, M. Giorgetti, M. Berrettoni, S. Zamponi, R. Marassi: J. Synchrotron Rad., **6**, 384 (1999)
22. A. Yoshiasa, T. Nagai, O. Ohtaka, O. Kamishima, O. Shimomura: J. Synchrotron Rad. **6**, 43 (1999)
23. D. Bazin, H. Dexpert, J. Lynch: [3, p. 113]
24. An interesting and informative history of XAFS is given by Lytle, J. Synchrotron Rad. **6**, 123 (1999)
25. S.J. Gurman, J. Synchrotron Rad. **2**, 56 (1995)
26. E.D. Crozier: Nucl. Instrum. Methods **B133**, 134 (1997)
27. D. Koningsberger, B. Moget, J. Miller, D. Ramaker: J. Synchrotron Rad. **6**, 135 (1999)
28. D.E. Ramaker, W.E. O'Grady: J. Synchrotron Rad. **6**, 800 (1999)
29. E.A. Stern: Phys. Rev. B **10**, 3027 (1974)
30. G. Bunker: Nucl. Instrum. Methods **207**, 437 (1983)
31. E.D. Crozier: [1, Chap. 9]
32. C. Bourder: J. Phys. Condens. Matter **2**, 701 (1990)
33. J.J. Rehr, R.C. Albers, S.I. Zabinsky: Phys. Rev. Lett. **69**, 3397 (1992)
34. S.I. Zabinsky, J.J. Rehr, A. Ankudinov, R.C. Albers, M.J. Eller: Phys. Rev. B **52**, 2995 (1995)

35. J.J. Rehr, R.C. Albers: Rev. Mod. Phys. (2000)
36. Workshop report: "Molecular Environmental Science: Speciation, Reactivity and mobility of Environmental Contaminants. An Assessment of Research Opportunities and the Need for Synchrotron Radiation Facilities." SLAC-R-95-477, 1995
37. P.J. Ellis, H.C. Freeman: J. Synchrotron Rad. **2**, 190 (1995)
38. XAFS International Workshop Report in "X-ray Absorption Fine Structure", ed. S.S. Hasnain, Ellis Horwood, New York, 1990.
39. J.W. Cook, D.E. Sayers: J. Appl. Phys. **52**, 5024 (1981)
40. N. Binstead, R.W. Strange, S.S. Hasnain: Biochemistry **31**, (1992) 12117
41. A. Filipponi, A. Di Cicco, T.A. Tyson, C.R. Natoli: Solid State Commun. **78**, 265 (1991)
42. E.A. Stern: J. Vac. Sci. Technol. **14**, 461 (1977)
43. T.L. Einstein: App. Surf. Sci. **11/12**, 42 (1982)
44. R. Fox, S.J. Gurman: J. Phys. C **13**, L249 (1980)
45. S.M. Heald, E. Keller, E.A. Stern: Phys. Lett. **103A**, 155 (1984)
46. S.M. Heald, H. Chen, J.M. Tranquada: Phys. Rev. B **38**, 1016 (1988)
47. M. Shirai, Y. Iwasawa: [3, p. 332]
48. B.A. Bunker, A.J. Kropf, K.M. Kemmer, R.A. Mayanovic, Q. Lu: Nucl. Instrum. Methods **B133**, 102 (1997)
49. T.L. Hanley, V. luca, I. Pickering, R.F. Howe: to be published.
50. T.L. Hanley, Y. Krisnandi, A. Eldewik, V. Luca, R.F. Howe: Ionics, submitted
51. C.J. Glover, M.C. Ridgeway, K.M. Yu, G.J. Foran, C. Clerc, J. Hansen, A. Nylandsted Larsen: J. Synchrotron Rad. **8**, 2001, in press
52. M.C. Ridgway, C.J. Glover: private communication

Part III

Processes and Applications

16 Minerals, Ceramics and Glasses

R.St.C. Smart

In materials science and technology applied to minerals, ceramics and glasses, surface analysis is used in three principal modes: *problem solving* in quality control for existing processes and materials; *materials characterisation* after adsorption, surface coating, reaction or modification; and *development* of new materials or processes. We can illustrate each of these modes with a few examples. In problem solving, difficulties with control of contamination, coating or adsorption chemistry, adherence (e.g. delamination), discolouration and changes in surface reactivity are common. Characterisation includes the rapidly expanding industry of surface engineering of ceramic and glass layers for corrosion and wear resistance, alteration of surfaces for composite (e.g. polymer) compatibility and mineral surface weathering. The last area encompasses long-term projects in processes as diverse as minerals separation to bioceramic design for materials as diverse as clays, nuclear waste solids and superconductors.

Much of the information on mineral, ceramic and glass surfaces has common features in both the categories of information required and the surface analytical techniques most applicable to extraction of that information. In general, the types of information include:

- for minerals and ceramics, definition of the crystalline phases in terms of their structure, composition (range), size and distribution;
- surface structure from atomic level to macroscopic features;
- definition of surface sites related to structure including defects, impurities and their reactivity;
- for minerals and ceramics, grain boundaries and intergranular films including their structure (width) and composition (segregation, triple points, precipitates etc.);
- variation of structure and composition with depth including depth profiles.

In addition to characterisation of the mineral, ceramic or glass and its surface properties, information on surface processes may be required in relation to:

- adsorption including the molecular structure, form and bonding, surface coverage, surface layer development and lateral distribution;
- surface reactions including oxidation, hydrolysis, weathering, dissolution (leaching), surface phase transformation;

- surface modification by, for instance, calcination, plasma reactions, surface layer deposition (e.g. sizing), hydrophobicity (e.g. flotation of minerals), surface conductivity (e.g. electrostatic separation).

The study of these properties of minerals, ceramics and glasses in general requires similar surface analytical techniques and methodological approaches. The choice of methodological approach and analytical techniques to obtain this information is usually based on a strategy of analysis beginning with the simplest, readily available techniques. They provide general or averaged information on surface properties The analysis may then proceed to more detailed, specific information with higher resolution in chemical or structural information using more advanced techniques.

Table 16.1 summarises an analysis strategy designed to give information from macroscopic through microscopic to atomic scale on chemical and structural properties using a sequence of analytical techniques. Structural information is most usefully obtained starting with the low-cost, accessible technique of optical microscopy, often used with image analysis enhancement, at the supra-micron scale. For minerals and ceramics, phase distributions can be readily identified using polarised light and standard mineralogical procedures on polished sections. Correlation with X-ray diffraction (XRD) will usually provide the essential information on crystalline phases that may be subject to adsorption, reaction or modification. The intermediate level of structure can be imaged and analysed using scanning electron microscopy (SEM) most recently in high resolution field emission (high brightness) imaging with resolution down to tens of nms. Near-surface ($\approx 1\ \mu m$) chemical composition or information can be obtained in the SEM with energy dispersive spectroscopy (EDS or EDAX). Scanning Auger microscopy/spectroscopy (SAM), again recently with field emission electron beam sources, can similarly be used for direct imaging coupled with surface analysis (2–5 nm) of chemical composition, lateral distribution and some chemical state information (Chap. 6). Finally, structural information at the atomic and molecular level is now relatively routinely available from the scanned probe microscopies (Chap. 10). Atomic force microscopy (AFM) for insulating samples, scanning tunnelling microscopy (STM) for conducting samples, or transmission electron microscopy (TEM) for thin sections or the edges of particles can be used. These techniques provide both imaging of lattice structure and local information on atomic sites in crystalline lattices or glass materials. For crystalline materials, i.e. minerals and ceramics, surface phases can be examined using glancing angle XRD and low energy electron diffraction (LEED) (Chap. 13). For these materials and glass surfaces, bonding in the surface layers can be derived in terms of nearest neighbour elemental identification, bond distances and (less reliably) coordination number from X-ray absorption fine structure XAFS (Chap. 15). Detailed information on surface sites in terms of chemical identification and structure is also the preserve of low energy ion scattering (LEIS) (Chap. 11).

Table 16.1. Analysis strategy for application of techniques to give information from macroscopic through microscopic to atomic scales

Structural Information		
Imaging		**Latices:Sites**
Optical Microscopy		XRD
↓		↓
SEM:SAM		XRD (glancing angle)
↓		↓
FESEM		XAFS:LEED
↓		↓
AFM; STM; TEM		LEIS
Chemical Information		
Near Surface	**Surface**	**Molecular**
EDS:XRF	XPS	FTIR:Raman
↓	↓	
XAFS:RBS:NRA	SAM	
	↓	
	SIMS (ToF)	
	↓	
	STS(STM):LEIS	

Surface chemical information is now relatively routinely obtained from X-ray photoelectron spectroscopy (XPS) (Chap. 7) and SAM with depth resolution of the order of 2–5 nm. Monolayer sensitivity can be obtained using static SIMS particularly from the relatively recent introduction of high resolution time of flight mass spectrometry (i.e. ToF-SIMS) (Chap. 5). The highest resolution chemical information on individual atomic and molecular sites is probably obtainable from STM in the scanning tunnelling spectroscopy (STS) mode. In STS, work functions of individual sites are measured (Chap. 10). The well-established molecular spectroscopies of FTIR and Raman have also been applied to obtain information on molecular structure in surface-sensitive modes such as attenuated total reflection, specular reflectance and surface enhanced Raman spectroscopy (Chap. 8).

In this chapter, a variety of case studies have been chosen to illustrate some of the more important industrial mineral, ceramic and glass materials and the type of information available from different techniques. It will be necessary for readers to assess the power of these techniques in relation to processes and materials with which they are familiar by extrapolating from the evidence presented in each of these case studies.

16.1 Minerals

16.1.1 Iron Oxides in Mineral Mixtures

Removal of iron oxides is of major concern in many mineral operations with, for instance, bauxite, kaolin, quartzite and titanium-based mineral sands. The iron in these three systems takes three main forms: free particles ($> \approx 1\,\mu m$) of oxides (e.g. hematite, magnetite), hydroxides (usually amorphous) and oxyhydroxides (e.g. goethite); as colloidal ($< 1\,\mu m$) "slimes") particles; and iron substituted into crystal lattices of other minerals (e.g. boehmite, gibbsite, kaolinite). It is usually possible to separate the free particles by flotation. The lattice-substituted iron cannot be separated but is usually < 0.5wt%, a level acceptable for most processing. The amount and location of the colloidal iron oxides is critical.

For example, after selective sedimentation, XPS of a bauxite product (principally gibbsite, boehmite and kaolinite) found a surface composition (atomic %) of: 3.0% Fe; 18.2% Al; 3.6% Si; 57.8% O, 16.1% C. TEM/EDS examination (Fig. 16.1) revealed gibbsite and boehmite particles larger than

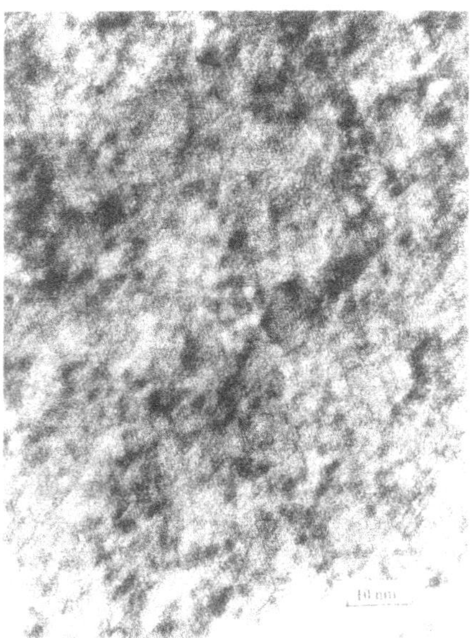

Fig. 16.1. High resolution TEM micrograph of part of a large gibbsite ($Al(OH)_3$) crystal, from bauxite pisolites, showing lattice fringes (striations). Some fringes are fragmented due to either disorder in the lattice or surface roughness (or both). The surface shows pits (Fig. 16.4) and protusions over distances of order 2–3nm consistent with the high surface area of this material. The dark patches (ca. 2nm diameters) show high Fe/Al ratio consistent with iron oxide particles (as hematite and goethite) attached to the gibbsite surface. From [1]

200nm with very small (i.e. 1–10 nm) iron-containing regions dispersed across their surfaces. Colloidal particles of hematite and goethite adhere to the surface of the larger platey minerals. They are not removed by ultrasonic agitation but reductive dissolution in dithionite dissolves them (at different rates for hematite and goethite). XPS confirms removal from the surface [1].

The efficiency of methods for separation of iron oxides can thus be monitored using XPS with electron microscopy.

16.1.2 Surface Layers on Minerals

In very many cases (perhaps even a majority), the surface of a mineral has neither the same composition nor structure as its bulk. This can be due to reaction with air (e.g. oxidation of sulfides), groundwater (e.g. silicate, carbonate depositions), contamination (e.g. humics, hydrocarbons), intergranular layers (e.g. graphitic) exposed on grinding, or adsorbed species (e.g. sulfates). These alterations to the surface can profoundly influence the behaviour of the mineral in a process.

For instance, a glacial sand deposit to be used for glass-making had unacceptable dewatering and melting properties very different from the dune sand it replaced. XRD and SEM/EDS found only quartz with very little clay and no significant impurities in the quartz after washing. XPS consistently revealed 2.5–5.0 at.% Al, with an aluminosilicate binding energy (not adsorbed Al^{3+}) in the surface. Ion etching suggested that this layer was no more than 200 Å thick. It could not be removed by any practical base-hydrolysis reaction but choice of a surfactant suitable for an aluminosilicate surface reduced dewatering to an acceptable time and adjustment of melting conditions for the presence of the layer allowed acceptable processing [2].

16.1.3 Mineral Processing of Sulfide Ores

Valuable sulfide minerals are usually separated from their waste (or gangue) minerals by froth flotation. The process involves chemical conditioning of the valuable mineral surface by selective adsorption of collector molecules designed to render the particle sufficiently hydrophobic to attach to a bubble and collect in the froth concentrate. Minerals separated in this way include pyrite FeS_2, chalcopyrite $CuFeS_2$, galena PbS, sphalerite ZnS and pentlandite $(Fe, Ni)_9 S_8$.

The process is complicated by such considerations as:

- chemical alteration of the surface layers of the sulfide minerals induced by oxidation reactions in the pulp solution;
- the presence of a wide range of particle sizes in the ground sulfide ores;
- galvanic interactions between different sulfide minerals to produce different reaction products on the mineral surfaces;
- interaction between particles in the form of aggregates and flocs;

- the presence of colloidal precipitates arising from dissolution of the sulfide minerals and grinding media;
- the mechanism of adsorption of reagents to specific surface sites;
- competitive adsorption between oxidation products, conditioning reagents and collector reagents.

Hence, the complex mixture of minerals and their interactions via dissolved species in most real ores makes the measurement of surface changes difficult to interpret using any technique. Nevertheless, a respectable body of literature has now been developed (e.g. [3–12]) in which some useful observations from surface analysis can be found. Studies, particularly using XPS and FTIR, of the effects of conditioning agents (e.g. metabisulphite, sulfide, cyanide), activators (e.g. Cu^{2+}) and collectors (e.g. alkyl xanthates, dithiophosphinates) on samples derived from processing plants and from synthetic minerals and ores have elucidated some of the mechanisms of flotation separation. A recent review [13] has summarised the application of surface analytical techniques to studies of sulfide mineral surface oxidation and collector adsorption.

An example can be seen in the effect of xanthate collector on an ore comprised predominantly of chalcopyrite, $Cu(I)Fe(III)S_2$, pyrite $Fe(II)S_2$, quartz, dolomite and iron oxides [14]. Table 16.2 shows the composition of the surface at $E_h + 200\,\mathrm{mV}$ before (Test 2) and after (Test 4) addition of sodium butyl xanthate ($NaC_4H_9OCS_2$) at 0.04 kg/tonne. Analysis of the flotation concentrates and tails (non-floating) is compared before and after ion etching to a depth of 25 nm. In the concentrates, several changes are evident after xanthate addition:

- the exposure of S (as sulfide) is increased roughly six-fold (some of which can be attributed to the adsorbate xanthate);
- exposure of Cu (as Cu(I) only) is markedly increased;
- exposure of Fe is significantly increased and the charge-corrected Fe(II) – $S(2p_{3/2})$ signal near 707 eV appears whereas only Fe(III) hydroxides were found before xanthate addition (Fig. 16.2);
- reduction of the amount of oxygen (as hydroxide) on the surface;
- the presence of carbon (as saturated hydrocarbon and carbonate) on both surfaces.

The tails show little evidence of S (and only as sulfate species) or Cu but Fe(III) (as hydroxides only) is found in both cases. Much more carbonate is found in the tails samples. The xanthate appears to have a cleaning action partly removing surface hydroxides and exposing the underlying mineral surfaces. Other studies, at different E_h values in solution (i.e. −400, −200, 0, +400 mV) show that $E_h + 200\,\mathrm{mV}$ is the most favourable condition for this effect of the xanthate. More detailed examination of the elemental regions in the XPS spectra can identify, for instance, the xanthate contributions to the C, O and S regions, sulfide/sulfate ratios, and different forms of Fe.

Table 16.2. Quantitative surface analysis of major species (mol%)

Samples	Graphite		CO_3^{2-}		SiO_2		Cu_2S	
	i[a]	25[b]	i	25	i	25	i	25
Test 2 Conc.	23.1	18.8	15.4	10.7	27.3	33.0	1.0	1.5
Test 2 Tail	6.1	0.8	24.4	15.2	23.1	30.6	0.3	0.9
Test 4 Conc.	24.0	16.4	13.7	7.3	12.3	14.1	12.2[c]	12.2[c]
Test 4 Tail	8.7	2.0	19.9	13.7	30.6	34.8	0.0	0.0

Samples	SO_4^{2-}		FeS_2		$Fe(OH)_3$	
	i	25	i	25	i	25
Test 2 Conc.	0.0	0.0	0.4	2.1	21.6	27.2
Test 2 Tail	0.5	1.5	0.0	0.0	21.6	28.4
Test 4 Conc.	0.0	0.0	30.6[c]	43.2[c]	4.7	6.8
Test 4 Tail	0.0	0.5	0.0	0.0	26.4	32.4

[a] Survey on sample without etch

[b] Survey on sample with 25nm etch

[c] Assignment of copper and sulfur is difficult due to presence of Cu xanthate and attenuation of signal

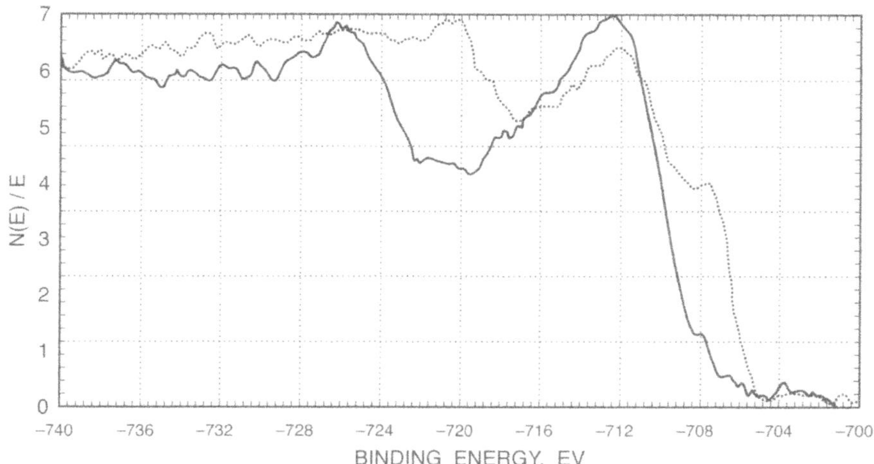

Fig. 16.2. Fe2p (4 at.% abundance) XPS spectra from two copper ore flotation concentrates from Mt. Isa (Australia) conditioned for 10min. at $E_h - 400\,\mathrm{mV}$ with no collector addition (*full line*) and $E_h + 200\,\mathrm{mV}$, 0.04 kg/t butyl xanthate collector addition (*dotted line*). The low E_h sample shows only broad iron hydroxide species on the surface. (Fig. 16.5) The collector-modified ore reveals new signals near 707 eV and 720 eV from clean chalcopyrite ($CaFeS_2$) and pyrite (FeS_2) surfaces.

SAM, SIMS and STM have been used to image and analyse oxidation products in a number of different forms and processes, namely:

- overlayers of ferric hydroxide, sulfate and carbonate on iron-deficient sulfide underlayers. In some cases, polysulfides S_n^{2-} and/or elemental sulfur S_n^0 are formed on the iron-containing sulfide substrate [3,5,6];
- oxidised fine sulfide particles adhering to larger particle surfaces [13];
- colloidal hydroxide (e.g. Fe(III), Pb(II), Cu(II)) precipitates and floccs [13,15,16];
- lateral distributions of isolated, patchwise, face-specific development of oxidation products of significantly varying depth under the same process conditions [13].

SAM elemental distribution maps from individual mineral particle surfaces after xanthate addition have illustrated strong association between Fe and O signals on pyrite surfaces (and in colloidal precipitates) with a converse, strong association between Cu and S from chalcopyrite surfaces [13].

STM has been used to follow the processes of oxidation in air and solution together with the action of collector adsorption on galena surfaces. Figure 16.3 illustrates some of these processes and actions. The time-dependent development of nm-scale sites of oxidation as protusions on galena in air is seen in Fig. 16.3a [17,18]. The development of pits due to dissolution of galena in air-purged water at pH 6 is illustrated in Fig. 16.3b [19]. The appearance of relatively large patches (100–200nm) of elemental sulfur after conditioning at pH 4.9 and $E_h > 200$ mV SHE (Fig. 16.3c) [20], and the appearance of colloidal particles of lead ethyl xanthate (confirmed by XPS and FTIR) after addition of the collector (10^{-4} M) at pH 7 (Fig. 16.3d) [15] are also shown. Atomic level imaging of the (001) surface of galena has been achieved by Eggleston, Hochella et al [21–27] including observation of the oxidation of a single S site at which the tunnelling current has been effectively extinguished [28].

Results of this kind from the different surface analytical methods have already provided major advances in understanding the chemistry of minerals processing. For instance, the actions of collector molecules have been shown [13] not to be confined to adsorption and increased hydrophobicity but also to act in several other modes, namely:

- detachment of fine oxidised sulfide particles from larger particle surfaces;
- detachment of fine oxide/hydroxide gangue particles;
- colloidal precipitation from solution (as for the lead ethylxanthate example above);
- partial removal of adsorbed and amorphous oxidised surface layers (Fig. 16.2);
- inhibition of oxidation (at higher concentrations);
- disaggregation of larger particles, and;
- patchwise or face-specific coverage.

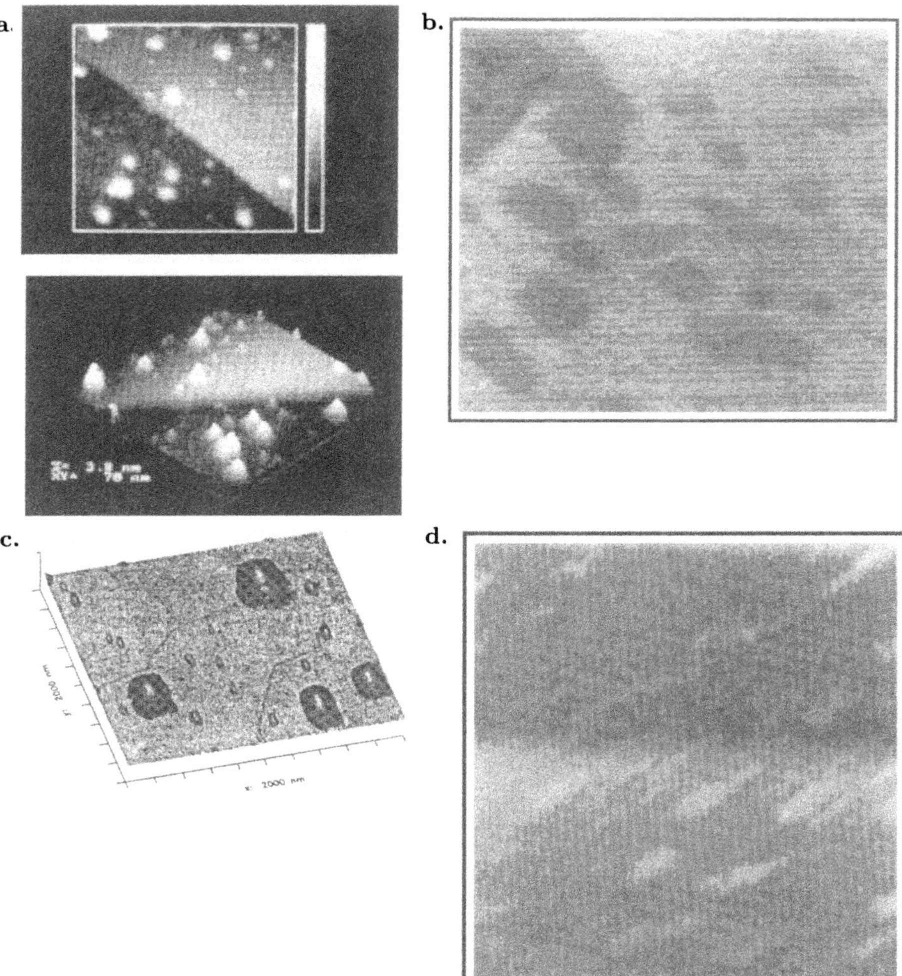

Fig. 16.3. STM images from cleaved galena (PbS) surfaces after **(a)** 270 min in air (70×70 nm area; upper image, top view; lower image 3D rotated; z scale 3.8nm). **(b)** 60 min air-purged pH 6 solution (500 × 500 nm area; top view image; z scale 2.6 nm). **(c)** in acetate buffer pH 4.9, applied potential +281 mV SHE (2000 × 2000 nm area; side view image; z scale 450 nm). **(d)** 30 min in 10^{-3} M Pb^{2+} at pH 7 (500 × 500 nm area; top view image; z scale 13.7 nm).

16.1.4 Adsorption and Reaction of Oxide and Clay Minerals

The adsorption/desorption reactions of anions on clay minerals is fundamental to plant nutrition and the mobility of environmental pollutants. Surface concentrations, chemical states and the structures of adsorbed ions can be determined with a combination of IR and XPS studies (e.g. [29,30]). In particular, goethite (α-FeOOH), as the major iron oxide soil mineral, is responsible for retention of phosphate in soils. IR shows that it is bonded as a binuclear HPO_4^{2-} complex, displacing two surface OH groups, across a [001]-directed surface trench. XPS (Table 16.3) shows that it is more strongly coordinated than sulfate (also binuclear) or selenite anions across the pH range from 3 to 12.

Surface analytical methods can monitor these surface species in soil environments in a relatively routine way.

Table 16.3. Atomic concentrations of surface atoms on goethite after adsorption of the anions shown at 5×10^{-2} *M* concentration, pH 3 and pH 12.

Atomic concentrations [atom %], pH 3					
Anion†	C	O	Fe	X†	X/Fe
PO_4^{3-}	19.7	59.9	18.2	2.2	0.12 ± 0.008
SeO_3^{2-}	15.8	59.8	21.7	2.6	0.12 ± 0.008
SO_4^{2-}	13.1	64.4	21.0	1.5	0.071 ± 0.008
Atomic concentrations [atom %], pH 12					
Anion†	C	O	Fe	X†	X/Fe
PO_4^{3-}	7.9	68.4	21.5	2.2	0.10 ± 0.008
SeO_3^{2-}	14.8	61.3	23.0	0.90	0.039 ± 0.008
SO_4^{2-}	16.2	60.4	22.8	0.57	0.025 ± 0.008

X† central atom of anion, i.e. P, Se and S respectively

Ratio of atomic % of the central atom X of the adsorbed anion to that of Fe and O at different pH of adsorption on goethite

Anion concentration $= 5 \times 10^{-2}$M		
pH	X	Ratio X/Fe/O
3	PO_4^{2-}	1 : 8 : 27
3	SeO_3^{2-}	1 : 8.3 : 23
3	SO_4^{2-}	1 : 14 : 43
12	PO_4^{2-}	1 : 9.8 : 31
12	SeO_3^{2-}	1 : 24.7 : 69
12	SO_4^{2-}	1 : 109.5 : 310

For pure goethite Fe/O = 1:2.5

The use of surface analytical techniques to study the surface chemistry during the dissolution of oxide and silicate minerals now has an extensive literature (e.g. [31]). Leaching reactions of feldspar (e.g. plagioclase) minerals in acid solution have been followed using a combination of TEM, XPS and SIMS depth profiles [32]. Leached layers of porous silica with thicknesses from 40 nm to 100 nm have been imaged and analysed revealing the loss of specific cations (e.g. Ca^{2+}, Mg^{2+}) to solution and the reduction of the Al/Si ratio in the leached layer. The thickness and composition of the layers is dependent on both the composition of the plagioclase feldspar and the pH of the solution with altered layers, under the same conditions of reaction, increasing in thickness in the order albite $<$ oligoclase $<$ labradorite $<$ bitownite. Changes in the structure and composition of leached minerals can be followed in detail with this combination of techniques.

Naturally weathered materials are quite often selectively leached and etched in a similar fashion to laboratory-simulated reacted minerals although, in the natural environment, the effect of contact with dissolved reaction products in relatively high solution concentrations over long periods of time can result in different mechanisms than those simulated in large-volume, short-term laboratory experiments. Topotactic secondary minerals, such as smectite, hematite and goethite can form on weathered olivine surfaces [33,34]. These mineral phases can be identified with a combination of SEM, electron diffraction in TEM and XPS, SAM or SIMS.

16.1.5 Surface Modification of Minerals

Modifying the surfaces of minerals by adsorption, reaction or surface coatings has now assumed considerable technological importance. Their uses include fillers in pulp, paper and polymer matrices, waste remediation by toxic and heavy metal element removal from solution, pigment materials in paints and coatings, composite materials including biodental and biomedical composites and in slurries like drilling muds. Hence, the objectives in their surface modification can be quite varied from alteration of hydrophobicity/hydrophilicity, chemical compatibility with the composite matrix (e.g. bonded methacrylate layers for PMMA incorporation), aggregation, optical properties (e.g. opacity, reflectivity, scattering), dispersion, absorptivity, wear, corrosion resistance or insulation (thermal, electrical).

Silane and siloxane reagents are commonly used to produce tailored surfaces ranging from fully hydrophilic to fully hydrophobic. With many minerals, particularly plate-like clay minerals (kaolinite, talc), the surfaces are relatively inert to adsorption of the reagents from solution. Often, only the edge sites of the platelets provide the primary sites for adsorption. The basal planes of the mineral can be activated for adsorption by partial dissolution in acid or alkaline solution or by the use of low-temperature (glow discharge) water vapour plasma reaction [35]. The extent of plasma activation of basal plane sites of kaolinite can be monitored by FTIR based on the absorbance of a new peak at 1400 cm^{-1} apparently due to a highly constrained, hydrogen-bonded OH group. The activated surface is completely wettable (i.e. hydrophilic) and gives complete dispersion in aqueous solution. It can be subsequently reacted with reagents, such as monochlorodimethyloctadecylsilane, to produce an irreversibly hydrophobic surface that will resist dispersion in water for periods of months. The uptake of the silane can be monitored by measuring the two resolvable contributions to the Si 2*p* XPS spectrum corresponding to the surface and adsorbed silane molecule. FTIR can also follow quantitatively this uptake through the νCH absorptions in the 2800–3000 cm^{-1} region. SEM and TEM images did not reveal any observable changes to the morphology of the kaolinite platelets resulting from plasma reaction or silane adsorption but

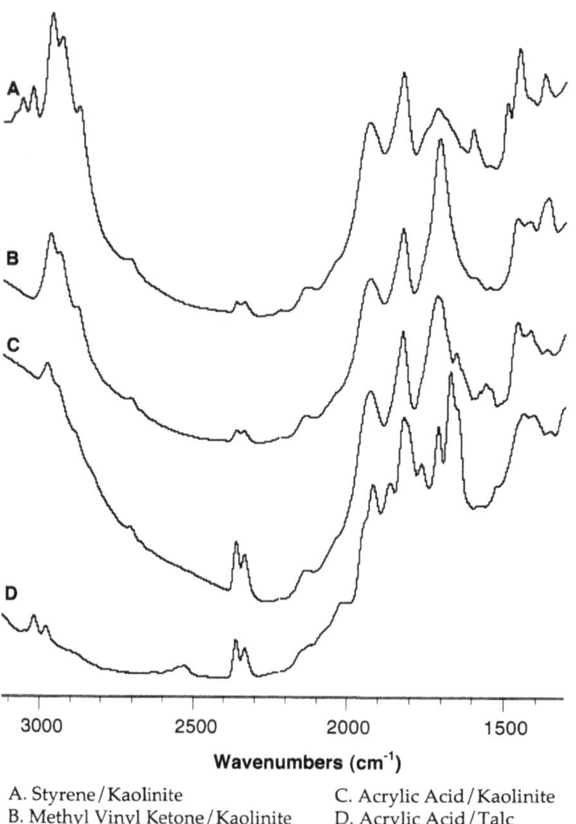

Fig. 16.4. FTIR spectra from polymer-coated clay particle surfaces (i.e. kaolinite, talc) activated in a low-temperature, low-density Ar plasma (10^{-1} torr; 50W) for 5–30 mins followed by admission of the monomer vapour (i.e. styrene, methylvinylketone, acrylic acid).

AFM imaging of the lattice structure of the kaolinite surface at the molecular level has revealed details of these sites of attachment of the silane molecules.

Improvement of the bonding of mineral particles into polymer matrices is reliant on improvements in the mechanical and chemical properties at the particle/polymer interface. With unmodified clay particles, this interface is often subject to disbonding (or "pull-out"), hydrolysis resulting in weaker bonding, preferential fracture and, in some cases, reduced brightness or even discolouration. The low-temperature plasma process can also be used to polymerise a variety of monomeric species and induce improved bonding to the activated mineral surface. The presence of the polymer films on the mineral surfaces can again be confirmed using both XPS and FTIR. Figure 16.4 gives some examples of FTIR spectra from polymer coatings applied to kaolinite and talc surfaces. Polystyrene νCH absorptions in the range 2800–3150 cm^{-1} and corresponding chain vibrations below 1800 cm^{-1} confirm the presence of the surface layer and can be correlated with C 1s XPS spectra from this layer. The polymer can be chosen to provide chemical compatibility with the corresponding polymer matrix, e.g. polyacrylate, polymethylvinylketone [35].

16.2 Ceramics

In ceramic technology, some of the main surface features requiring definition for control of the properties of the ceramics (see also Chap. 3) are:

- grain size and distribution;
- the location of matrix elements and impurities (e.g. in new phases, lattice-substituted in major phases, in intergranular regions);
- the structure and size of intergranular films and pores;
- degradation via leaching/dissolution/recrystallisation of the surface

Properties of concern in optimisation of ceramic products include mechanical strength, thermal conductivity, corrosion and wear resistance, all of which depend on the microstructure of the ceramic. Optimisation of materials and fabrication conditions to achieve the correct microstructure is the aim of this analysis.

Grain sizes and their distribution can usually be adequately mapped with a combination of optical microscopy (with image analysis), SEM/EDS (lateral resolution down to $\approx 0.5\,\mu m$) and SAM (down to $\approx 0.1\,\mu m$ if moderately conductive) as illustrated in Chapters 2 and 3. Comparison of results from each technique applied to the same material is often useful because the different analysis depths, i.e. $\approx 1\,\mu m$ and $\approx 5\,nm$ respectively for EDS and SAM, can reveal surface segregation of particular elements. Reactions of the air-exposed surface, and surface residues from dies or grinding agents are also seen. If the grain size is very small, i.e. $< 100\,nm$, it may be necessary to use STEM/EDS to study this distribution. This method has the advantage of allowing structural determination of individual grains by electron diffraction so that compositional inference of phase identification can be confirmed. The disadvantage of STEM/EDS is that it normally requires ion beam etching for thin sections with associated damage to surface layers. Some examples of the use of these techniques applied to a fine-grained ceramic can be found in [36,37].

Locating elements in ceramic microstructures generally relies on the same three techniques, i.e. SEM/EDS, SAM and STEM/EDS, but XPS and SIMS can also be useful in certain cases. New phases can be identified from X-ray diffractometry but, since individual grains can vary widely in composition, STEM/EDS or SAM is needed to locate and analyze a representative set of grains. Substitution of particular elements into grains of the major ceramic matrix phases can also vary considerably. For instance, Cs contents of Ba hollandite (i.e. $Ba_xCs_yTi_8O_{16}$) grains in the Synroc ceramic are found to cover the range $y = 0$ to $y = 1.5$. More extensive examples of determination of lattice substitution and new phase formation can be found in [38–40]. Homogeneity in mixing of the precursor materials and any additives is often essential to the performance of the material in practice.

The structure, size and location of elements in intergranular films and pores has been discussed and illustrated in Chapters 1 and 3. The use of XPS and static SIMS in studies of fracture faces of ceramics can be particularly useful in these applications [37].

16.2.1 Leaching and Dissolution

Resistance to surface degradation is now a major area of research and development in ceramics as diverse as Y-Ba-Cu oxide superconductors, perovskites and nuclear-waste disposal materials like Synroc [41]. Characterisation of loss to solution of particular elements by ion-exchange (leaching) or dissolution (reaction of the matrix) and the formation of protective layers of reprecipitated or recrystallised (in-situ) products form the basis of this research. Depth profiling the reacted surface layer using XPS, AES or SIMS gives important information on these processes but is often difficult to interpret on a multiphase ceramic surface with redeposited crystallites. Figure 16.5 illustrates elemental XPS profiles from such a surface [41,42]. It is clear that Ba, Cs, Mo and Ca are lost from the surface on reaction whilst Ti, Zr and Al are retained. Further study with SEM and TEM (Fig. 16.6) revealed, however, that one particular phase, i.e. $CaTiO_3$ perovskite with substituted Sr and rare earth elements, was dissolving preferentially leaving a recrystallised TiO_2 (brookite and anatase) product in the original perovskite sites. Hence, Ti loss was not significant in the XPS profiles despite the extensive reaction in these sites. Work on the single-phase ceramic perovskite $CaTiO_3$ [43,44] has shown that the mechanism for surface degradation is: loss of only 1–2 monolayers of Ca by ion-exchange; base-catalyzed hydrolysis reaction of the perovskite lattice to a depth of ≈ 100 mm (ultimately diffusion-limited) releasing the majority of the Ca to solution; and at temperatures above 90°C, recrystallisation

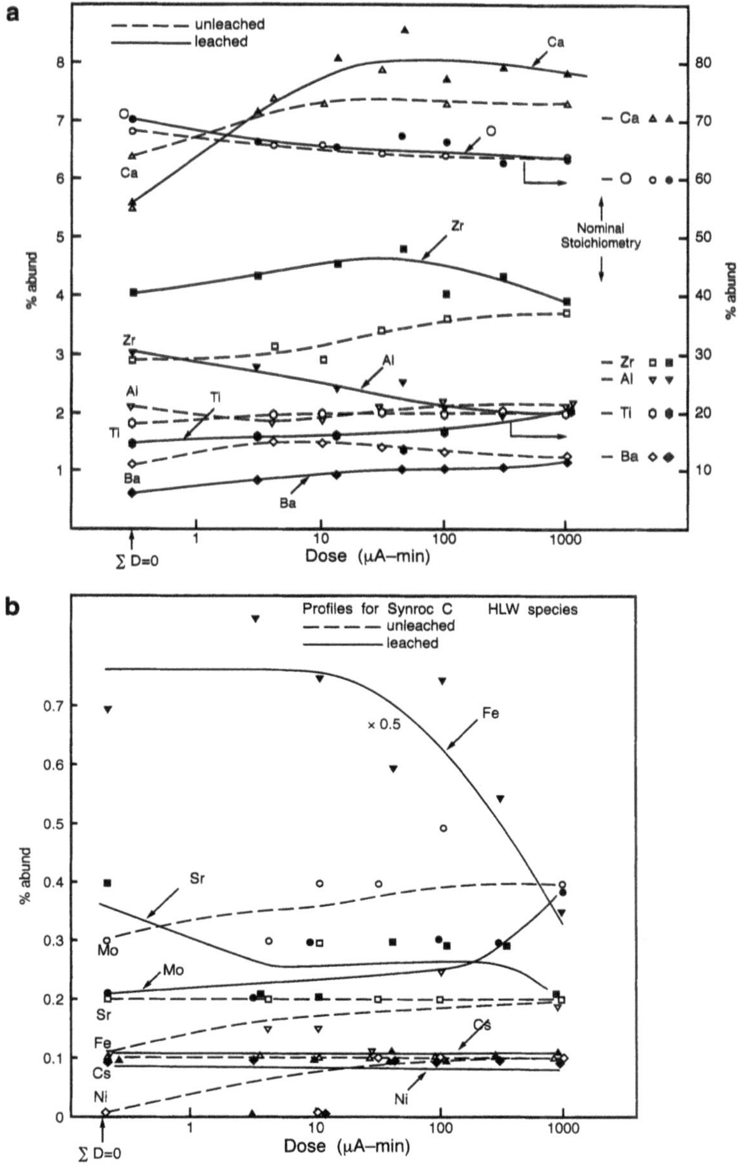

Fig. 16.5. **(a)** XPS depth profiles from the ceramic Synroc C matrix (i.e. major phase) elements before (*open symbols*) and after (*closed symbols*) hydrothermal attack for 25 days at 350°C. Nominal average bulk stoichiometries are marked as bars on the right with % abundance scales for Ba, Al, Zr, Ca at left and Ti, O at right. 1μA.min ion etch is ca. 0.2nm removal. **(b)** XPS depth profiles for simulated waste species in the same specimen as Fig.16.5a recorded under the same conditions. Uncertainties in % abundance are of order 0.2at.%. From [42].

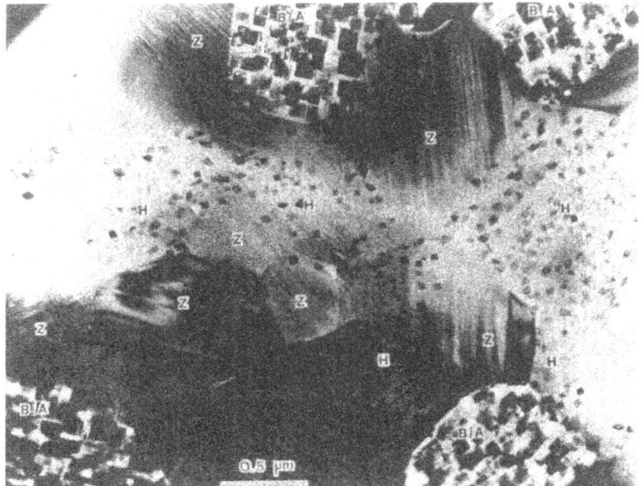

Fig. 16.6. TEM micrograph of TiO_2 (anatase A and brookite B) crystals formed *in-situ* from perovskite ($CaTiO_3$) grains in the surface of a Synroc B ceramic. An ion beam-thinned disc of Synroc B was subjected to hydrothermal attack in water at 190°C for 1 day. The original perovskite grain size was ca 1 μm in diameter. The other less-reacted phases are hollandite (H, $BaAl_2Ti_6O_{16}$) and zirconolite (Z, $CaZrTi_2O_7$).

of the initially-amorphous, reacted layer to crystalline TiO_2. On multiphase ceramic surfaces, a variety of crystalline products can form from solution [e.g. $CaCO_3$, $Ca_3(PO_4)_2$, molybdates etc.] or from recrystallisation in situ. Some of the precipitates can be multilayered on the same nucleation site as different products reach saturation in the solution. Mechanisms and their definition using surface analytical techniques are illustrated in [35,41,44].

16.2.2 Ceramic Surface Layers: Bioceramics

Surface layers of ceramic materials are often applied to metal, oxide and polymer surfaces to impart corrosion resistance, thermal or electrical insulation and, in bioceramic applications, layers that protect the substrate, harden the interface and induce biocompatibility for tissue or bone growth. A variety of methods for applying these ceramic surface layers are now used including: precipitation or sol-gel deposition followed by calcination; physical vapour deposition; chemical vapour deposition; and plasma reactions either in low density, low temperature processes or as high temperature plasma sprays. The composition, particularly stoichiometry and impurity levels, structure, integrity, continuity, conformity, interfacial bonding, interfacial flaws and porosity determine the performance of the ceramic layer and, in turn, the coated device for material in application. For this reason, detailed information is required from surface and analytical techniques in order to optimise the processing parameters and their control.

An example of this characterisation can be found in the coating of metal substrates with silica layers using low temperature plasma deposition from tetraethoxysilane: water vapour mixtures [45,46]. The ceramic layer can be designed to be compositionally and functionally graded through the metal/metal oxide/orthosilicate/pyrosilicate/chain silicate/layer silicate/silica to bulk ceramic material. The plasma-deposited layers are generally < 50 nm thick, continuous, conforming, fully bonded to the substrate and deformable without disbonding. The objectives in the design of these layers were initially to produce metal surfaces protected from oxidation, corrosion and acid attack with improved metal-ceramic bonding but have now been extended to bioceramic titanium-based interfaces for deposition of bioactive hydroxyapatite in improved dental and medical implants.

The Si modified-Auger parameter derived from XPS spectra has been used to determine this graded layer structure and the changing stoichiometry of the SiO_x bonding in the layers. This parameter combines the kinetic energy of the most intense Auger electron emission (KE_A) and the binding energy of the main photoelectron from the same elements (BE_{PE}) in the expression:-

$$\alpha' = KE_A + BE_{PE} . \tag{16.1}$$

This parameter has the advantage that surface charging is automatically removed by the opposing shifts in the two components and it is a very sensitive measure of energy changes due to extra-atomic relaxation related to bonding to next nearest neighbours. An example is given in Fig. 16.7 where the information is presented as a chemical state plot similar to that developed by Wagner [47] giving each of the KE_A and BE_{PE} components as well as the α' values. With increasing Si at.%, as the TEOS:water plasma reaction proceeds, reactive SiO species from the plasma are incorporated into the oxide layer on the nickel substrate forming, initially, orthosilicate (SiO_4^{4-}) and pyrosilicate ($Si_2O_7^{6-}$) structures. Increasingly-linked siloxane structures develop as the reaction proceeds through to complete coverage of bulk anhydrous silica with no evidence of exposed nickel. Once full coverage has been achieved, the dissolution rates of the nickel in pH 2 acid are reduced by two orders of magnitude over periods of 700 h [45].

High resolution TEM examination of the plasma-reacted layer has added important information on the structure and composite nature of this functionally graded layer [48]. TEM specimens were prepared from a thin nickel disc dimpled from one side using a conventional disc grinder followed by chemical polishing of the nickel substrate. The plasma-assisted chemical vapour deposition (PACVD) TEOS: water procedure was then applied to the flat, non-dimpled side of the disc and the treated disc acid etched from the dimpled side until a hole of $\approx 500\,\mu m$ appeared. This thin section was examined using TEM in combination with EDS analysis. Figure 16.8 shows that the outer surface consists of an ≈ 10 nm layer of amorphous silica (i.e. no electron

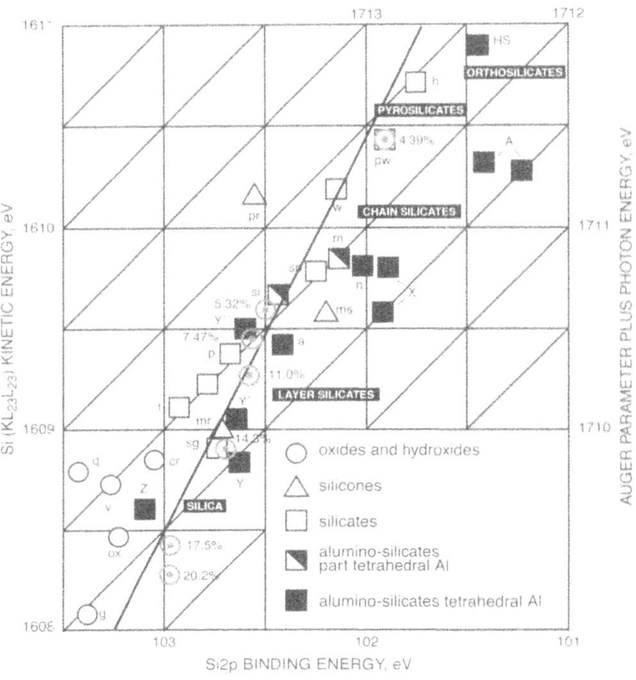

Fig. 16.7. Modified Auger parameter plot for a variety of silicates (from [47]) and TEOS plasma-reacted oxidised nickel surfaces.

diffraction pattern) overlying a darker grey contrast region with embedded distinct crystallites (≈ 10 nm) giving electron diffraction for NiO and EDS analysis showing both nickel and silicon. The reacted region between the nickel substrate and this overlayer shows interpenetration of the oxide and silicate structures between the grains of the original nickel substrate. Hence, the composite nature of the layer locks the reaction products into the substrate to give gradation in physical and chemical properties from the metal substrate to the ceramic surface layer. The silica is shown by both XPS and FTIR to be, somewhat surprisingly, anhydrous but this surface layer can be modified by reaction with the plasma by incorporation of H_2O_2 with the TEOS and water to produce a hydrolysed silica film.

Essentially the same procedure can be applied to a variety of metal substrates including iron, mild steel, stainless steel, brass and titanium. The last substrate is the subject of recent research using the metal-ceramic bonded

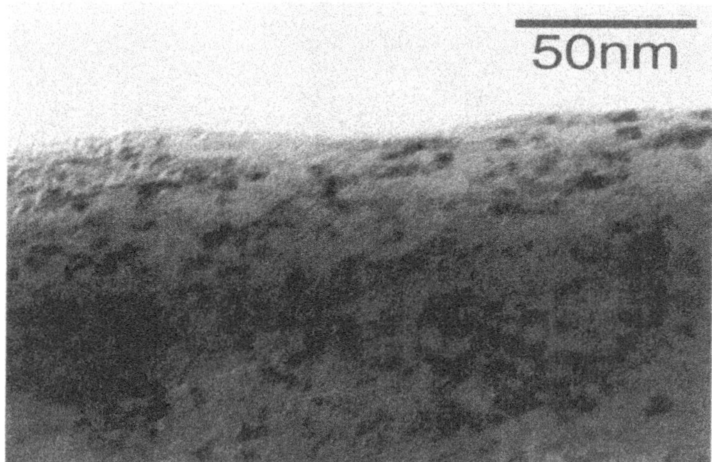

Fig. 16.8. TEM photograph of edge of silica thin film produced on nickel by plasma assisted chemical vapour deposition of TEOS:Water (Etch Type Specimen).

layer as the substrate for further deposition of ceramic hydroxyapatite layers from solution by sol-gel or precipitation methods.

This example may serve to illustrate that a complete understanding of the physical and chemical nature of the PACVD layer could only be obtained by the combination of several techniques; in this case, XPS, TEM, EDS and FTIR.

16.3 Glasses

Surface analysis has been used extensively, particularly SEM, XPS, SIMS and FTIR, to characterise glass surfaces before and after weathering or aqueous attack (e.g. [49–51]). Like ceramics, attack by water at glass surfaces proceeds by four processes: ion exchange in the first few atomic layers; the formation of an amorphous, so-called siliceous layer from base-catalyzed hydrolysis disrupting the network bonding; surface segregation of particular elements into the hydrogen-bonded, hydroxylated siliceous layer; and recrystallisation in-situ or precipitation from solution [41]. These reactions are important in control of glass technology such as surface powdering, mould transfer, opacity, surface hardening and corrosion.

16.3.1 Leached and Recrystallised surfaces

An example of this attack on a zinc-containing glass is presented here to illustrate the type of information leading to elucidation of these mechanisms [50]. A simulated nuclear waste borosilicate glass containing 20.9 wt.% Zn

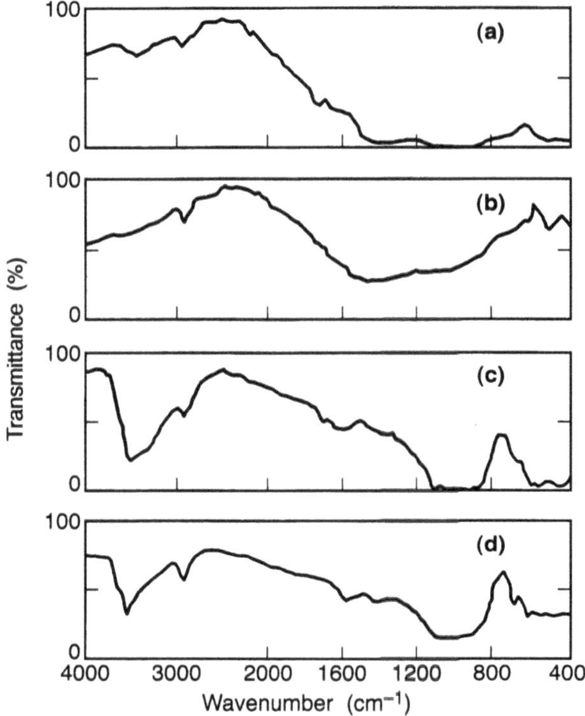

Fig. 16.9. Infrared spectra from the surface of a nuclear waste glass **(a)** initially; **(b)** after evacuation ($\approx 10^{-5}$ Torr) at 170°C for 1 h; **(c)** after hydrothermal reaction at 200°C for 18 h, drying at 200°C for 1 h in air and cooling in a vacuum dessicator; **(d)** sample (c) evacuated at 170°C for 1 h. Note the changes in the hydrogen-bonded $Si(OH)_x$ vibrations in the broad absorptions near 3500–3300 cm^{-1}. The sharp shoulder near 3700 cm^{-1} arises from free (not H-bonded) OH groups. The broad absorption in (a) and (d) near 1640 cm^{-1} is the bending vibration of trapped H_2O molecules in the reacted layer. From [50].

(or 11.7 at.%) with 14 other cations, including Na, Mg, Ca, Fe and Ni, was reacted at 170–200°C in water. Figure 16.9 illustrates IR spectra, from both in-situ (ATR) and detached (transmission) samples of the altered hydrolyzed layer. The initial ν(OH) absorption on the unattacked surface near 3500 cm^{-1} is entirely removed by evacuation at 170°C. After aqueous reaction at 170°C for 60 h, strong, broad, hydrogen-bonded ν(OH) (3200–3500 cm^{-1}) and $\gamma(H_2O)$ (near 1600 cm^{-1}) absorption is found. Evacuation removed some of the $\gamma(H_2O)$ absorption but not the ν(SiO) region around 1000 cm^{-1} [52] indicating possible crystalline products within the siliceous layer.

Figure 16.10 shows XPS profiles for the same glass before and after a relatively short duration of aqueous attack at 200°C for 1 hour. Note that

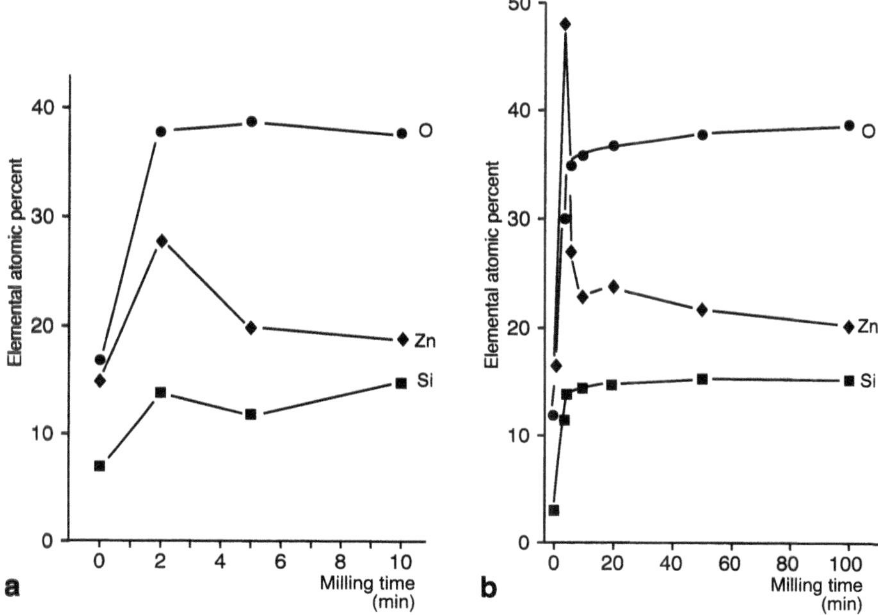

Fig. 16.10. XPS depth profiles ($\approx 1\,\mathrm{nm\,min^{-1}}$) of O, Zn and Si in the surface of a polished disc of a zinc-containing nuclear waste glass **(a)** before and **(b)** after hydrothermal attack in water at 200°C, 1 h. Note change of depth scale from (a) to (b). From [50].

the Si and O signals are relatively unaltered in the siliceous layer but there is a strong accumulation of Zn in the first 10 nm of the surface. SIMS profiles of longer duration attack have shown Zn accumulation in the surface to depths greater than 100 nm. Other elements, notably Mg, Fe and Ni, are also found segregated into the siliceous layer in this and other glasses after aqueous attack whilst the alkali metal cations (e.g. Na^+, K^+, Rb^+) are almost always lost from the reacted layer.

SEM studies of the surface after extensive reaction show a relatively uniform (conducting) surface layer with small crystallites embedded in it as illustrated in Fig. 16.11. TEM with electron diffraction, however, established that more than 90% of the layer is amorphous. The crystalline forms were indexed to the mineral zinc hemimorphite $Zn_4(OH)_2Si_2O_7.H_2O$. In other zinc-containing glasses, different crystal forms: e.g. Zn_2SiO_4 have been identified. No crystalline forms containing the segregated elements of Fe, Mg or Ni were found so that it appears that they are at least partly preferentially bound in the siliceous layer instead of dissolving in the solution.

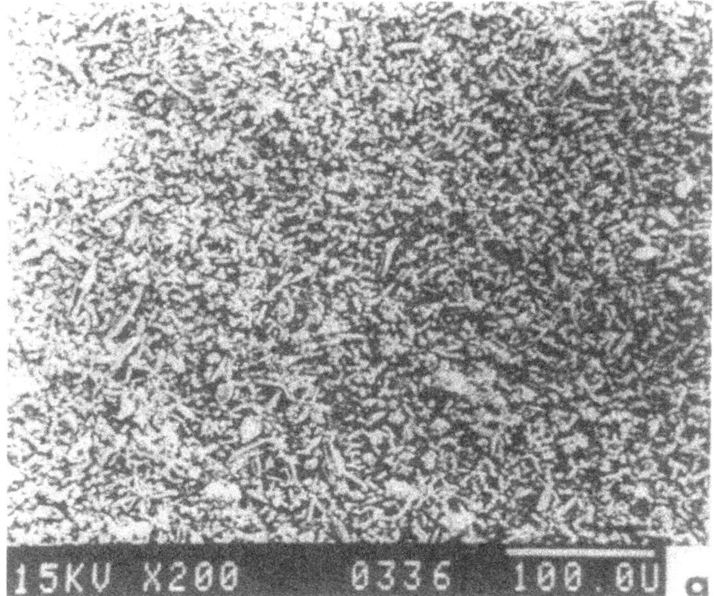

Fig. 16.11. SEM micrograph of the surface of a polished disc of a zinc-containing glass after hydrothermal attack in water at 170°C for 60 h, dried 170°C, 1 h. The white bar in the micrograph is 100 μm.

Even more complex behaviour of different elements can be found in other glass systems. This combination of techniques, however, can give a full description of the changes in composition and chemistry of the surface layers.

16.3.2 Surface Modification of Glass Surfaces

So-called "fibre-glass" generally uses E-glass fibres to reinforce polymer composites. The glass fibres are subjected to a surface modification step to enhance bonding between the fibre and the polymer since, without this surface modification, diffusion and reaction of water at the glass-matrix interface may lead to delamination and degradation of the mechanical properties of the composite. A variety of coupling agents, usually based on silane or siloxane reagents, are commonly applied from dilute aqueous suspensions, partial hydrolyzates or from organic solvents. Even before reaction with the coupling agents, however, the glass surface undergoes some reaction in processing or hydrolysis in air causing the surface composition to be significantly different from the nominal composition of the glass (Table 16.4). For the fibres, a silica-rich surface layer was measured with decreased atomic concentrations of all elements, particularly boron, sodium, magnesium, potassium and iron which were of negligible concentration on either glass fibre or plate surfaces compared with the bulk glass.

Table 16.4. The composition of E-glass as determined from the components used in the manufacturing process and by XPS. The XPS percentages were adjusted so that the carbon contamination was taken as zero [53].

Element	Composition (At.%)		
	Manufacture	XPS	
		Fibre	Plate
O	61.7	66.2	72.7
Si	18.8	21.9	17.9
Ca	7.9	7.1	2.6
Al	5.9	4.3	4.7
Na/K	1.1	0.5*	0.4*
B	4.4	< 0.05	1.7
Mg	0.3	< 0.05	< 0.05

Atomic percentages of sodium only; no potassium was found on the glass fibre or plate surface

Combined studies, using XPS, DRIFT and contact angle measurements, have been made on surface modification of E-glass using silanes, siloxanes and long-chain alcohol reagents [35]. The thickness of the overlayer can be determined by a relatively simple XPS measurement of the intensity of the Ca $2p$ emission before and after reagent adsorption. The calcium present in the E-glass fibre surface can be used to determine this thickness (t) using the expression

$$t = -\lambda_{\mathrm{Ca}} \ln\left[I_{\mathrm{Ca}}/I^*_{\mathrm{Ca}}\right] 2/\pi \qquad (16.2)$$

where I_{Ca} is the measured calcium concentration on the treated sample, I^*_{Ca} is the calcium concentration for the E-glass surface before treatment and λ_{Ca} is the attenuation length of the Ca $2p$ photoelectron. The factor $2/\pi$ accounts for the curvature of the fibre which can increase the effective path length for photoelectrons in the surface layer. A variety of functionalised siloxanes with different numbers of polymer repeat units have been studied in this way producing layers varying in thickness from 1.2 to 4.1 nm [35]. Confirmation of the layer structure can be obtained from the inverse correlation of the C $1s$ and Ca $2p$ elemental concentrations in static depth profiling using an argon ion beam.

Long-chain (i.e. C_{12}, C_{14}, C_{18}) alcohols have been shown to strongly adsorb on glass surfaces and may offer a cheaper alternative to silanes for some applications. Reacted with E-glass fibres at 130°C, it is found that these alcohols adhere to the glass surface after more than 10 extraction processes using hexane or cyclohexane. Advancing contact angles in the range 85–97° are measured on these surfaces corresponding to hydrophobic methylene or polyethylene-like surfaces. Quantitative DRIFT measurements correspond to surface coverage of 1–5 monolayers from comparison of methylene peak intensities with known alcohol concentrations on standards [35]. However, angle resolved XPS analysis of C 1*s* and Si 2*p* signals shows that full coverage of the glass surface is unlikely and that instead disordered multilayers in islands is more likely.

A similar study [54] of a non-ionic alcohol, namely alkylphenylpoly(oxyethylene), frequently used in sizing blends for glass fibre processing, reached similar conclusions from a combination of XPS, streaming potential and Wilhelmy measurements. The surface properties are substantially modified with only a fraction of the surfaces covered by these surfactants. Surface exposure of Si in XPS spectra reaches a plateau at a level corresponding to 30% of the clean surface concentration corresponding to a less negative (but now unchanging) streaming potential. The advancing contact angle increases only slightly from 43° to 48° and there is significant hysteresis in the wetting. AFM studies of similar systems of adsorbed layers of silanes, siloxanes and alcohols on glass surfaces have also supported patchwise adsorption in relatively thick, multilayered regions. This is in sharp contrast to adsorption of the same long chain alcohols on silica where < 3 molecules nm^{-2} at sub-monolayer coverage are adsorbed.

References

1. K. Kuys, J. Ralston, R.St.C. Smart, S. Sobieraj, R. Wood, P.S. Turner: Surface characterisation, iron removal and enrichment of bauxite ultrafines. Miner. Eng. **3**, 421 (1990)
2. R.St.C. Smart, B.G. Baker, P.S. Turner: Aspects of surface science in Australian industry. Aust. J. Chem. **43**, 241 (1990)
3. A.N. Buckley: The application of X-ray photoelectron spectroscopy to flotation research. Colloids and Surf. **93**, 159–172 (1994)
4. I. Kartio, K. Laajalehto, E. Suoninen: Application of electron spectroscopy to characterization of mineral surfaces in flotation studies. Colloids and Surf. **93**, 149–158 (1994)
5. J.R. Mycroft, H.W. Nesbitt, A.R. Pratt: X-ray photoelectron and Auger electron spectroscopy of air-oxidized pyrrhotite: distribution of oxidized species with depth. Geochim. Cosmochim. Acta. **59**, 721–733 (1995)
6. R.St.C. Smart: Surface layers in base metal sulfide flotation. Miner. Eng. **4**, 891–909 (1991)
7. S.L. Chryssoulis, K.G. Stowe, F. Reich: Characterisation of composition of mineral surfaces by laser-probe analysis. Trans. Inst. Min. Metall. Sect. C **101**, c1–6 (1992)
8. J.S. Brinen, S. Greenhouse, D.R. Nagaraj, J. Lee: SIMS and SIMS imaging studies of adsorbed dialkyl dithiophosphinates on PbS crystal surfaces. Int. J. Miner. Process. **38**, 93–109 (1993)
9. E. Suoninen, K. Laajalehto: Structure of thiol collector layers on sulfide surfaces. Proc. Int. Miner. Process. Congr., 15th (Sydney, Aust.). Australian Institute of Mining and Metallurgy, Melbourne **3**, 625–629 (1993)

10. J.M. Cases, P. de Donato, M. Kongolo, L. Michot: An infrared investigation of amyl xanthate adsorption by pyrite after wet grinding at natural and acid pH. Colloids and Surf. **36**, 323–338 (1989)
11. A.N. Buckley, I.C. Hamilton, R. Woods: Investigation of the surface oxidation of sulphide minerals by linear potential sweep voltammetry and X-ray photoelectron spectroscopy, in *Flotation of Sulphide Minerals*, ed. K.S. Forssberg (Elsevier, Amsterdam 1985) pp.41–60
12. J. Mielcarzski: XPS study of ethyl xanthate adsorption on oxidized surface of cuprous sulfide. J. Colloid Interface Sci. **120**, 201–209 (1987)
13. R.St.C. Smart, J. Amarantidis, W. Skinner, C.A. Prestidge, L. LaVanier, S. Grano: Surface analytical studies of oxidation and collector adsorption in sulfide mineral flotation, in *The Solid-Liquid Interface: A Surface Science Approach*, (Eds K. Wandelt, J. O'Connor and S.M. Thurgate) (Springer Verlag, Berlin 2001), in press
14. S. Grano, J. Ralston, R.St.C. Smart: Influence of electrochemical environment in the flotation behaviour of sulphide ores. Int. J. Miner. Proc. **30**, 69 (1990)
15. B.S. Kim, R.A. Hayes, C.A. Prestidge, J. Ralston, R.St.C. Smart: In-situ scanning tunneling microscopy studies of galena surfaces under flotation related conditions. Colloids and Surf. A, **117**, 117–129 (1996)
16. A.R. Gerson, A.G. Lange, K.P. Prince, R.St.C. Smart: The mechanism of copper activation of sphalerite. Appl. Surface Sci., **137**, 207–223 (1999)
17. K. Laajalehto, R.St.C. Smart, E. Suoninen. STM and XPS investigation of reaction of galena in air. Appl. Surf. Sci. **64**, 29–39 (1993)
18. B.S. Kim, R.A. Hayes, C.A. Prestidge, J. Ralston, R.St.C. Smart: Scanning tunneling microscopy studies of galena: the mechanism of oxidation in air. Appl. Surf. Sci. **78,** 385–397 (1994)
19. B.S. Kim, R.A. Hayes, C.A. Prestidge, J. Ralston, R.St.C. Smart: Scanning tunneling microscopy studies of galena: the mechanisms of oxidation in aqueous solution. Langmuir **11**, 2554–2562 (1995)
20. G. Wittstock, I. Kartio, D. Hirsch, S. Kunze, R. Szargan: Oxidation of galena in acetate buffer investigated by atomic force microscopy and photoelectron spectroscopy. Langmuir, **12**, 5709 (1996)
21. C.M. Eggleston, M.F. Hochella, Jr: Scanning tunneling microscopy of sulfide surfaces. Geochim. Cosmochim. Acta. **54**, 1511 (1990)
22. C.M. Eggleston, M.F. Hochella, Jr: Scanning tunneling microscopy of galena (100) surface oxidation and sorption of aqueous gold. Science **254**, 983 (1991)
23. C.M. Eggleston, M.F. Hochella, Jr: Scanning tunneling microscopy of pyrite (100) surface structure and step reconstruction. Am. Mineral. **77**, 221 (1992)
24. C.M. Eggleston, M.F. Hochella, Jr: The structure of hematite (001) surfaces by scanning tunneling microscopy. Am. Mineral. **77**, 911 (1992)
25. C.M. Eggleston, M.F. Hochella, Jr: Scanning tunneling microscopy applied to PbS (001) surfaces. Am. Mineral. **78**, 877 (1993)
26. C.M. Eggleston, M.F. Hochella, Jr: in *The Environmental Chemistry of Sulphide Oxidation*, ACS Symposium Series, 550, eds C.N. Alpers and D.W. Blowes, American Chemical Society, Washington DC, 201 (1994)
27. M.F. Hochella, Jr, C.M. Eggleston, V.B. Elings, G.A. Parks, G.E. Brown Jr, C.M. Wu, NS k. Kjoller: Mineralogy in two dimensions: scanning tunneling microscopy of semiconducting minerals with implications for geochemical reactivity. Am. Mineral. **74**, 1235 (1989)

28. M.F. Hochella, Jr: in *Mineral Surfaces*, eds. D.J. Vaughan and R.A.D. Pattrick, Min. Soc. Series No. 5, (Chapman and Hall, London 1995) p.17
29. R.J. Atkinson, R.L. Parfitt, R.St.C. Smart: Infrared study of phosphate adsorption on goethite. J. Chem. Soc., Faraday **I 70**, 1472 (1974)
30. R.R. Martin, R.St.C. Smart: X-ray photoelectron of anion adsorption on goethite. Soil Sci. Soc. Am. J. 51, 1 (1987)
31. D.J. Vaughan and R.A.D. Pattrick (eds). Mineral Surfaces, Min. Soc. Series No. 5, (Chapman and Hall, London 1995)
32. W.H. Casey: ref. [31], p185ff
33. J.F. Banfield, B.F. Jones, D.R. Veblen: An AEM-TEM study of weathering and diagenesis, Abert Lake, Oregon II: diagenetic modification of the sedimentary assemblage. Geochim. Cosmochim. Acta, **55**, 2795–810 (1991)
34. R.A. Eggleton: Formation of iddingsite rims on olivine: a transmission electron microscope study. Clays Clay Miner., **32**, 1–11 (1984)
35. R.St.C. Smart with P. Arora, B. Braggs, H.M. Fagerholm, T.J. Horr, D.C. Kehoe, J.G. Matisons, J.B. Rosenholm, J. Ralston: Modification of Oxide, Glass, Mineral and Metal Surfaces Using Adsorption, Plasma and Sol-Gel Reactions, in *Interfaces of Ceramic Materials: Impact on Properties and Applications*, eds. K. Uematsu, Y. Moriyoshi, Y. Saito and J. Nowotny, Trans Tech Publications, 361–404 (1996)
36. R.L. Segall, R.St.C. Smart, P.S. Turner, T.J. White: Microstructural characterisation of Synroc C and E by electron microscopy. J. Am. Ceram. Soc. **68**, 64 (1985)
37. J.A. Cooper, D.R. Cousens, J. Hanna, R.A. Lewis, S. Myhra, R.L. Segall, R.St.C. Smart, P.S. Turner, T.J. White: Intergranular films and pore surfaces in Synroc C: structure, composition and dissolution characteristics. J. Amer. Ceram. Soc. **69**, 347 (1986)
38. T.J. White, R.L. Segall, P.S. Turner: Radwaste immobilisation by structural modification the crystallochemical properties of Synroc, a titanate ceramic. Angew. Chemie (Int. Ed. Engl.) **24**, 357 (1985)
39. P.E. Fielding, T.J. White: Crystal chemical incorporation of high-level waste species in aluminotitanate-based ceramics: Valence, location, radiation damage and hydrothermal durability. J. Mater. Res. **2**, 387 (1987)
40. W.J. Buykx, K. Hawkins, D.M. Levins, H. Mitamura, R.St.C. Smart, G.J. Stevens, K.G. Watson, D. Weedon, T.J. White: Titanate ceramics for the immobilization of sodium-bearing high-level nuclear waste. J. Am. Ceram. Soc. **71**, 678 (1988)
41. S. Myhra, R.St.C. Smart, P.S. Turner: The surfaces of titanate minerals, ceramics and silicate glasses: surface analytical and electron microscope studies. Scanning Microsc. **2**, 715 (1988)
42. S. Myhra, A. Atkinson, J.C. Riviere, D. Savage: A surface analytical study of Synroc subjected to hydrothermal attack. J. Am. Ceram. Soc. **67**, 223 (1984)
43. T. Kastrissios, M. Stephenson, P.S. Turner, T.J. White: Hydrothermal dissolution of perovskite: implications for Synroc formulation. J. Amer. Ceram. Soc. **70**, C144 (1987)
44. S. Myhra, D.K. Pham, R.St.C. Smart, P.S. Turner: Surface reaction and dissolution of Ceramic and high temperature superconductors, in *Science of Ceramic Interfaces*, ed. J. Nowotny Mat. Sci. Monographs, **75**, (Elsevier, Amsterdam 1991) p.569

45. P.S. Arora, R.St.C. Smart: Formation of silicate structures in oxidised nickel surfaces using low temperature plasma reaction. Surf. Interface Anal. **24**, 539–548 (1996)
46. M. Steveson, P.S. Arora, R.St.C. Smart: XPS studies of low temperature plasma-produced graded oxide-silicate-silica layers on titanium. Surf. Interface Anal **26**, 1027–1034 (1998)
47. C.D. Wagner: Faraday Discuss. Chem. Soc., **60**, 291 (1975)
48. R.St.C. Smart, P.S. Arora, M. Steveson, N. Kawashima, G.P. Cavallaro, H. Ming, W.M. Skinner: New approaches to metal-ceramic and bioceramic interfacial bonding, in *Ceramic Interfaces: Properties and Applications II*, ed. S-J Kang, H-I Yoo and J. Nowotny, Inst. Materials Press (UK), 293–326 (2000)
49. D.E. Clark, L.L. Hench: An overview of the physical characterisation of leached glass surfaces. *Nucl. Chem. Waste Manage.* **2**, 93 (1981)
50. R.A. Lewis, S. Myhra, R.L. Segall, R.St.C. Smart, P.S. Turner: The surface layer formed on zinc-containing glass during aqueous attack. J. Non-Cryst. Solids **53**, 299 (1982)
51. N.S. McIntyre, G.C. Strathdee, D.F. Phillips: SIMS studies of the aqueous leaching of a borosilicate glass. Surf. Sci. **100**, 71 (1980)
52. V.C. Farmer (ed.): The infrared spectra of minerals. Mineral. Soc. Monograph 4, London 285–303 (1974)
53. H.M. Fagerholm, J.B. Rosenholm, T.J. Horr, R.St.C. Smart: Modification of E-glass fibres by long chain alcohol adsorption. Colloids and Surf. **110**, 11–22 (1996)
54. H.M. Fagerholm, C. Lindsjo, J.B. Rosenholm, K. Rokman: Colloids and Surf. **69**, 79 (1992)

17 Characterization of Catalysts by Surface Analysis

N.K. Singh and B.G. Baker

Solid catalysts are the basis of many important industrial processes. Reaction of gases or liquids to form particular products occurs at specific sites on the catalyst surface. The structure and composition of the catalyst surface is critical in determining the reactivity and selectivity of a catalyst. The techniques of surface analysis provide the means of characterizing a catalyst in terms of the actual composition and structure of the surface rather than by its bulk properties. The objective of such studies is to provide a scientific basis for improving catalyst formulations and understanding the processes of activation and deactivation which the catalyst undergoes. Supported catalysts, the type most widely used in industry, consist of an active component dispersed on the internal surface of a porous inorganic oxide. High area solids and light loadings very highly dispersed are often employed to maximize the catalytic activity of expensive components. Metals and oxides may be formed on a support by decomposing or reducing a salt which has been introduced by solution impregnation. For studies of model catalysts under UHV conditions metals can be deposited in situ on to oxide surfaces usually formed on metal substrates. The preparation, pre-treatment and actual catalytic reaction conditions may result in reaction between the components including the support. It is the purpose of the surface analysis to reveal these processes.

Specific objectives of the surface analysis of catalysts are:

1) to determine the surface composition of the catalyst in reactive form;
2) to identify the valence state of elements present at the surface of the active catalyst, and to identify the catalytically active sites;
3) to monitor the extent of interaction between the catalyst components and the support and possibly explain the effects of varying the support;
4) to determine the effects on surface composition of various catalyst preparation and pretreatment procedures.

The most useful techniques are X-ray photoelectron spectroscopy, XPS (Chap. 6), Auger electron spectroscopy, AES (Chap. 6), and ion scattering spectroscopy (ISS). Others include SIMS (Chap. 5), FT–IR (Chap. 8), RBS and NRA (Chap. 9), LEIS (Chap. 11), LEED (Chap. 13) and UPS (Chap. 14). The application of these surface techniques to the study of catalysts poses a number of experimental problems:

i) The active components of the catalyst are generally present in low concentration and are at the surface of a high area porous solid. Most of this surface is internal and therefore not directly accessible to photon, ion or electron beams.

ii) The catalyst support is generally an insulating solid which tends to charge electrostatically during the analysis. This has the effect of displacing peaks and makes the determination of chemical shift effects more difficult.

iii) The active form of the catalyst persists only in a controlled environment and will not stand exposure to the atmosphere.

To overcome these problems the spectrometer for catalyst studies should have a sample preparation chamber which provides for heating, reactant gas exposure and vacuum transfer to the analysis chamber. The method of mounting the sample needs to be compatible with these processes and particular attention must be given to the effects of electrostatic charging in order to quantify chemical shift effects.

Whilst all techniques noted above are excellent for characterising catalysts materials, a combination of XPS and FT–IR is particularly powerful, as they provide complementary information. XPS provides information on the chemical environment of the catalyst components. Infrared spectroscopy, by virtue of its inherent sensitivity to chemical functional groups and their conformations within a molecule or particular sites in framework structure, is ideally suited for identifying adsorption sites on catalyst surfaces, and following catalytically controlled reactions. A UHV spectrometer, equipped to perform XPS, AES UPS and ISS and which has been modified for catalyst studies is shown in Fig. 17.1. Based on the Leybold LHS10 electron spec-

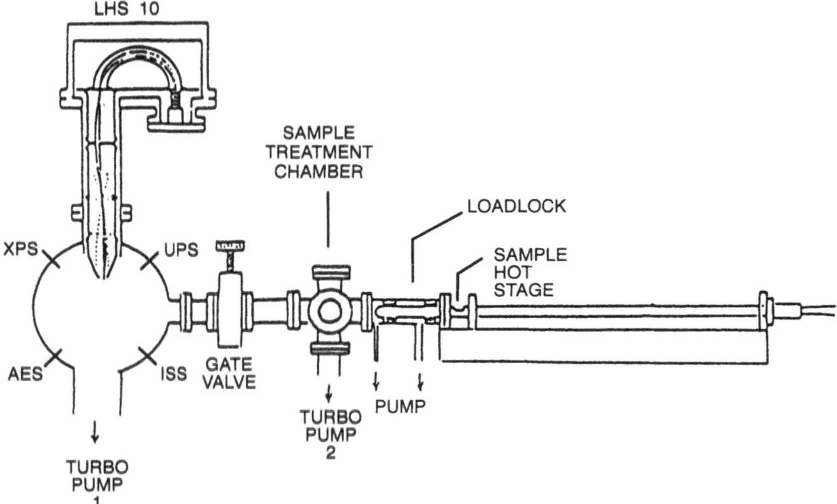

Fig. 17.1. Electron spectrometer (Leybold) with sample treatment chamber

trometer, this system has a rod with a heated stage [1]. The sample may be heated in the treatment chamber in either a gas flow or static atmosphere to create conditions comparable to practical service of the catalyst. High vacuum is then achieved before opening the gate valve and transferring to the spectrometer chamber. The sample stage can be heated during analysis if required. Other systems provide for transfer of a sample holder and allow the treatment chamber to be isolated during analysis. Systems combining a micro-reactor with facilities for kinetic studies and electron spectroscopies have also been employed [2]. Vibrational spectroscopy requiring UHV conditions, especially monitoring of reactions on single crystal transition metal and semiconductor surfaces, are also performed in similar spectrometers but incorporating oppositely mounted differentially pumped infrared transparent windows. One such spectrometer is described in reference [3], which is set up for Fourier transform reflection absorption infrared spectroscopy (FT–RAIRS) (also known as infrared absorption spectroscopy, IRAS), LEED and AES. In RAIRS the infrared beam is focussed from a source, using focussing mirrors, at a glancing angle onto a surface held inside an ultrahigh vacuum chamber through a KBr window. The reflected beam from the surface is focussed, again using mirrors, through another KBr window onto an external liquid nitrogen cooled MCT detector. The KBr windows, the input and the output optics, including the detector, are all contained within cells. The two cells are continually purged with dry nitrogen gas to minimise absorption of infrared beam by atmosphere. The reader is referred to articles by Greenler [4,5], Pritchard [6] and Horn [7] for a more detailed description of this particular mode of vibrational spectroscopy. High surface area catalysts do not require UHV environment, and hence these studies are usually performed in high vacuum utilizing glass cells / compartments or under an inert gas environment, as discussed in Chap. 8.

The objective in each experimental set-up is to examine catalysts in their reactive form. The mounting of powder or granular samples of catalysts for analysis must allow for good thermal and electrical contact with the holder and ensure mechanical stability so that sample is not spilled during evacuation. An effective method is to press a pellet of sample in a small stainless steel cup. The top of the pellet can be cleaved off to expose fresh surface which has not contacted the plunger of the press. The cup is then mounted on the stage of the spectrometer providing good electrical and thermal contact. The latter is not a problem when oxide thin films, e.g. Al_2O_3, are formed on a metal or semiconductor substrates in vacuum. A fresh film is formed prior to XPS/RAIRS measurements which then eliminates the cleaning, cleaving requirements. In XPS, an insulating sample tends to acquire a positive charge which decreases the observed kinetic energy of the photoelectrons. This shift is in the same direction as a chemical shift where the element has a more positive valence state. Catalysts frequently show both effects and in addition often exhibit differential charging, i.e., some parts of the sample charge

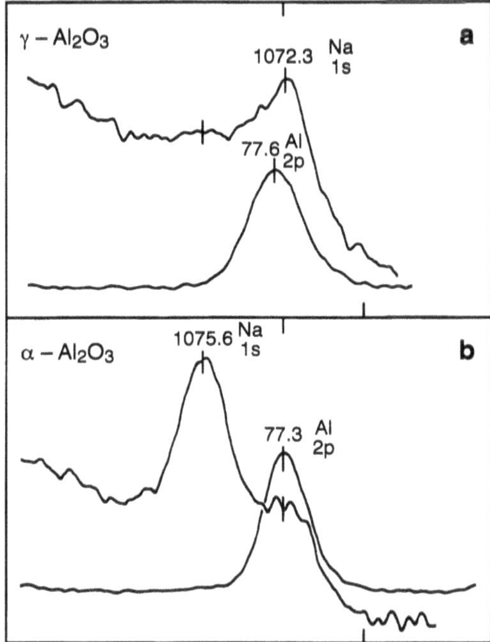

Fig. 17.2. X-ray photoelectron spectra for sodium ($1s$) and aluminium ($2p$) in (**a**) γ-Alumina (Merck) and (**b**) heat treated (1200° C, 30 min) alumina. The apparent binding energies are displaced $\sim$ 3.4 eV except for the sodium peak in the initial γ-alumina. Sodium in this material is present in a β-alumina surface phase which conducts. Heat treatment destroys the β-alumina leaving sodium compounds on the surface

while others are uncharged. This can be of use in identifying particular elements as being in combination in a separate phase. Charging is detected by comparison of observed energies with values tabulated for reference peaks in the spectrum. A low energy electron "flood gun" may also be used to directly neutralize charge at a surface. This is particularly useful in cases of differential charging.

17.1 Examples of Catalytic Systems Studied by XPS

17.1.1 Alumina

A variety of forms of alumina are used as catalyst supports. The valence state of aluminium is invariant at +3 and the oxide is insulating. Analysis by XPS provides a good measure of the extent of charging. This analysis also reveals surface impurities which are likely to be incorporated into a catalyst prepared on this support. Gamma alumina prepared by the Bayer process has an area of $\sim 70\,m^2\,g^{-1}$ and typically contains about 0.2 percent sodium as impurity. This sodium is readily detected by XPS showing that it must be concentrated at the alumina surface [8]. The spectra in Fig. 17.2a show Al(2p) and Na(1s) on separate scales. The binding energy for Al(2p) at 77.6 eV indicates charging of $\sim$ 3.4 eV, typical of Al in alumina. The binding energy for Na(1s) at 1072.3 eV equals the reference value for uncharged sodium. This extraordinary example of differential charging is explained by structural identification of β-alumina ($Na_2O.11Al_2O_3$) by electron diffraction. This exists as a surface phase in this type of γ-alumina. It is an electrolytic conductor and thus no local charging of sodium occurs. The Al(2p) spectrum is dominated by the more abundant bulk Al_2O_3 and so is charged. Heat treatment of this alumina causes a phase change to produce α-alumina and decompose the β-alumina. Spectra after heating, Fig. 17.2b, show both Al(2p) and Na(1s) charged by $\sim$ 3.4 eV. The sodium in the heat-treated alumina exists now as islands of soluble sodium oxide on the reconstructed alumina surface.

17.1.2 Tungsten Oxide Catalysts

Catalysts containing tungsten oxide in a partially reduced state have been shown to have activity for the skeletal isomerization of alkenes [9]. Conditioning and operation of the catalyst requires a controlled reducing atmosphere achieved by hydrogen and water vapour. The chemistry of tungsten in this catalyst has been studied by following the $4f$ peaks in XPS [1]. Unsupported WO_3 is semiconducting and gives no charging problems. The binding energy of the $4f_{7/2}$ peak is 35.2 eV compared with 30.2 eV for tungsten metal. The chemical shift of 5.0 eV results from the change from zero to +6 valence state. Reduction of WO_3 in wet hydrogen results in a shift of the $4f$ peaks to lower binding energies (Fig. 17.3). Intermediate stages of reduction show mixed spectra indicating sample heterogeneity but a number of intermediate oxides of tungsten $W_{20}O_{58}$, $W_{18}O_{49}$ are known and the blue colour of these oxides is observed. The reduction by this method is limited to WO_2.

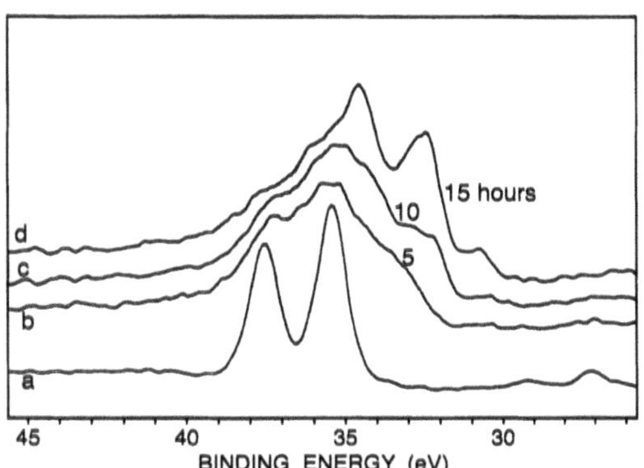

Fig. 17.3. XP spectra of unsupported WO_3 reduced in wet hydrogen ($pH_2/pH_2O = 40$) at 405°C. (a) WO_3 before reduction: (b) after 5 h reduction: (c) after 10 h reduction: (d) after 15 h reduction

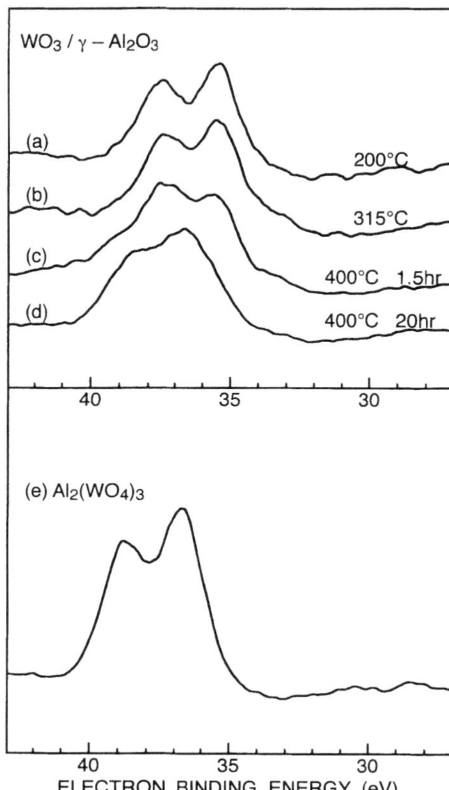

Fig. 17.4. XP spectra of a sample of 6% $WO_3/\gamma\text{-}Al_2O_3$ heated in air. (*a*) Heating to 200°C; (*b*) heating to 315°C: (*c*) heating at 400°C for 1.5 h: (*d*) heating at 400°C for 20 h: (*e*) XP spectrum of $Al_2(WO_4)_3$

The supported form of the catalyst consists of 6 percent loading of WO_3 on alumina. The final statestage of the preparation involves heating in air to decompose tungstic acid. This preparation step has also been followed by successive XPS analyses (Fig. 17.4). Prolonged heating at 400°C results in a shift of the $4f$ peaks to even higher binding energy (Fig. 17.4d). This chemical shift is not due to a change of valence state but to the change of tungsten from the octahedral sites of WO_3 to the tetrahedral sites of aluminium tungstate $Al_2(WO_4)_3$. The tungsten oxide which has reacted with the support is not readily reduced. Spectra in Fig. 17.5 show that reduction in dry hydrogen at 400°C reduces little of the sample from Fig. 17.4d whereas the sample (Fig. 17.4a) can be reduced to metallic tungsten. Aluminium tungstate is not a catalyst for the reaction and catalysts overheated in preparation had poor activity.

Some charging effects were observed in spectra from supported tungsten catalysts. In Fig. 17.6, the spectrum (*i*) from unreduced WO_3 on alumina shows three peaks. Operation of an electron flood gun shows that only the two peaks at lower kinetic energies are from charged regions of the sample.

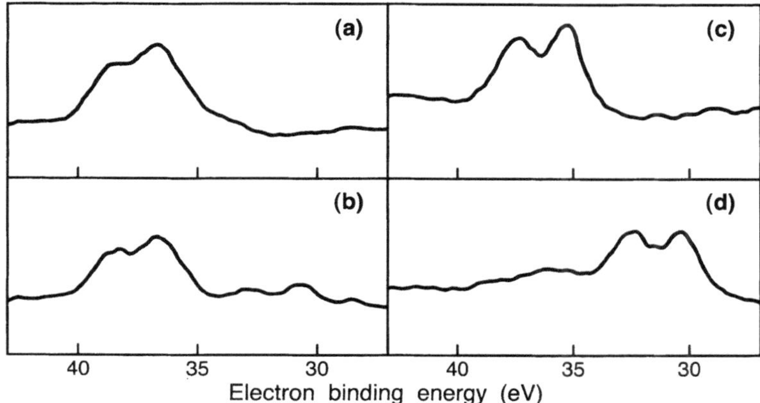

Fig. 17.5. XP spectra (**a**) of sample from Fig. 17.4d after exposure to wet hydrogen at 400°C. (**b**) Subsequent exposure to dry hydrogen at 400°C (15% reduction). (**c**) of sample from Fig. 17.4a heated in air to only 200°C. (**d**) Subsequent exposure to dry hydrogen at 400°C (75% reduction)

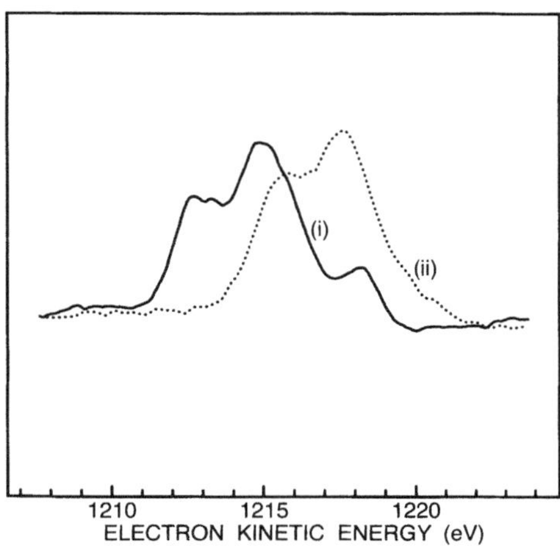

Fig. 17.6. A sample of unreduced WO_3/HT-Al_2O_3 (air, 450°C) serves as an example of how charging by X-rays can complicate a spectrum. (*i*) Spectrum recorded without using the flood gun. (*ii*) Flood gun used to discharge sample: 2.7 eV charge (see text)

The recognition of multiple charging effects is obviously important before interpreting spectra in terms of elemental valence states. However, spectra recorded with the flood gun operating are broadened and, as the setting of the flood gun is arbitrary, peak positions are indefinite.

Ion scattering spectroscopy may also be applied to the study of catalysts. The technique requires the analysis of the energies of positive ions and so requires the analyzer to operate with potentials the reverse of those used for electron spectroscopy. The incident beam from an ion gun is typically 1 keV helium ions. The loss of energy of the ion in a single binary elastic collision depends on the mass of the surface atom. The spectrum is recorded

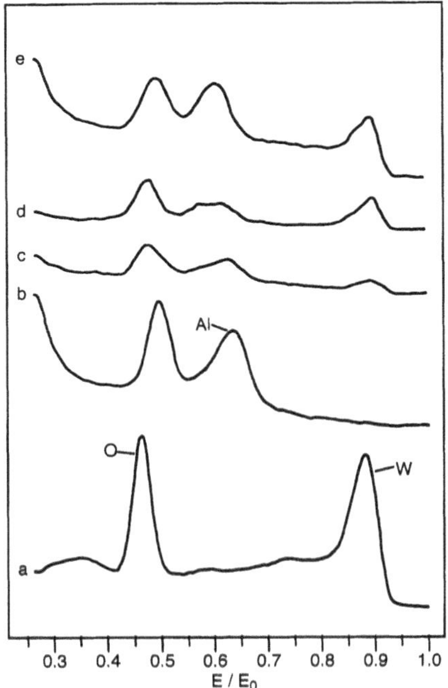

Fig. 17.7. He^+ ($E_0 = 1\,kV$) ion scattering spectra of (*a*) bulk WO_3; (*b*) HT-Al_2O_3: (*c*) 6% WO_3/HT-Al_2O_3, untreated; (*d*) 6% WO_3/HT-Al_2O_3, heated in air at 400°C for 20 h: (*e*) bulk $Al_2(WO_4)_3$

as E/E_0, where E is the measured kinetic energy of the scattered ion and E_0 the incident energy. Electrostatic charging of insulators is a problem in applying the technique to catalysts and there are difficulties in quantifying the data [10]. The technique is however more surface sensitive than XPS and is of particular use in defining the location of a component on a catalyst support.

Analysis of the tungsten oxide catalyst system using ion spectroscopy is shown in Fig. 17.7. Oxygen, aluminium and tungsten are identified. The relative intensities of tungsten and aluminium in the prepared catalyst indicate that tungsten oxide is well dispersed on the support effectively covering the surface.

17.1.3 Palladium on Magnesia

The following experiment to test the reactivity of palladium metal on magnesium oxide shows the nature of the metal–support interaction for a catalyst based on these materials. A thin film of palladium was evaporated on to a single crystal magnesium oxide surface. The sample was heated in vacuum at 400°C and then argon ion milled to the metal-oxide interface. Analyses

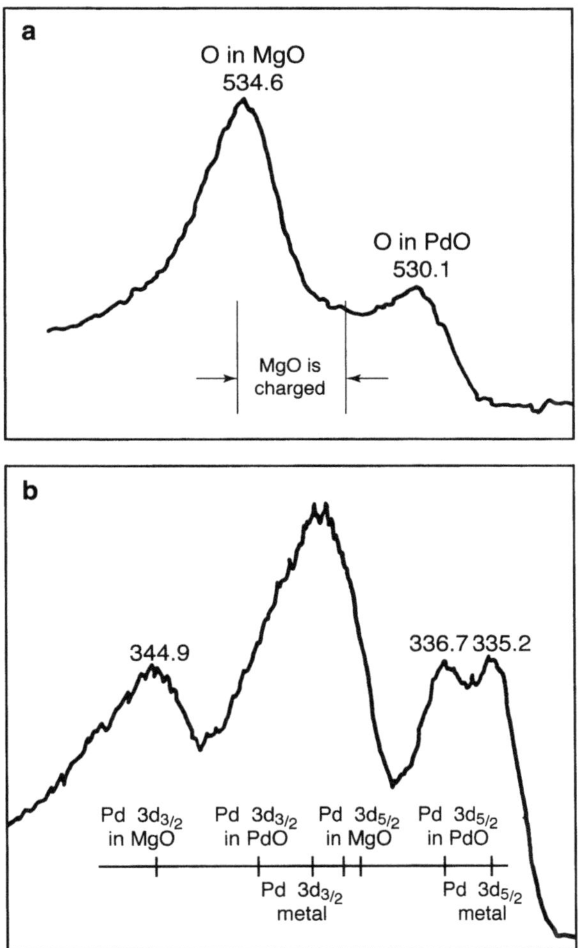

Fig. 17.8. X-ray photoelectron spectra of a palladium film deposited on single crystal magnesium oxide after heating and after argon ion milling to the metal-oxide interface. (**a**) Oxygen in MgO distinguished from oxygen in PdO by chemical shift and by electrostatic charging. (**b**) Palladium $3d_{3/2}$ and $3d_{5/2}$ peaks showing chemical shift due to oxide formation and identifying Pd in the charged MgO

by XPS (Mg $K\alpha$) are shown in Fig. 17.8. Oxygen 1s peaks from PdO and MgO differ in binding energy by 1.8 eV for chemical reasons. The observed spectra in Fig. 17.8a differ by 4.5 eV due to electrostatic charging of the MgO only. The complicated palladium 3d spectrum in Fig. 17.8b is interpreted as a superposition of three sets of the $3d_{3/2}$ $3d_{5/2}$ doublet; the metal, palladium oxide which is conducting and palladium in the insulating magnesium oxide.

17.1.4 Cobalt on Kieselguhr Catalysts

This type of catalyst is used for the Fischer-Tropsch synthesis of hydrocarbons from carbon monoxide and hydrogen. Under these reducing conditions, cobalt may be reduced to the metallic state. However, promotors such as MgO and ThO_2 are present and may interact with cobalt. The purpose of a surface analysis by XPS was to determine the state of cobalt under various reducing conditions and to define the role of the promotors [11]. The unreduced catalyst containing both promotors was found to be electrically conducting and exhibited no charging. The XPS analyses (by Al, $K\alpha$) are shown in Fig. 17.9. All of Mg, Th, Si and Co are detected and the oxygen $1s$

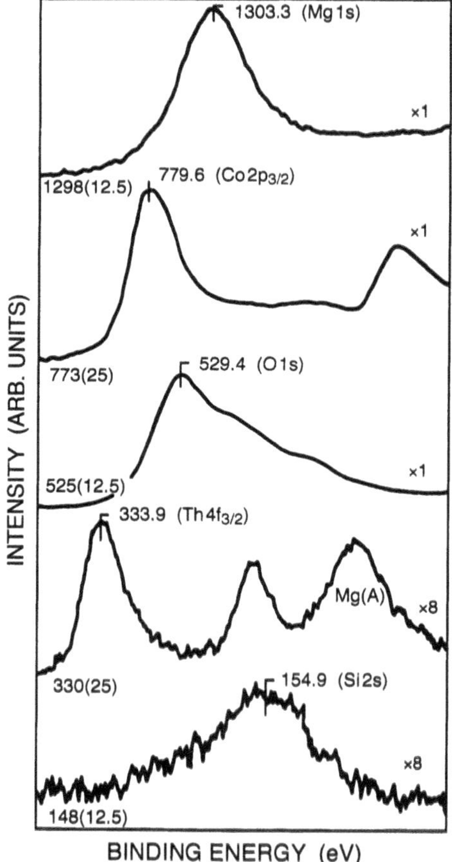

Fig. 17.9. Representative XPS spectra of a Co-ThO_2-MgO-Kieselguhr catalyst (Al $K\alpha$) in the unreduced (calcined) state. Data are shown for Mg $1s$, Co $2p_{3/2}$, O $1s$, Si $2s$, and Th $4f_{7/2}$ regions. This catalyst was electrically conducting and exhibited no charging

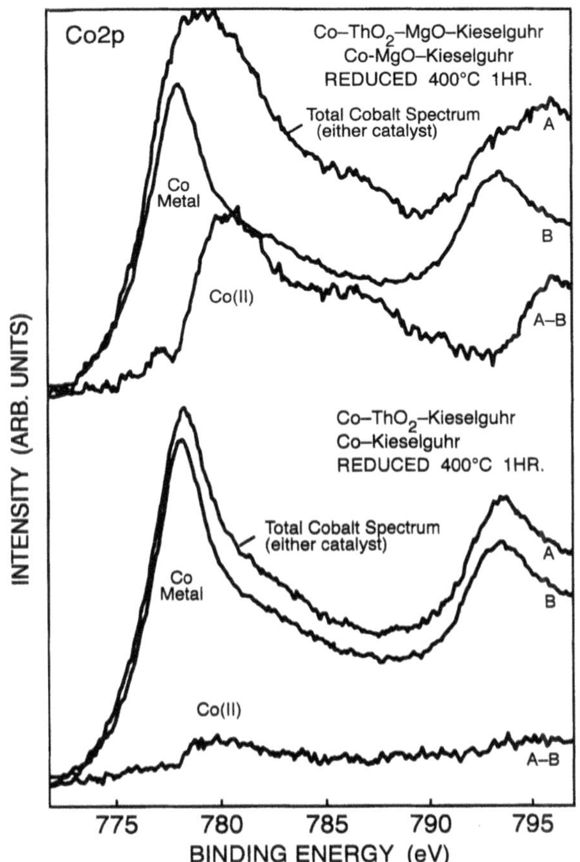

Fig. 17.10. Difference Co $2p_{3/2}$ spectra for magnesia-promoted and other cobalt catalysts after reduction at 400°C for 1 h in H_2. The cobalt metal lineshape (B) was deconvoluted from the total spectrum (A) to give the residual Co(II) unreduced component (A–B)

region shows a complex structure due to the overlapping of the various oxide components. The binding energy and shape of the Co $2p_{3/2}$ peak identify the presence of Co_3O_4. The largest component of the O $1s$ peak is also identified with Co_3O_4.

After reduction of the catalyst, the Mg $1s$ and Th $4f$ lines indicated no reduction of Mg^{2+} and Th^{4+} whereas the cobalt and oxygen spectra were markedly changed. Four catalysts, reduced at 400°C for 1 hour gave cobalt spectra as in Fig. 17.10. Difference spectra are plotted to reveal that the amount of unreduced cobalt is large in catalysts containing MgO, but very small when MgO is absent. The oxygen spectra showed overlapping peaks and they required a curve fitting procedure to resolve the contributions. An

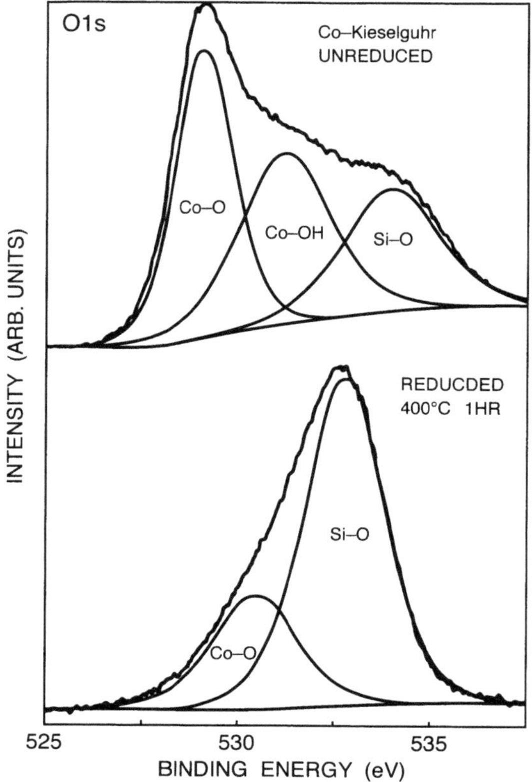

Fig. 17.11. Oxygen 1*s* spectra of unreduced and reduced Co-KG catalyst at 400°C including curve-fitting analysis (three-peak fit). Both the Co-O and Co-OH components are considerably attenuated after reduction, due to conversion of cobalt to the metallic phase. All peaks are normalized to the peak maximum so the change in Si-O intensity is only apparent. For most of the catalysts the Si-O intensity is a measure of the exposed KG surface and is approximately constant

example is shown in Fig. 17.11 for the unpromoted catalyst. The reduced form contains only small contribution of CoO to the O 1*s* peak, consistent with the observation that cobalt is reduced to metal in this catalyst. The corresponding result for MgO-promoted catalysts shows CoO present after reduction. It is concluded that there is a strong interaction of MgO with cobalt and that CoO possibly forms a solid solution which is resistant to reduction.

17.1.5 Iron Catalysts

Catalysts containing iron are used in the synthesis of ammonia and for the Fischer-Tropsch synthesis of hydrocarbons from carbon monoxide and hydrogen. Under these reducing conditions it might be expected that metallic iron is the active component. However, structural stabilizers or supports and a variety of promotors are present and the possibilities of reaction with iron are many. The control of the chain growth process in the Fischer-Tropsch synthesis and selectivity to the formation of olefins can be achieved by a lightly loaded supported iron catalyst [12]. In this type of catalyst, promotors and support are present in excess so that the selective reaction may well depend on compounds of iron rather than on the metal. Such a catalyst is, however, very difficult to study by XPS because the active components are highly dispersed and of very low concentration. The alternative is to study model systems designed to test the feasibility of reaction which may occur in the conditioned catalyst.

In the first type of experiment, a catalyst was prepared by precipitating iron, aluminium and praseodymium from solution in the atomic ratio 0.7 : 0.2 : 0.2. After drying and calcining, a sample was reduced and tested and found to have modest Fischer-Tropsch activity. This material is an unsupported version of the catalyst formed by supporting iron on alumina with a rare earth promoter. The difference, apart from the lower surface area and activity, is that alumina is present as a minor component. Analysis by XPS should then reveal the behaviour of iron without the problems of the dominant insulating support. As a preliminary experiment, a sample of Fe_2O_3 was tested by reducing in hydrogen at 320°C in the sample preparation chamber of the electron spectrometer. Analysis of the Fe $2p$ peaks showed that complete reduction to zero-valent iron occurs. The precipitated catalyst was then analyzed at various stages of reduction under the same conditions. The spectra in Fig. 17.12 show effects due to both electrostatic charging and chemical shifts. Initially Al $2p$ is charged while Fe $2p$ and Pr $3d$ are not charged. Oxygen $1s$ shows evidence of two states. Evidently at this stage separate phases exist. As the reduction proceeds the multiple charging effect on the Al $2p$ and O $1s$ spectra at first increases, then disappears. The final result is that both Al and O are in a single uncharged state. The Pr $3d$ signal shows no charging. The Fe $2p$, by position and shape, is present in the +2, +3 valence states. This condition persists after treatment under reaction conditions with CO and H_2. It is concluded that iron has reacted under the reducing conditions to form a conducting compound with Al and O. Reaction could also have involved Pr but since the oxide of this element is also conducting, there is no direct evidence. The possible products of reaction are aluminates within the composition range Fe_3O_4 to $FeAl_2O_4$. Surface phases of this composition would be expected to have considerable conductivity.

A second approach to the investigation of the reactivity of oxide supports towards iron is to directly introduce the metal to the external oxide surface by vacuum evaporation. Experiments of this kind have been conducted on alumina, silica, titania and praseodyminium oxide [13]. The objective is to observe any reaction of the metal with the oxide support under the conditions

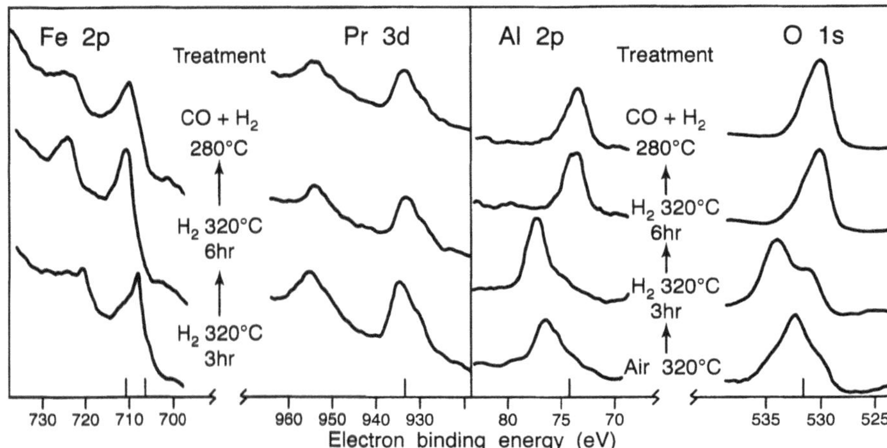

Fig. 17.12. Precipitated iron catalyst. XPS analyses at various stages of reduction. The reference marks on the energy scales indicate the peak position for an uncharged sample. In the final state of reduction the sample is uncharged but iron is not in the metallic state (see text)

for catalyst service. Very light loading of metal is necessary to ensure that excess metal does not obscure the analysis of the interface. The XPS analyses for iron on alumina are shown in Fig. 17.13. Trace (*b*) after deposition of the iron film shows two charge states; Fe $2p_{3/2}$ at 706.7 eV and 709.6 eV. The former, probably represents a conducting, continuous iron film and the latter represents islands or clusters of discontinuous Fe on the alumina. After reaction with H_2 at 320°C for 3 hours, the spectra are markedly changed [trace (*c*)]. The binding energies of Fe($2p_{3/2}$) XPS spectra can then be attributed to Fe(II) and Fe(III) at 709.5 eV and 711.0 eV, respectively, with charging $\sim$ 4 eV. The iron has reacted with alumina under conditions known to reduce iron oxide to metal. This result is confirmed by analysis by electron-induced Auger electron spectroscopy. Auger spectra, recorded before and after the XPS spectra, (Fig. 17.14a, b) show a strong Fe peak at 47.5 eV corresponding to Fe in the metallic state. After the treatment in hydrogen the Auger spectrum shows complete loss of the 47.5 eV Fe(O) peak and appearance of weak, broad oxide peaks at 51.0 eV and 37.5 eV. This is characteristic of oxidized iron [14].

A similar experiment for iron on titania showed that iron remains metallic on TiO_2 after the treatment in hydrogen. The XPS Fe $2p$ peaks by position and shape indicate zero-valent iron and the Auger analyses (Fig. 17.14) show the peak characteristic of metallic iron is present after the hydrogen treatment. The peak has moved to 46.5 eV due to charging and is of decreased intensity but there is no evidence of oxide peaks at 51.0 and 37.0 eV. It is concluded that titania is unreactive towards iron at 320°C in hydrogen and

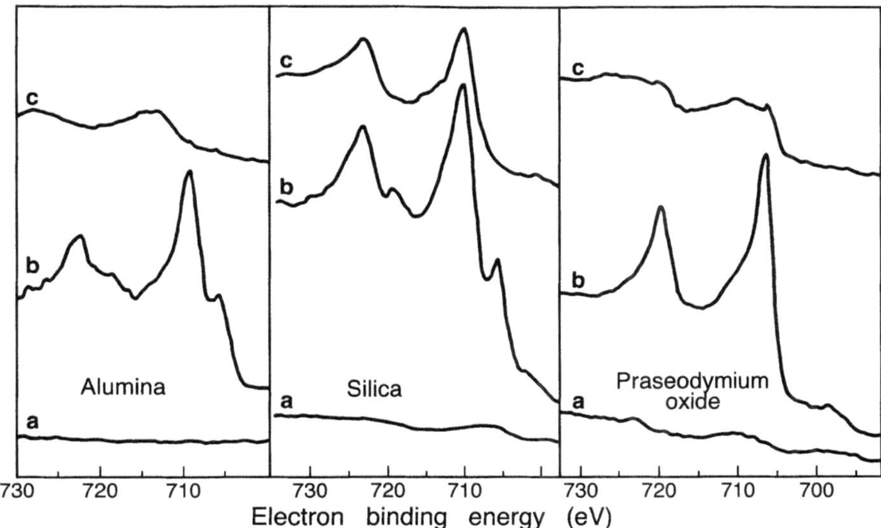

Fig. 17.13. XPS Fe 2*p* peaks from an iron film on an oxide support. Traces (*b*) as deposited; (*c*) after heating in hydrogen at 320° C

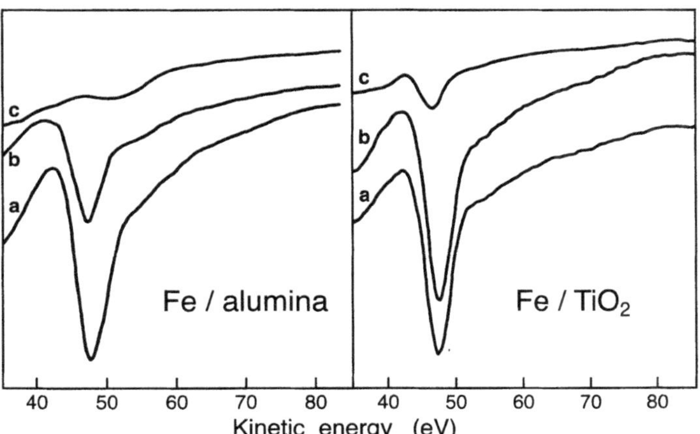

Fig. 17.14. Iron on alumina and on TiO_2 Auger spectra from the thin iron film. (*a*) before, and (*b*) after taking the XP spectra, and (*c*) after heating in H_2 at 320°*C*

that it may be classed as an inert support in contrast to alumina which was found to be reactive.

An iron on silica experiment was conducted under the same conditions. The XPS analyses (Fig. 17.13) show metallic iron, charged at 710.8 eV and uncharged at 706.3 eV (trace *b*). After heating in H_2 at 320°C the sharp peak at 710.8 eV is consistent with zero-valent iron but there is a broad oxide emission near 714 eV. This suggests limited reaction of iron with silica has occurred. Auger analysis of the reduced sample was difficult due to electrostatic charging but a weak broad peak at 56 eV and a peak near 48 eV are consistent with the conclusion that part of the iron is reacted.

Reference was made previously to the effect of praseodymium oxide as an additive to iron Fischer-Tropsch catalysts. To test the reactivity of this oxide towards iron, a sample of the oxide (Pr_6O_{11}) was reacted with an iron film under reducing conditions. The XPS results in Fig. 17.13 are free of the effects of electrostatic charging: Pr_6O_{11} is a semiconductor. The iron $2p$ peaks are well defined for the metal film (trace *b*). After heating in hydrogen at 320°C most of the iron is in a higher valence state. Extensive reaction of iron has occurred. Possibly the compound formed is $PrFeO_3$ however the amount present as a surface phase is too small for structural identification.

This series of results, after the reduction of the deposited iron layers in H_2 at 320°C, can be summarized as follows:

- Fe on Al_2O_3 reacts totally to form a new surface phase in which the Fe is in oxidized form.
- Fe on TiO_2 does not react and remains as separate Fe metal particles in islands on the TiO_2 surface;
- Fe on SiO_2 is partly oxidized suggesting that the role of this oxide as a catalyst promoter is in the formation of a compound with iron.

The extreme differences in reactivity of the metal with different oxide substrates correlates with the behaviour of the supported metal catalysts where alumina is the preferred substrate for selective Fischer-Tropsch activity. Silica is less effective and titania is not useful as a substrate.

17.2 Examples of Catalytic Systems Studied by FT-Infrared Spectroscopy

In this section we present case studies of catalyst characterization using surface vibrational spectroscopy performed in a number of different experimental modes. The results obtained in each case have also been confirmed by at least one other spectroscopic technique.

17.2.1 ZSM-5 Zeolites

Zeolites exist in many forms and exhibit a variety of structures and compositions [15]. They find applications as catalysts in numerous industrial reactions such as cracking, dehydrogenation, alkylation, isomerization, oxidative addition and dehydrocyclization. These solids, which could be microporous or mesoporous, possess three-dimensional structures, replete with cages and channels which may or may not intersect. The catalytically active sites are the bridging hydroxyl groups (the Brønsted acid centres). These link the pores and are distributed uniformly throughout the solid. The nature of the active sites can be modified during the synthesis of the material, and indeed it is this attribute of these solids that make them such versatile catalysts for a multitude of catalytic reactions.

One such zeolite is a high silica aluminate, ZSM-5, composed of linked SiO_4 and AlO_4 tetrahedra. The presence of aluminium in the lattice makes the framework negatively charged, which is compensated by mobile extra-framework cations, such as hydrogen and metal ions. In the former case the resulting zeolite is then referred to as HZSM-5, and in the latter case it would be M-HZSM-5. The acid centres in these zeolites can be easily characterised by transmission FT-IR using zeolite samples pressed into self-supporting wafers and weighing approximately 10–20 $mg\,cm^{-2}$.

In the following example we will show how IR spectroscopy can be used to identify acid sites in four proton-exchanged HZSM-5 zeolites [16]. These zeolites, labelled Z12, Z16, Z27 and Z121 vary in their Si/Al ratios, having values of 12, 16, 27 and 121, respectively. The values of bulk aluminium (framework aluminium) per unit cell, again respectively, are 7.7 (7.6), 6.0 (2.3), 3.5 (1.4) and 0.8 (0.8), and the difference between these two sets of values yields the values of extra-framework aluminium per unit cell. In Fig. 17.15 the IR spectra in the ν(OH) region of the four zeolites, measured at 423 K after heating in flowing nitrogen at 673 K, are shown. The spectra for the samples have

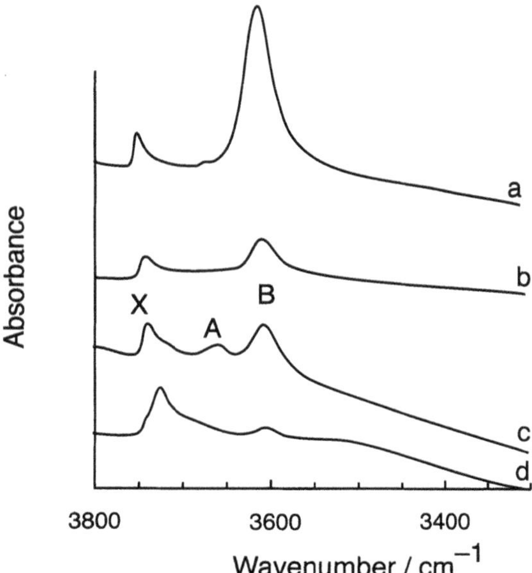

Fig. 17.15. Infrared spectra in n(OH) region of HZSM-5 samples: *(a)* Z12, *(b)* Z27, *(c)* Z16, *(d)* Z121. X, A and B refer to SiOH, AlOH and Brønsted acid hydroxyls, respectively. (Reprinted from [16]; copyright 1999 Elsevier Science Publishers, B. V.)

been normalised to equalize the intensities of the zeolite framework overtone band at $\sim 2000\,\mathrm{cm}^{-1}$. All show a band at $3610\,\mathrm{cm}^{-1}$ (labelled B), which is due to Brønsted acid sites noted above. The intensity of this band provides an indication of the framework aluminium concentration. Hence, the Z12 sample, with the highest concentration of framework aluminium has the most intense peak at $3610\,\mathrm{cm}^{-1}$. The samples show an additional ν(OH) band at $3740\,\mathrm{cm}^{-1}$ (labelled X), attributed to non-acidic silanol group (SiOH), terminating the zeolite lattice. In the case of sample Z121 with the lowest framework aluminium concentration, the $3740\,\mathrm{cm}^{-1}$ band is actually a shoulder on the more intense $3720\,\mathrm{cm}^{-1}$. The latter peak is attributed to a second non-acidic silanol group associated with internal defects in the zeolite framework. Hydrogen bonding between these internal silanol groups are responsible for the broad band observed at $\sim 3500\,\mathrm{cm}^{-1}$ for Z121 zeolite. An additional ν(OH) band is observed in zeolites Z27 and Z16, which is particularly pronounced in the spectrum of Z16. This band is attributed to extra-framework aluminium, and Z16 with the highest extra-framework aluminium, shows the most intense band.

In summary, the IR spectra in Fig. 17.15 have shown that the four zeolites contain different relative concentrations of Si–OH Brønsted acid sites and extra-framework Al–OH species. They contain, if at all, very few Lewis acid sites. However, to confirm the presence of Lewis sites, infrared studies involving pyridine adsorption need to be carried out.

With metal exchanged zeolites, e.g. Fe-ZSM-5, it is usual to use a probe molecule such as NO to study the interaction of iron species with the Brønsted sites of the zeolite. The NO molecule allows that because the NO stretching vibrations will be different when adsorbed at different iron sites. We will use a recent work of Joyner and Stockenhuber [17] on Fe-ZSM-5 to illustrate how infrared spectroscopy can be used to determine the degree of interaction between iron and the zeolite framework.

Fe-ZSM-5 was prepared from the parent zeolite by a number of ion exchange methods, including a novel method proposed by Feng and Hall (discussed in detail in reference [17] which was effective in achieving an 80% exchange. The infrared spectrum of the pressed wafer, taken in the transmission mode, in the ν(OH) region, Fig. 17.16 shows the expected bands, as observed previously for ZSM-5 zeolites in Fig. 17.15. The band at $3670\,\mathrm{cm}^{-1}$ is due to extra-framework species, those at $3600–3610\,\mathrm{cm}^{-1}$ to bridging acidic hydroxyl stretching vibrations, and that at $3740\,\mathrm{cm}^{-1}$ due to stretching vibrations of terminal Si–OH groups.

The redox behaviour and the oxidation state of the exchanged iron was studied by NO adsorption. The NO adsorption characteristics provides information on the *heterogeneity* of the iron-oxo species within the zeolites, and the occurrence of *isolated* iron species, which averaging techniques such as XPS and EXAFS (Extended X-ray Absorption Fine Structure) cannot explore. In Fig. 17.17 the difference infrared spectra following NO adsorption on a

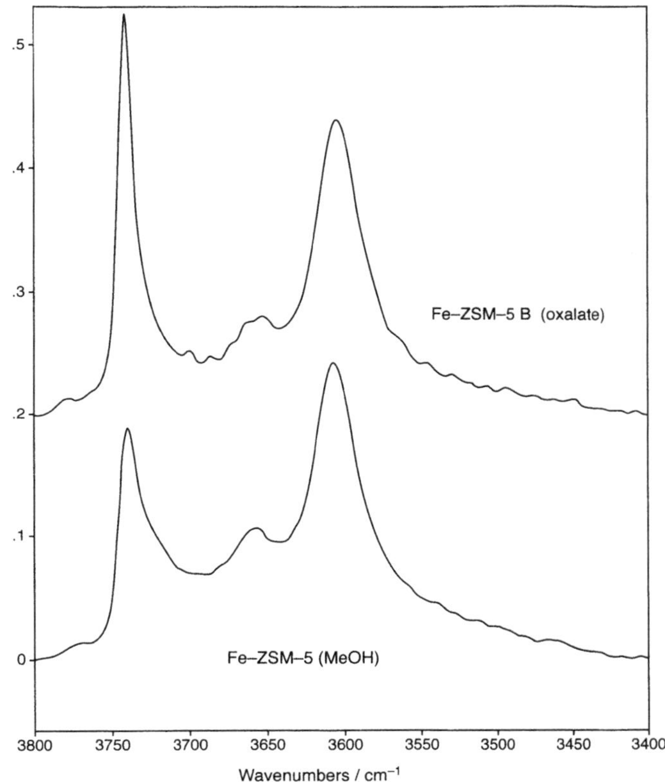

Fig. 17.16. Infrared spectra measured at 425K after pretreatment in vacuo at 773 K: *(lower curve)* Fe-ZSM-5 prepared by exchange in methanol; *(upper curve)* Fe-ZSM-5 prepared by method of Feng and Hall (see text). (Reprinted from ref. 17.17; copyright 1999 American Chemical Society)

vacuum treated sample of Fe-ZSM-5 are shown. At 10^{-3} mbar NO pressure, the NO bands appear at 1880 cm^{-1} (intense), 1841 and 1816 cm^{-1} (shoulders) and a weak band at 1786 cm^{-1}. Discrete bands at 1816 and 2250 cm^{-1} appear when the NO pressure is increased to 10^{-2} mbar. At 1 mbar pressure the 1815 and 2250 cm^{-1} bands increase in intensity, and a new band at 2135 cm^{-1} appears. From 1 to 10 mbar the bands at 1815, 1860 and 2250 cm^{-1} bands show a decrease in intensity while the feature at 2135 cm^{-1} increases dramatically. It should be noted that the 1880 cm^{-1} band stays constant in intensity at all NO pressures. NO adsorption at 10^{-1} mbar on samples with different pre-treated conditions (spectra shown in Fig. 17.18) show changes in intensities in 2250, 2135, 1880, 1815 and $\sim$ 1780–1741 cm^{-1} bands to varying degrees. The important point to note is that the bands at 2135 and

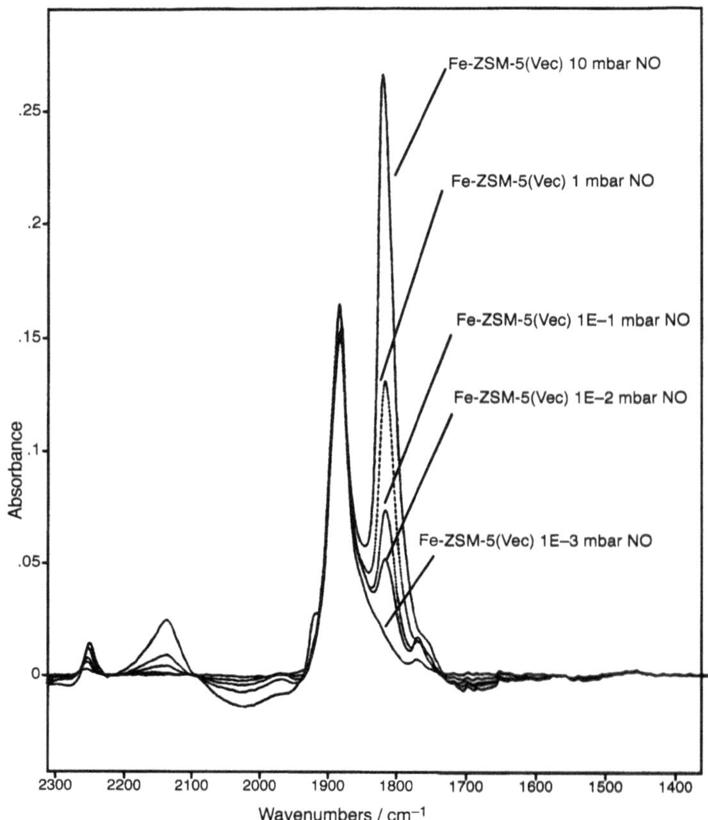

Fig. 17.17. Infrared spectra measured at 303 K for increasing NO pressure from 10^{-3} mbar to 10 mbar over Fe-ZSM-5. (Reprinted from [17]; copyright 1999 American Chemical Society)

1635 cm^{-1} show a marked increase in intensity after calcination in oxygen. Using results presented here and EXAFS/literature information the bands are assigned in the following manner. The observed bands are due to high-spin Fe(II)-NO complex (1880 cm^{-1}), low-spin NO complex involving isolated Fe(II) ions ($\sim$ 1780 cm^{-1}) and to NO adsorption on more covalent Fe–O–Fe species (1815 cm^{-1}). The intensity of the 1815 cm^{-1} band only starts increasing when the 1880 cm^{-1} band becomes saturated, suggesting that this band is due to adsorption on the iron-oxo nanocluster sites with pre-adsorbed NO. The intensity of the 1841 cm^{-1} NO peak was found to show a linear relationship (not shown here) with the number of acid sites destroyed by iron exchange. It is therefore attributed to adsorption of NO on isolated iron ions. The band at 2135 cm^{-1}, sensitive to calcination in oxygen, is either due to NO^{2+} species hydrogen-bonded to the acidic OH groups of zeolite, or to NO^{+}

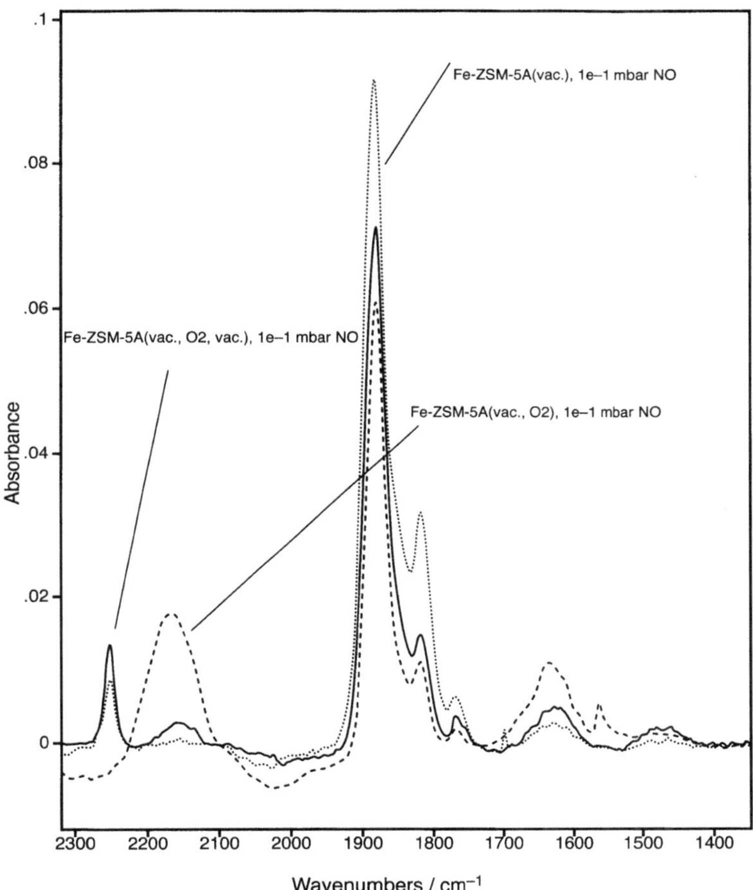

Fig. 17.18. Infrared spectra showing the influence of pretreatment conditions on Fe-ZSM-5. (Reprinted from [17]; copyright 1999 American Chemical Society)

species acting as a charge balancing cation. The interaction of iron with the acidic protons are quite complex and not fully understood, as yet.

In summary, the infrared data presented above, in conjunction with XPS/EXAFS data show that the exchange of iron with zeolite ZSM-5 leads to the formation of both *isolated iron atoms*, and *iron-oxo nanoclusters* of typical size Fe_4O_4 with very short Fe–Fe inter-atomic distance of ca. 2.50 Å. These iron species exist predominantly within the pore structure of the zeolite and not on the external surfaces.

17.2.2 Thin Alumina Films

Thin oxide films of aluminium, formed by oxidation of aluminium on a metal substrate can be used as a model catalyst support, since films grown this way are known to mimic high surface area supported systems such as γ-alumina. Infrared spectroscopy can be used to scrutinize the properties of these thin oxide films, for example, to examine the nature and the type of surface acid sites, to monitor changes in the extent of hydroxylation, and to determine the Lewis acid sites by titration by Lewis bases such as ammonia. In this section we will describe how RAIRS data has been used to characterise thin alumina films grown on Mo(100) substrate by the oxidation of surface aluminium by water in ultrahigh vacuum [18]. Using this method the alumina formed (thickness > 20 Å) has a stoichiometry of Al_2O_3, as confirmed by AES.

In Fig. 17.19 infrared spectra of alumina films formed under different pretreatment conditions are displayed. Spectrum (a) is that of the film formed using H_2O. The intense asymmetric feature centred at $\sim 935\,cm^{-1}$ corresponds to the longitudinal optical (LO) vibrations of the Al–O bonds in alumina. The corresponding transverse optical (TO) mode is not observed because it is forbidden by surface selection rules. The OH groups on the hydroxylated sample should be characterised by the O–H stretching frequency at $\sim 3750\,cm^{-1}$, which is not observed here because this region could not accessed by the MCT detector used in the study. Spectrum (b) is that of a similar film formed after heating the freshly prepared alumina in vacuo to 850 K, then hydroxylating with D_2O, followed by heating to 700 K to desorb any molecular water. The Al–O vibrations are now centred at $960\,cm^{-1}$ with the asymmetric curve having a tail to lower frequencies. This is attributed to the alumina film being a more-ordered, more γ-like Al_2O_3. The feature centred at $2700\,cm^{-1}$ is assigned to the O–D stretching mode due to hydroxyl species present on the surface of the oxide film. The corresponding frequency of the bending mode of gas phase D_2O, expected at $\sim 1178\,cm^{-1}$ is not observed, indicating the complete absence of any molecular water on the surface. Spectrum (c) was obtained by exposing a dehydroxylated alumina film (formed by heating to 900 K in vacuo), where the oxide was grown using H_2O, to gaseous ammonia. The feature at $1260\,cm^{-1}$ is assigned to deformation mode of NH_3 adsorbed at Lewis sites of alumina. This ν_2 mode is sensitive to the acid strength of the oxide to which it is co-ordinated: the greater the electron-acceptor power the greater the shift of ν_2 to higher frequencies. The large shift in frequency observed for this mode on the alumina surface from the gas-phase value ($\sim 950\,cm^{-1}$) indicates the high strength of the Lewis acid sites present on the surface. It should be noted that the ν_4 mode, that accompanies the ν_2 mode, expected at $\sim 1628\,cm^{-1}$, is not evident. The reason is simple. While the ν_2 mode has a_1 (z symmetry) the ν_4 mode has $e(x, y)$ symmetry, making this mode symmetry forbidden by the experimental arrangement used in the investigation.

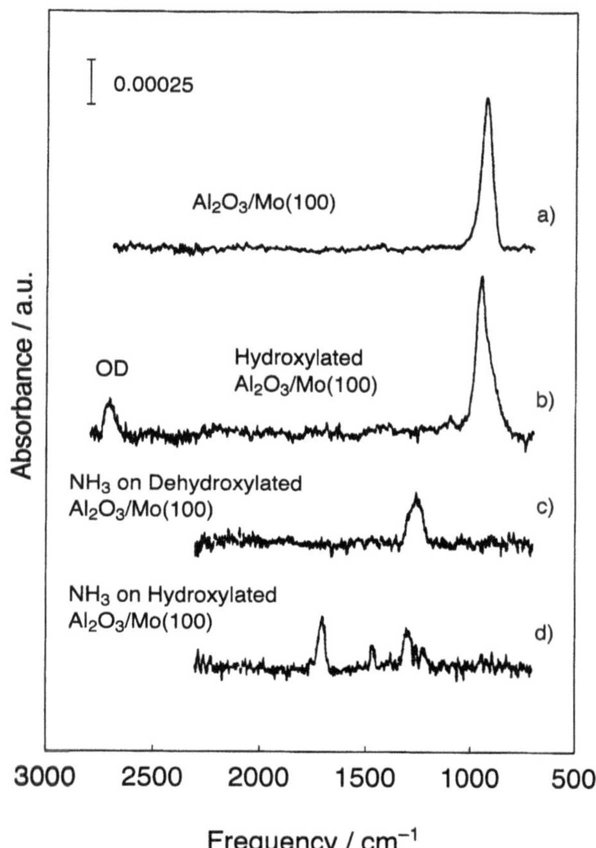

Fig. 17.19. RAIRS of *(a)* alumina grown on Mo(100) by reaction between aluminium and water (H_2O), *(b)* alumina grown on Mo(100) by reaction between aluminium and deuterium oxide (D_2O), *(c)* alumina grown on Mo(100) by reaction between aluminium and water (H_2O), annealed at 900 K to remove surface hydroxyls, exposed to ammonia at 80 K and then annealed to 260 K before data acquisition at 80 K, *(d)* alumina grown on Mo(100) by reaction between aluminium and water (H_2O), then exposed to ammonia at 80 K, annealed to 225 K before data acquisition at 80 K. (Reprinted from [18]; copyright 1999 Elsevier Science Publishers, B. V.)

Finally, spectrum (d) is that of ammonia adsorbed on hydroxylated alumina grown with reaction with H_2O at 80 K and subsequently annealed to 225 K. Note that, two new features are observed compared to spectrum (c), a new feature at $\sim 1710\,cm^{-1}$ and a small peak at $1450\,cm^{-1}$. Also note that the feature at $\sim 1260\,cm^{-1}$ is not a single broad feature any more but has more structure. This feature becomes a single peak when the surface is annealed to 225 K and disappears completely by 350 K. The band at $\sim 1260\,cm^{-1}$ again indicates the presence of Lewis acid site. The $\sim 1450\,cm^{-1}$ band indicates the presence of NH_4^+ species formed by NH_3 adsorption at Brønsted sites. Alumina is not generally considered to have strong Brønsted sites and hence the very weak peak observed here. The assignment of the $\sim 1710\,cm^{-1}$ band is not as straight-forward, and is beyond the scope of this book. The reader is referred to the original source [18] for more information.

17.2.3 Rhodium Supported Alumina Films

RAIRS with CO as a probe molecule constitutes an ideal tool to study nucleation behaviour of metal deposits on oxide substrates. In this section we illustrate how changes in carbonyl stretching frequency following adsorption of CO on rhodium supported alumina [20] can be used to determine the types of rhodium nucleation sites and their particle sizes. Rhodium was evaporated onto a thin well-ordered alumina film grown on NiAl(100) at 90 K. At this deposition temperature rhodium is deposited at the rotational and antiphase domain boundaries separating long-range ordered domains of the aluminium oxide film. The average number of atoms per particle was determined from the number of metal atoms deposited, and the island density was observed by STM. The data reported here are for rhodium deposits on alumina with average particle size of 9 atoms.

The infrared spectrum of CO saturated rhodium on alumina deposit is shown in Fig. 17.20(a) (top spectrum). The most prominent feature at $2117\,cm^{-1}$ is due to symmetric C=O stretching frequency of terminally bound CO molecule on *isolated* rhodium atoms bound to oxide defects. The asymmetric stretch expected at $\sim 2035\,cm^{-1}$ is not observed. This is because its dynamic dipole is oriented parallel to the oxide surface and the mode is thus forbidden due to surface selection rule. The bottom spectrum shown in Fig. 17.20(a) is that of rhodium deposit saturated with carbon monoxide containing an equimolar mixture of ^{12}CO and ^{13}CO. Three bands at 2117, 2101 and $2068\,cm^{-1}$, with relative intensity ratio of 1:2:1, are observed. These bands signal the formation of the rhodium gem-dicarbonyl species, $Rh(CO)_2$ and not RhCO. The latter would have given rise to a second infrared band, due to 13CO, $47\,cm^{-1}$ lower than the intense $2117\,cm^{-1}$ peak.

Spectra presented in Fig. 17.20(b) and the accompanying STM images can be used to determine the type of nucleation sites at which the $Rh(CO)_2$ species is formed. CO was adsorbed to saturation coverage on rhodium deposits formed under three different experimental conditions. For deposition at 300 K, when STM shows that metal deposits are formed at *oxide domain boundaries* giving rise to *line defects*, no band at $2117\,cm^{-1}$ is seen (top spectrum), that is, no $Rh(CO)_2$ species is formed. When rhodium is deposited at 90 K, and rhodium atoms are associated with *oxide point defects*, then the $2117\,cm^{-1}$ band is observed (middle spectrum). Rhodium deposition first at 300 K followed by deposition at 90 K results in rhodium nucleation at both line defects and oxide point defects on the alumina surface. CO adsorption on

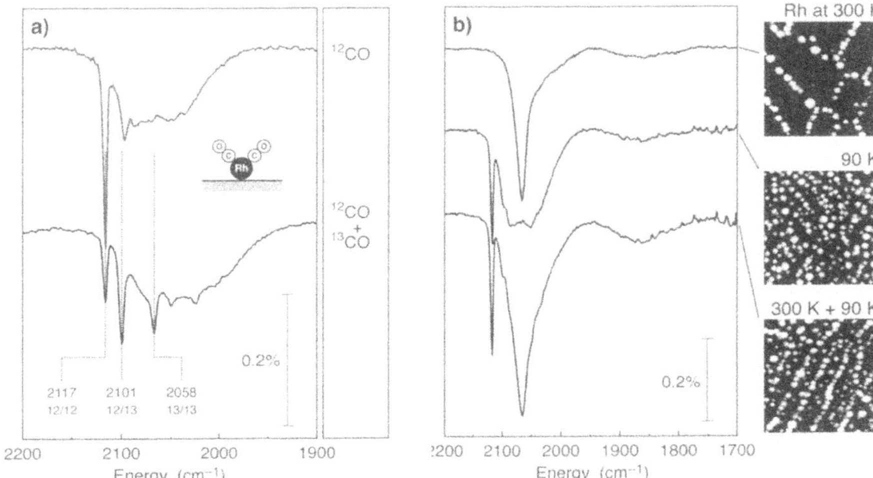

Fig. 17.20. **(a)** Infrared reflection-absorption spectra taken after deposition of 0.028 ML of rhodium and subsequent saturation with ^{12}CO *(top)*, and an approximately equimolar mixture of ^{12}CO and ^{13}CO *(bottom)* at 90 K. The isotopic compositions giving rise to the three dicarbonyl bands are indicated below the corresponding wavenumbers, **(b)** Infrared spectra recorded after CO saturation of rhodium deposits at 90 K, along with room temperature STM images (500 Å× 500 Å). *(Top)*: 0.057 ML of rhodium deposited at 300 K, *(Middle)*: 0.057 ML of rhodium deposited at 90 K, *(Bottom)*: 0.057 ML of rhodium at 300 K, followed by the same exposure at 90 K. (Reprinted from [19]; copyright 2000 Elsevier Science Publishers, B. V.)

such a deposit shows a band at 2117 cm^{-1}, implying that Rh(CO)$_2$ species will form as long as rhodium nucleation sites exist at *oxide point defects.*

17.2.4 Copper Supported Silica Films

In this section we provide another study which utilizes RAIRS for the characterization of a metal supported catalyst, copper on silica [20]. CO was used as a probe molecule to elucidate the structures of copper deposits on silica support. Thermal desorption spectroscopy was used to support the infrared data. Copper was evaporated onto a planar silica thin film ($\sim$ 100 Å) supported on Mo(110) surface at 100 K, while the CO adsorption on the resulting model catalyst was carried out at 90 K.

Fig. 17.21 shows the infrared spectra of CO saturated silica deposit annealed to temperatures indicated. At 100 K an intense peak at 2099 cm^{-1} with a shoulder on the low frequency side is observed. A new band appears at 2070 cm^{-1} when heated to 300–500 K, which gradually increases in intensity as the catalyst is heated to 900 K. The 2099 cm^{-1}, on the other hand, splits into two peaks at 2108 cm^{-1} and 2094 cm^{-1}. Using literature values of CO on Cu(111), the observed bands are attributed to CO adsorbed on several

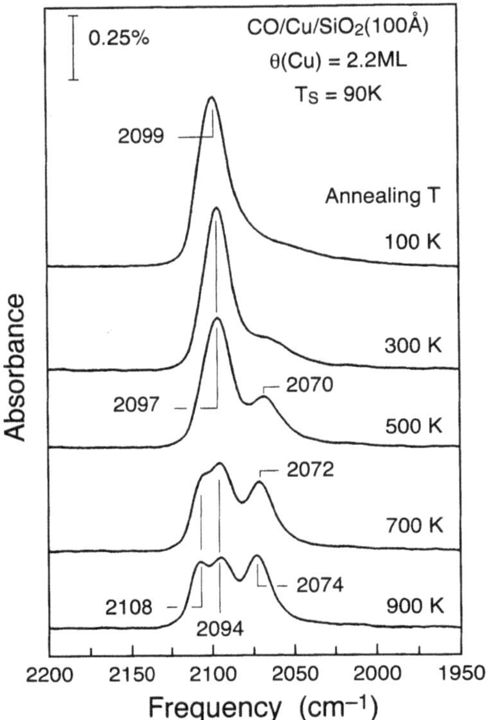

Fig. 17.21. Infrared reflection-absorption spectra of CO on silica-supported copper as a function of the preannealing temperature. Copper was deposited at 100 K and annealed to the indicated temperatures, followed by CO adsorption to saturation at 90 K. (Reprinted from [20]; copyright 1993 American Chemical Society)

distinct *atop* sites on copper; CO adsorbed on *bridging* and *hollow* sites on Cu(111) produce bands in the 1814–1834 cm^{-1} range.

The structure of the supported copper also depends on the copper coverage and the temperature to which the deposit is annealed. In Fig. 17.22 spectra of CO saturated catalyst is shown for increasing copper coverage. In each case, following copper deposition at 100 K, the film was annealed to 900 K, then cooled to 90 K and saturated with CO. At low copper coverages (0.5 ML) two bands at $\sim$ 2110 cm^{-1} and 2066 cm^{-1} are evident. The 2110 cm^{-1} band corresponds to (111)-facets of copper particles. Atomic force microscopy (AFM) shows that the particles at this low coverage consist of small clusters of average size $\sim$ 200 Å. The 2066 cm^{-1} band is lower in frequency than previously observed for CO stretching on any single crystal surface, and is attributed to a particle size effect; this phenomenon is explained fully in reference [20]. With increasing copper coverage a new band at $\sim$ 2094 cm^{-1} appears, which grows with copper coverage till the coverage

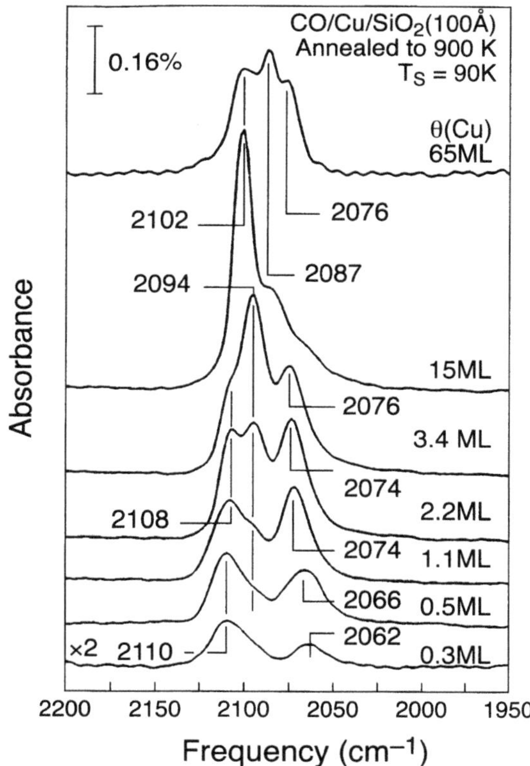

Fig. 17.22. Infrared-reflection spectra of CO on silica-supported copper as a function of copper coverage. Copper was depsoited at 100 K and annealed to 900 K, followed by CO adsorption to saturation at 90 K. (Reprinted from [20]; copyright 1993 American Chemical Society)

reaches 3.4 ML. This band indicates the presence of (110)-facets of copper deposits. The 2110 cm^{-1} band remains constant till this coverage, but with a further increase of copper coverage (15 ML) it becomes the main feature of the spectrum. The 2094 cm^{-1} band is not evident for copper coverages $>$ 15 ML. The dominance of the 2110 cm^{-1} mode in the whole CO spectral region suggests that at high copper coverages *continuous films* with majority (111) orientations form during deposition. The high index (211) and (311) planes may also be present for these coverages as the spectral range for these planes fall in the 2108–2110 cm^{-1} range. At even higher copper coverages ($>$ 30 ML) CO exhibits three well-resolved absorption features at 2102, 2087 and 2076 cm^{-1}. The 2087 and 2076 cm^{-1} bands are due to CO adsorption onto the (111) and (100) facets, respectively, and the 2101 cm^{-1} band is attributed to adsorption at step/edge sites of the polycrystalline copper film that forms at this coverage.

It should be noted that the results presented here do not provide any evidence of interactions of reaction of copper with the support material silica. However, a later paper of the same researchers [21] shows, using XPS and temperature programmed desorption, that a small fraction of copper in the copper–silica interface is partially oxidized and this phase desorbs at 1300 K.

In summary, this infrared study of a model copper catalyst shows that at low effective copper coverages, the un-annealed copper deposit has at least two types of small copper islands on the silica surface. As the copper coverage increases the small clusters are merged into an extended smooth film. On the basis of the CO stretching frequencies it has been proposed that the copper atoms are initially aggregated into (111)-like islands and high index islands. Annealing the copper to 900 K following copper adsorption at 100 K causes the formation of three-dimensional clusters.

17.3 Conclusion

The above examples show that surface techniques, such as XPS and vibrational spectroscopy (FT–IR, FT–RAIRS), can reveal much of the complex chemistry of a catalyst surface. The reactions which occur during the conditioning or activation of the catalyst involve only materials contacting at a surface. The surface phases which form are not readily characterized by bulk techniques nor do they necessarily follow the thermodynamics of bulk preparations. Direct analysis of the surface is the only way to reliably characterize the catalyst surface.

References

1. P.J. Chappell, M.H. Kibel, B.G. Baker: J. Catalysis **10**, 139 (1988)
2. D.J. Dwyer: In *Catalyst Characterization Science*, ed. by M.L. Deviney and J.L. Gland (American Chemical Society 1985) pp. 124–132
3. R. Raval, M.A. Harrison, D.A. King: J. Vac. Sci. Technol. **A9 (2)**, 345–349 (1991)
4. R.G. Greenler: J. Chem. Phys., **44**, 310 (1966)
5. R.G. Greenler: J. Vac. Sci. Technol. **12**, 1410 (1975)
6. J. Pritchard, M.L. Sims: Trans. Faraday Soc., **66**, 427 (1970)
7. A. Horn: In *Spectroscopy for Surface Science*, ed. by R.J.H. Clark and R.E. Hester (John Wiley & Sons Ltd., 1998), pp. 273–339
8. B.G. Baker, N.J. Clark: New Fischer-Tropsch Catalysts. Report NERDDP EG85/470 (Australian Govt. Printing Service 1985)
9. B.G. Baker, N.J. Clark, H. McArthur, E. Summerville: Catalysts. U.S. patent 4610975, New Zealand patent 216337, Australian patent application No. 73419/87
10. J.C. Carver, S.M. Davis, D.A. Goetsch: In *Catalyst Characterization Science*, ed. by M.L. Deviney and J.L. Gland (American Chemical Society 1985) pp. 133–143

11. B.A. Sexton, A.E. Hughes, T.W. Turney: J. Catalysis **97**, 390 (1986)
12. B.G. Baker, N.J. Clark, H. McArthur, E. Summerville: U.S. patents 4610975, 4666880, 4767792; New Zealand patent 207355; Australian patent 565954
13. R.St.C. Smart, P.S. Arora, B.G. Baker: Proc. Aust. X-ray Analytical Association (AXAA-88) 219–228 (1988)
14. C. Klauber: Ph.D. Thesis, Flinders University (1984) p. 7.42
15. J.M. Thomas, W.J. Thomas: Principles and Practice of Heterogeneous Catalysis, (VCH Verlasgesellschaft, 1997), 347–364
16. S.M. Campbell, X.-Z. Jiang, R.F. Howe: Microporous and Mesoporous Materials, **29**, 91–108 (1999)
17. R. Joyner, M. Stockenhuber: J. Phys. Chem. B, **103**, 5963–5976 (1999)
18. M. Kaltchev, W.T. Tysoe: Surface Science, **430**, 29–36 (1999)
19. M. Frank, R. Kuhnemuth, M. Baumer, H.-J. Freund: Surface Science, 454–456, 968–973 (2000)
20. X. Xu, D.W. Goodman: J. Phys. Chem., **97**, 683–689 (1993)
21. X. Xu, J.-W. He, D.W. Goodman: Surface Science, **284**, 103–108 (1993)

18 Application to Semiconductor Devices

P.W. Leech and P. Ressel

Semiconductor devices form the basis of nearly all applications in electronics and optoelectronics. Silicon integrated circuits (ICs) containing a range of advanced logic devices are dominant in the electronics industry [1]. Other commercial electronic devices based on substrates of III-V compound semiconductors include high electron mobility transistors, heterojunction bi-polar transistors and integrated circuits. Examples of commercially available optoelectronic devices in III-V semiconductors include light emitting diodes and multi-quantum-well lasers. For military and industrial markets, arrays of infrared diodes and detectors are fabricated using $Hg_{1-x}Cd_xTe$ or equivalent bandgap II-VI semiconductors. Recent developments in semiconductor devices include the field of silicon nanoelectronics in which $Si/SiGe/SiO_2$ structures are used to fabricate single electron metal-oxide-semiconductor transistors with nano-scale gate lengths. The performance of all of these devices is critically dependent on the integrity of the fabrication process in terms of their dimensional accuracy, the control of interfacial reactions and high purity of the component materials. While electrical measurements on the devices can provide indirect data on these material parameters, their characterisation also requires direct information from the range of surface analytical techniques. Each of the techniques has unique capabilities in the analysis of the structural features in the upper 1–2 μm of the surface within which are located the devices and circuits. Hence, many of the problems encountered in the development of semiconductor devices are sufficiently complex to require a number of the techniques used in combination.

The first part of this Chapter is a review of analytical techniques which are applied in silicon integrated circuits and their component materials. This section outlines the role of the techniques in both the diagnosis of failure in manufactured integrated circuits and in the monitoring of the many steps in the processing of a circuit. In view of the introductory nature of this Chapter, emphasis is placed on the more conventional methods with details of other techniques available in the referenced texts. The second section reviews the characterisation of ohmic metal to semiconductor contacts as an example of the analysis of a complex, layered structure of the type encountered in modern device technology.

18.1 Micro and Nano-Analysis of Integrated Circuits

Semiconductor technology continues to move on a path towards an increasing density of devices per unit area and a shrinking size of minimum features. The miniaturisation of devices as described in Moore's Law represents a doubling of the number of transistors on an integrated circuit every ≈ 18 months [1]. This trend in integration is driven by the requirement in the marketplace for increased performance and functionality at a reducing cost and seems likely to continue into the near future. In current commercial circuits, 10^8 or more electronic devices are interconnected by 6,7 or more multiple-levels of wiring with separation between the layers by an oxide/nitride dielectric material [2]. The minimum size of the "interconnects" or thin film wires patterned in Al or Cu-based metallurgy are currently (2001) as small as 180 nm and are predicted to reduce to 70 nm by the year 2010 [3–5]. Other examples of nano-scale features in modern circuits include the ultra-shallow source/drain junctions (with a depth of 36–72 nm) formed by implantation with B^+ ions and the thermally grown gate oxides of 3.5–4.0 nm thickness [3]. The scale of these dimensions requires the application of an increasing level of sensitivity and accuracy in the probing of the structures by analytical techniques. A further challenge in the analysis of very large scale integrated (VLSI) circuits is the increasing emphasis on the properties of surfaces, interfaces and junctions in the thin films used in fabrication. Examples of interfacial regions include a) the contact layers of reacted silicide such as $TiSi_2$, WSi or $CoSi_2$ [4], b) the barrier layers of TaN, WN, $TiSi_2$ or Ti/TiN [6], and c) gate structures of polysilicon/silicon dioxide/silicon. The fabrication of the circuit from these elements typically involves more than 100 sequential steps. Each step must be optimised in terms of process parameters and performed in a non-contaminating environment. A very high level of purity must be maintained in all of the materials, with concentrations of contaminant metals of $< 10^{10}$ atoms/cm^2 (or less than 10^{-5} of a monolayer) [7]. At these levels of contamination, many of the available analytical techniques are approaching their limit of resolution and thereby require the use of a variety of methods for the unambiguous identification of features. Figure 18.1 illustrates the areas of application of key analytical techniques in examination of a complimentary metal-oxide-metal (CMOS) device.

Analytical techniques perform an important role in both the development of the processing steps and in the diagnosis of defects in the manufactured circuit. Both of these functions are reviewed here, commencing with the role in failure analysis. The various classification systems of failure in ICs are based on either the source of the defect [8], the physical nature of defects/mechanisms [8] or the time dependence of failure [9,10]. Those failures which occur early in the lifetime of integrated circuits are commonly the result of defects generated during the manufacturing steps [8,10]. A summary of common modes of failure of ICs during manufacturing and their associated causes has been given by Richards and Footner [8] and Grovenor [9].

a) 
b) 

Fig. 18.1. Example of a display in SEM-VC dynamic mode showing images with 500 ns between states. The images have a **(a)** 1.2 μs and **(b)** 1.7 μs delay at a clock speed of 4 MHz. (Courtesy of J. Rogers, Telstra Research Laboratories).

These modes of failure include breaks or displacement in the metal patterns, cracking in the semiconductor or passivation and the corrosion or lack of adhesion of the metallization due to contamination [8–10]. Following the initial running-in period, an extended regime of failures in service occur at a much lower rate, usually by degradation mechanisms in the metallisation or semiconductor. Common defects arising in the service of ICs include the bridging of metal tracks by electromigration, the interdiffusion of multilayered structures, void formation, hillock growth and the corrosion of interconnects [8–10].

The sequence of examination in the failure analysis of ICs proposed by Richards and Footner has been summarised in Table 18.1 [8]. The ranking of the techniques in Table 18.1 is based on their ease of accessibility and economy of cost. An important factor in determining relative cost is the time of a straightforward analysis which increases as 0.2 h (SEM), 0.5 h (RBS, AES, SIMS and ESCA) and 16 h (TEM) [11]. In general, the examination of an IC commences with stereo-microscopy which allows the detection of gross damage to the package or connecting pins and X-ray radiography which can determine the presence of broken or incorrectly connected wires. Any examination by subsequent techniques requires the de-packaging or de-capsulation of the chip. Metal transistor outline cans or ceramic dual-in-line packages may be opened by mechanical methods. Alternatively, plastic dual-in-line packages require the chemical dissolution of the encapsulant in a jet of hot sulphuric acid. Following removal of the encapsulating plastic, the brightfield mode of optical microscopy is used in for a preliminary examination of features such as the metal tracks, the contact windows and coverage of steps by patterned films as described in more detail in [8,11]. As a supplementary method, the dark field mode of the optical microscope may enhance the visibility of non-planar features on the surface of the chip such as particles, scratches or chemical residues. A recent development has been the use of UV microscopy operating at 350 nm to reduce the lateral resolution of op-

Table 18.1. Sequence and selected capabilities of analytical tools in investigation of semiconductor devices (after [8])

	Typical Applications	Lateral resolution	Depth resolution
1. Visual / optical inspection			
2. Radiography			
3. Electrical tests			
4. Package tests for hermicity			
5. De-packaging			
6. **First rank analytical techniques**			
Optical microscopy	Imaging	0.5 μm	0.01 μm
SEM-EDX	Plan view and cross-sectional imaging / microanalysis of thin films	45 Å(SEM) 0.3–1 μm (EDX)	1–5 μm
Metallography-sectioning			
7. **Second rank analytical techniques**			
Atomic force microscopy (AFM)	Quantitative, 3-dimensional surface imaging with near atomic resolution	50–100 Å	0.001 μm
SEM-VC: SEM-EBIC, SEM-OBIC			
Auger electron spectroscopy (AES) Scanning Auger Microprobe (SAM)	Major and minor elemental identification and mapping at surface / depth profiling	150–1000 Å	< 20 Å
X-ray photoelectron spectroscopy (XPS)	Elemental identification and chemical bonding at surface of organic / inorganic molecules / selected element profiling	25–1000 μm	1–100 Å
Scanning acoustic microscopy (SAM)	Imaging of internal interfaces	1.0 μm at high frequency	0.001 μm
X-ray diffraction / topography	Phase identification, crystal size / orientation		

Table 18.1. (cont.)

8. **Third rank analytical techniques**			
Secondary ion mass spectroscopy (SIMS)	Dopant / impurity depth profiling	0.15–1.0 μm (imaging) 30 μm (profiling)	50–300 Å
Rutherford backscattering spectroscopy (RBS)	Quantitative thin film composition / thickness	2 mm	20–200 Å
Thermal imaging (IR microscopy)			
Transmission electron microscopy (TEM)	High resolution imaging / elemental micro-analysis	1.8–50 Å	
9. Comparison with literature			
10. Simulation tests if appropriate			

tical microscopy from 0.5 μm to 0.25 μm [12]. However, many of the defects which can cause failure in ICs cannot be resolved even at this level by optical microscopy. Hence the scanning electron microscope (SEM) has become a standard tool in the analysis of failure in advanced ICs [8,13,14]. The secondary electron (SEM-SE) mode is used to reveal the topography of devices with a lateral resolution down to ≈ 4 nm while the backscattering electron (SEM-BSE) imaging mode depicts the contrast between different materials due to variation in electron density. Another useful mode is electron beam induced current (SEM-EBIC) microscopy which can determine minority carrier lifetime and diffusion length at p-n doped junctions. Alternatively, the optically beam induced current (SEM-OBIC) mode creates pairs of charge carriers by bombardment with light to give visibility to the p-n junctions (sources of current) or defects (sinks of current).

However, the most important mode of the SEM for the simple dc analysis of circuits and the measurement of time response is voltage contrast (SEM-VC) microscopy [8,13,14]. In the static mode of voltage contrast at its simplest level, a potential difference of +5 V is applied between the grid of the detector and a conductor track on the chip. Any discontinuity in the conductor results in a local electric field with fewer secondary electrons reaching the detector and a darker appearance in that region. This technique is particularly effective in the location of faulty connections and defects which create open circuit on the chip surface. In the dynamic mode of VC, the SEM beam is chopped stroboscopically at exactly the same frequency as the modulation of voltage in an operational device. Measurements of time-resolved waveform can be performed from any part of the IC using a probe to explore the delays

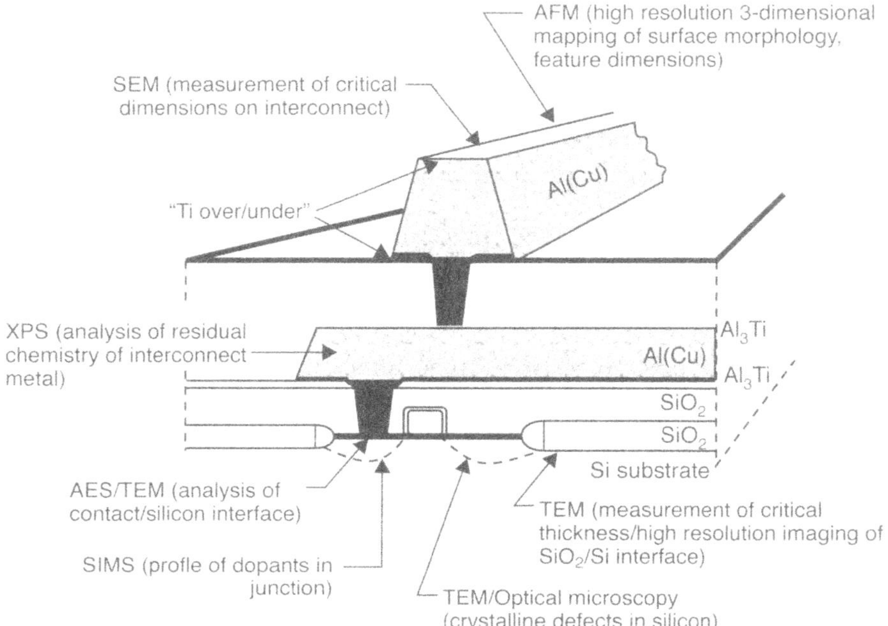

Fig. 18.2. Schematic illustration of a cross-section of a CMOS device showing examples of the use of analytical techniques to acquire data at specific features (after refs. 18.7 and 18.19).

in signal propagation and allow comparison with the logic states in the design of the circuit. This dynamic mode of SEM-VC is very useful in locating faults in very large scale integrated circuits and examples of stroboscopic images are shown in Fig. 18.2. The different logic states and state transitions can be displayed in a colour voltage contrast (CVC) mode [15]. The techniques of stroboscopic microscopy and waveform point analysis are collectively known as "e-beam testing". These testers are increasingly integrated into CAD systems for design validation as well as inspection for the validation of process steps and the failure analysis of circuits.

Another technique increasingly used in conjunction with the SEM-VC is the focused-ion beam (FIB). A finely focused beam (250 Å probe size) of energetic ions is scanned over a specific area of an IC. The beam can be used in the dual roles of imaging/analysis and modification of features in the circuit with very precise cross-sectioning or micro-milling on-chip and rapid analysis at the point of a defect [7,16]. The location of FIB within the fabrication line enables the repair or redesign of interconnects within a circuit while remaining in production.

The atomic force microscope (AFM) with capabilities of atomic level resolution and 3-dimensional mapping can perform measurements on the nm-scale features encountered in ICs. This level of precision overcomes the difficulty of the SEM in making accurate vertical measurements without the necessity for sectioning of wafers. Hence, AFMs are increasingly used in the fabrication line to provide a calibration standard for other equipment. Examples of the application of AFM include monitoring of the planarity of oxide following anisotropic plasma etching and the measurement of surface roughness, morphology and grainsize in sputtered films of TiN/TiW [17].

Some analyses require a higher resolution than is possible with SEM (≈ 4 nm) or seek additional information on the crystal structure of phases, the distribution and types of lattice defects or the structure of interfacial layers. These details are often the basis of understanding the usefulness of materials in the development of a process or in the causes of failure in devices. The transmission electron microscope (TEM) is extensively used in the high resolution imaging and chemical analysis of thin films, the features of patterned structures (including porosity, voids) and grain or crystallographic structure [8]. For example, a critical application of TEM is in the examination of the structure of the silicon/polysilicon interface which determines the attainable speed in bi-polar transistors [18]. The continued reduction in feature sizes of circuits increases the importance of defects and interfaces with further emphasis on the use of TEM to resolve the structural features. However, the lengthy and often difficult specimen preparation and the problem of locating a defect in an IC combine to limit the use of this technique.

In addition to information on morphology and structure, the elemental/compositional details of the features of ICs are of key importance in the development of process and in the diagnosis of degradation. A listing of the major techniques of compositional analysis useful in the development of semiconductor process together with a summary of lateral resolution and the depth resolution are included in Table 18.1.

Auger electron spectroscopy (AES) instruments are available with focused electron beam to ≈ 50 nm and depth resolution of 2–3 nm under optimised conditions. The highly focused electron beam enables a maximum sensitivity but with a beam current density below the limit to induce damage. The beam is often rastered over an area to minimise heating and damage and hence the name Scanning Auger microprobe (SAM). Analysis by the Auger technique is combined with sputter etching to allow profiling into the depth of the specimen. However, in the sputtering of the non-planar surface topology of ICs, it is often very difficult to take advantage of the small probe size and excellent depth resolution. The surfaces tend to become roughened during sputtering with the further potential for distortion of the profile resulting from an enhanced Auger emission from protrusions and a shadowing from recessed features. It is also necessary to use Auger Sensitivity Factors to adjust for the different probabilities of sputtering of the respective elements in a compound material. However, by careful optimisation of the parameters, highly accurate results are reported in the analysis of device structures [20].

Useful applications of AES include the characterisation of ultra-thin films of nitrided gate oxide [21], the monitoring of concentrations of contamination in barrier metals and the analysis of corrosion or contamination failures in packaged integrated circuits [22]. As a complimentary technique, X-ray photoelectron spectroscopy (XPS) is capable of gathering chemical bonding information in addition to the identification of atomic species and concentrations in the near surface region (≈ 100 Å). However, the fine depth resolution (10 nm) of XPS at small spot size is accompanied by a spatial resolution of 25 µm–2 mm. Among the many useful examples of application of XPS is the evaluation of polymer deposits and of the chemical state of SiO_2 after the reactive ion etching in fluorine based plasmas [23,24].

Secondary ion mass spectroscopy (SIMS) has a very high sensitivity to all elements in the ppm to ppb range with depth resolution of < 1 nm, making this an essential technique in elemental profiling. SIMS has been routinely used in the profiling of free carrier and impurity distributions, particularly in the implanted and annealed regions of a semiconductor [25]. It is the only major technique capable of detecting levels of dopant as low as 10^{12} atoms/cm^3 by dynamic SIMS and surface metallic impurities in Si at levels of 10^7–10^{10} atoms/cm^3 by time-of-flight or static SIMS [26]. Static-mode SIMS is usually performed with a very low beam current in the picoamp to nanoamp range and pulsed sputtering. In the analysis of dopants by dynamic SIMS, it is necessary to take samples at a comparatively large beam diameter of > 1 µm in order to obtain the highest sensitivities. However, for the probing of semiconductor devices on a nano-scale, it has been proposed that ultra-high intensity post-ionisation of all sputtered species by Nd:YAG laser is combined with a kHz pulsed micro-focused ion beam [25]. Beam diameters of 100 nm are now commercially available using liquid ion sources. Together with time-of-flight mass spectrometry, this technique is promising in the characterisation of nano-scale structures [25]. Applications of SIMS include the depth profiling of high purity semiconductor materials, the verification of concentration profiles in implanted Si substrates [27] and identification of B as a contamination in gate oxides [28].

Rutherford backscattering spectrometry (RBS) has become the primary non-destructive technique for the quantitative analysis of stoichiometry and thickness of layer structures up to 1 µm in depth [29]. A major advantage of RBS is the ability to determine absolute values of concentration of heavy elements on light substrates which are insensitive to the effects of chemical bonding. However, the interpretation of data obtained by RBS can be difficult because both mass and depth information are projected on the same scale. Although capable of depth resolution of very thin layers in the range of ≈ 100 Å, the lateral resolution of the beam (100 µm–1 mm) limits the usefulness of RBS to specially prepared samples of larger area. A useful application of RBS is in the channeling of crystals to assess the degree of crystalline perfection. In the channeling mode, RBS is very sensitive to atoms on off-lattice sites and

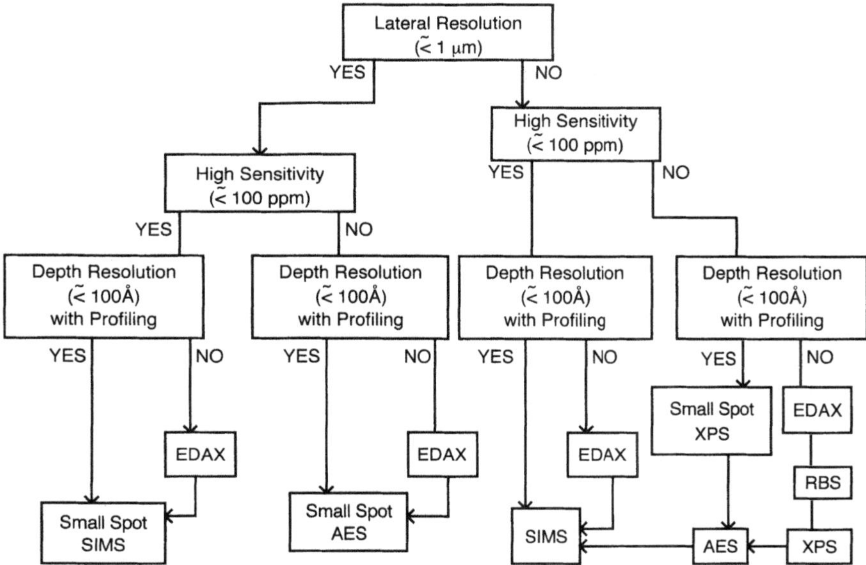

Fig. 18.3. Decision tree for the application of compositional analysis techniques in a device or circuit [30].

to defects in the crystal. While RBS has a poor sensitivity to light elements on heavy substrates, the measurement of x-rays by particle induced x-ray emission (PIXE) in conjunction with RBS provides complimentary information about heavy elements. In plan-view and cross-sectional imaging, energy dispersive x-ray (EDX) analysis provides a rapid method of multi-elemental identification with good resolution (1 μm) and a visual correlation with the SEM in the region of the sample. The depth resolution of 1–5 μm means that this is essentially a bulk technique with sensitivity of 100–2000 ppm and a limited ability to identify chemical states depending on the element.

An alternative form of protocol for the analysis of devices or circuits using these preceding techniques is shown in Fig. 18.3 [30]. Here, the choice of technique is based on the required lateral resolution, sensitivity and depth in profiling. Where the techniques are listed in a horizontal series, the first shown in the sequence is preferred on the basis of a number of factors including non-destructivity, quantitative character and cost.

18.2 Analytical Techniques in the Characterisation of Ohmic Metal/Semiconductor Contacts

An integral feature of all electronic and optoelectronic devices based on semiconductors are the metal/semiconductor contacts. These "contacts" consist of single or multiple layers of metal which are lithographically patterned into fine geometries on the surface of an n- or p-type region of semiconductor. They provide the electrical linkage for voltage and current signals between the active region of the semiconductor and the external circuit. An interaction between the metal and semiconductor is often encouraged by annealing to obtain the required metallurgical and electrical characteristics. The two specific types of metal/semiconductor contacts which are required in the design of a device are either ohmic, low resistance contacts or Schottky rectifying contacts.

Ohmic contacts must exhibit a low resistance in order to realise the full performance level of devices. Any parasitic resistance across the metal/semiconductor interface will limit the maximum frequency and cause excessive Joule heating, potentially leading to the failure of devices. Ohmic contacts therefore require a resistance which can be essentially ignored during the operation of the device and a near-linear relationship between the measured current and applied voltage. The electrical parameter most commonly used to evaluate the quality of an ohmic contact is the specific contact resistance, $\rho_c = V(x)/J(x)$ where $V(x)$ is the voltage drop across the interfacial region and $J(x)$ is the current density through the layer. A variety of methods has been used for the measurement of ρ_c [31] although the most widely adopted has been the Transmission Line Model (TLM) [32]. The measured values of ρ_c in ohmic contacts to III-V and II-VI compounds and Si depend very strongly on the details of structure and the density of defects at the metal/semiconductor interface and hence the importance of data from analytical techniques. Although the primary consideration in the choice of the metallization in an ohmic contact is a low value of ρ_c, the other requirements may be summarised as follows:

a) Thermal stability during subsequent processing of the device (up to $\approx 400°C$)
b) Negligible diffusion in a lateral or vertical direction
c) Smooth surface morphology
d) Good adhesion at the metal/semiconductor interface
e) High ductility to allow bonding with Al or Au wires
f) Minimum internal stress
g) Controlled metallurgical reaction with the semiconductor

Because of the difficulty in meeting all of the above criteria using a single layer of metal, it is common to adopt a system of 3 or more layers which attempt to obtain an optimal combination of properties. The first layer produces the required metal/semiconductor reaction and determines the electrical characteristics while providing good adhesion. The intermediate layer acts as a barrier to the diffusion of the outer film of Au into the interfacial region. A ductile metal such as Au is used as an outer layer to enable the formation of wire bonds to the contact. A further option is an intermediate layer of a dopant element which becomes incorporated into the semiconductor during the interfacial reaction.

Schottky barrier theory predicts that to produce a good ohmic contact the interfacial metal requires a work function which is smaller than the electron affinity of the semiconductor i.e. $\varphi_m < \chi_s$. However, in the case of wide band gap semiconductors including III-V compounds, the properties are dominated by the pinning of the Fermi level by interface states. This is due to the high densities of surface states which appear in the band gap of the semiconductor and effectively pin the Fermi level. The two main methods used to reduce the barrier height involve additional processing steps either ex-situ or in-situ to the epitaxial growth chamber. Ex-situ contacts are formed by the deposition of a layer or the ion implantation of a dopant metal. Practical types of ex-situ contacts to III-V compound semiconductors include a melting/alloying reaction such as in the Ni/Ge/Au system [33], a solid phase regrowth such as in Pd/Ge/Au or Pd/Si/Au systems [34] or sintering as in Ti/Pt/Au [35]. In-situ schemes use very heavy doping to $> 10^{19}\text{cm}^{-3}$ in the growth of the either the substrate or of a narrow bandgap overlayer.

An example of an epitaxial layer with a low barrier height on either GaAs or InP is $In_{1-x}Ga_xAs$. The $In_{1-x}Ga_xAs$/InP heterojunction has been widely used in the fabrication of photodiodes, multi-quantum well lasers, and high electron mobility transistors because of the high values of electron mobility and electron saturation velocity of $In_{1-x}Ga_xAs$. Furthermore, the composition of $In_{1-x}Ga_xAs$ with a mole fraction of InAs = 0.53 can be grown as lattice matched to InP and exhibits a low Schottky barrier height to both n- and p-type InP. Over the range of applications of $In_{0.53}Ga_{0.47}As$, one of the most challenging examples of an ohmic contact is formed to the thin, p-type base layer ($< 100\,\text{nm}$) on InP-based Heterojunction bi-polar transistors (HBTs). This dimension of the base layer severely limits the available depth of the metallurgical reaction allowing only a moderate level of doping (5–10 $\times 10^{18}\text{cm}^{-3}$) in the p-$In_{0.53}Ga_{0.47}As$ while requiring a low specific contact resistance of $\approx 1 \times 10^{-6}\ \Omega\text{cm}^2$. For the purposes of this section, a review is made of an ex-situ contact scheme, the Pd (20 nm) / Zn (10 nm) / Au (10 nm) / LaB_6 (100 nm) / Au (100 nm) metallisation to p-$In_{0.53}Ga_{0.47}As$/InP. In this scheme, the interfacial layer of Pd reacts with the p-$In_{0.53}Ga_{0.47}As$ by a process of regrowth in the solid phase accompanied by the release of the dopant Zn. This reaction also allows the penetration of native oxides at the interface. The intermediate film of LaB_6 is inserted in the contact as a barrier to the indiffusion of the outer layer of Au. An understanding of the reactions which occur in this complex, multi-layered structure requires a range of analytical techniques, each contributing information on the microstructure and composition of the interfacial region. In the remainder of this section, the Pd/Zn/AuLaB_6/Au to p-$In_{0.53}Ga_{0.47}As$ contacts are used as an example of the analysis of thin layered structures as encountered in modern semiconductor technology. Initially, the Pd/Zn/Pd/Au system without a diffusion barrier is examined by AES followed by characterisation of the Pd/Zn/Au/LaB_6/Au system containing a LaB_6 barrier using several other techniques.

AES is widely applied in metal/semiconductor contacts to obtain rapid, quantitative profiles of the surface and underlying layers and to determine the extent of interdiffusion during the metallurgical reaction [35–37]. For depth profiles using AES, a controlled sputtering with an ion gun is used with alternate recordings of Auger spectra. However, in multi-layered structures the ion bombardment may produce surface roughening, a preferential sputtering of some species or inhomogeneous mixing. These effects may degrade the resolution of a profiling experiment unless means such as rastering of the ion beam are adopted to minimise the heating and damage. For the analysis of Pd (10 nm) / Zn (5 nm) / Pd (20 nm) / Au (150 nm) contacts to p-$In_{0.53}Ga_{0.47}As$/InP, the relative Auger sensitivity factors were determined from pure films of Au, Pd, In and Zn and GaAs while beam current was low at 1.5 $mAcm^{-2}$ [38]. AES depth profiles of Pd/Zn/Pd/Au contacts to $In_{0.53}Ga_{0.47}As$ (Zn doped to $2 \times 10^{19} cm^{-3}$) both as-deposited and after an annealing treatment at 350 and 450°C are shown in Fig. 18.4 [38]. For the as-deposited contacts, the AES profiles demonstrate a relatively sharp interface between the Au and inner Pd layer with little evidence of diffusion or intermixing. However, the presence of Pd within the $In_{0.53}Ga_{0.47}As$ indicates an interfacial reaction even at room temperature for the unannealed contacts, as consistent with a low value of ρ_c. In comparison, at 350°C and more particularly at 450°C, the AES profiles reveal large-scale intermixing of the metal layers and $In_{0.53}Ga_{0.47}As$ with the width of the interface estimated as several hundred nm. In particular, evidence is seen of an extensive outward diffusion of both the In and Ga cationic components through the Pd and into the Au layer. Further evidence of extensive indiffusion from the profiles at 450°C was the high concentration of Au in the $In_{0.53}Ga_{0.47}As$. The distribution of Pd at the interface indicates that the extent of reaction with $In_{0.53}Ga_{0.47}As$ was marginally increased during annealing at 450°C while accompanied by a decrease in ρ_c by a factor of ≈ 2 to 5.

RBS is extensively used to determine the stoichiometry, depth and thickness of multilayered films in metal/semiconductor contacts as well as monitoring the progress of interfacial diffusion or reactions [35–37]. The RUMP code (an RBS analysis package) allows the calculation of the thickness and depth of the layers by a comparison of the experimentally determined spectra with a simulated plot. Figure 18.5 shows examples of the RBS spectra obtained using a 1.8 MeV He^{++} beam for a Pd (20 nm) / Zn (10 nm) / Au (10 nm) / LaB_6 (100 nm) / Au (100 nm) contact to p-$In_{0.53}Ga_{0.47}As$/InP [39–42]. Comparison is made between as-deposited and annealed samples. The spectra show separate peaks corresponding to the top layer of Au (1.8 MeV), the La in the film of LaB_6 at 1.6 MeV, the interfacial layer of Pd (1.5 MeV) and the plateau for $In_{0.53}Ga_{0.47}As$ at the low energy end (1.3 MeV). The peak for La was overlapped at the high energy side by the inner layer of Au so that essentially only one peak is visible. Figure 18.5 shows that annealing at temperatures of either 425 or 500°C produced only a minor reduction in height

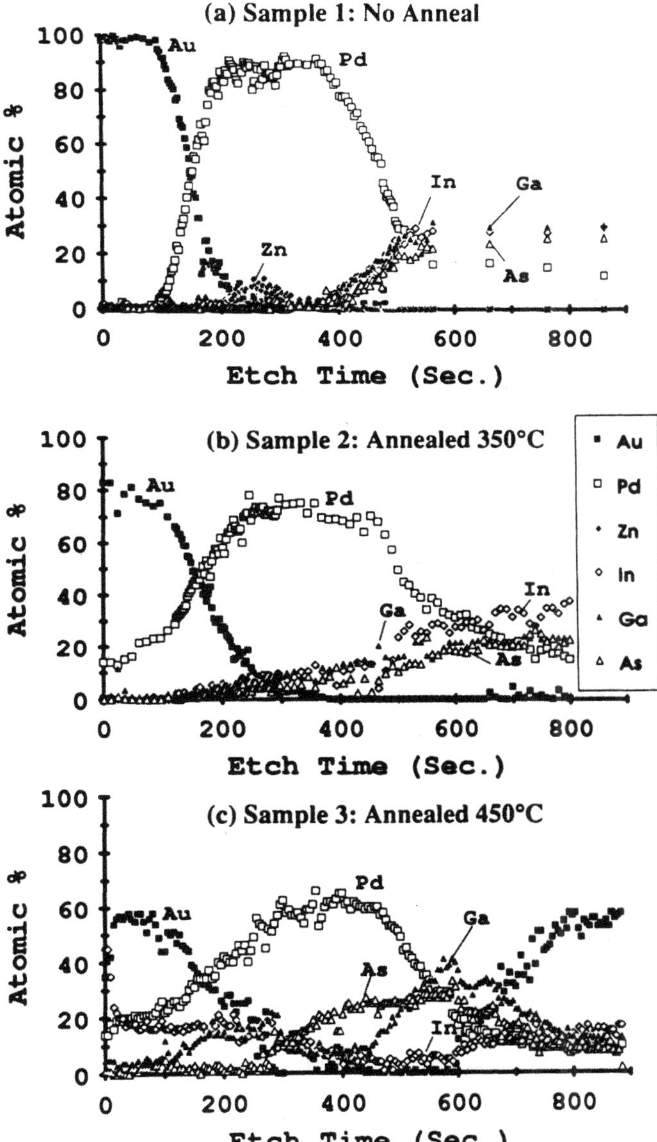

Fig. 18.4. AES depth profile of the Pd/Zn/Pd/Au to $In_{0.53}Ga_{0.47}As$/InP contacts (**a**) as-deposited and after annealing at (**b**) 350°C and (**c**) 450°C [38].

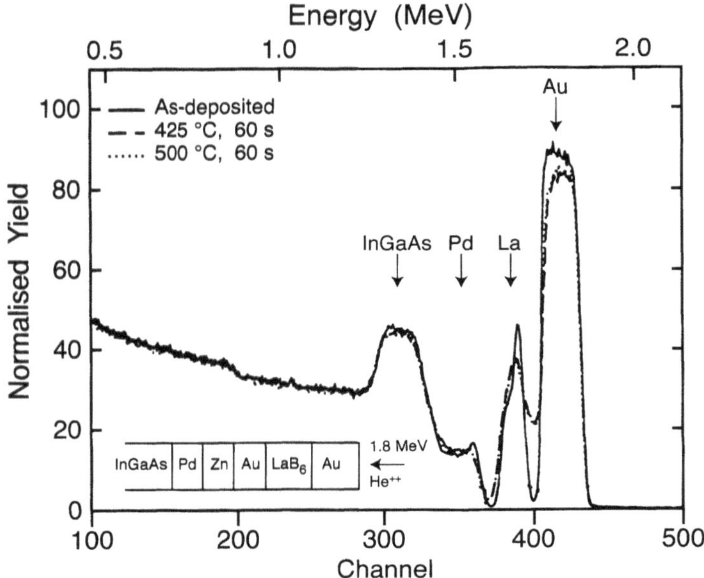

Fig. 18.5. RBS spectra of the Pd/Zn/Au/LaB_6/Au metallisation on $In_{0.53}Ga_{0.47}As$ as-deposited and after annealing at 425 and 500°C [39].

of the outer Au peak, indicating only a very limited indiffusion. The spectra also show that during annealing there was little or no change in the In, Ga and As plateau or Pd peak, with the interfacial reaction confined to a narrow width of below 100 nm without extensive intermixing. In general, these RBS spectra show the effectiveness of the LaB_6 layer in preventing an indiffusion of the outer Au layer and in enhancing the stability of the Pd/$In_{0.53}Ga_{0.47}As$ interface without a significant increase in ρ_c.

SIMS is particularly well-suited to the analysis of metal/semiconductor contacts because of its high (ppm) sensitivity for trace concentrations such as the dopant elements released into the semiconductor during annealing. This technique also provides a good resolution of the compositional cross-section of the contact, giving information on the depths of penetration of the metals. However, the depth resolution of SIMS in multi-layer contacts may be degraded by surface roughness which can be induced either by the sputter process or by annealing of the metallisation. In addition, the ion bombardment may cause a mixing or knock-on of metallic species into the semiconductor. As a consequence, it is difficult to accurately detect by SIMS the profile of a thin, low concentration region of dopant beneath a metal layer of high concentration.

These effects can be circumvented by sputtering from the substrate or backside of the contact following a chemical thinning of the sample [43,44].

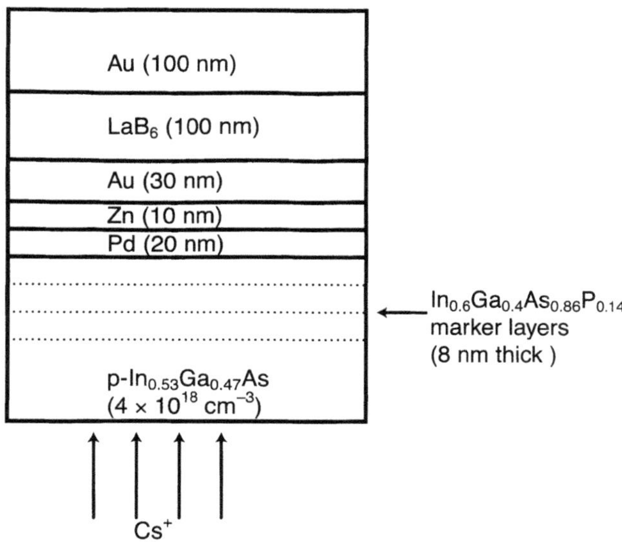

Fig. 18.6. Schematic illustration of the structure of Pd/Zn/Au/LaB_6/Au metallisation on $In_{0.53}Ga_{0.47}As$ showing method of analysis by Backside SIMS [45].

Backside SIMS gives a maximum depth of resolution for a sharp transition, moving from the low to high concentration in the profiling direction. For SIMS analysis of the Pd/Zn/Au/LaB_6/Au contacts to $In_{0.53}Ga_{0.47}As$, Fig. 18.6 shows the configuration of layers in the chemically thinned sample. Figure 18.7 shows examples of SIMS profiles using a 5.5 keV Cs^+ ion beam at 42° incidence for samples which were a) implanted with Zn ions prior to contact formation or b) contained an evaporated layer of Zn in the metallisation [45]. These samples were annealed at 400 and 425°C respectively. The distributions of In, Ga and As were normalised to the concentrations of these elements in $In_{0.53}Ga_{0.47}As$. In these profiles, the sputter time on the horizontal axis equals the depth in nm within the region of semiconductor as determined from the known location of the P-based markers. The maximum depth affected by metallurgical reactions is determined by the onset of deviation in the semiconductor from its bulk level at the interface. Diffusion is evident beyond the depth of reaction of the semiconductor. Also evident in Fig. 18.7 in both of the types of contact is a reaction of Pd with $In_{0.53}Ga_{0.47}As$ to a depth of ≈ 100 nm. Analysis by SIMS has revealed different profiles of Zn for the implanted and evaporated contacts. In the ion implanted contacts, Fig. 18.7(a), an accumulation of Zn is shown at the interface while in the evaporated contacts, Fig. 18.7(b), there is diffusion of Zn away from the Zn layer towards the interface. The depth of diffused Zn from the evaporated contact exceeds the reaction depth by ≈ 70 nm while the implanted Zn remains within the reaction depth. In addition, SIMS has shown that the

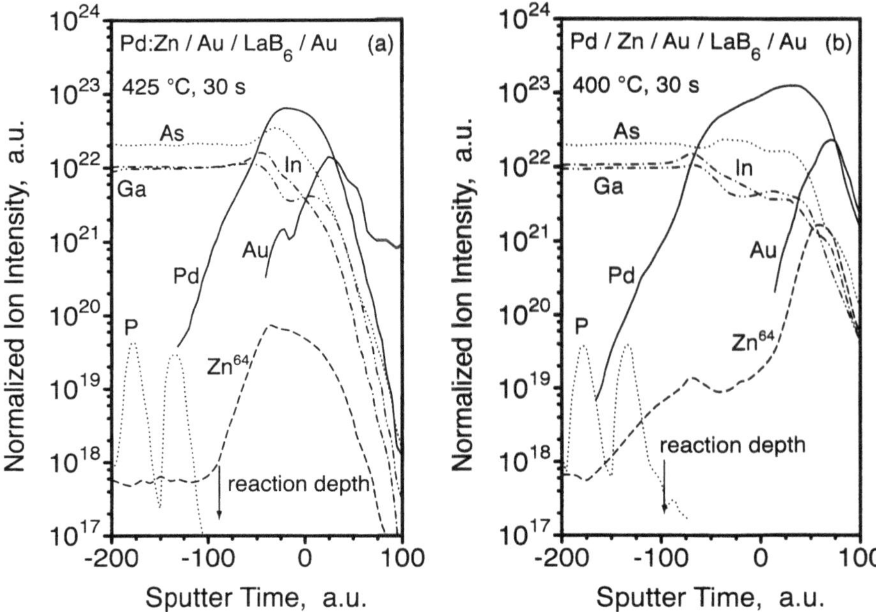

Fig. 18.7. Normalised backside SIMS profiles of Pd/Zn/Au/LaB_6/Au metallisation on $In_{0.53}Ga_{0.47}As$ with the Zn either a) implanted or b) evaporated. The sputter time equals depth in nm within the semiconductor [45].

interfacial concentration of Zn was $> 10^{19}cm^{-3}$ at the interface for evaporated Zn contacts and $\approx 10^{18}cm^{-3}$ for the implanted contacts. The onset of high interfacial concentrations of Zn as determined by SIMS coincided with the onset of a low resistance behaviour for annealing temperatures $\geqslant 375°C$. Higher concentrations of Zn in the interfacial region achieved in evaporated contacts produced a lower specific contact resistance $< 1 \times 10^{-6}\Omega cm^2$ on p-$In_{0.53}Ga_{0.47}As$ than the comparatively low concentrations by implantation of Zn. Using SIMS profiles, it is possible to accurately determine the depth of reaction as a function of the implant dose or the thickness of the evaporated Zn layer and annealing temperature. Data from SIMS profiles indicates that the incorporation of a layer of Zn in the metallisation acts to catalyse a reaction with the $In_{0.53}Ga_{0.47}As$.

TEM combined with EDX can provide information on the distribution and composition of phases and the lateral roughness at the interface. These techniques are frequently combined with XRD analyses to obtain details of the quality of crystal. The combined TEM, EDX and XRD are used extensively in fundamental studies of ohmic contacts as well as more generally as a tool for the development and optimisation of metallisation as a processing step. Prior to examination by TEM, the structure is thinned to $< 3\,\mu m$ using mechanical

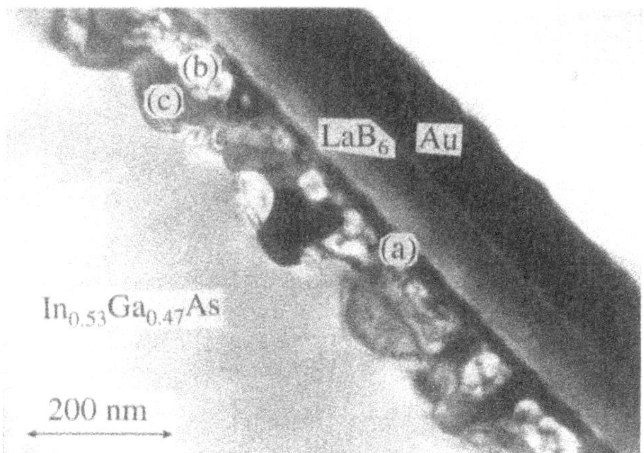

Fig. 18.8. Cross-sectional transmission electron micrograph of a Pd/Zn/Au/ LaB_6/Au on $In_{0.53}Ga_{0.47}As$ contact annealed at 425°C for 30 s [45].

polishing, dimpling and Ar ion milling. Figure 18.8 shows a cross sectional TEM micrograph of a Pd/Zn/Au/LaB_6/Au contact to $In_{0.53}Ga_{0.47}As$ annealed at 425°C [45]. The technique allows a clear identification of the Au, LaB_6 and $In_{0.53}Ga_{0.47}As$ regions and of the reacted interfacial region formed during the thermal treatments. Electron diffraction has shown the LaB_6 as amorphous. Within the reacted interfacial region, TEM reveals 3 zones: (a) a thin (≈ 7 nm) almost continuous band adjacent to the LaB_6 (b) areas of pronounced contrast which were amorphous with a 3-dimensional morphology resembling a sponge and (c) large crystallites with a mean size of ≈ 95 nm. EDX analysis with a local resolution of < 20 nm is able to determine the composition of the patterns except for region (a) which was below the resolution of the apparatus. The contrast-rich regions contain 31 at% Pd, 8 at% Au, 3 at% Zn, 6 at% In and 27 at% Ga and 23 at% As while the areas of large crystallites comprise 25 at% Pd, 14 at% Au, 3 at% Zn, 14 at% In, 28 at% Ga and 13 at% As. Both types (b) and (c) regions thereby contain all elements from the metallisation and the semiconductor.

XRD analysis provides further information on the phases formed during thermal treatment. Prior to analysis, the top layer of Au is etched off to avoid a possible overlap of peaks. The spectra in Fig. 18.9 allow identification of two compounds, cubic $PdAs_2$ and hexagonal $Pd_{12}Ga_5As_2$. The identification of the phases involves a comparison of the XRD peak family with calculated values assuming $a = 0.9468$ nm and $c = 0.372$ nm and analysis of single films of Pd on $In_{0.53}Ga_{0.47}As$ [45]. These spectra, combined with EDX data, suggest that the large crystallites in areas c) in Fig. 18.9 are $Pd_{12}Ga_5As$ where part of the Pd and Ga are replaced by Au and In. Combined TEM and XRD data show the formation of $(Pd, Au)_{12}(Ga, In)_5As_2$ phase during annealing at > 360°C. The formation of this solid solution phase during

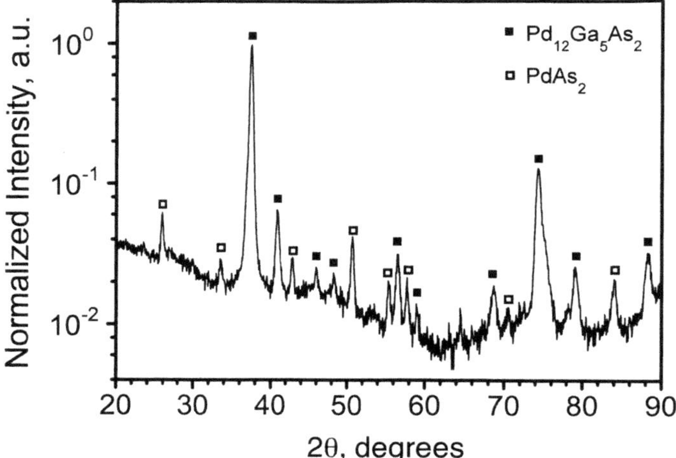

Fig. 18.9. X-ray diffraction spectrum of a Pd/Zn/Au/LaB_6/Au metallisation on $In_{0.53}Ga_{0.47}As$ annealed at 425°C for 30 s [45].

annealing correlates with high interfacial concentrations of dopant Zn at the interface and with the onset of ohmic behaviour producing a minimum in the value of ρ_c of $< 1 \times 10^{-6}\ \Omega cm^2$.

18.3 Summary

Analytical techniques fulfil a critical role in both the development of the processing steps and in the diagnosis of failure in integrated circuits. The selection of the most suitable technique(s) depends on requirements of the level of sensitivity, the lateral resolution, the depth resolution and the cost of analysis. Trade-offs are required between a small sampling volume, high resolution and an enhanced sensitivity. The characterisation of the Pd/Zn/Au/LaB_6/Au metallisation to a p-type $In_{0.53}Ga_{0.47}As$ substrate is used as an example of the analysis of a complex, multi-layered structure. The application of a range of techniques (AES, RBS, SIMS, TEM, EDX and XRD) provides details of the interfacial structure, the extent of interdiffusion between layers and the distribution of the dopant element.

References

1. C.R. Barrett, MRS Bulletin, **18(7)**, 3, (1993)
2. J.G. Ryan, R.M. Geffken, N.R. Poulin and J.R. Paraszczak, IBM J. Res. Develop., **39(4)**, 371, (1995)
3. L. Peters, Semiconductor International, **21(1)**, 61, (1998)
4. M. Lerne, Proc. ESSDERC, Ed. H. Grunbacher, Stuttgart, 42, (1997)

5. *The National Roadmap for Semiconductors (Semiconductor Industry Association)*, San Jose, CA, (1994)
6. T. Ohba., MRS Bulletin, **20(11)**, 46, (1995)
7. H. Fatermi: in *Electronic materials chemistry*, ed. by H.B. Pogge, (Marcel Dekker Inc., New York, 1995), 367
8. B.P. Richards, P.K. Footner, *The role of microscopy in failure analysis*, (Oxford University Press, 1992), 2
9. C.R.M. Grovenor (Ed.): *Microelectronic Materials 1. Microelectronic Devices. Design and Manufacture*, (IOP Publishing, Bristol, 1992), 482
10. C.M. Bailey, and D.H. Hensler: in *ASM Metals Handbook, 9th Ed.*, **11**, 766, (1989)
11. H.W. Werner: in *Microelectronic Materials and Processes*, (Kluwer Academic Publishers, Netherlands), 845, (1989)
12. W. Hunn, Semiconductor International, **21(8)**, 277, (1998)
13. B.P. Richards, P.K. Footner, Metals and Materials, 75, (1989)
14. S. Gorlich and E. Wolfgang: in *Analysis of microelectronic materials and devices*, ed. By M. Grasserbauer and H.W. Werner, (John Wiley and Sons, UK, 1991), 811
15. M. Dax, Semiconductor International, **20(2)** , 60, (1997)
16. A. Braun, Semiconductor International, **20(12)**, 105, (1997)
17. D. Praminik, M. Weling and Li Zhou, Solid State Technology, **37(7)**, 79, (1994)
18. J.M. Stork, E. Gavin, J.D. Cressler, G.L. Patton and G.A. Sai-Halasz, IBM J. Res. Develop., **31(6)**, 617, (1987)
19. J.M.E. Harper and K.P. Rodbell, J. Vac. Sci. Technol., **B15(4)**, 763, (1997)
20. W. Palmer, Surface and Interface Analysis, **22**, 331, (1994)
21. G. Yoon, and Y. Epstein, J. Electr. Soc., **145(5)**, 1679, (1998)
22. S. Wolf and R.N. Tauber, *Silicon processing for the VLSI era- Vol.1 Process Technology*, (Lattice Press, Sunset Beach, CA, 1986), p.603
23. P. Czuprynski and O. Joubert, J. Vac. Sci. Technol., **B16(3)**, 1051, (1998)
24. M.H. Kibel and P.W. Leech, Surface and Interface Analysis, **24**, 605, (1996)
25. S.W. Downey: in *Handbook of Compound Semiconductors*, ed. by P.H. Holloway, G. E. McGuire, (Noyes Publications, NJ, 1995), 653
26. P.K. Chu, B.W. Schueler, F. Reich and P.M. Lindley, J. Vac. Sci. Technol., **B15(6)**, 1908, (1997)
27. J. Sun, R. Bartholomew, K. Bellur, A. Srivastava, C. Osburn and N. Masnari, J. Electr. Soc., **144(10),** 3659, (1997)
28. K. Kawagishi, M. Susa, T. Maruyama and K. Nagata, J. Elect. Soc., **144(9)**, 3270, (1997)
29. A.E. Morgan: in *Characterisation of semiconductor materials: principles and methods* ed. by G.E. McGuire, (Noyes Publications, NJ, 1989), 49
30. C.R. Helms, Mat. Res. Soc. Symp. Proc., **69**, 3, (1986)
31. A. Scorzoni and M. Finetti, Materials Science Reports, **3**, 79, (1988)
32. G.K. Reeves and H.B. Harrison, IEEE Electron Device Letts, **EDL-3(5)**, 111, (1982)
33. M. Murakami, K.D. Childs, J.M. Baker and A. Callegari, J. Vac. Sci. Technol., **B4(4)**, 903, (1986)
34. L.C. Wang, X.Z. Wang, S.N. Hsu, S.S. Lau, P.S. Lin, T. Sands, S.A. Schwartz, D.L. Plumpton and T.F. Kuech, J. Appl. Phys., **69**, 4364, (1991)
35. A. Katz: in *Handbook of compound semiconductors*, ed. by P.H. Holloway, and G.E. McGuire, (Noyes Publication, NJ), 170, (1995)

36. C.J. Palmstrom, and D.V. Morgan, in *Gallium Arsenide, Materials, Devices and Circuits*, Eds. M.J. Howes and D.V. Morgan, (J. Wiley and Sons), 195, (1985)
37. T.C. Shen, G.B. Gao, and H. Morkoc, J. Vac. Sci. Technol., **B10(5)**, 2113, (1992)
38. P.W. Leech, G.K. Reeves and M.H. Kibel, J. Appl. Phys., **76(8)**, 4713, (1994)
39. P. Ressel, P.W. Leech, G.K. Reeves, W. Zhou and E. Kuphal, Appl. Phys. Letts., **68(13)**, 1841, (1996)
40. P.W. Leech, P. Ressel, G.K. Reeves, W. Zhou and E. Kuphal, Mater. Res. Soc. Symp. Proc., **406**, 419, (1996)
41. P.W. Leech and G.K. Reeves, Thin Solid Films, **298**, 1, (1997)
42. P.W. Leech, G.K. Reeves, W. Zhou and P. Ressel, J. Vac. Sci. Technol., **B16(1)**, 227, (1998)
43. J. Herniman, J.S. Yu and A.E. Staton-Bevan, Appl. Surf. Science, **52**, 289, (1991)
44. S.A. Schartz, M.A. Pudensi, T. Sands, T.J. Gmitter, R. Bhat, M. Koxza, L.C. Wang and S.S. Lau, Appl. Phys. Lett., **60(9)**, 1123, (1992)
45. P. Ressel, P.W. Leech, P. Veit, E. Nebauer, A. Klein, E. Kuphal, G.K. Reeves and H.L. Hartnagel, J. Appl. Phys., **84(2)**, 861, (1998)

19 Characterisation of Oxidised Surfaces

J.L. Cocking and G.R. Johnston

Metals, in general, critically depend on their surface oxide scales for environmental stability, particularly in aggressive oxidising atmospheres at high temperatures. The protective capabilities of oxides are dependent on many physical and chemical properties, as well as on their mechanical adherence to the metal surface. In summary, an "ideal" protective oxide would be:

- physically and chemically stable. An ideal oxide would not dissociate nor melt at the temperatures and pressures of interest;
- mechanically stable. The scale would be capable of maintaining intimate contact with the surface of the metal, particularly when sudden temperature changes occur;
- a barrier to diffusion. The function of a protective oxide is to separate the metal from the oxygen in the gas phase. The ideal oxide would, therefore, have a low diffusion rate for both oxygen and metal ions, otherwise the oxidation reaction would proceed at the oxide/metal or oxide/gas interfaces respectively;
- continuous and dense. When pores or cracks are present in the oxide scale the protective capabilities of the oxide are lost.

All of these properties that are essential for oxidation protection are dependent on the chemistry and morphology of the oxide scale, the oxide/metal interface and the near-surface region of the metal.

So that alloys can be developed with improved oxidation resistance, it is essential to know the mechanisms whereby oxidation occurs. Such knowledge can only be obtained through characterisation of oxidised surfaces, particularly of surfaces in the initial stages of the oxidation process. The characterisation of thin oxide scales, particularly for oxides less than a micron thick, has only been possible in recent years with the development of the array of sophisticated surface analytical equipment described elsewhere in this book. A number of these techniques have been used by the authors to characterise oxidised surfaces of metals [1–3]. In this review, application of three techniques, namely scanning Auger microscopy (SAM), microbeam Rutherford backscattering spectrometry (μ-RBS) and extended X-ray absorption fine structure (EXAFS), to the study of the oxidation of selected cobalt- and nickel-based alloys is described. SAM and μ-RBS are powerful tools, particularly when used together, for analysing multi-phase surfaces where spatial resolution of better than 20 μm is required, whereas EXAFS has great potential for the non-destructive determination of atomic structure in thin surface layers.

19.1 The Oxidation Problem

Nickel- and cobalt-based superalloys are used extensively in demanding high temperature applications such as hot end components in gas turbine engines. The alloy chemistry of the superalloys has been formulated to give maximum strength and not maximum oxidation resistance. Protective coatings are therefore applied to the surfaces of turbine components such as blades, vanes and combustion chamber liners, to increase their oxidation resistance and improve their durability. Optimum oxidation protection of alloys and coatings for high temperature use is provided by oxide scales of alumina (Al_2O_3). Chromia (Cr_2O_3) and silica (SiO_2) are also protective oxides, but Cr_2O_3 has limited use at high temperatures because of its volatility in oxidising atmospheres, while SiO_2 has a thermal expansion coefficient which is incompatible with most high temperature alloys and hence its adherence is poor. In practice, the protective oxide scales that form are never exclusively a single oxide, but instead are mixtures of oxides and spinels.

The generic formulation for one class of oxidation resistant coatings used both in gas turbine engines (particularly those used in shipboard applications) and in thermal barrier coatings (as the oxidation-resistant layer beneath the outer ceramic layer) is MCrAlX, where M is either Ni, Co or Fe and X is an element such as Hf, Y or Ce, which has a high affinity for oxygen. The disproportionate and beneficial effects produced by minor additions of these oxygen-reactive elements on the adherence of the oxide to the surface was the principal reason that this present work was undertaken. Many mechanisms have been proposed to explain their role in improving the adherence and hence the oxidation resistance of MCrAl alloys, but no consensus has been reached. The mechanisms proposed include:

i) mechanical "keying" or "pegging" of the scale to the substrate [4];
ii) sulphur "gettering" by the oxygen-reactive element, which is also sulphur-reactive, thus tying up the sulphur impurities in the alloy [5]; and
iii) a combination of (i) and (ii).

Almost all previous mechanistic studies of the oxidation of high temperature alloys have been reported for long oxidation times, where the thickness of oxide film is typically much greater than a micron. Similarly where more than one phase is present, little or no attempt has been made to resolve the oxidation behavior of each phase. In the work briefly reviewed here, which was a collaborative study involving Materials Research Laboratory, Melbourne, the US Naval Research Laboratory, Washington DC, the University of Pittsburgh PA, the Royal Melbourne Institute of Technology and the University of Melbourne, the following programme was undertaken:

- the initial stages of oxidation were studied, i.e. oxide films much less than 1 μm in thickness were characterised;
- the oxidation mechanisms for both phases of a two-phase alloy were investigated; and

- the effects of specific elements on the initial oxidation mechanisms of both phases was studied using ion implantation.

The experimental procedures for sample preparation, ion implantation and surface analysis techniques employed have been detailed previously [6,7]. With the exception of EXAFS, the analytical techniques from which the results were obtained have been described elsewhere in this book.

19.2 Oxidation of Co-22Cr-11Al

The first example presented in this brief review of the application of microanalytical techniques to oxidation studies is the oxidation of Co-22Cr-11Al (nominal composition, wt%) cast alloys. This alloy has a two-phase structure consisting of a matrix of β-CoAl with α-Co solid solution precipitates ranging in size from 5 to 20 µm. The oxidation experiments were performed, in air, at 700°C for times between 10 min and 96 h. One surface of each sample was metallographically polished to a 0.25 µm diamond finish. Half of this polished surface was implanted with 2×10^{16} Hf (or Y) ions/cm^2 with an energy of 150 keV, while the other half was shielded with a tantalum mask. This technique allowed the in situ comparison of the oxidation of non-implanted and implanted alloys on the same surface.

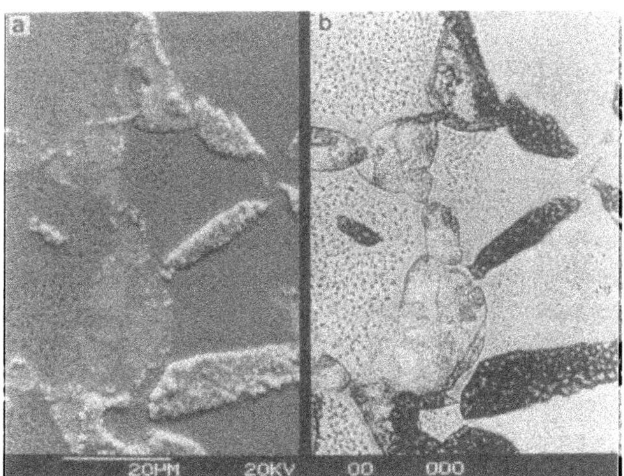

Fig. 19.1. Secondary electron (**a**) and backscattered electron (**b**) micrographs of the boundary region between the non-implanted (left of boundary) and Hf-implanted (right of boundary) areas of a Co-22Cr-11Al surface oxidised for 9.5 h at 700°C. These micrographs show the presence of voids in the non-implanted alloy and their suppression following Hf implantation

The effects of specific elements on the oxidation mechanisms are visibly apparent with this experimental arrangement. After oxidation, the two halves of the specimens appear different, even to the naked eye. The most striking effect of ion implantation with Y and Hf is the marked suppression of void formation in the metal at the metal-oxide interface, predominantly at the α/β phase boundaries [7]. The scanning electron micrograph in Fig. 19.1 clearly shows the voids in the metal in the non-implanted alloy seen through the thin, transparent oxide film on the β-phase. Across the sharp boundary, it is obvious that Hf implantation has altered the oxidation mechanism and suppressed void formation. The critical contribution of this work is the determination of the chemical composition and thickness of the oxide films, which is now described.

19.2.1 Chemical Characterisation

The two most powerful microanalytical tools for analysing the oxides on both the α and β phases are scanning Auger microscopy (SAM) and microbeam Rutherford backscattering spectrometry (μ-RBS). These two techniques are extremely powerful tools when used in combination. High resolution SAM gives chemical composition data in the top 4 to 10 atomic layers of the surface. For depth information from SAM, ion beam milling must be used. This technique, however, is at best semi-quantitative with respect to depth because of many experimental problems including:

(i) preferential sputtering of certain elements with respect to others;
(ii) different milling rates through different phases; and
(iii) redeposition of sputtered material onto the site under analysis.

With the present experimental technique where both the implanted and non-implanted surfaces are on the same alloy face (and in some cases at the boundary, on the same grain), SAM depth profiles give accurate relative depth profiles for the oxides.

RBS and μ-RBS on the other hand are non-destructive techniques that give quantitative depth information from the backscattered spectra. The two techniques (SAM and μ-RBS) therefore provide complementary information for very detailed surface characterisation.

19.2.2 Scanning Auger Microscopy

One of the great strengths of SAM is its ability to determine quantitatively the concentration of species present on the surface of a sample in the first 4 to 10 atomic layers. Its relative weakness is in the assigning of depth from the original surface following ion beam milling. To overcome this potential problem, all thickness determinations in this work are reported as relative ion beam milling times rather than as units of thickness. As mentioned above, RBS is used to confirm relative thicknesses. Ion beam milling was performed on areas that included the interface between the implanted and non-implanted regions. Information obtained using the SAM technique is now illustrated with results obtained for oxidised yttrium-implanted alloys.

Figure 19.2a shows a scanning electron micrograph of the area of interest for a sample which was oxidised for 60 min in air at 700°C. The Y-implanted region is on the right hand side of the micrograph. A number of analysis points were chosen in both the α and the β phases on both the implanted and non-implanted regions. The five elements of interest – Co, Cr, Al, Y and O – were monitored before, during and after ion milling through the oxide into the underlying alloy. Milling was interrupted at selected times to allow elemental area maps to be made. These times were chosen at, or near, the interface between the oxide and the alloy for each phase, and examples are shown in the remainder of Fig. 19.2. The left hand series are the area maps obtained after 5 minutes of milling (near the oxide/alloy interface for Y-implanted β-phase) and the right hand series after 14 minutes milling (near the oxide/alloy interface for the Y-implanted α-phase). Depth profiles of the five elements, at each analysis point, were also constructed, using sensitivity factors to convert the Auger signal to elemental concentrations. Examples of depth profiles for non-implanted β-phase and Y-implanted β-phase are given in Fig. 19.3a and b respectively. The sensitivity factor used to construct Fig. 19.3 pertain to the elements in the oxide and therefore the concentrations in the metal are not accurate. It is possible to construct composite concentration profiles using sensitivity factors for elements in both the oxide and the metal, but this was not done in Fig. 19.3.

The depth profiles are strikingly similar to the concentration profiles generated from RBS spectra (see next section). Preliminary correlations between the two techniques have been published [2], with a more detailed publication in preparation. The area maps in Fig. 19.2 show quite distinctly the very uniform thicknesses of the oxides on the same phases. The area maps also show the positions of the implanted layer in each of the α and β phases after oxidation. The Y Auger signal was too weak to monitor during depth profiling – it was only possible to detect it accurately during the compiling of the area maps. In agreement again with RBS results, the Y is in the oxide scale near the oxide/alloy interface. Because of the considerable difference in the thickness of the oxide which formed on the two phases, the Y-rich layer on the α-phase is located at a vastly different depth beneath the outer surface than the corresponding layer on the β-phase.

19.2.3 Rutherford Backscattering Analysis

The μ-RBS technique, which has a spatial resolution of about 10–20 μm and therefore gives information on the individual phases, has been used to characterise both implanted and non-implanted, α and β phases of the alloys. Spectra for the β-phase of the oxidised Co-22Cr-11Al, both Hf-implanted

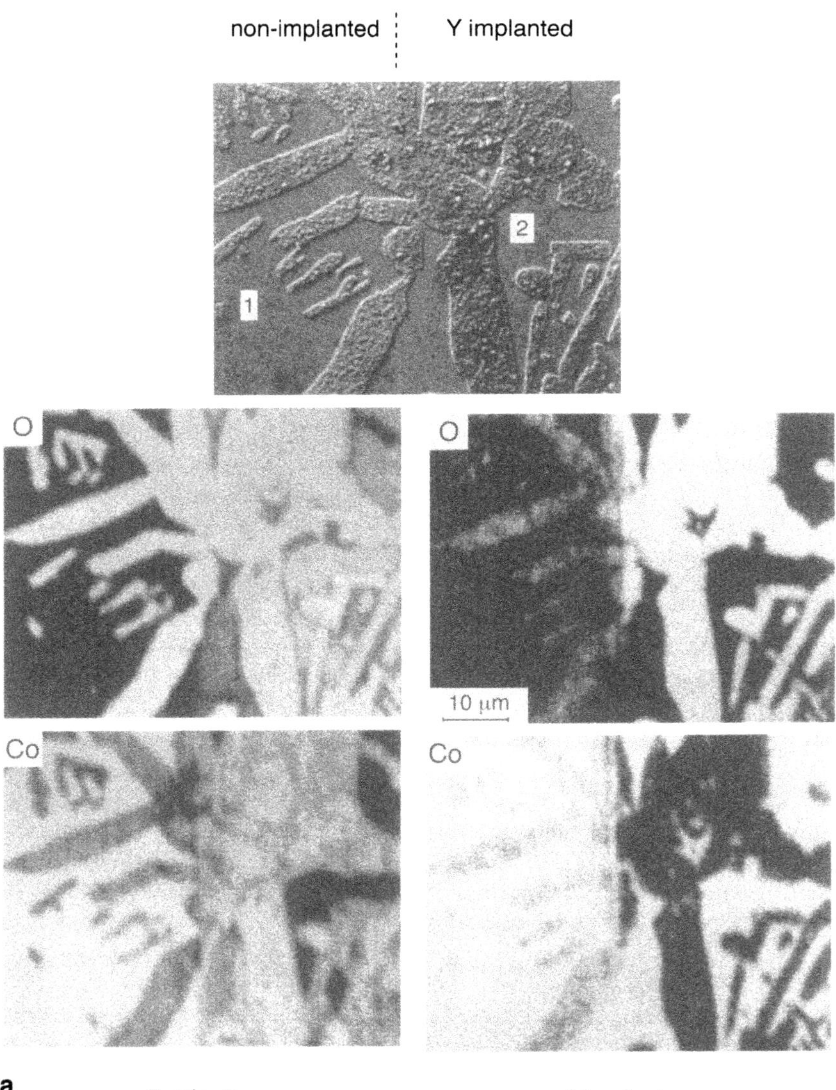

Fig. 19.2. For caption see the opposite page

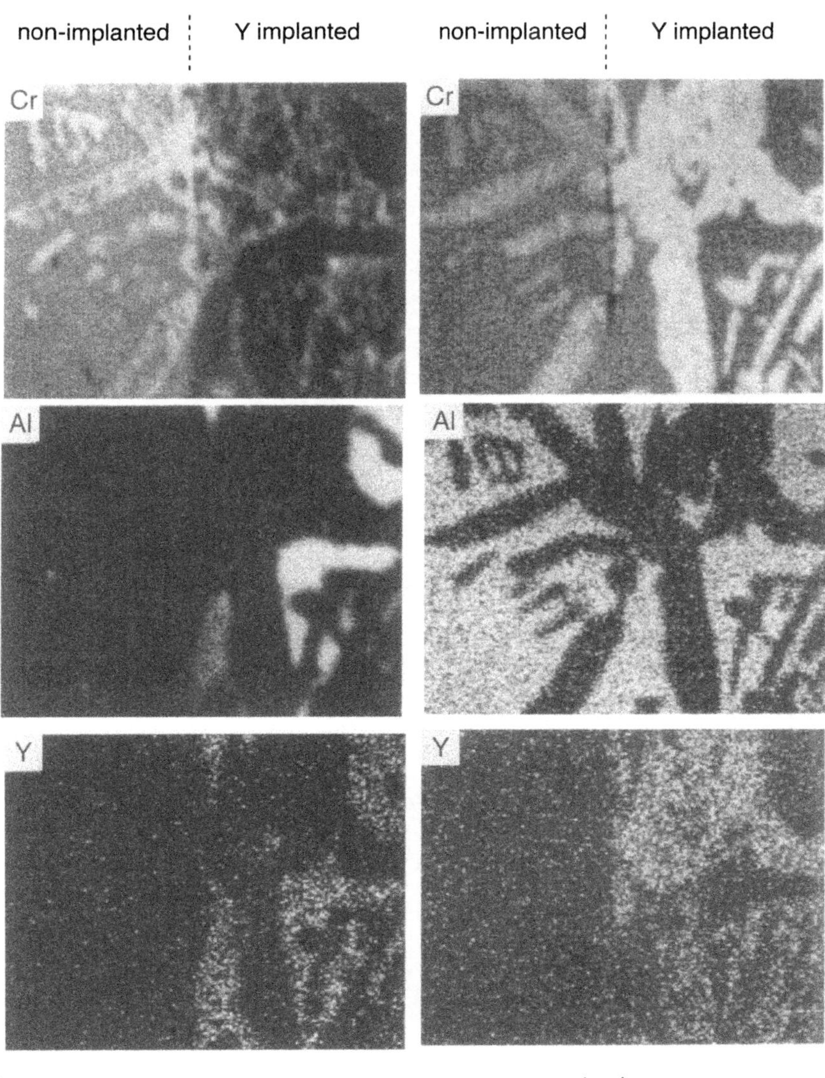

Fig. 19.2. Scanning electron micrograph and Auger area maps at the interface region between non-implanted (left of boundary) and Y-implanted (right of boundary) Co-22Cr-11Al oxidised in air for 1 h at 700°. The Auger maps on the left side were recorded after 5 min of ion beam milling, while the maps on the right side were obtained after 14 min of milling. Depth profiles at points 1 and 2 are shown in Fig. 19.3

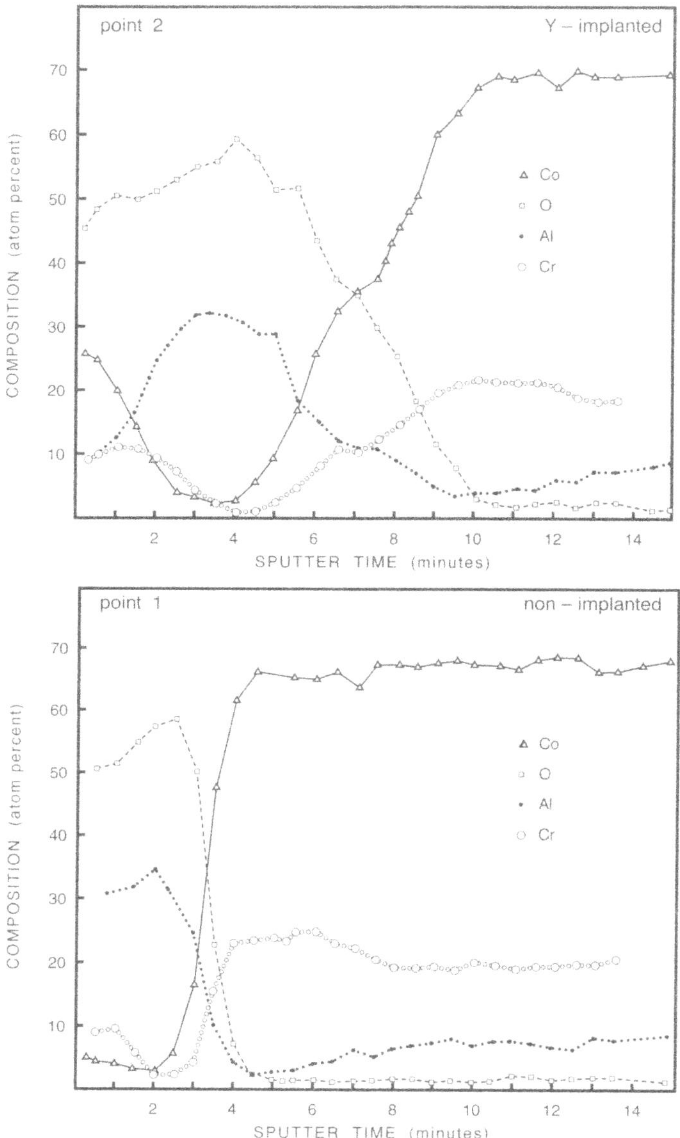

Fig. 19.3. Auger element depth profiles for Co, Cr, Al and O (Y concentration too low to monitor in this mode) at the two points designated in Fig. 19.2, i.e. point 1 – β-phase, non-implanted; point 2 – β-phase, Y-implanted

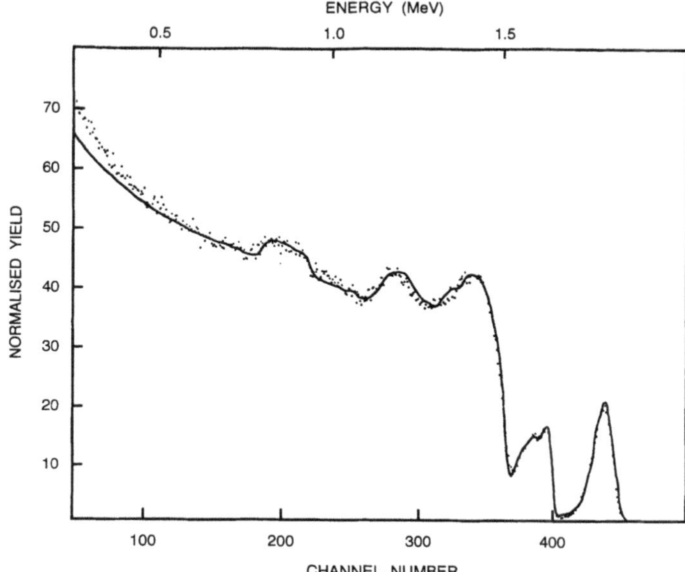

Fig. 19.4. Experimental Rutherford backscattering spectrum (*dots*) of Hf-implanted, β-phase Co-22Cr-11Al oxidised in air for 1 h at 700°C, together with the theoretical spectrum (*solid line*) iteratively generated from the concentration profile shown in Fig. 19.6

and non-implanted, are given in Figs. 19.4 and 19.5 respectively. Figures 19.6 and 19.7 give the respective elemental concentration profiles obtained by deconvolution of the spectra using the RUMP programme [3]. The important results obtained from Figs. 19.4 to 19.7 are:

i) The oxide on the Hf-implanted β-phase is more than twice as thick as the oxide on the non-implanted β-phase, in close agreement with the result illustrated above for the Y-implanted alloy;
ii) Cobalt is enriched at the outer surface of the Hf-implanted β-phase, where its concentration is at least five times greater than on the corresponding non-implanted phase, again in agreement with the SAM result. Both β-phase oxides are predominantly Al rich (presumably Al_2O_3) as would be expected for the oxidation of CoAl;
iii) the final position of the ion implanted layer of Hf, after oxidation, is near the oxide/metal interface. This is also true for the oxidised α-phase, even though the α-phase oxide is appreciably thicker than the β-phase oxide [2], again in agreement with the SAM results.

Perhaps the most important contribution of RBS and μ-RBS to this surface characterisation exercise is that the thickness of the various oxide films are quantitatively defined without recourse to milling techniques and without the

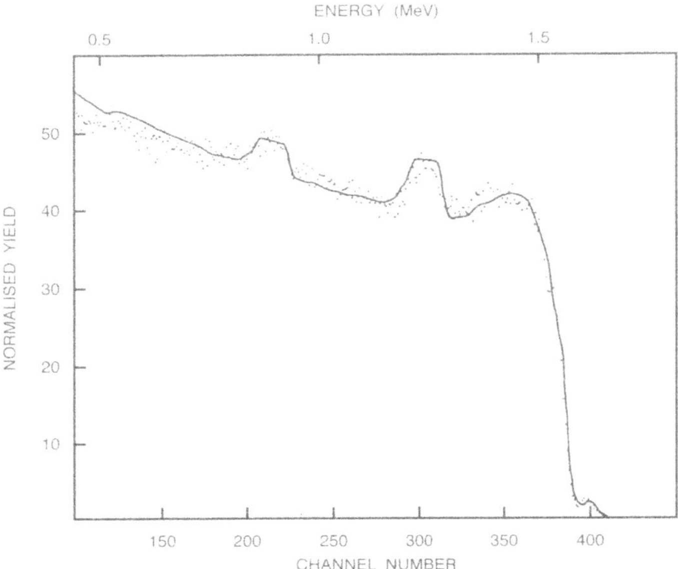

Fig. 19.5. Experimental Rutherford backscattering spectrum (*dots*) of non-implanted, β-phase Co-22Cr-11Al oxidised in air for 1 h at 700° C, together with the theoretical spectum (*solid line*) iteratively generated from the concentration profile shown in Fig. 19.7

inaccuracies inherent in the use of sensitivity factors (Sect. 19.2.2). Milling techniques, which are essential for SAM analyses, introduce uncertainties in thickness measurements through physical processes that include preferential sputtering of elements from the surface, activated diffusion on the surface and redeposition of sputtered material. Detailed mechanisms proposed to account for the features will be addressed in a future publication. The aim in the present publication has been to illustrate the great capabilities of SAM and μ-RBS in completely characterising the oxidised surfaces of individual phases as small as about 20 μm in diameter.

19.3 Oxidation of Ni-18Cr-6Al-0.5Y

The second example presented in this review is the oxidation of Ni-18Cr-6Al-0.5Y (nominal composition, wt%) alloys, ion implanted with V ions. Some samples were produced as castings, which were subsequently polished, while other samples were produced by plasma spraying. The surfaces of the samples were implanted with 2×10^{16} (low dose) or 2×10^{17} V ions/cm^2 (high dose) with an energy of 85 keV. Both non-implanted and ion implanted samples were oxidised in a mixture of 80% Ar and 20% 0_2 for 10 min at 900°C. Photoelectron EXAFS (extended X-ray absorption fine structure) spectra

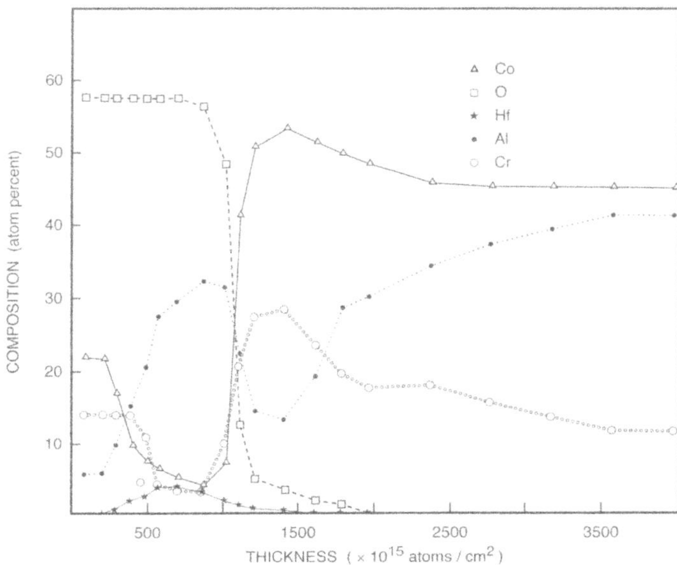

Fig. 19.6. Elemental concentration profiles of the oxide and near-surface region of oxidised, Hf-implanted β-phase Co-22Cr-11Al iteratively generated to produce the spectrum (*solid line*) in Fig. 19.4

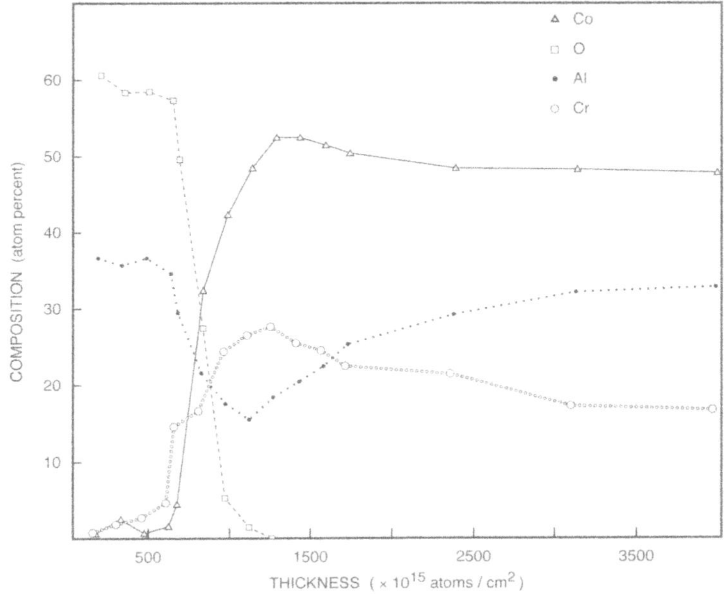

Fig. 19.7. Elemental concentration profiles of the oxide and near-surface region of oxidised, non-implanted β-phase Co-22Cr-11Al iteratively generated to produce the spectrum (*solid line*) in Fig. 19.5

were collected from all samples. The EXAFS data were used in conjunction with RBS, SAM and STEM data to monitor the changes induced by the implantation and oxidation processes. Only the EXAFS data will be discussed in this section.

19.3.1 Extended X-Ray Absorption Fine Structure

EXAFS is a technique which utilises synchrotron radiation. The technique involves measurement of small oscillations which are apparent in the region about 1000 eV above the K- (and L-) shell absorption edges in spectra of X-ray absorption edge coefficients versus X-ray photon energy. These small oscillations arise when an electron leaving the atom from whose shell it has been ejected backscatters from its nearest neighbor atoms, hence causing interference with the photoelectron wave emerging from the atom. Data obtained from the oscillations enable the average distance of a particular type of atom from its nearest neighbors to be determined. In addition, it is theoretically possible to determine the number and the kind of neighbors, as well as the distance to more distant neighbors with the EXAFS technique.

Because each element has different electron binding energies, individual elements in a polyatomic material can be selectively excited, thus enabling the local environment of the atoms in a material to be characterised in a way that is not possible by conventional diffraction techniques. An additional advantage of this technique is that the escape depth of the photoelectrons is of the same order as that of the implantation depth, i.e. around 100 nm.

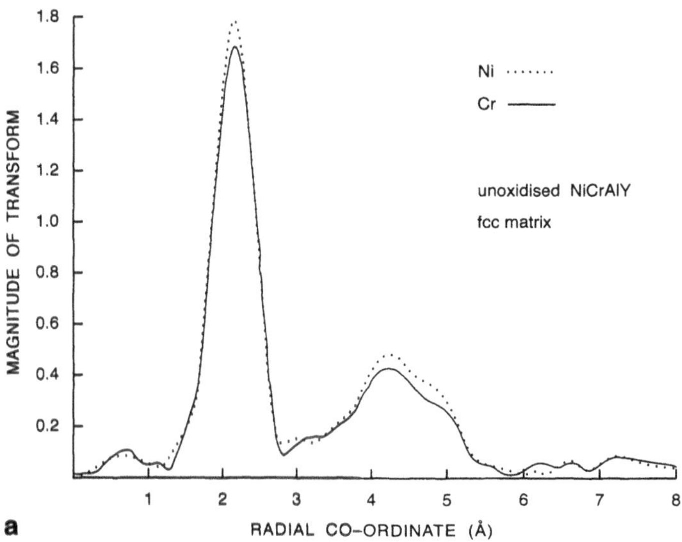

Fig. 19.8. See opposite page for caption

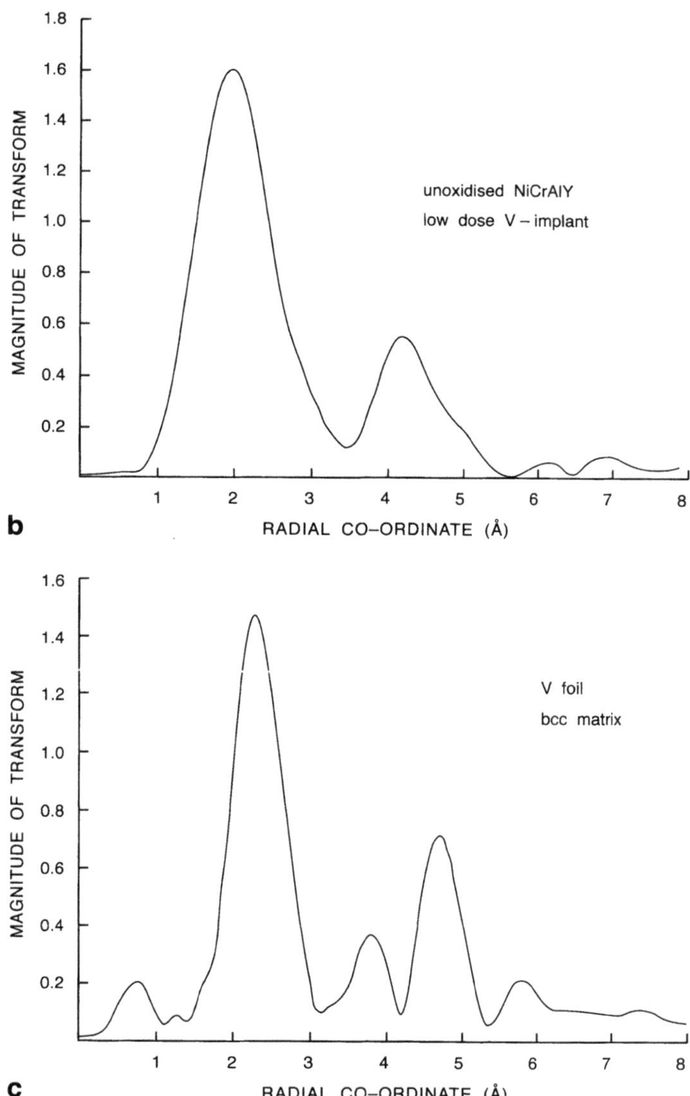

Fig. 19.8. Fourier transforms of (**a**) the Ni and Cr edges of unoxidised Ni-18Cr-6Al-0.5Y; (**b**) the V edge of V-implanted, unoxidised NiCrAlY; and (**c**) the V edge of pure vanadium

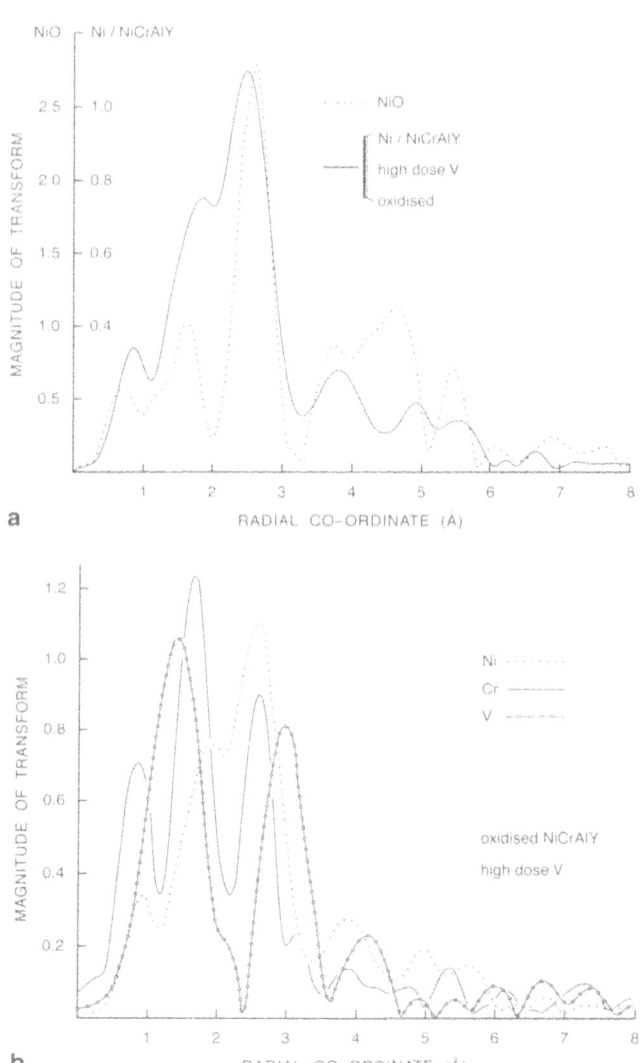

Fig. 19.9. For caption see the opposite page

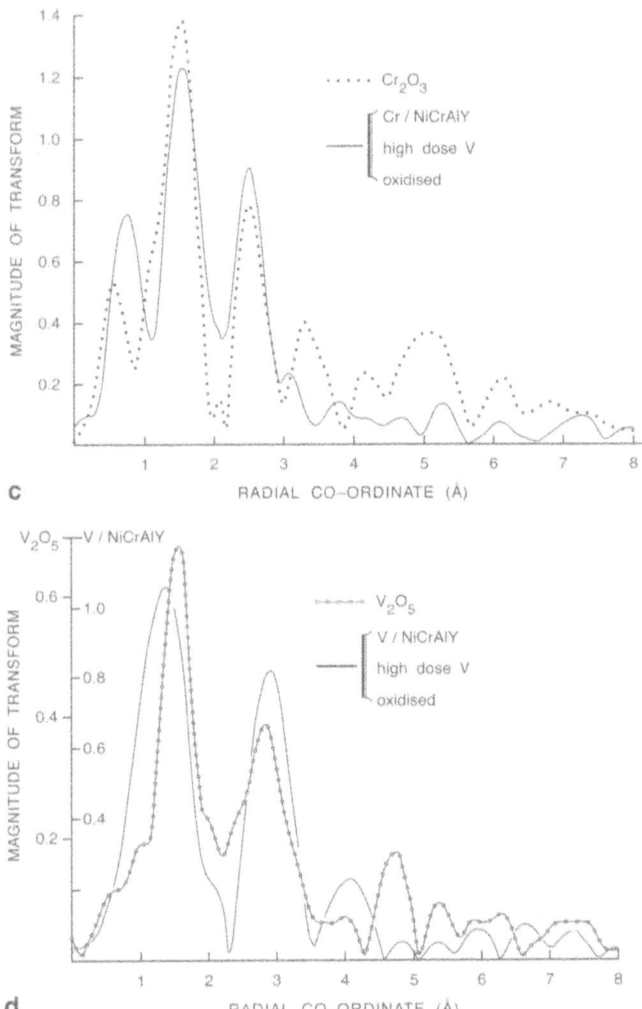

Fig. 19.9. Fourier transforms of (**a**) the Ni, Cr and V edges of V-implanted Ni-18Cr-6Al-0.5Y oxidised for 10 min at 900°C; (**b**) the Ni edge of oxidised, V-implanted NiCrAlY (*solid line*) compared with the Ni edge of pure NiO; (**c**) the Cr edge of oxidised, V-implanted NiCrAlY (*solid line*) compared with the Cr edge of pure Cr_2O_3; (**d**) the V edge of oxidised, V-implanted NiCrAlY (*solid line*) compared with the V edge of pure V_2O_5

The EXAFS spectra of the K-edges for Ni, Cr and V in non-implanted, implanted, unoxidised and oxidised samples were collected on the Naval Research Laboratory's beam line at the National Synchrotron Light Source. Fourier transforms of the spectra above the absorption edges were made to obtain information about the coordination of these elements in the NiCrAlY alloy. The Fourier transforms of the Ni edge of the non-implanted, unoxidised samples, such as the one shown in Fig. 19.8a, showed that the Ni had a fcc structure. The large peak near 2 Å is the contribution from the first shell of neighbors. The other peaks correspond to more distant neighbors. The Fourier transform of the Cr edge in the same sample, Fig. 19.8a, showed essentially the same profile. This signifies that the Cr atoms have the same local environment as the Ni atoms, i.e. Cr substitutes directly into the Ni lattice. This result was confirmed by examination and analyses of the same samples using STEM [8].

The Fourier transform of the V edge in the low dose, unoxidised sample (illustrated in Fig. 19.8b) is similar to that of Ni and Cr, rather than that of pure V, the transform of which is shown in Fig. 19.8c. This means that during implantation V ions substitute into the fcc lattice rather than forming clusters or discrete precipitates. If V had formed clusters it could be expected that the near-neighbor environment would be similar to that of pure V, which has a bcc lattice. Alternatively, if V had formed discrete precipitates with Cr and/or Ni the transforms for these elements would have been modified, which did not occur. This example clearly demonstrates that EXAFS is a powerful nondestructive technique, particularly for the analysis of implanted metals.

After 10 min oxidation at 900°C a number of different oxides form on both the non-implanted and implanted alloys. Analyses using STEM showed that the oxides present included NiO, Al_2O_3, $NiAl_2O_4$ and $NiCr_2O_4$ [8]. The STEM analyses also showed that the oxides were layered, with NiO richer at the outermost surface and Al_2O_3 richer at the metal/oxide interface. The oxides formed on the V-implanted samples were similar to those formed on the non-implanted samples. No discrete V-containing oxides were resolved in the selected area diffraction patterns, however, EDS analyses of the implanted, oxidised samples performed in STEM, suggested that V was associated with the Cr-rich oxides. SAM analyses confirmed that oxides tended to be layered. SAM, however, was not able to confirm an association of V with Cr in the oxide scale.

The Fourier transforms obtained from the EXAFS spectra for the Ni, Cr and V edges in an oxidised, high-dose V-implanted NiCrAlY are shown in Fig. 19.9a. An examination of the transforms shows that the local environment of atoms of each of these elements is different. Much microstructural information is contained in these transforms. Comparisons of the transforms from the implanted, oxidised alloy with those of simple, pure oxides (Figs. 19.9b, c and d) confirm the proposal that the oxides formed are complex oxides and not a physical mixture of simple oxides, because of the considerable differences that are readily apparent. Detailed analyses of these results are yet to be performed, but it is possible to characterise the oxides scale on the implanted alloy by following one or both of two paths. Firstly transforms of mixtures of oxide standards may be obtained experimentally and compared with the oxide formed on the oxidised alloys. Alternatively, computer generated transforms may be obtained by modelling the oxides. In order to do this, information provided by other analytical techniques is necessary, particularly at this stage in the development and application of the EXAFS technique. The information provided by STEM will enable comparisons of the transforms obtained experimentally with transforms generated from models of oxides to assist in the characterisation of the oxides. The present results highlight the potential of the EXAFS technique for characterising surfaces.

References

1. G.R. Johnston, J.L. Cocking, W.C. Johnson: Oxid. Metals **23**, 237 (1985)
2. G.R. Johnston, P.L. Mart, J.L. Cocking, J.W. Butler: Materials Forum **9**, 138 (1986)
3. J. Saulitis, G.R. Johnston, J.L. Cocking: Thin Solid Films **166**, 201 (1988)
4. C.S. Giggins, B.H. Kear, F.S. Pettit, J.K. Tien: Met. Trans. **5**, 1685 (1974)
5. J.G. Smeggil, A.W. Funkenbusch, N.S. Bornstein: Met. Trans **17A**, 923 (1986)
6. J.A. Sprague, G.R. Johnston, F.A. Smidt, S.Y. Hwang, G.H. Meier, F.S. Pettit: In *High Temperature Protective Coatings*, ed. by S.C. Singhal (Warrendale, PA, TMS of AIME 1983) pp. 93–103
7. F.A. Smidt, G.R. Johnston et al.: In *Surface Engineering*, ed. by R. Kossowsky, S.C. Singhal (NATO ASI Ser., Ser. E 1984), pp. 507–523
8. J.L. Cocking, J.A. Sprague, J.R. Reed: Surf. Coatings Technol. **36**, 133 (1988)

20 Coated Steel

R. Payling

Metal manufacturing and metal goods figure prominently in the industrial applications of surface analysis [1]. This is not surprising when one considers the enormous surface areas being produced each year, for example, in the sheet metal industry. Such sheet metal products undergo complex multistage industrial processing much of which interacts with the particular surface exposed during each process; and the performance and appearance of these products are judged, at least in part, by surface features, such as surface flatness, corrosion resistance, color, gloss, etc.

Coated steel, as the name suggests, is composed of a steel substrate (typically 0.4 to 1.5 mm thick) with various metallic or organic coatings (typically 20 μm thick). The four major areas for the consumption of coated steel are building products, home appliances, office furniture, and the automotive market. The choice of which coating to use in each area depends on the particular application, since coatings differ in their cost, their corrosion resistance, their suitability for forming, welding and painting, in their high temperature stability, and their suitability for containing food stuffs, etc.

The production of coated steel sheet is a multistage process and the changes in surface composition at each stage of processing must be determined if a knowledge of the complete operation is to be obtained [2]. Such information is useful when trying to trace a particular product feature, such as surface segregation, to a particular processing stage(s). Since each processing stage will leave telltale chemical traces which are then overlaid by subsequent processing, knowledge of the composition of near surface layers of the product is often more useful than the immediate surface composition. Such information can often separate the manifestation of a problem, such as a corrosion product, from the underlying processing variables which caused the problem.

A typical processing sequence (popularly called a route, or routing) for painted, metallic coated, steel strip, starting with the steel coil from a hot strip mill, is [3]:

a) pickling, to remove iron oxide scale in either hydrochloric or sulphuric acid;
b) oiling, immediately after pickling to prevent coil rusting and assist in cold reduction;

c) cold reduction, to reduce the steel thickness to the desired level – here, an emulsion of oil in water is applied to the strip to aid lubrication and heat dissipation;
d) metallic coating by immersion in a liquid metal bath of controlled composition;
e) gas-jet stripping, to control metallic coating thickness;
f) temper rolling and tension levelling, to improve base steel properties and surface flatness;
g) a paint pretreatment process, involving alkaline cleaning, hot and cold water rinsing, a proprietary conversion coating, and a final chromate rinse;
h) a multistage painting process, including the application of primer and top coats, with stoving and water quenching.

Each of these 8 stages acts on the exposed surface, which is changing throughout the sequence from uncoated steel, to a metallic coating, and finally to the surface of the polymer coating. The following series of applications illustrates this journey, and some of the potential hazards along the way, beginning inside the steel and ending with the polymer surface.

20.1 Applications

20.1.1 Grain Boundaries in Steel

Auger analysis of the fracture surface of steel is normally concerned with detecting the presence of segregating elements such as carbon, sulphur, phosphorus or oxygen [4,5] or aluminium and nitrogen [6]. The segregation of these elements is of most general interest when the fracture is inter-granular rather than trans-granular, since the latter may indicate only the existence of inclusions while the former may provide an explanation for the fracture failing at the embrittled grain boundary. Before commencing the Auger analysis, the presence of inter-granular fracture must first be confirmed with scanning electron microscopy, i.e., by the presence of its characteristically smooth secondary electron image.

The detection of segregating elements alone does not always provide the complete answer to fracture analysis, as the mere detection of elements does not, of itself, determine the chemical species of these elements. This information is contained in the position and shape of the individual Auger peaks. In a particular steel which was failing in a through thickness tensile test, it was suspected that segregation of boron nitride was weakening the grain boundaries. Scanning electron microscopy showed that the fracture faces had regions of inter-granular fracture, with the exposed grain faces typically being 20 μm across, large enough to make Auger analysis relatively easy. In the inter-granular regions approximately 25 atomic % boron and nitrogen were detected. The peak positions of both elements were consistent with boron nitride, namely 170 and 382 eV, respectively, [7–9] and the peak shape of the

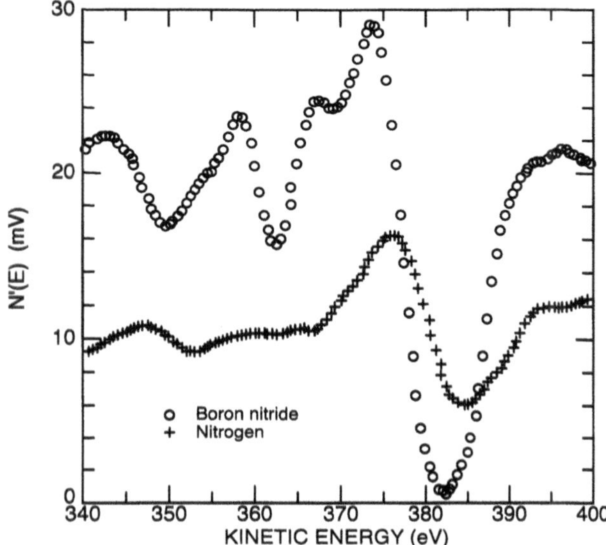

Fig. 20.1. Change in N_{KLL} peak shape between boron nitride and adsorbed nitrogen on steel

nitrogen, in particular, was characteristic of boron nitride [10] (Fig. 20.1). It was therefore concluded that boron nitride was present on the inter-granular fracture surface.

20.1.2 Steel Surface

Auger and GD-OES analysis of uncoated steel sheet, even after solvent degreasing, shows a complex surface, in which the exposed iron content may typically account only for 5–20 atomic % [3,11]. Other elements detected and their likely sources are: carbon and zinc, from residual oils, particularly extreme pressure additives; manganese, silicon, aluminum, phosphorus, and chlorine from the steel base, by diffusion during annealing; silicon and phosphorus from residual alkaline cleaning solutions; sodium from rolling lubricants; and calcium and magnesium from industrial rinse water. Of these, residual carbon has been of special concern for some years, especially in the automotive industry, because of a strong correlation between residual carbon levels on uncoated steel and the corrosion of painted steel [3]. Consequently, sheet steel manufacturers have put considerable effort into determining the amount, type, and sources, of this residual carbon. Scanning Auger mapping shows that the carbon may form a uniform layer or be distributed in patches, and AES depth-profiling provides information on the depth distribution and overall mean thickness of the residual carbon layer. If this thickness is greater than 7–8 nm it can be anticipated that subsequent processing, such as painting, will give problems, providing of course that the sample(s) chosen is representative of the whole surface. Common to any off-line analysis, one is always concerned whether the small area analyzed is representative of the whole 12 tonne steel coil, especially since only the leading or trailing ends of the coil are readily accessible.

Various surface studies have shown the complexity of residual carbon forms on steel. Authors differ in their interpretation of the dominant form [12–15] but carbon may be present as organic molecules, graphite, amorphous soot or smut, or iron carbide. One common problem with coil annealing of oiled steel strip is redeposition of carbon near the edge of the strip, so-called "snaky" edge. While surface analysis is useful here, ultimately the solution to such problems is found in the proper control of the production process, in controlling oil levels and furnace conditions, for example.

In XPS studies of oiled steel surfaces, the carbon 1*s* peak is typically found at a binding energy of 286 eV. Degreasing the sample, to remove excess oils, shifts this peak to 284.7 ± 0.4 eV. This peak is present on most real surfaces and corresponds to a form of tenacious hydrocarbon, often called adventitious hydrocarbon, which has a reference value of 284.9 eV [16]. Unfortunately, the carbon 1*s* peaks for graphite, amorphous carbon, and iron carbide, all occur around 284.3 eV; so that positive identification of carbon on steel by XPS is uncertain. Since ion-bombardment of all these forms of carbon on steel results in the carbon 1*s* peak shifting to 284.2 ± 0.2 eV [3], information on any variation in the type of carbon with depth is therefore also unobtainable.

An alternative XPS method for characterizing oxygen-containing hydrocarbons, such as those found in lubricants used for rolling steel, involves a comparison of the binding energy of the oxygen 1*s* peak with the kinetic energy of the X-ray excited oxygen Auger *KLL* peak [3]. Both peaks appear in the same XPS spectrum and the results are called XPS chemical state plots, or Wagner plots [17]. One possible source of residual carbon is the formation of iron soaps between the acid groups in rolling oils and the steel. The Wagner method has been used to demonstrate that organic residues can survive closed coil annealing at 700°C, though energy shifts indicate some structural modification does occur.

Auger analysis is capable of distinguishing carbides from other forms of carbon due to a characteristic peak shape [18]. The C_{KLL} (272 eV) peak shape changes markedly from hydrocarbon (or graphite) to carbide, and the sensitivity increases by a factor of about 3 [18–20]. When an uncoated steel sample is ion-bombarded to remove most of the oxide film (typically to a depth of 10–15 nm), the surface carbon *KLL* peak shape, typical of either graphite or hydrocarbon, is replaced by a carbide peak shape [13]. But this does not necessarily indicate an underlying iron carbide. The measured carbon level ($\sim 6\%$) is 20 times the bulk value and diffusion of carbon from the surface or bulk and incorporation of oil during iron oxide formation have been proposed to explain this apparent carbide enrichment [21]. Steel samples abraded to mid-thickness, however, show exactly the same behavior and the explanation may lie in a combination of other factors. First, in differential spectra a narrow peak gives a greater apparent peak-to-peak height, so that the peak shape change to the narrow carbide shape exaggerates the amount of carbide present, by a factor of about 3; secondly, carbon has a very low sputtering rate and the AES technique analyzes what is left on the surface during or after sputtering, which exaggerates the total carbon concentration measured below the surface; and thirdly, carbon mixed with iron – possibly as a result of the sputtering process – may give a carbidic AES peak shape without its being a true carbide, because of short-range Coulomb interactions.

In the enamelling of steel, for the manufacture of enamelled stoves, bath tubs, hot water systems, etc., a frit is applied to the steel and then fired to form the enamel surface. Good adhesion of the enamel requires a sufficient reaction between the frit and the steel surface. In an early study of the factors affecting the adhesion of enamel to steel, it was found that better adhesion was related to a thinner surface oxide [3] (Fig. 20.2). Other factors which have been found to be relevant to enamel adhesion are surface carbon (also associated with blistering of the enamel coating), silica, and steel grade. Titanium killed steels, for example, have exceptionally thin oxides (typically $\leq$ 4 nm) but may give enamel adhesion problems, presumably because the surface oxide formed on titanium killed steels inhibits the bonding reaction between the steel and the frit.

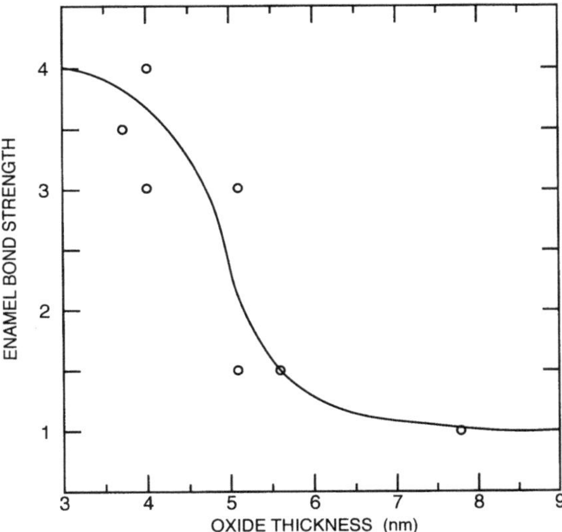

Fig. 20.2. Variation in enamel bond strength on steel in one study as a function of surface oxide thickness

20.1.3 Alloy Region

When the steel strip enters the liquid metal bath, alloying begins between the liquid and the steel surface. Control of this alloy growth is important for proper coating properties and for coating adhesion [22,23]. An understanding of the alloying process requires a knowledge of the alloy composition. This is not always straightforward [3]. For chemical analysis of the separate alloy phases, the individual layers must be isolated through chemical etching techniques which may alter the composition of the layer of interest. To complicate matters, the layers are generally too thin for definitive electron microprobe analysis and depth-profiling Auger analysis may introduce artefacts through preferential sputtering.

In a particular 54 mass % aluminum, 44 mass % zinc and 1.5 mass % silicon coating studied, the alloy region was a two phase layered structure with a total thickness of 2 μm. Table 20.1 lists the AES, electron microprobe, and chemical analyses of the two major alloy phases. The Auger sensitivity factors were obtained by setting the AES analysis of the top phase equal to that of the microprobe analysis, where the microprobe and chemical analysis were in good agreement, with the result that the AES analysis matched more closely the chemical analysis of the bottom phase.

Table 20.1. Estimates of alloy composition by various methods (in mass %)

Method	Composition							
	Top Phase				Bottom Phase			
	Al	Fe	Zn	Si	Al	Fe	Zn	Si
Electron microprobe	54.2	31.9	7.7	6.2	43.1	48.6	6.8	1.4
Chemical	55.1	32.5	7.2	5.2	52.3	39.3	4.8	3.6
AES	54.2	31.9	7.7	6.2	56.3	36.0	4.4	3.2

20.1.4 Metallic Coatings

Metallic coatings may be broadly divided into three categories: hot-dipped, electrolytic (or electroplated), and vapor deposited coatings. The first two of these broad categories may be further subdivided: hot-dipped coatings into zinc (or galvanized), zinc-aluminum alloy, zinc-iron (or galvanneal), and lead-tin (or terne); and electrolytic coatings into zinc, zinc alloy, and tin (or tinplate). Only hot-dipped coatings and tinplate will be discussed here.

The surfaces of hot-dipped metallic coatings on steel generally differ considerably from the bulk compositions of the coatings [24]. The dominant metallic species present on the surface of all coatings reported here may be explained simply by consideration of the relative oxygen affinities of the metals in the coatings [25]. As the metallic coating solidifies the constituents of

Table 20.2. Standard free energy of oxide formation, ΔF, at 400°C (673 K)

Oxide	ΔF_{673} [kJ/mol]
Ce_2O_3	−1066
MgO	−1050
Al_2O_3	−974
ZnO	−563
SnO_2	−444
PbO	−306

the coating compete at the surface for oxygen. Various mechanisms, such as minimum surface energy, charge diffusion, or rejection by the solid coating, may all influence surface composition; but, because of the relatively large energies involved in oxidation, the oxygen reaction would appear to dominate [25]. Table 20.2 lists the standard free energy of formation for a number of oxides, referenced to 400°C [26]. This temperature was chosen as representative of the high oxidation temperatures of the coatings after exiting the pot. Within the simple theoretical framework considered, the more negative the free energy the more likely the species will be found at the surface. For example, the large negative value for Al_2O_3 compared with ZnO explains the

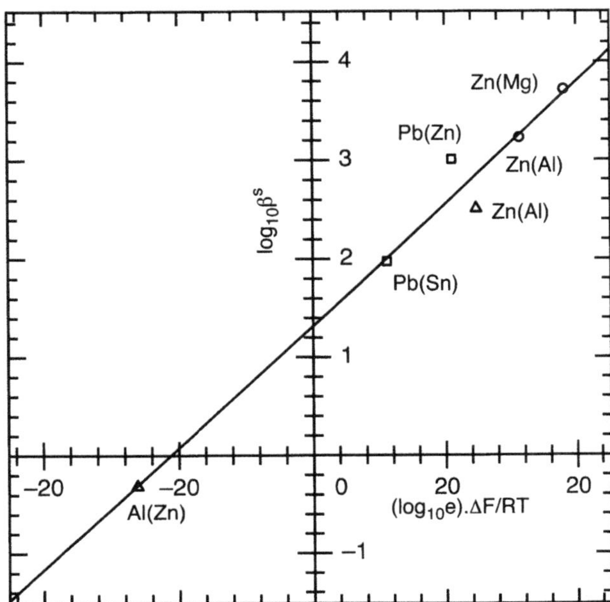

Fig. 20.3. Surface enrichment, β^s, for a range of hot-dipped metallic coatings on steel as a function of the difference in free energy of oxide formation, ΔF, between solute (*inside brackets*) and solvent metals (*outside brackets*)

aluminum oxide on galvanized coatings. The large negative value for MgO justifies current interest in magnesium additions. Defining the surface enrichment factor, β^{s}, by:

$$\beta^{\mathrm{s}} = \frac{X^{\mathrm{s}}/Y^{\mathrm{s}}}{X^{\mathrm{c}}/Y^{\mathrm{c}}} \tag{20.1}$$

where X denotes the solute concentration and Y the solvent concentration and where superscripts s and c refer to surface and bulk respectively, and assuming the segregation mechanism may be described by the Gibbs adsorption isotherm, then the following equation may be derived:

$$\log_{10} \beta^{\mathrm{s}} = \log_{10} e \, (\Delta F_y - \Delta F_x)/RT \, . \tag{20.2}$$

Equation 20.2 describes the segregating effect of oxidation, where ΔF is the difference in free energy of oxide formation between the two metallic species, x and y, solute and solvent, respectively and where R is the universal gas constant and T is the absolute temperature. Figure 20.2 shows good agreement of (20.2) with all four categories of hot-dipped metal surfaces considered; except for the slope of the line of best fit which is only 0.07, rather than unity. This discrepancy in slope may indicate the oxidation process is not in thermodynamic equilibrium. Nevertheless, the general energy dependence of (20.2) appears to be supported empirically by the data in Fig. 20.3.

a) Galvanized Steel. The solid surface formation on hot-dipped metallic coatings is a complex process. A steel strip is passed through a liquid metal bath (or pot). The thickness of the resultant liquid coating, typically 20 μm, is determined by the combined effects of gravity runoff and air jet stripping. The liquid coating, exposed to ambient air, cools rapidly below its solidification temperature. For galvanized coatings this degree of supercooling is about 8°C [27]. Nucleating at or near the steel surface, the coating then solidifies rapidly. Inside the galvanized coating, the initial rapid solidification releases sufficient latent heat of fusion that the remaining liquid is brought back to the melting temperature, 418°C [27]. Final solidification of the outer coating may then be quite slow, possibly over several seconds, until the remaining latent heat is dispersed, and those characteristic granular structures popularly called spangles are formed.

The chemistry of a galvanized surface is complex even though the galvanizing process only involves passing the steel through a molten bath of zinc and a few minor alloying elements, chiefly aluminum at about 0.2 mass %. During the slow final solidification and while the temperature remains high after solidification until water quenching, significant surface segregation of the metals occurs. Prior to passivation, the surface of the galvanized coating is predominantly aluminum oxide with some zinc oxide and a varying amount of hydrocarbon contamination [25] (Fig. 20.4a). Unpassivated galvanized surfaces typically have 20–30 atomic % aluminum and 5–10 atomic % zinc, with

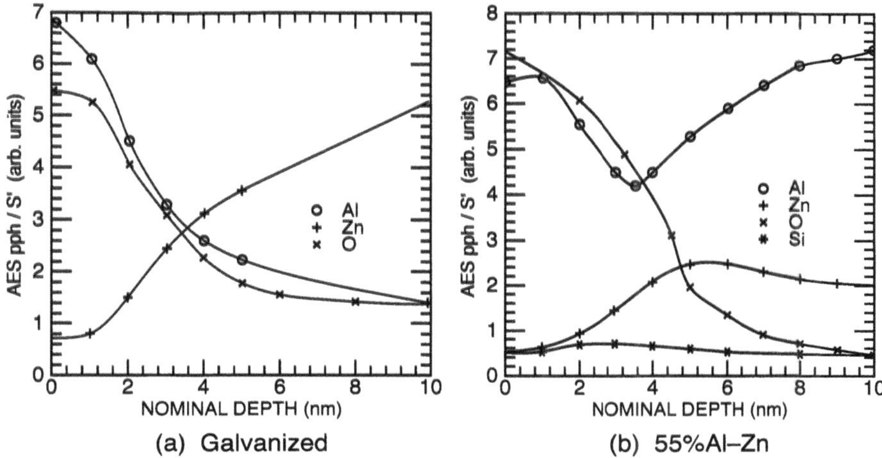

Fig. 20.4. AES depth profiles of the surfaces of (**a**) galvanized steel, and (**b**) 55% Al-Zn coated steel. The AES signals (pph) have been normalized with modified elemental sensitivity factors (S')

the variation largely depending on surface carbon levels which are typically 10–30 atomic %. Though significant levels of hydrocarbons are present on all industrial surfaces, the hydrocarbons on fresh metallic coatings are generally very thin, typically less than 1 nm.

The aluminum oxide surface on galvanized steel could be expected to form an excellent barrier to oxygen diffusion and hence a barrier to oxidation of the zinc coating beneath. Unfortunately, if galvanized steel is left unpassivated and stored in a wet environment, the aluminum oxide soon converts to aluminum hydroxide and an unseemly "white" corrosion of the zinc forms [25].

The lead content of galvanized coatings varies greatly (typically from 0 to 1.5 mass %), depending on the manufacturing process. Lead has an extremely low solubility in solid zinc and so is rejected during solidification both towards grain boundaries and towards the surface of the coating. This rejection mechanism is separate from the oxidation mechanism presented above, as during solidification the lead coalesces within the remaining liquid to form discrete metallic lead particles at the final solid interfaces rather than a continuous thin oxide layer on the surface. The absence of lead at the immediate surface, however, indicates the lead is under the surface aluminum oxide layer [3].

Since aluminum oxide is a tightly bound, stable oxide in dry conditions, its stability is likely to be affected by defects in the oxide and these in turn be affected by trace impurities. Trace levels are beyond the sensitivity limits of Auger analysis (typically limited to concentrations above about 0.5% atomic). SIMS does not share this limitation and elements detected by SIMS on the surface of a series of galvanized steel samples but not detected by AES were arsenic, indium, lithium, gallium, and fluorine, though copper, tin and bismuth could not be detected at low levels with the experimental conditions used because of peak overlap. Clearly, knowledge of the surface composition is vital in such work.

b) 55% Al-Zn Coating. Commercial 55%Al-Zn coatings contain 54 mass % (73 atomic %) aluminum, 44 mass % (25 atomic %) zinc, and 1.5 mass % (2 atomic %) silicon [3]. A typical depth-profile from a commercial sample is shown in Fig. 20.4b. The commercial coatings are dual phased with aluminum-rich or zinc-rich regions typically 2–8 μm in size. The depth-profile in Fig. 20.4b therefore represents an average of these two phases. The separate phases would be expected to passivate differently and are therefore of separate interest.

Besides knowledge of the immediate surface, it is essential to know how the composition of the coating varies through its full thickness. Such information not only helps explain the composition of the immediate surface (because of segregation during solidification) but how the coating will perform in long term corrosion. The GD-OES technique is outstanding in its ability to depth-profile metallic coatings, profiling at typically 1 μm/minute, with high sensitivity (around 5 ppm), for virtually the whole Periodic Table, including hydrogen. Figure 20.5 shows a depth-profile of a 23 μm thick 55%Al-Zn coating on steel. The depth-profile shows the thin oxide covered surface, the

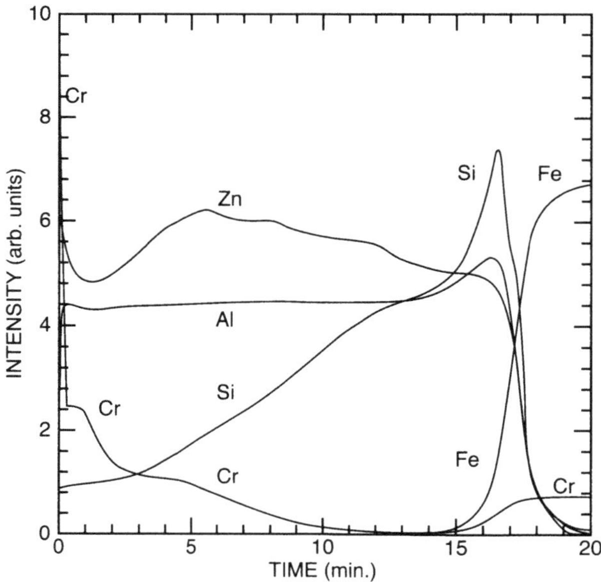

Fig. 20.5. GD-OES depth profile of 55% Al-Zn coated steel

nearly uniform centre of the coating and the complex zinc, aluminum, silicon, iron alloy region, described earlier, joining the steel to the coating. Such coatings are now often covered with a very thin (0.5 μm) non-conductive polymer coating to aid final fabrication and are best analysed with rf GD-OES.

20.1.5 Treated Metallic Coating Surface

Commonly applied passivation treatments for hot-dipped coatings contain either chromic acid (hexavalent chromium, or chromium VI) or mixed hexavalent/trivalent chromium (chromium VI/chromium III) based solutions [2]. When chromate solutions were first formulated, that is before surface analysis, it was not known that the surface of hot-dipped galvanized steel was predominantly aluminum oxide. While chromium VI reacts with zinc it does not react with aluminum oxide, so the acid must first disrupt the surface layer to reach the zinc beneath. The chromium III component then forms an inert protective barrier while the chromium VI is reduced to chromium III by reaction with the zinc. Alternatively, a fluoride may be added to the chromate solution to attack the aluminum oxide and thereby increase the reactivity of the surface [14].

A valuable method of AES analysis for monitoring total surface chromium levels is integration of the area under the chromium depth-profile [28]. This method is of special interest in recording the way chromium concentrations vary over small distances, for example, across grains or between grains, when the standard chemical analysis for chromium requires an area of approximately $67\,cm^2$.

20.1.6 Metal–Polymer Interface

When studying the adhesion of paint or polymer systems to metallic coatings, it is useful to know the location at which adhesion failure occurs [3]. Failure often occurs at the interface between the paint and the metallic coating which would indicate a problem exists either with the paint formulation or with surface cleanliness. But in one investigation, paint was removed from a galvanized surface by adhesive tape following a reverse impact test and XPS showed the underside of the removed paint contained aluminum, zinc, and chromium which had been detached from the metallic coating surface. Auger analysis of the exposed metallic coating showed mostly zinc, with the expected mixed chromium, aluminum oxide surface no longer present. Apparently the paint had adhered so well that separation occurred at the more loosely bound oxide in the metallic coating surface. This result helped focus further work into the nature of the oxide rather than into the paint. Indeed, samples taken before painting indicated the surface oxide was 9 nm thick rather than the usual 4 nm for that coating.

20.1.7 Polymer Surface

The natural weathering of paint or polymer surfaces involves complex processes of chemical and biological attack and thermal and photo degradation, resulting in subtle to gross physical and chemical changes. Specific areas where XPS can provide information on the weathering of paint surfaces are: a) changes in surface composition, for both organic and inorganic paint constituents [29]; b) oxidation and hydrolysis, especially from wet storage, by monitoring the oxygen to carbon ratio [30–32] and the carbon peak width [29]; c) changes in cross-linking, by monitoring the nitrogen peak, when cross-linking agents are used [33]; d) bond saturation, via shake-up satellites [29,31,34]; and e) variations in pigment concentration with depth, through combined XPS and ion-bombardment. The titanium depth-profile may be particularly useful, because most paints contain titanium (as titanium dioxide, or rutile) and on unexposed paint surfaces the titanium concentration

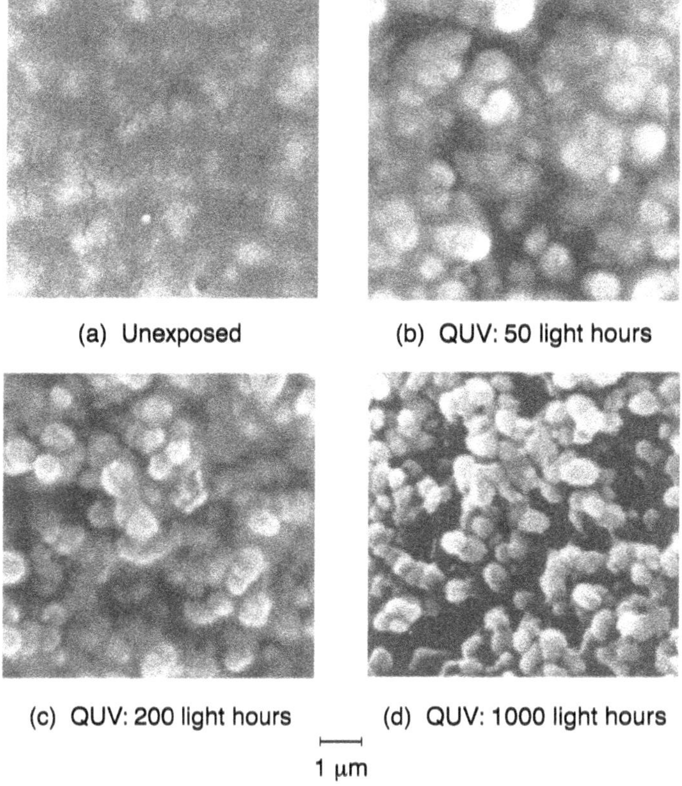

Fig. 20.6. SEM micrographs of a commercial polymer showing the progressive revealing of pigment particles following exposure in a QUV weatherometer

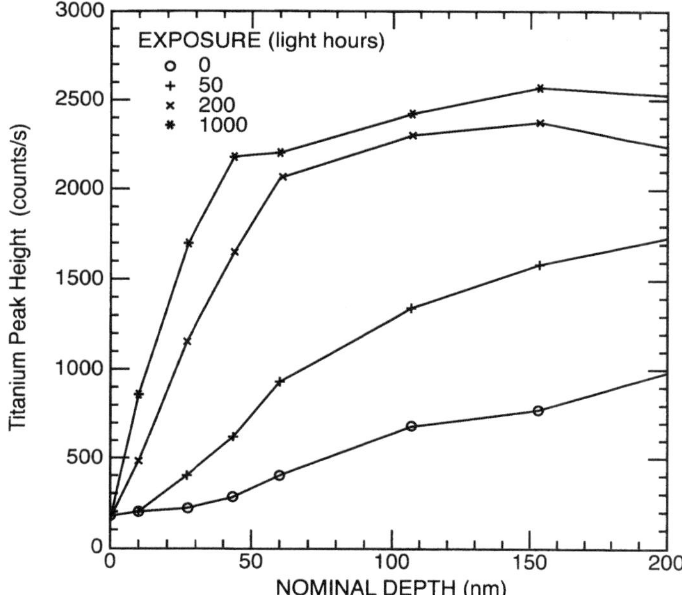

Fig. 20.7. Changes in XPS depth profiles for titanium from a commercial polymer following exposure in a QUV weatherometer

steadily increases with depth, thus providing a possible depth marker for paint film loss during weathering [32]. The ion-bombardment of polymers is known to create compositional artefacts under certain conditions, so special care is required with the technique [35].

As part of a larger program, described in [36], metal panels were coated with a commercial white silicone modified polyester (SMP) and kept unexposed or subjected to 50, 200 or 1000 light hours (2000 total hours) exposure in a cyclic accelerated weathering instrument (QUV). White was chosen because it is a relatively simple pigmented system based on rutile (titanium dioxide). Unexposed and exposed samples were then examined by scanning electron microscopy (SEM) and XPS. The SEM micrographs, in Fig. 20.6, show that the surface morphology changed markedly with exposure, with the smooth unexposed surface becoming quite rough following erosion of the polymer resin and exposure of the pigment particles. XPS analysis combined with argon ion etching indicated the presence of a pigment-free layer (about 20 nm thick) on the unexposed polymer surface which quickly disappeared during accelerated testing. The elements detected by XPS were carbon, nitrogen, oxygen, silicon and titanium. The surface carbon and nitrogen concentrations both dropped during testing, their loss representing the loss of the polymer and of the melamine cross-linking agent in the polymer, respectively. The nitrogen/carbon ratio decreased during testing, with rapid loss of nitrogen in the first 200 hours; whereas, the oxygen/carbon ratio increased rapidly in the first 200 hours. An increase in oxygen/carbon ratio could indicate increased oxidation or hydrolysis of the polymer, however, the carbon peak width decreased slightly (from 3.6 eV unexposed to 3.3 eV at 1000 horus) which does not support an increase in the number of C–O bonds. The increasing oxygen signal was therefore more likely to be associated with the exposed pigment particles. This exposure of pigment is best illustrated by changes in the titanium depth-profile in XPS (Fig. 20.7) or GD-OES [37].

20.2 Conclusion

Surfaces play a vital role in problems of corrosion, adhesion, and appearance of industrial products. A painted, coated stell sheet, for example, contains upwards of nine distinct interfaces, or surfaces, all of which at some stage influence manufacturing and final product performance. The surface composition of such industrial coatings often differ radically from their bulk composition. XPS, AES, SIMS and GD-OES provide essential information on surface composition and the changes to surface composition during such complex, multistage industrial processing.

References

1. M.P. Seah: Surf. Interface Anal. **2**, 222 (1980)
2. P.D. Mercer, R. Payling: Nuclear Instrum. Meth. **191**, 283 (1981)
3. P.D. Mercer, R. Payling: Proc. 4th Australian Conf. on Nuclear Techniques of Analysis, AINSE, Lucas Heights, NSW, 6–8 Nov. 1985, p. 132
4. M.T. Thomas, R.H. Jones, D.R. Baer, S.M. Bruemmer: The PHI Interface **3**(2), 3 (1980)
5. A.J. Garratt-Reed, G. Cliff, G.W. Lorimer and R. Pilkington, Interfacial Engineering for Optimized Properties Symp., Boston, pp 103-108 (1997)
6. R.H. Edwards, F.J. Barbaro, K.W. Gunn: Metals Forum **5**(2), 119 (1982)
7. A.P. Coldren, A. Joshi, D.F. Stein: Metall. Trans A **6**a, 2304 (1975)
8. M.P. Seah: Surf. Interface Anal. **1**, 86 (1979)
9. G. Hanke, K. Muller: Surf. Sci. **152/153**, 902 (1985)
10. R.H. Stulen, R. Bastasz: J. Vac. Sci. Technol. **16**(3), 940 (1979)
11. J. Angeli, K. Haselgrbler, E.M. Achammer and H. Burger, Fresenius J. Anal. Chem. **346**, 138 (1993)
12. J.A. Slane, S.P. Clough, J. Riker-Nappier: Metall. Trans. A **9**, 1839 (1978)
13. P.L. Coduti: Metals Finishing **78**, 51 (1980)
14. V. Leroy: Mater. Sci. Eng. **42**, 289 (1980)
15. R.A. Iezzi: PhD Thesis, Lehigh University (1979)
16. L.E. Davis, N.C. MacDonald, P.W. Palmberg, G.E. Riach, R.E. Weber: *Handbook of Auger Electron Spectroscopy*, 2nd ed. (Physical Electronics Div., Perkin-Elmer Corp, Eden Prairie, MN 1978)
17. C.D. Wagner, D.A. Zatko, R.H. Raymond: Anal. Chem. **52**, 1445 (1980)
18. S. Craig, G.L. Harding, R. Payling: Surf. Sci. **124**, 591 (1983)

19. K. Ishikawa, Y. Tomida: J. Vac. Sci. Technol. **15**, 1123 (1978)
20. R. Payling, G.L. Harding, S. Craig: Appl. Surf. Sci. **24**, 11 (1985)
21. R. Bastasz, G.J. Thomas: J. Nucl. Mater. **76–77**, 183 (1978)
22. S.J. Makimattila, E.O. Ristolainen, M. Sulonen, V.K. Lindros: Scripta Metall. **19**, 211 (1985)
23. P. Karduck, T. Wirth and H. Pries, Fresenius J. Anal. Chem. **358**, 135 (1997)
24. V. Leroy, B. Schmitz: Scand. J. Metall. **17**, 17 (1988)
25. R. Payling, P.D. Mercer: Appl. Surf. Sci. **22/23**, 224 (1985)
26. C.R. Weast (Ed.): *Handbook of Chemistry and Physics*, 55th ed. (CRC Press, Cleveland, OH 1974) D-45–50
27. D.I. Cameron, G.J. Harvey: 8th Int. Conf. on Hot Dip Galvanizing, London (1967)
28. R. Payling: Appl. Surf. Sci. **22/23**, 215 (1985)
29. D. Briggs: In *Practical Surface Analysis by Auger and X-ray Photoelectron Spectroscopy*, ed. by D. Briggs, M.P. Seah (Wiley, Chichester 1983) p. 359
30. D.T. Clark: In *Physicochemical Aspects of Polymer Surfaces*, Vol. 1, ed. by K.L. Mittal (Plenum, New York 1983) p. 3
31. A. Dilks: J. Polymer Science: Polymer Chemistry Edition **19**, 2847 (1981)
32. S. Skeldar, A. Zalar, R. Zavasnik: 16th FATIPEC Congress, 335 (1982)
33. E. Takeshima, T. Kawano, H. Takamura: Trans ISIJ **23**, 652 (1983)
34. L.C. Lopez, D.W. Dwight, M.P. Polk: Surf. Interface Anal. **9**, 405 (1986)
35. D.E. Williams, L.E. Davis: In *Characterization of Metal and Polymer Surfaces*, Vol. 2, ed. by L.H. Lee (Academic, New York 1977) p. 53
36. C.J. Aldrich, E.M. Boge, J.T. Murada, R. Payling, J.R. Bird: *EUREM 88*, Proc. 9th European Congress on Electron Microscopy, York, England, 4–9 September 1988
37. D.G Jones and R. Payling: In R. Payling, D.G. Jones & A. Bengtson (Eds.) Glow Discharge Optical Emission Spectrometry, John Wiley & Sons, Chichester (1997), p. 661

21 Thin Film Analysis

G.C. Morris

The increasing importance of thin films for new technologies has encouraged fundamental and applied research on their physical and chemical structures and on the interfaces made with them [1]. Their physical structures (e.g. morphology, topography, crystallite properties, extent and type of defects) are explored by diffraction and microscopic techniques including X-ray and electron diffraction, scanning tunnelling microscopy and ultrasonic microscopy. Their chemical structures (e.g. element type, concentration and spatial distribution) are explored by microanalytical techniques such as Fourier transform infra-red spectroscopy, secondary ion mass spectrometry (SIMS), X-ray photoelectron spectroscopy (XPS), Auger electron spectroscopy (AES), ion scattering spectroscopy (LEIS, HEIS), as well as dispersive X-ray analysis and electron energy loss spectrometry with scanning and transmission electron microscopy. As an ultimate objective, researchers desire a three-dimensional elemental map on an atomic scale for the thin film and its interfaces. Some progress towards that aim has been made, but achievement is some time away.

Thin film analysis may be considered as characterizing materials about one micron thick in which the outer few atomic layers probably have a different structure and composition for a number of reasons especially atmospheric contamination. A wide diversity of application areas, some of which are noted in Table 21.1, need such analysis to solve problems and understand phenomena which occur. The analytical techniques used for the particular application must be carefully selected using criteria such as information depth probed, lateral resolution required, detection limit needed, elemental type and chemical state investigated. Complementary methods are recommended to confirm or deny conclusions made using one technique alone. Thus, a typical analysis might include e.g. X-ray diffraction data, a physical image from electron microscopy, AES and XPS analyses of specific areas exposed by ion sputter removal of successive layers (depth profile analysis). The compositional data from the two techniques, AES and XPS, could be compared. Analytical electron microscopy could also give useful comparative data. Elsewhere in this book, each of the main analytical techniques have been reviewed so that their advantages/disadvantages for a particular application can be assessed.

The aim of this chapter is to illustrate how surface sensitive techniques provide a data base to understand processes and phenomena which occur with particular thin films. Example from several of the different application areas

Table 21.1. Some application areas benefiting from thin film analysis

Materials Science/Technology	*Film Technology*
Corrosion	Adhesion problems
Passivation	Coating thickness
Segregation phenomena	Failure analysis
Failure analysis	Optical problems
Inclusion analysis	Electrical problems
Diffusion analysis	Decorative problems
Alloy composition	Segregation problems
Surface modification	Lubrication phenomena
Metallurgy	*Semiconductor/Electronic Materials*
Fatigue failures	Dopant distribution
Stress corrosion cracking	Interface widths
Chemical reaction problems	Purity analysis
Diffusion phenomena	Processing phenomena
Segregation phenomena	Diffusion phenomena
Composition studies	Failure diagnosis
Quality assurance	Ageing processes
Temper brittleness	Cleaning processes
Others	
Catalyst distribution	Quality assurance
Catalyst deterioration	Medical prosthesis
Mineral composition	Packaging problems
Materials signature	Adhesion phenomena
Tribology studies	Delamination
Surface-bulk differences	Magnetic tape irregularities
Dental materials	Opto-electronic films

shown in Table 21.1 could be used. However, it is more instructive to use a single system even though it will have its own complex of problems because the general principles of characterization remain the same, e.g. structure and composition profile are determined, 'good' systems are compared with 'faulty' systems to probe defects, etc. Thin film photovoltaics based on II-VI semiconductors is the system chosen for illustrating how surface analytical methods are an essential probe for examining the materials and their interfaces.

21.1 Thin Film Photovoltaics

21.1.1 Use for Solar Electricity

Photovoltaic devices made from thin films have a future as a viable solar energy alternative if module costs are lowered and device stabilities of 20 to 30 years can be achieved [2]. To establish those aims, efficient, stable photovoltaic systems must be made inexpensively in industrial quantities. The criteria of efficiency and stability require that the effects which arise from contaminations at surfaces, in the bulk and at interfaces, must be minimized, chemical reactions and compositional changes induced by humidity, light and temperature must be inconsequential, and time dependent processes such as impurity segregation and interdiffusion must be insignificant. Surface sensitive analytical methods (SIMS, XPS, AES), combined with structural, morphological (XRD, SEM) and microelectrical characterization (electron beam induced current – EBIC) play a key role in probing the processes which occur during preparation of the thin films and when fabricating them into photovoltaic devices. This will be illustrated using one of the most promising devices for solar energy conversion, viz. that based on the heterojunction cadmium sulphide/cadmium telluride (*n*CdS/*p*CdTe) [2].

21.1.2 The Thin Film Solar Cell: Glass/ITO/*n*CdS/*p*CdTe/Au

The thin film solar cell glass/ITO/*n*CdS/*p*CdTe/Au is fabricated in several steps:

(i) The base material is soda glass coated with a 200 nm layer of indium tin oxide (ITO), (ii) a layer of about 80 nm CdS is deposited on the glass/ITO out of a chemical solution [3] or by an electrodeposition method using a solution containing Cd^{2+} and S_2O_3=ions [4], (iii) a layer of about 1.5 µm CdTeis electrodeposited on the CdS from a solution containing Cd^{2+} and $HTeO_2^+$ ions [5,6], (iv) the as-deposited CdS,CdTe films are *n*-type and the heterojunction *n*CdS/*p*CdTe is formed during a conversion step in which the film is heated to about 400°C in air, (v) the desired low resistance contact to the *p*CdTe is accomplished by a chemical etching of the surface followed by metallization e.g. using a 60 nm thick layer of gold (Au).

Single junction solar cells which convert up to 13% of sunlight into electrical energy have been made [4,6–8]. To fabricate laboratory cells reproducibly and to transfer the technology to industry, variables which affect each process must be recognized and controlled. Development proceeds in an iterative fashion with the knowledge gained at each processing step fed back to define the processes more precisely. Progress in understanding scientifically what is the influence of each variable and what occurs in each process requires surface analysis. This will be illustrated by discussing some typical examples of use in e.g. (i) purity control of the films, (ii) doping profiles in the films, (iii) analysis of layered structures, (iv) analysis of surfaces after processing steps.

Figure 21.1 represents the thin film system and illustrates the main techniques discussed. The primary ion beam is used for SIMS and for sputter profiling away surface contamination and successive layers to expose the underlying surfaces for compositional analysis by XPS and AES. It should be remembered that the uncertainty in such analyses is determined by bombardment induced atomic mixing and by the measured signals being convolutions of species concentrations over the electron escape depth. These points are discussed in Chap. 4.

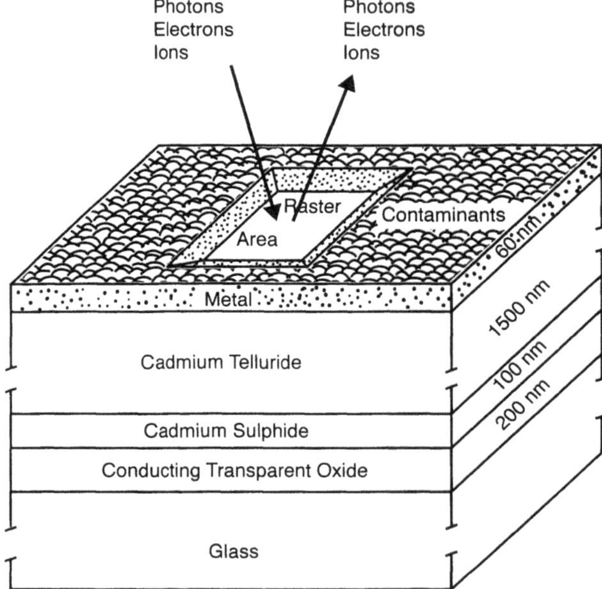

Fig. 21.1. Representation of surface analysis of a thin film photovoltaic device

21.2 Film Purity

21.2.1 Low Level Impurities – Qualitative

Material purity is an important characteristic of photovoltaics. Some desired data include (i) the types, extent and location of impurities in the semiconductor films, (ii) the limitations imposed by those impurities on the final cell's performance, (iii) the minimum purity of starting material to produce adequate cells, (iv) the alterations in various impurity profiles during the fabrication of cells from the films.

Because of the low concentration of impurities and especially because these films (e.g. 1.5 µm CdTe) are on a thick (1 mm) glass substrate, SIMS is an essential rapid and routine analytical tool for multi-element detection. In Fig. 21.2 the positive SIMS spectrum from an electrodeposited CdTe film is compared with that from a single crystal purchased as 5N purity from a commercial source. Spectra were obtained using a Perkin Elmer PHI Model 560 Multitechnique System (XPS, SAM, SIMS). The ion gun was a differentially pumped electron impact Ar^+ source. SIMS data for both the film and crystal were recorded under similar experimental conditions. It is clear that the electrodeposited film had fewer impurities than the commercial 5N crystal used for the data of Fig. 21.2 and, indeed, two other 5N crystals from different commercial suppliers [5].

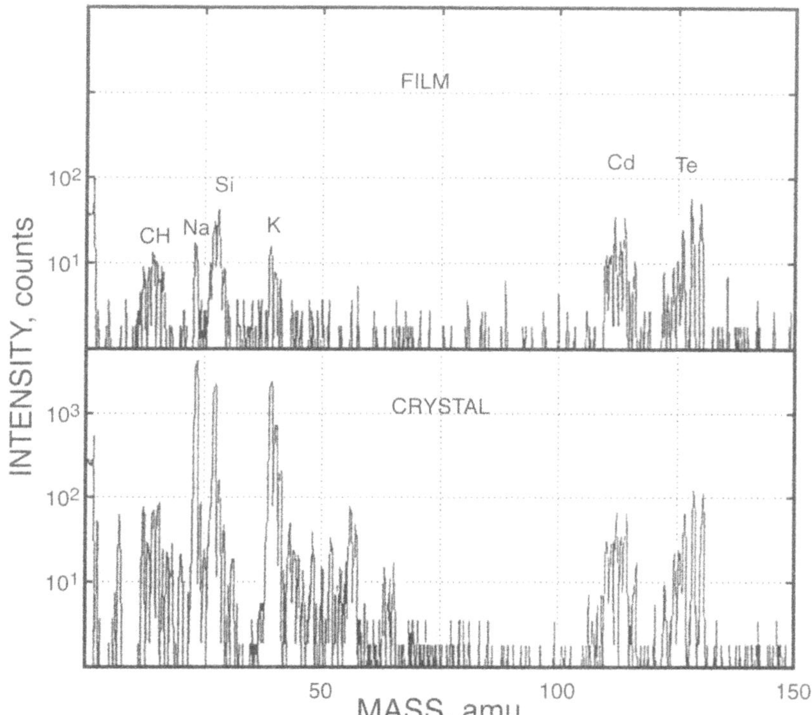

Fig. 21.2. Positive SIMS spectra for an electrodeposited CdTe film and a CdTe single crystal supplied as 5N pure. From [9]

Cross Contamination and Fringe Neutral Problems in SIMS. In such studies, in addition to the obvious precautions of keeping constant the type of primary ion, its energy and current density, secondary lens settings, sample geometry and environment, the 'fidelity' of data for trace element analysis required identification and elimination of transported material from surrounding areas onto the sample analysis area [9,10]. This transported material which was left from a previously sputtered sample contaminated the sample surface producing secondary ions intermixed with 'true' sample signals. To minimize cross-contamination the sample chamber must be carefully designed to avoid large area metal surfaces near the sample.

Of equal concern for reliable SIMS results are the secondary ions produced by energetic unfocused fringe neutrals bombarding the region around the analysis area. Secondary ions produced by fringe neutrals striking the sample mount and attachments, e.g. spring clips or cover plates, invalidate results of trace element studies [9]. The extent of this effect has been quantified as a function of distance from the beam centre and of beam current and beam energy [11]. For example, with an Ar^+ current in the range 1 µA–4 µA and a beam energy of 4 or 5 keV, the secondary ion signals which arose from neutrals at distances greater than 3 mm from the beam centre was 1.6% of those generated by the primary ion. At distances greater than 6 mm and 9 mm from the beam centre, the values were 0.3% and 0.1% respectively. Hence, to minimize the contribution to the 'true' SIMS signal of fringe neutral signals, two strategies should be used. Firstly, sample mounting hardware such as cover plates, screws, etc., on the front of the sample should be avoided. A small stub projecting about 2 cm from the sample mount covered by the sample has been found useful [9]. Secondly, a large crater such as 9 mm × 9 mm should be etched during initial profiling and then the analysis taken over a smaller area such as 3 mm × 3 mm so that data representative of the film's bulk composition could be collected. Thus the SIMS signal whether from the primary ion or the fringe neutrals would originate initially from an area at a similar depth and with similar composition.

The use of an ion gun with a curved ion path can significantly reduce fringe neutral effects. Analysis of the CdTe films with such a gun (using Cs^+ and O^- primary ions) detected each element observed with the simpler in-line ion gun thus validating the procedures used.

21.2.2 Low Level Impurities – Quantitative

The thin film used to provide the SIMS data of Fig. 21.2 was made using an A.R. grade (2N) source of cadmium sulphate in the deposition solution although this was subsequently electrolytically purified. Purer films can be made with better starting material as illustrated in Fig. 21.3 where SIMS data for films prepared by A.R. grade (2N) and a 4N grade of cadmium sulphate are compared. Although the data of Fig. 21.3 show that the film made from the 2N grade was 'dirtier' over all regions of the mass spectrum, cells as efficient as those made with the purer starting material were obtained. That result suggested that the cleaning procedures [5] removed the electrically active impurities.

Although SIMS is primarily a qualitative technique, it is useful for comparative work provided similar matrix samples are used and the precautions stated above are taken. Useful information as illustrated by the data of Figs. 21.2 and 21.3 can be obtained. However, quantitation is often needed and other techniques must be used to complement the SIMS data. Quantitative XPS and AES data can be useful, although restricted to elements with concentrations above 0.1%. Inductively coupled plasma-atomic emission and atomic-absorption spectrophotometry have been used to determine the magnitude of impurities in the components of the plating bath solution and six different supplies of cadmium sulphate [12]. The origin of an impurity and in some cases an upper limit to its concentration in the films has been determined.

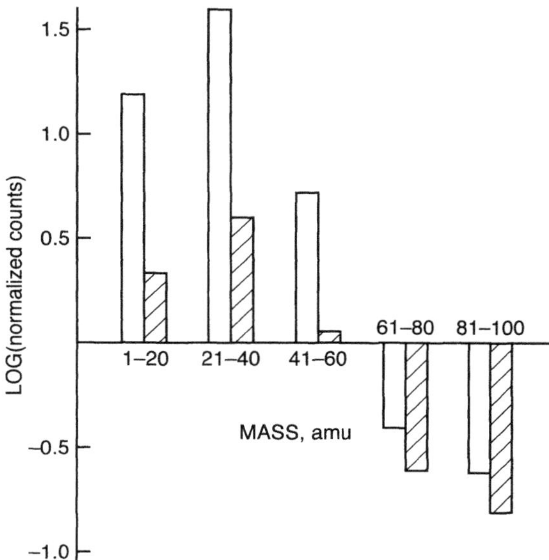

Fig. 21.3. Intensities of positive SIMS signals in various mass regions for electrodeposited CdTe films prepared using a 2N grade (□) and a 4N grade (▨) of cadmium sulphate. The SIMS counts have been normalized to the combined Cd + Te count

With the type of studies outlined above, the purity of the films produced under varying conditions could be monitored to ensure that the films were purer than 5N single crystals purchased from commercial suppliers.

21.2.3 Doping Profiles in Thin Films

One of the common uses of SIMS is to look at doping profiles e.g. as a consequence of ion implantation of ^{11}B or ^{31}P in Si to alter the electrical character. Provided that the rate of sputter profiling of the host materials was known, quantitative implantation profiles of ^{11}B in Si have been tracked by SIMS [13] over six orders of magnitude to levels of $\sim 10^{13}\,\mathrm{cm}^{-3}$.

The sputter rate of the host material needs to be determined from other experiments, usually with reference to standard materials such as Ta_2O_5 film on Ta. For electrodeposited CdS and CdTe films, thickness of various films were determined by SEM , by a surface profiler, by near infra-red interference spectrophotometry, and by the charge passed during deposition. The rate of Ar^+ sputtering of these films under defined ion beam conditions was compared with the rate of sputtering standard Ta_2O_5/Ta film [14]. The ratios of sputtering rates were CdS:CdTe:Ta_2O_5 as 6.3:3.9:1 [12.11].

The data in Fig. 21.4 illustrate a typical doping profile formed in this case by heating n CdTe in an ampoule with phophorus and tellurium. The SIMS signal intensity from the species of $^{31}P^+$ and $^{31}PO^+$ were determined as a

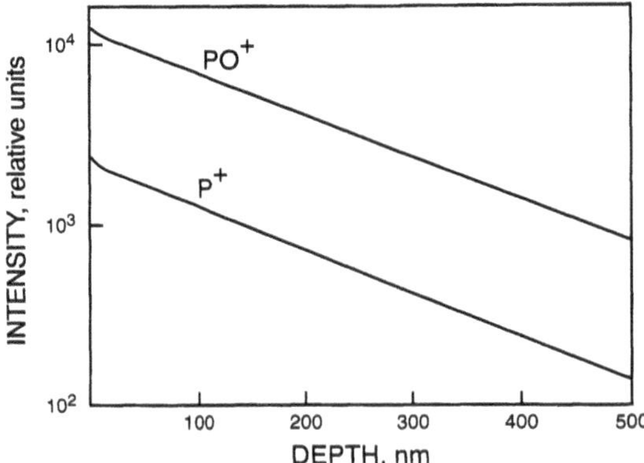

Fig. 21.4. Depth profile of phosphorus doped into an electrodeposited CdTe film measured by SIMS signals from PO^+ and P^+. Rate of depth profiling was 20 nm min^{-1}

function of the depth into the cadmium telluride. Note that the PO^+ species had a similarly shaped profile to the P^+ species and was considerably more intense, making it the preferred ion species to monitor the phosphorus level.

21.3 Composition and Thickness of Layered Films

A common problem in thin film analysis is determining the composition and thickness of adventitious or deliberately added thin layers on surfaces or at interfaces. When the layers have a varying composition profile, this can be a complex problem. Examples will illustrate how some types of problems can be solved.

21.3.1 Composition Gradation in Films, e.g. $Cd_xHg_{1-x}Te$ Films on Platinum

Films with composition $Cd_xHg_{1-x}Te$ have been cathodically electrodeposited onto platinum [6,15]. There was a possibility because of the deposition method used that the film composition could vary as the film thickness increased. Measurement of the atomic concentrations by XPS with ion sputter profiling was impractical because mercury may preferentially sputter from the film surface. Instead, a series of successively thicker films were made using the same deposition conditions and solutions of similar composition. The surface atomic concentration of mercury expressed as (atom percent Hg)/atom percent (Hg+Cd) was obtained by XPS. These data are shown in Fig. 21.5 which illustrates that a composition gradient would exist in the deposited films.

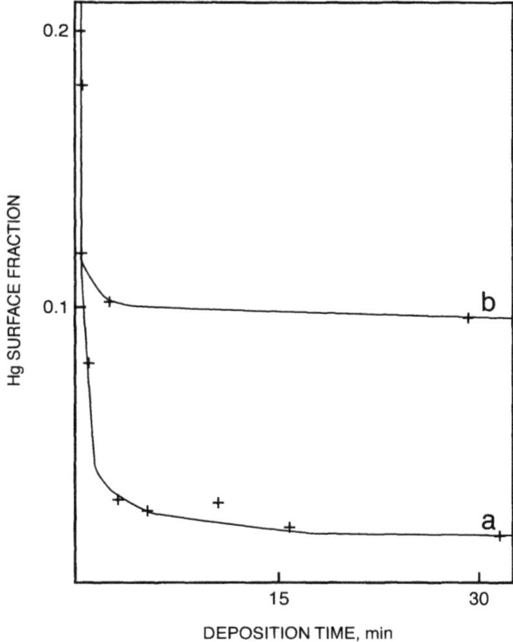

Fig. 21.5. Mercury content expressed as (Hg/Hg+Cd) on the surface of electrodeposited $Cd_xHg_{1-x}Te$ films of various thicknesses prepared from solutions containing Hg concentrations of (*a*) 5 ppm (*b*) 10 ppm

21.3.2 Thin Overlayers on Films

a) Growth of Oxide/Hydroxide layers on CdTe Films. Electrodeposited CdTe films exposed to the atmosphere grew oxide/hydroxide layers which SIMS depth profiling showed to be about 2 nm thick. Angular resolved XPS which provided the data of Table 21.2, confirmed that the oxygen species were concentrated at the surface layers. The binding energy for the XPS O 1*s* peak differed for the surface and bulk species. The surface oxygen was probably present as an hydroxyl species [16].

Table 21.2. Angular resolved XPS data from CdTe films exposed to the atmosphere

	Atomic concentration [Atom %]			O 1*s* binding energy [eV]
	Cd	Te	O	
Grazing angle	32.0	37.5	30.5	531.9
Normal	45.8	49.3	4.9	530.4

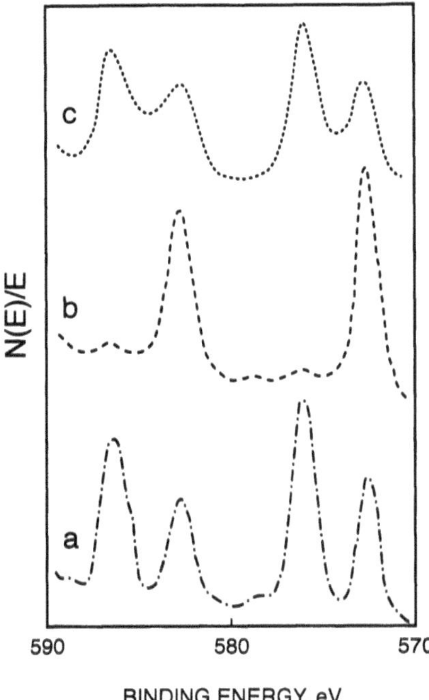

Fig. 21.6. XPS spectrum in the Te 3*d* region of the surface of a CdTe crystal (*a*) after air anneal 10 min, 350°C, (*b*) after 20 s etch with 80°C KOH, 30% (w/w), (*c*) after ion sputtering the oxidized film for 2 min

When the *n*-CdTe was type converted by heating the film in air at 350–400°C for about 10–15 minutes, an oxygen-rich film, readily removed by a basic etch, was formed on the surface as shown by the XPS spectrum for the Te 3*d* region (shown in Fig. 21.6) [17]. The oxygen-rich layer, said to be $CdTeO_3$ [18] was shown to be a few nm thick by depth profiling until the atomic concentrations of Cd and Te were 50%. The change in the Te 3*d* peak position as a function of ion sputter time was another useful way of displaying the surface nature of the oxygen-rich region as shown in Fig. 21.6c.

b) Surface Modification of Thin Films. Surface modification of materials is finding increasing importance in technology. Probing the chemical changes which surface modifications introduce is essential to suitable control of the modifications. For example, the surface of a *p*-CdTe film must be modified for a low specific resistance contact with a metal and this has been accomplished with a thin, very heavily doped layer which allowed carrier tunnelling. A favorite recipe to produce the layer has been to use one of several chemical etchants such as bromine in methanol (BM), orthophosphoric acid with nitric acid (OPN) and acidified potassium dichromate (KD). XPS studies have shown the chemistry of what is happening on the surface as will be illustrated by data [19] obtained using a BM etch (0.1% Br_2 in methanol). Figure 21.7 shows the XPS survey spectra of the film after etch-

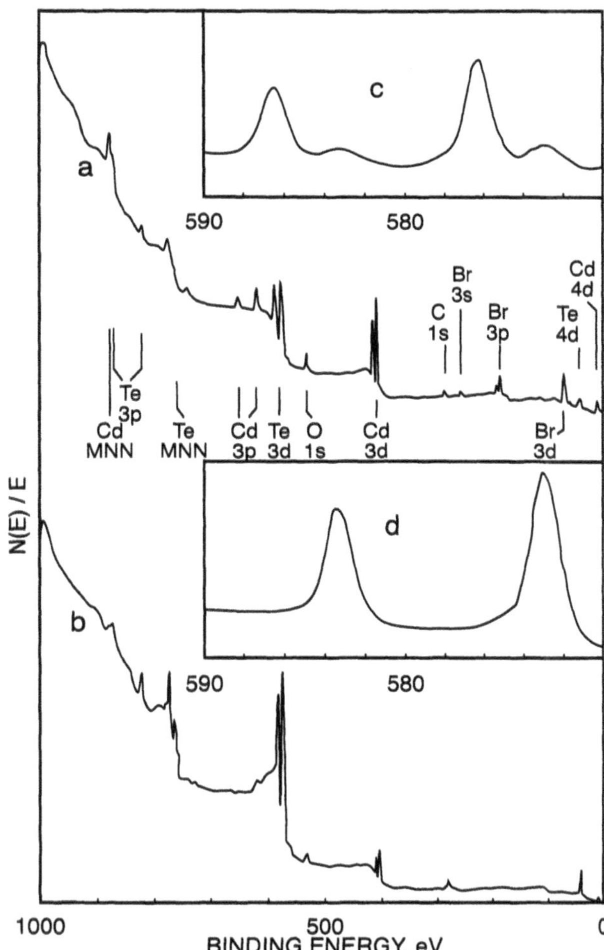

Fig. 21.7. XPS survey spectra of CdTe film after statically etching for 5 s in 0.1% bromine-methanol then **(a)** blow-dried with nitrogen; **(b)** washed each in methanol (1 min) and Milli-Q water (1 min) then blow-dried with nitrogen; **(c)** Te 3*d* region from spectrum (a); **(d)** Te 3*d* region from spectrum (b)

ing and before and after washing with water. The bromine 3*d* peaks were observed near 69 eV when the film was not properly washed. Washing also altered both the intensity and position of the cadmium 3*d* and tellurium 3*d* peaks. The unwashed film showed a strong cadmium 3*d* signal consigned to cadmium bromide on the surface and also showed a split tellurium 3*d* signal assigned to Te^0/Te^{2-} and tellurium bromide. Washing removed the bromide species as evidenced by the disappearance of the Te^{4+} peaks (Fig. 21.7d) and

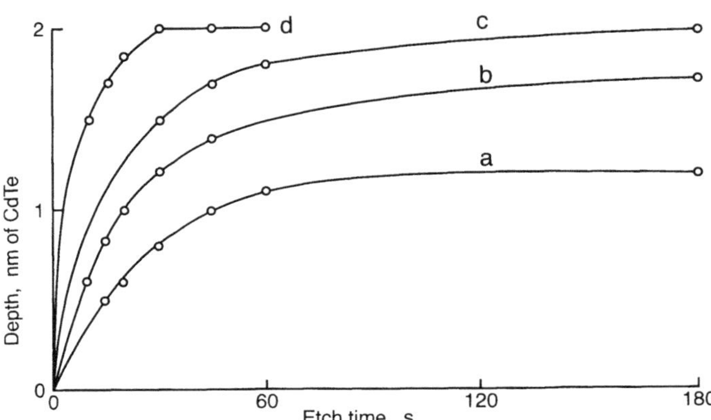

Fig. 21.8. Depth of Cd depleted surface with etch time for CdTe single crystals statically etched by BM of different concentrations: **(a)** 0.1%; **(b)** 5%; **(c)** 10%; **(d)** 50%. From [19]

a change in the relative intensity of the Cd:Te $3d$ peaks from 0.7 (unwashed) to 0.5 (washed).

The depth of the cadmium depleted region was determined using XPS with ion profiling as a function of chemical etch time and concentration. A sputter rate of about $0.5\,\text{nm min}^{-1}$ was used with XPS analyses after successive 1 min sputter times. The atomic concentrations of Cd and Te were plotted as a function of sputter time with the endpoint reached when the Cd:Te ratio was 1.00, the value it was in the crystal bulk. Figure 21.8 shows the result of such a study. The amount of material removed depended on the time and the concentration of the chemical etch but even though the crystal was etched at a rate of $14 \pm 4\,\text{nm s}^{-1}$ by a static 0.1% BM solution, the depth of the Cd depleted region was limited to about 2 nm. Note that even though the surface material was not CdTe the depth of the Cd depleted region has been expressed as if the surface species sputtered at the same rate as CdTe. This is a convenience, but not accurate e.g. elemental Te sputters about twice as fast as CdTe.

c) Contaminant Modification of Surfaces e.g. Carbon on Gold. As discussed previously, thin layers of adventitious materials can grow on exposed surfaces. For an example, the effects of contaminants on the gold layer evaporated to form a metal contact to the etched p-CdTe films are discussed. The contact resistance increased twenty fold and the cell's efficiency decreased after the cell was left exposed to the atmosphere for one year. A freshly evaporated gold surface showed only the presence of gold at and near the surface. However, as shown by the AES data of Fig. 21.9, the gold contact deteriorated with age because of the accumulation with time of carbonaceous materials on the surface. Suitable encapsulation and permanent contacts would nor-

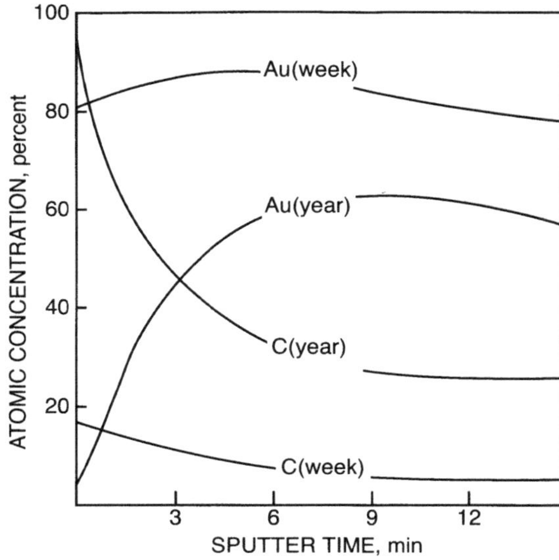

Fig. 21.9. Depth profiles of the atomic concentrations of C and Au in Au deposited on CdTe films

mally be used to avoid such problems. Note that the atomic concentrations of carbon and gold in the data of Fig. 21.9 are less than 100% after several minutes of ion beam etching. Tellurium species are observed in this region, but a detailed study of the diffusion of the near surface species has not been published.

21.4 Beam Effects in Thin Film Analysis

Beam induced effects such as preferential sputtering of one element in a compound, preferential diffusion of a species, surface oxidation, adsorbate dissociation, redeposition, etc., are a constant problem in thin film analysis. The use of a multi-technique facility can help to pinpoint the source of a problem. For example, the magnitude of the atomic concentration of oxygen on the surface of either CdS or CdTe was measured by AES to be about ten times larger than the value measured by XPS. This variation was traced to oxidation of the compounds under the electron beam probably by a thermal reaction with residual water vapour in the system [9,20]. Under simultaneous ion beam sputtering the atomic concentration of oxygen measured by AES decreased as the inverse of the ion beam current and at high ion beam currents was the same as measured by XPS. Clearly simultaneous rapid sputtering was needed to provide reliable AES data for oxygen.

The experienced researcher generally recognizes when beam effects are altering the 'true' data. Readers are referred to the article by *Kazmerski* [21] for further examples.

21.5 Conclusion

Surface analytic methods are a fruitful way of learning some details of thin film structures as shown by examples in this chapter mainly dealing with chemical structures of thin film photovoltaics. The usual methods used to investigate these devices are by current-voltage-time-temperature relationships to probe device parameters and current generation and trapping mechanisms. Various spectroscopies such as time- and energy-resolved luminescence, quantum yield, admittance provide further details, including the type and extent of traps. Ideally, data from these experiments should be correlated with details of the physical and chemical structures available from surface analytic methods. That task is currently too complex for completion for even one type of thin film device made by one preparative method. However, some progress has been made and further development of analytic methods of near atomic resolution will hopefully lead to inexpensive, durable, efficient thin film photovoltaics.

References

1. H. Oechsner (ed.): *Thin Films and Depth Profile Analysis*, Topics Curr. Phys. Vol. 37 (Springer, Berlin, Heidelberg 1984)
2. U.S. Department of Energy, National Photovoltaics Program Five Year Research Plan, 1987–1991, DOE/CH1000093-7, 1987
3. W.J. Danaher, L.E. Lyons, G.C. Morris: Solar Energy Materials **12**, 137 (1985)
4. G.C. Morris, P.G. Tanner, A. Tottszer: 21st Photovoltaic Specialist Conference, IEEE (1990)
5. L.E. Lyons, G.C. Morris, D.H. Horton, J.G. Keyes: J. Electroanal. Chem. **168**, 101 (1984)
6. B.M. Basol: Solar Cells **23**, 69 (1988)
7. G.C. Morris, L.E. Lyons, P. Tanner, C. Owen: Technical Reports of the 4th International Photovoltaic Science and Engineering Conference, Sydney (1989), p. 487
8. A 10% efficient solar array covering about 5% of the area of Australia would generate more electrical power than the world's power stations
9. G.C. Morris, L.E. Lyons, R.K. Tandon, B.J. Wood: Nucl. Instrum. B **35**, 257 (1988)
10. C.W. Magee, R.K. Honig: Surf. Interface Anal. **4**, 35 (1982)
11. G.C. Morris, B.J. Wood: Materials Forum **15**, 44 (1991)
12. L.E. Lyons, G.C. Morris, R.K. Tandon: Solar Energy Materials **18**, 315 (1989)
13. D.S. Simons, P. Chi, R.G. Downing, J.R. Ehrstein, J.F. Knudsen: Proc. Sixth International Conference on Secondary Ion Mass Spectrometry, ed. by A. Benninghoven, A.M. Huber, H.W. Werner (Wiley, Chichester 1988) p. 433
14. National Physics Laboratory, Certified Reference Material NPL No. S7B83, BCR No. 261
15. G.C. Morris, M. Marychurch: Materials Forum **15**, 143 (1991)
16. C.T. Au, M.W. Roberts: Chem. Phys. Lett. **74**, 472 (1980)
17. W.J. Danaher, L.E. Lyons, G.C. Morris: Appl. Surf. Sci. **22/23**, 1083 (1985)

18. F. Wang, A. Schwartzman, A.L. Fahrenbruch, R. Sinclair, R.H. Bube, C.M. Stahle: J. Appl. Phys. **62**, 1469 (1987)
19. W.J. Danaher, L.E. Lyons, M. Marychurch, G.C. Morris: Appl. Surf. Sci. **27**, 338 (1986)
20. A. Ebina, K. Asano, Y. Suna, T. Takahashi: J. Vac. Sci. Technol. **17**, 1074 (1980)
21. L.L. Kazmerski: Solar Cells **24**, 211 (1988)

22 Identification of Adsorbed Species

B.G. Baker

Adsorption at a solid surface is the initial step in many heterogeneous processes. The reactivity of solids, corrosion, inhibition, catalytic reaction and some methods of separation depend on adsorption. The process may involve specific chemical interaction with surface sites and, in many cases, results in dissociation of the adsorbate molecule. The formation of the initial monolayer needs to be understood in detail in order to explain the behaviour of adsorbent materials in contact with gas or solution.

The methods of surface analysis are inherently suited to the study of the atomic detail of adsorbates. Measurement of surface concentration, element identity and valence state can be made by X-ray photoelectron and Auger spectroscopy. When these techniques are applied to particulate or polycrystalline materials, the results may represent a combination of the adsorption behaviour of a variety of surfaces. Different crystal planes of pure substances provide surfaces which may have markedly different adsorption properties. Techniques which aim to define the spatial arrangement of adsorbates or the details of chemical bonding are applied to prepared single crystal surfaces. The behaviour of practical materials must then be interpreted in terms of a series of studies of several different crystal planes.

The following techniques have been employed to study adsorbate layers: low energy electron diffraction (LEED) to detect ordered spatial arrangements; X-ray photoelectron (XPS) and Auger electron spectroscopies (AES) to measure surface concentration and states of chemical combination; ultraviolet photoelectron spectroscopy (UPS) for detecting molecular surface species and changes in electron densities resulting from adsorption; electron energy loss spectroscopy (EELS) and reflection infrared spectroscopy for measuring vibrational states of adsorbed species. The application of these techniques is contained in the examples of adsorption studies that follow. Others are discussed in Chaps. 1, 5, (SIMS), Chap. 10 (STM), Chap. 11 (LEIS) and Chap. 12 (RHEED).

22.1 Examples of Adsorption Studies

22.1.1 Nitric Oxide Adsorption on Metals

The practical objective of most of the scientific interest in nitric oxide adsorption is the control of emissions of this gas into the atmosphere. The control of automotive emissions by catalysts containing noble metals is now established technology. The high cost of Pt, Pd and Rh and the question of reliable supply provides an ongoing incentive to discover alternative materials. Furthermore the exhaust catalyst to control nitric oxide requires the reaction of NO with carbon monoxide. This is only possible when oxygen levels are controlled at a very low level. Emissions from diesel engines and industrial furnaces contain too much oxygen to allow control by existing catalyst technology. Empirical testing of large numbers of materials has failed to solve the problem. Perhaps a more detailed understanding of the atomic and molecular features of adsorbed nitric oxid will reveal new catalytic strategies.

The decomposition of nitric oxide and the reaction of nitric oxide with carbon monoxide have been investigated on polycrystalline iron and nickel [1]. The initial reaction, on both metals, at 500–550 K is the decomposition of nitric oxide to nitrogen and nitrous oxide and the incorporation of oxygen into the film to form an oxide layer. The rate of the decomposition reaction decreases as the thickness of the oxide layer increases and becomes immeasurably slow at an oxide thickness > 50 nm. The oxidized surface catalyzes the reduction of NO by CO to form N_2 and CO_2 but with relatively low activity.

The initial stages of the action of nitric oxide on nickel has been studied in a number of experiments on single crystals. The chemisorption of nitric oxide on (110) nickel has been investigated by Auger electron spectroscopy, LEED and thermal desorption [2]. It was found that NO adsorbs irreversibly at 300 K forming a faint (2×3) LEED pattern. At 500 K this pattern intensifies. The process of dissociation is revealed by Auger spectroscopy. The data in Fig. 22.1 were obtained by successive excursions of the temperature followed by measurement of the Auger peak heights for oxygen and nitrogen. Comparison of the measured peak heights with those calculated by theory for a monolayer of atoms shows agreement only for oxygen not for nitrogen. This is interpreted as being due to NO being bound to the surface via nitrogen so that oxygen is outermost. Above 500 K the nitrogen signal increases and oxygen decreases indicating that the molecule has dissociated leaving the surface covered with atomic N and O. By measuring the rate of the decomposition as a function of temperature the dissociation energy is calculated at 125 kJ mol^{-1}. At $\sim$ 860 K nitrogen desorbs. The rate of this desorption was measured by AES and by quantitative thermal desorption. It is shown that the desorption of N_2 is first order and that the binding energy is 213 kJ mol^{-1}.

The behaviour of nitric oxide on the (110) crystal plane is unlike that observed on the close packed (100) and (111) planes [3,4], The initial adsorption on these planes involved dissociation at temperatures < 250 K. At increased exposures of nitric oxide gas, a molecular form is found to adsorb. This form is stable only to about 400 K when it desorbs as NO gas. There is no dissociation process occurring at $\sim$ 500 K as found on the (110) plane.

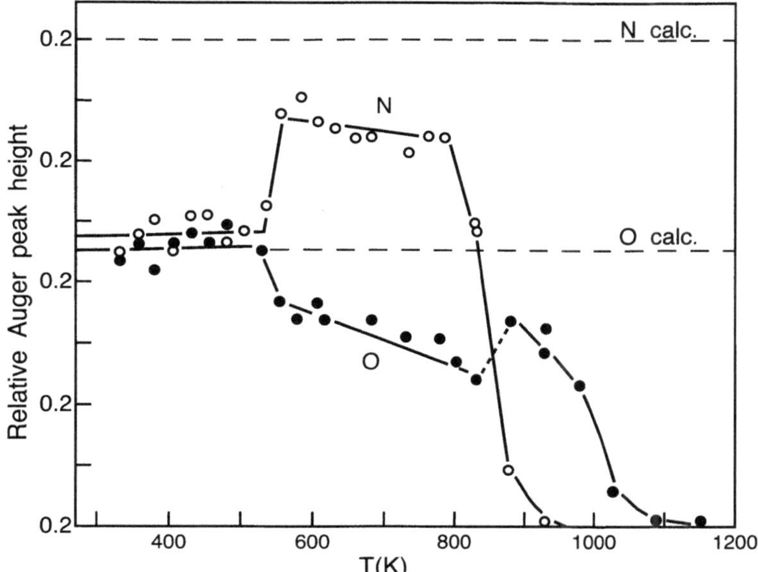

Fig. 22.1. Auger analyses for nitrogen (○) and oxygen (•) from nitric oxide adsorbed on (110) nickel. The analyses were recorded at room temperature after successively heating the crystal to the plotted temperatures at a rate of $10\,K\,s^{-1}$ and cooling quickly

Comparison of the single crystal work with the results on polycrystalline nickel, previously mentioned, shows that the (110) plane most closely corresponds to the observations on polycrystalline metal. Decomposition of an adsorbed nitric oxide molecule to form adsorbed oxygen and nitrogen at 500 K is the common feature. In the single crystal adsorption experiment, these atomic species persist. In the higher pressure reaction experiment, the oxygen diffuses to form a surface oxide while the nitrogen is displaced to the gas phase by incoming nitric oxide.

The polycrystalline surface is likely to expose various crystal planes including the (100) and (111). The behaviour of the (110) plane may well be typical of various rough or stepped planes. The reactive properties of the polycrystalline surface may derive from structural features which constitute only a minor fraction of the total surface.

Ultraviolet photoelectron spectroscopy (UPS) is an important technique in the study of adsorbate-surface bonding. In the case of nitric oxide, where there is obviously interest in both molecular and dissociated forms, the UPS technique is most important. A detailed identification of molecular orbitals is possible, adsorbed atomic oxygen and nitrogen are identified, and the change in electron density at the surface resulting from the adsorbate interaction is recorded.

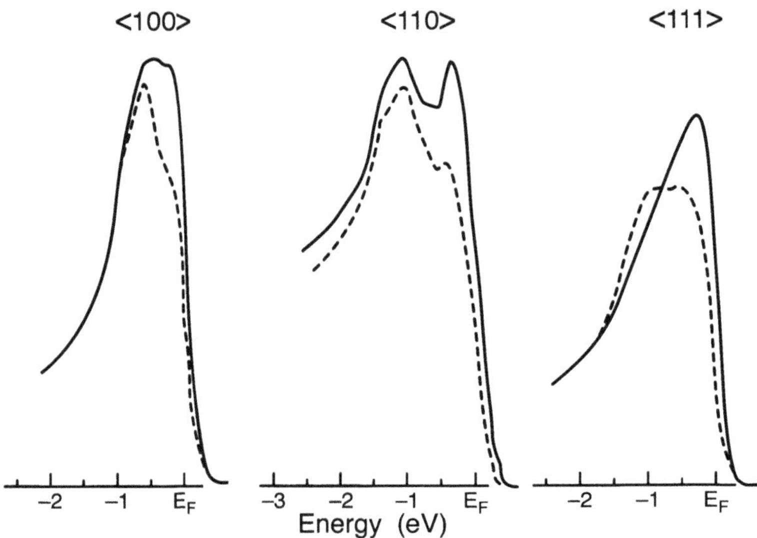

Fig. 22.2. Angle-resolved spectra from clean nickel (100) taken in three crystallographic directions (*full lines*); surfaces with adsorbed nitric oxide (*broken lines*)

The photon source for UPS is typically a helium discharge lamp providing HeI radiation at 21.2 eV (Chap. 14). This is introduced to the surface via a windowless port with differential pumping between the lamp and work chamber. The energy range of this form of UPS is $\sim$ 16 eV i.e. 21.2 eV less the work function. The Fermi energy is usually detected and is taken as a reference so that spectra have an energy scale relative to E_F. The transition metals important in nitric oxide adsorption have a high density of electrons near the Fermi level. Chemisorption of a gas results in a decrease in intensity in the spectrum near E_F. This is shown in Fig. 22.2 for the adsorption of nitric oxide on (100) nickel [5]. These spectra are taken in angle-resolved mode in the apparatus. The surface orientation is (100). Three directions of emission from this surface reveal differing d-band features. In fact, the shape for the directions $\langle 110 \rangle$ and $\langle 111 \rangle$ correspond closely to those observed from surfaces prepared with these orientations. The effect of adsorption of nitric oxide is shown in each case to decrease the intensity at close to E_F. This represents electrons from the metal becoming involved in the process of chemisorption.

Nitric oxide may be considered to be intermediate electronically between CO and O_2, two molecules capable of reaction with NO. The series CO, NO and O_2 represent occupancies of the outermost Π^* antibonding orbital of 0, 1 and 2 electrons. Back-donated charge from a metal surface is accommodated in the vacant Π^* orbital of CO without causing dissociation of the chemisorbed molecule. The corresponding process in O_2 results in dissociation which is always observed as the initial result of chemisorbing oxygen on

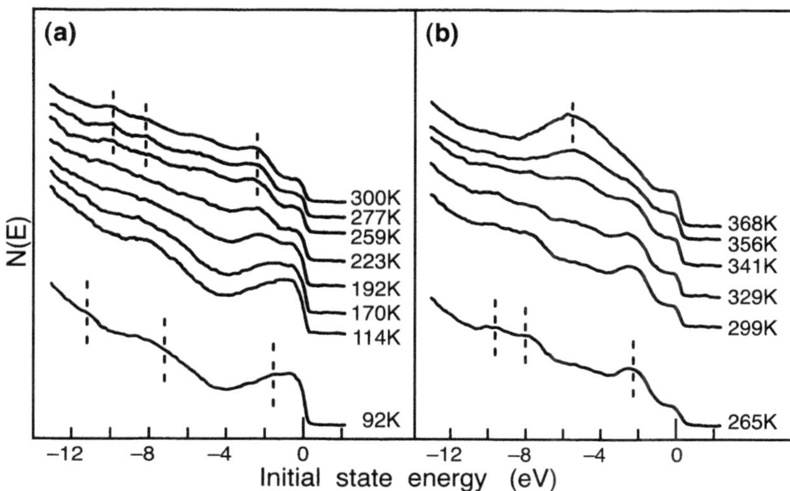

Fig. 22.3. (**a**) Initial adsorption of NO on Fe(110) at 92 K (α_1 state) followed by a sequence of temperature flashes to form the β state. (**b**) Initial adsorption of NO on Fe(110) at 265 K (β state) followed by a sequence of temperature flashes to form the dissociated (d) state

metals. But NO with one electron in the Π^* anti bonding may form molecular or dissociated adsorbate depending on the conditions.

A study of the adsorption of NO on (110) iron by UPS has revealed considerable complexity in the modes of molecular adsorption [6]. The initial adsorption of NO at 92 K results in the UPS spectra shown in Fig. 22.3. The broken lines mark features not present in the spectrum of the clean (110) iron surface (see also Fig. 22.4). They are attributed to a molecular adsorbed state of NO referred to as α_1. Repeated temperature flashes show a series of spectral changes until at temperatures above 330 K a single major feature is centred at −5.5 eV (Fig. 22.3b). This feature indicates photoemission from N 2p and O 2p and hence atomic adsorbates. The important characteristic spectra are summarized in Fig. 22.4 where they are represented as difference spectra by subtracting the spectrum of the clean Fe(110) crystal.

Thus the adsorption of nitric oxide on an Fe(110) single crystal surface at temperatures 90–350 K results in the formation of at least four distinguishable adsorption states, depending on the substrate temperature. The molecular adsorption states, α_1 over 90–110 K and α_2 over 110–170 K, are both thought to involve NO chemisorbed, N end down, possibly at an off-normal angle. Only slight differences exist between the valence electronic structure of the α states. The β adsorption state exists over a very wide temperature range of 170–290 K, although its concentration can be made to peak at 270 K. A model for the β state is that the initial dissociative adsorption occurs randomly and non-incorporatively, thereby allowing the formation of single vacant sites at

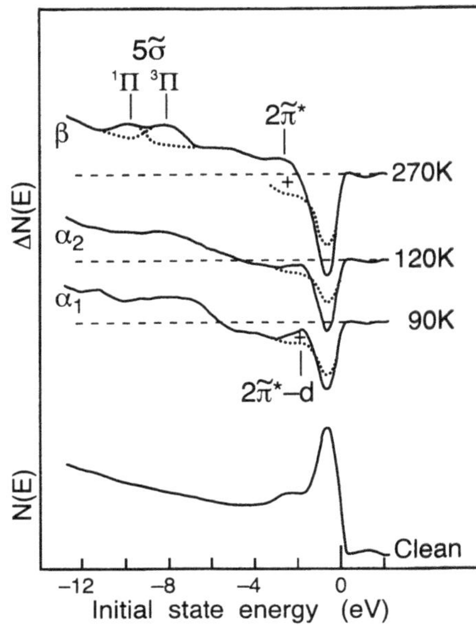

Fig. 22.4. HeI UPS difference spectra, weight averaged and smoothed, comparing the α_1, α_2 and β molecular adsorptions. An estimate of the inelastic scattering contribution is shown by dotted curves

which further dissociative adsorption cannot occur. Additional adsorption is thus restricted to molecular adsorption at these single sites. Schematically this can be represented:

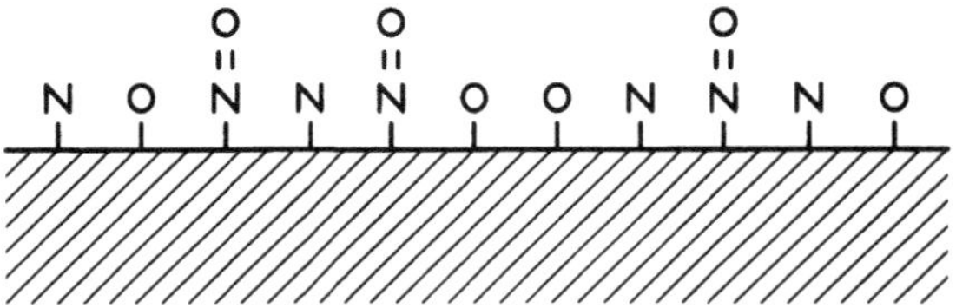

The β state is apparently stabilized at all substrate temperatures at which NO would dissociate given the opportunity, i.e. an adjacent vacant site.

Nitric oxide was found to initially dissociate at all substrate temperatures above 110 K. The extent of dissociation increased with temperature, being complete above 300 K (d state). No measurable molecular NO was found to exist above 330 K. The lower temperature adsorption states are shown to undergo an irreversible conversion to the next highest temperature state upon heating the substrate: $\alpha_1 \rightarrow \alpha_2 \rightarrow \beta \rightarrow d$. A LEED pattern showed an ordered C(2×2) structure for low doses at 350 K but all other adsorption conditions resulted in disorder. Auger analysis of the NO intensity ratio indicated that atomic N is readily incorporated into the sub-surface region.

Of the three molecular states the β state showed the closest correspondence to a state that would be expected for a weak chemisorption of NO. Stabilization of the β state at temperatures which caused NO dissociation on a clean surface may be due to the blocking of adjacent sites by adsorbed atoms or by an electronic stabilization effect in which adjacent adsorbed atoms reduce back-donation into the antibonding $2\pi^*$ orbital.

The complexity of nitric oxide adsorption and its dependence on the specific structure and composition of the surface might at first sight appear discouraging. However it is this diversity which offers hope of finding materials and conditions to hold nitric oxide in a reactive state for catalytic destruction.

UPS studies of molecularly adsorbed NO on various surfaces of Fe, Co, Ni, Cu, Ru, Pd, Pt, Rh, W, Re and Ir have been summarized in [7].

22.1.2 Aurocyanide Adsorption on Carbon

The selective adsorption of aurocyanide $Au(CN)_2^-$ onto the surface of activated carbons is of interest due to its widespread industrial use in enriching solutions of dissolved gold as part of the processing of gold ore. Some understanding of the process has been gained by observing the behaviour of adsorption from solution under varying conditions but the determination of the nature of the bonding and the structural identity of the adsorbate species requires the application of a spectroscopic technique. The system has been examined by X-ray photoelectron spectroscopy [8]. The detailed interpretation of the spectra provides a good example of the potential of the technique for adsorption studies.

An activated carbon sample was treated with a potassium aurocyanide solution buffered at pH 10. The adsorbate concentration which resulted was much greater than is usual in the industrial process. This was necessary to provide sufficient sensitivity in the XPS analysis but investigation over a range of coverages showed that a consistent adsorption state resulted. Excess gold solution was removed by washing and a series of samples prepared by subjecting them to acid washing of increasing severity. The dried samples were crushed and pressed into indium foil to provide a conducting holder for the XPS analysis. The samples in the spectrometer were maintained at 150 K during analysis to minimize possible radiation damage.

The XPS peaks of greatest importance in this study are the Au $4f$ and N $1s$ peaks. These serve to monitor the overall stoichiometry of the adsorbed aurocyanide and also indicate the nature of the bonding. The N $1s$ peak is taken as a monitor of cyanide since carbon is in excess as the adsorbent. Calibration samples showed that there are two distinctly different environments for nitrogen in these cyanides. Structurally $KAu(CN)_2$ consists of nearly linear $Au(CN)_2^-$ anions with both carbons bound to the central gold atom, the nitrogen atoms being in the terminal positions. In AuCN a linear polymeric chain exists with nitrogen bridging from carbon to gold. The observed core level binding energies (BE) for N $1s$ are 398.6 eV in the terminal position and 399.1 eV in the bridging position. This difference in BE provides the basis for distinguishing adsorbed cyanide species on carbon.

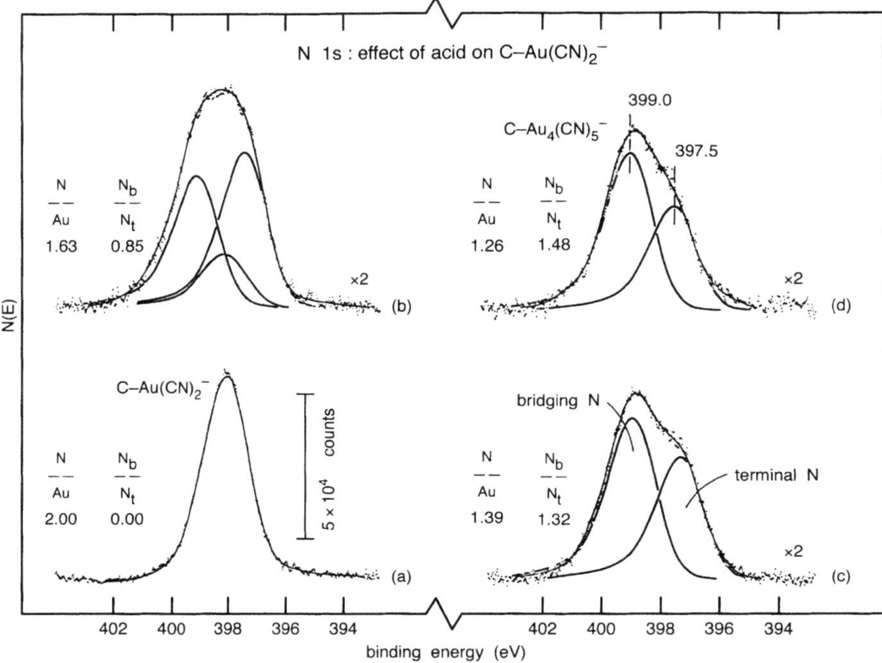

Fig. 22.5. N 1*s* photoelectron peak for (**a**) $Au(CN)_2^-$ adsorbed from alkali then subjected to increasingly severe acid (1M HCl) reaction, (**b**) 15 min at 298 K, (**c**) 180 min at 298 K and (**d**) 15 min at 373 K. Curves normalized to constant gold coverage

The series of N 1*s* spectra for the aurocyanide ion adsorbed on carbon are shown in Fig. 22.5. Adsorption from alkaline solution (trace (a)) shows nitrogen in the terminal position only and a ratio N/Au = 2 consistent with $Au(CN)_2^-$ as the adsorbate. The first acid treatment results in a broadening of the N 1*s* feature (trace (b)). This has been decomposed to reveal contributions from two other nitrogen states with binding energies 399.0 and 397.5 eV. The spectra in traces (b), (c) and (d) are then interpreted in terms of a diminishing coverage of the adsorbed $Au(CN)_2^-$ species and increasing coverage of the two other states. Taking the lower BE to represent the terminal N_t and the higher BE to represent N_b, the spectra are interpreted to determine total N/Au and N_b/N_t ratios. Note that the initial $Au(CN)_2^-$ species is not present in traces (c) and (d).

It is known that heating an acidified solution of $Au(CN)_2^-$ results in the formation of polymeric AuCN and the evolution of HCN. This reaction clearly has not gone to completion on the carbon surface. The limiting values of the measured ratios N/Au = 1.26 and N_b/N_t = 1.48 suggest that the polymerization process has resulted in a limited chain length anion. This oligomer

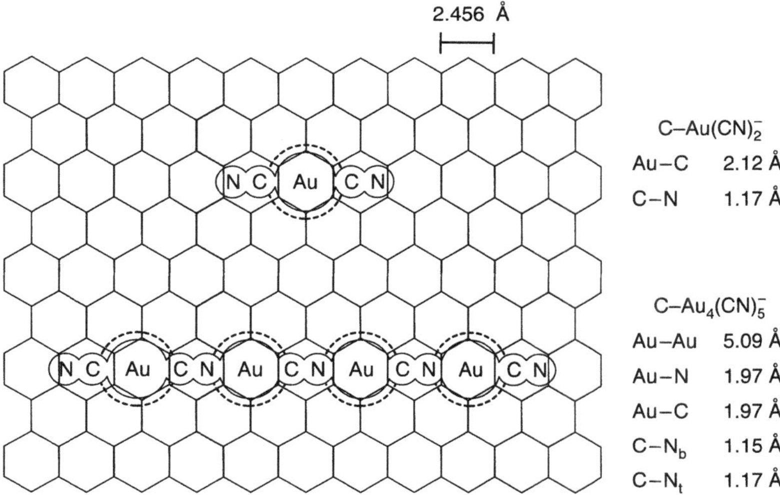

Fig. 22.6. Schematic depiction of the aurocyanide species $(Au(CN)_2^-$ and $Au_4(CN)_5^-$ adsorbed in a graphite plane. C and N atoms are scaled according to their covalent radii and Au atoms to the hard sphere ionic radius for Au(I) (*solid line*) and atomic radius (*broken line*)

would have both terminal and bridged N and a N/Au ratio greater than unity. The species $Au_4(CN)_5^-$ is proposed. This has N/Au = 1.25 and $N_b/N_t = 1.5$ in excellent agreement with the experiment.

The monomer and tetramer are represented in Fig. 22.6. The $Au(CN)_2^-$ species adsorbed on activated carbon from alkaline solution is adsorbed on and parallel to the graphitic planes. The bonding mechanism proposed is a π-donor bond from the graphite surface to the central cationic gold atom. The Au $4f_{7/2}$ peak has a binding energy 0.3 to 0.5 eV lower than that observed for the compound $KAu(CN)_2$ suggesting that the surface bond involves a net charge transfer of this magnitude. A shift of the same magnitude exists for the N 1s peak from the adsorbate suggesting that there is also a transfer of charge to the terminal nitrogens. This is not an indication of direct bonding of N to the surface but a consequence of the π-donor bond to the gold.

A similar bonding model is applied to the tetramer $Au_4(CN)_2^-$ formed by the acid induced oligomerization of C–$Au(CN)_2^-$. The geometry of the adsorbate on the graphitic carbon lattice is shown in Fig. 22.6. All four gold atoms are bound by π-donor bonds each resulting in the same BE shift for Au $4f_{7/2}$ and hence the same charge transfer. In this case, however, there are four gold atoms and only two terminal nitrogens: the observed BE shift in N 1s is larger.

The above study is remarkable in that a single technique, XPS, has been applied to samples prepared in solution to yield results comparable in detail to many in situ, single crystal, multitechnique experiments. The absence of electrostatic charging on activated carbon has facilitated precise measurement and interpretation of binding energy shifts.

22.1.3 Adsorbed Methoxy on Copper and Platinum

Methoxy species (CH_3O) play an important role as intermediates in the oxidation of methanol to formaldehyde. The reaction is catalyzed selectively by copper and silver whereas on platinum and other metals much CO_2, CO and water are formed along with formaldehyde. An explanation is sought in terms of the nature and stability of the reaction intermediate. For polyatomic adsorbates a surface vibrational spectroscopic technique is required to identify the chemical bonds and the mode of interaction with the surface.

Electron energy loss spectroscopy (EELS) has been developed as a technique for the identification of adsorbed species [9]. The vibrational modes are observed as energy losses in the range 0–500 meV in electrons reflected from the surface. The primary electron beam must have low energy, typically 1–15 eV, and be monochromatic usually within 5–10 meV. The spectra may be compared directly with infrared spectra by reporting the energy as wave numbers (cm^{-1}). The resolution of EELS, 40–80 cm^{-1}, is inferior to infrared spectroscopy but the range 0–4000 cm^{-1} is impressively broad. The technique requires ultra high vacuum and single crystal conducting samples. It is compatible with other surface techniques, UPS, XPS, AES and LEED, and has been incorporated into multitechnique systems.

The design of the spectrometer used in the study of the methoxy adsorbate is shown in Fig. 22.7 [9]. Two electron energy analyzers are located with a fixed scattering angle of 60° from the normal to the crystal. One acts as a monochromator running at a low pass energy of < 1 eV and directing a current of $\sim 10^{-9}$ A to the crystal. By adjusting the crystal position and deflectors, the reflected (0,0) beam is directed into the second analyzer. This analyzer is scanned through elastic and inelastic regions to detect the vibrational peaks. Vibrational energies are measured as the difference between elastic and inelastic energies. Identification of peaks is made by reference to infrared and Raman spectra of known molecules and by observing frequency shifts in deuterium-labelled molecules.

The spectrum in Fig. 22.8 was obtained by reacting methanol and atomic oxygen on a copper (100) crystal surface [10]. Water, a product of the reaction has been desorbed by heating to 370 K. The five fundamental vibrational modes associated with the methoxy species are assigned as follows: ν(Cu–O) 290 cm^{-1}, ν(C–O) 1010 cm^{-1}, $\delta(CH)_3$ 1450 cm^{-1}, ν_s(CH) 2830 cm^{-1} and ν_a(CH) 2910 cm^{-1}.

The intense stretching modes (Cu–O) and (C–O) and the absence of bending vibrations of the Cu–O–C chain indicate a perpendicular orientation of the methoxy species relative to the Cu(100) surface, bound through O to Cu. Although the spectrum was measured at 100 K, the preparation involved

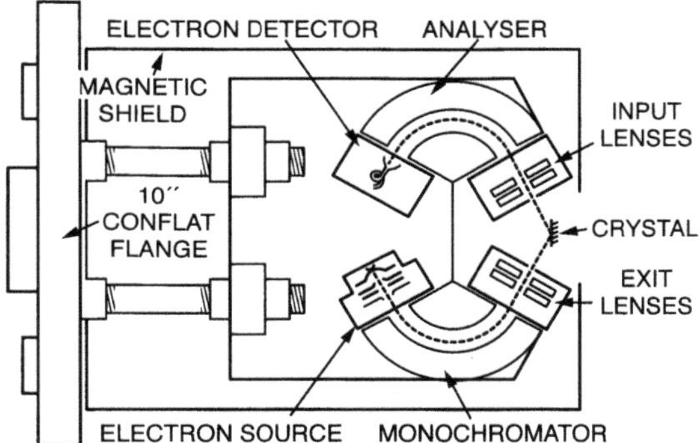

Fig. 22.7. Schematic of a 127° high resolution electron energy loss spectrometer for vibrational studies of surfaces. The incidence angle is 60° from the surface normal

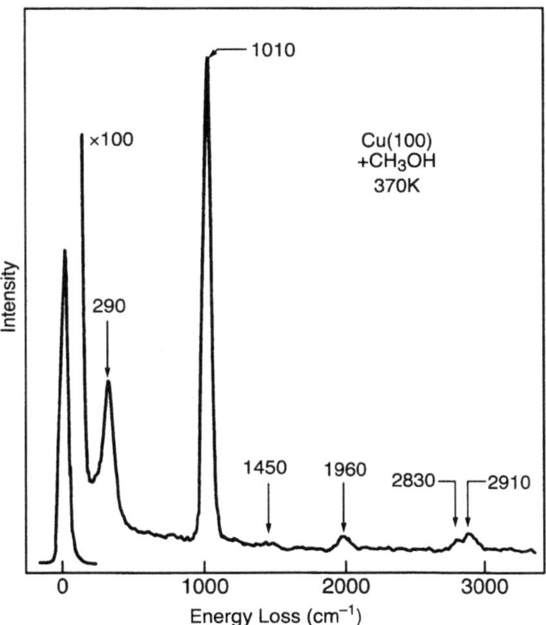

Fig. 22.8. Vibrational spectrum of the methoxy (CH_3O) intermediate chemisorbed on Cu(100) at 100 K. The beam energy was 5 eV. A monolayer of atomic oxygen was reacted with excess CH_3OH and annealed to 370 K to remove water

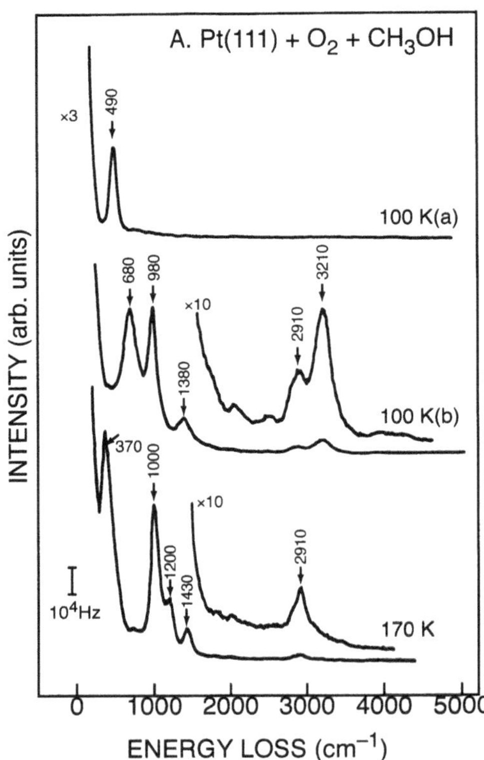

Fig. 22.9. Energy loss spectra of the reaction of methanol (CH_3OH) with atomic oxygen on Pt(111) to produce the methoxy intermediate (CH_3O). (**a**) Atomic oxygen $p(2 \times 2)$ structure. (**b**) Excess methanol condensed on (**a**). Annealing to 170 K produced the methoxy spectrum (*lower*). The beam energy was near 1 eV

heating to 370 K. The methoxy layer is stable to this temperature. At higher temperatures, CH_3O decomposes to formaldehyde and hydrogen.

An experiment on platinum (111) reveals a similar mechanism for the formation of methoxy from methanol and oxygen [11]. The EEL spectra in Fig. 22.9 show atomic oxygen on Pt(111) at 100 K with ν(Pt–O) = 490 cm^{-1} (trace (a)) followed by methanol added at the same temperature (trace (b)). At this stage reaction has not occurred. The methanol spectrum has obscured the Pt–O mode with strong features due to OH bending (680 cm^{-1}), CH stretch (980 cm^{-1}), CH_3 deformation (1380 cm^{-1}), CH stretch (2910 cm^{-1}) and OH stretch (3210 cm^{-1}). When the crystal was heated to 170 K the OH bending and stretching modes disappeared and a new Pt–O stretch at 370 cm^{-1} was evident. This is shown in the lower trace in Fig. 22.9 and is the evidence for the formation of the methoxy species. The shift to lower energy of the Pt–O as compared to that for atomic oxygen is due to the extra mass of the methyl group on the oxygen.

This methoxy-covered Pt(111) surface is stable only to 170 K. Above this temperature, the methoxy species is not stable on Pt(111). Decomposition in this case is to carbon monoxide and hydrogen; the Pt surface breaks the

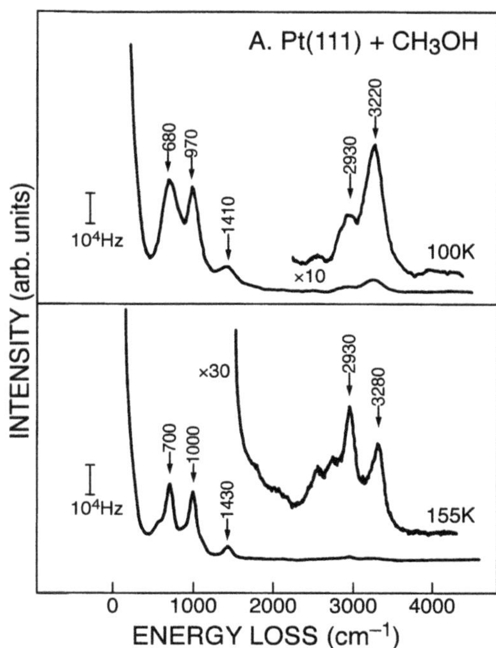

Fig. 22.10. Comparison of the energy loss spectra of multilayer methanol (100 K) and monolayer methanol (155 K) on Pt(111). In the monolayer spectra considerable "softening" of the methyl group CH stretching modes can be seen. The beam was near 1 eV

C–H bonds in the methyl group and the CO inverts to bind carbon to the surface, the usual mode of adsorption of CO.

The breakdown of the methyl group at low temperature clearly precludes the chance to produce formaldehyde. The mechanism may well occur on transition metals generally which strongly chemisorb hydrogen and carbon monoxide. Evidence for the effect of the platinum surface on methanol is shown by the EEL spectra in Fig. 22.10 [11]. The upper trace is for bulk methanol, i.e. a multilayer of methanol ice. No oxygen is added. The lower trace is for a monolayer of physically adsorbed methanol at 155 K. In this case vibrational modes are sharper and the OH stretch has moved to $3280\,\mathrm{cm}^{-1}$. This is an indication of decreased hydrogen bonding between molecules in the two-dimensional layer. Also in the two-dimensional layer, new frequencies are observed in the CH region near $2900\,\mathrm{cm}^{-1}$. These occur at lower energies and indicate that there is strong hydrogen bonding of the methyl group hydrogens to the platinum surface. Thus the methyl group in the physically adsorbed state is strongly influenced by the Pt(111) surface. When this monolayer was heated to $> 200\,\mathrm{K}$, complete decomposition to CO and hydrogen was observed by EELS and no methoxy groups formed.

The distinction between copper and platinum as catalysts for the selective oxidation of methanol to formaldehyde then depends on the stability of the methoxy species. The EEL spectra show that this species is more stable on copper and that, although it can form on platinum at low temperature, it decomposes to adsorbed CO and hydrogen.

The EELS techinque as applied in the above examples has one very important advantage over infrared spectroscopy. The bonds of an atom to the surface, Cu–O, Pt–O at energies $< 500\,\mathrm{cm}^{-1}$ are directly measured. These are out of the normal range of surface infrared methods. At higher energies, the superior resolution of IR spectra provides greater detail of the vibrational modes. The two techniques should be seen as complementary in their roles in surface science.

22.2 Conclusion

The examples discussed above have been chosen to illustrate the detailed nature of the information about adsorbates obtainable by surface analytical techniques. The application of LEED, AES, UPS, XPS and EELS to particular adsorption systems has been discussed. These techniques are best combined with other methods of detecting reaction or desorption by monitoring the gas phase by mass spectrometer. It is also generally found that more than one surface technique is needed to obtain a satisfactory description of an adsorbate. The emphasis on single crystal techniques is inherent in this area of surface science. A surface exposing multiple crystal planes would produce mixed spectra from multiple adsorbate states. The behaviour on practical surfaces is best deduced by measuring overall properties of adsorption and reactivity of the practical surface and comparing with the detailed studies on various crystal planes of pure substances.

References

1. B.G. Baker, R.F. Peterson: Proc. Sixth Int. Congress on Catalysis **2**, 988–996 (1977)
2. G.L. Price, B.A. Sexton, B.G. Baker: Surf. Sci. **60**, 506 (1976)
3. G.L. Price, B.G. Baker: Surf Sci. **91**, 571 (1980)
4. H. Conrad, G. Ertl, J. Küppers, E.E. Latta: Surf. Sci. **50**, 296 (1975)
5. G.L. Price, B.G. Baker: Surf. Sci. **68**, 507 (1977)
6. C. Klauber, B.G. Baker: Appl. Surf. Sci. **22/23**, 486 (1985)
7. C. Klauber, B.G. Baker: Surf. Sci. **121**, L513 (1982)
8. C. Klauber: Surf. Sci. **203**, 118 (1988)
9. B.A. Sexton: Appl. Phys. A **26**, 1 (1981)
10. B.A. Sexton: Surf. Sci. **88**, 299 (1979)
11. B.A. Sexton: Surf. Sci. **102**, 271 (1981)

23 Surface Analysis of Polymers

H.A.W. StJohn, T.R. Gengenbach, P.G. Hartley, and H.J. Griesser

The properties and composition of polymer surfaces play an important role in a number of modern applications of polymeric materials, such as wetting, printing, adhesive bonding, membranes, and biomedical devices [1]. Interfacial interactions govern, for instance, the adsorption and denaturation of proteins on the surface of a biomedical implant or a membrane; adverse interactions give rise to biocompatibility problems and membrane fouling [2]. The surface composition of polymeric materials and the resultant interfacial forces must be carefully optimized for such applications; surface analytical methods occupy a key role in such optimizations and in general in the research and development of novel polymeric materials designed to possess specific surface properties [1–4]. Many of the analytical methods detailed elsewhere in this book are well suited to such work since their probe depths are of a similar magnitude to the thickness of the surface layers that interact with the "environment" (e.g. water, ink, adhesives, and biological fluids)

The surface analysis of polymers requires, however, consideration of some of the marked differences in structure and properties of polymers compared with metallic and ceramic materials. Accordingly, analysis protocols need to be adjusted and results interpreted based on a thorough understanding of polymer materials science and, if available, also the specific properties of the polymer under investigation. Some analysis methods that are successfully applied to inorganic materials do not lend themselves to application to polymers. A prominent example is Auger Electron Spectroscopy: the electron beam used to excite Auger electrons leads to excessively rapid destruction of the analyte surface, thereby preventing meaningful analyses. On the other hand, other techniques, particularly static secondary ion mass spectrometry (SSIMS), offer a high detail of information and are particularly well suited to the analysis of polymeric materials. The main objective of this chapter is therefore not to give a comprehensive overview of the analysis of polymer surfaces but to highlight the very specific issues encountered by the analyst when dealing with polymeric surfaces as opposed to other materials such as metals or ceramics. For detailed descriptions of analysis techniques and methods the reader is referred to the relevant chapters in this volume.

The main specific issues that complicate the surface analysis of polymers are:

- the intrinsic mobility of polymer chains

- the fact that most polymers contain various additives and/or are processed with the aid of processing agents such as extrusion slip agents
- the marked susceptibility of most polymers to compositional alterations ("radiation damage") while the analysis is proceeding
- the inability to perform depth profiling by ion beam sputtering and
- the multifunctional nature of many polymeric surfaces

In addition, as for inorganic analyte surfaces, issues such as ready adsorption of contaminants and charging of insulating surface layers must be considered.

23.1 Specific Properties of Polymers

The two main issues of relevance are the covalent connectivities of polymers and the intrinsic mobility of polymer chains. The multitude of covalent bonds that connect the various atomic constituents of polymer materials lead to unique effects such as the preferential formation of specific fragments in SSIMS. The intrinsic mobility of polymer chains can lead to polymer surface layers displaying compositions and properties that are time-dependent and can vary with the environment that the polymer is exposed to.

While the range of elements that are typically found in polymers is quite limited, with C, H, N, O, and Si the most frequent and F, Cl, Br, and S less frequent, and a few other elements very occasionally found, the ways in which these elements can be assembled into polymeric materials is literally unlimited. Hydrocarbon polymers such as polyethylene, polypropylene, polybutadiene, polystyrene, and poly-para-xylylene possess very different properties as a result of the ways in which the C and H atoms are connected. Moreover, the properties also depend on the average molecular weight (there always arises a distribution of chain lengths and thus molecular weights of polymer chains as a result of the random nature of monomer addition processes), crosslinking degree, and processing conditions (for instance polyethylene is fabricated with different densities which affects mechanical properties). However, established surface analysis methods, particularly XPS, are best suited to elucidating elemental compositions and consequently are relatively insensitive to the long-range discrete structural connections that are so important for polymers. It is necessary to study primary and secondary chemical shifts, shake-up satellites, and valence band spectra in order to gain an appreciation of the composition and structure of polymer surfaces beyond the level of nearest-neighbour elemental bonds.

As for inorganic materials, the spatial arrangements of the atomic constituents of polymers also play a key role. Most polymers are composed of regular repeat structures that are added during synthesis in a linear fashion to assemble to long linear polymer chains. Some polymers also possess a branched structure which confers crosslinks between linear segments. Examples of polymer structures are shown in Fig. 23.1. The covalent connections

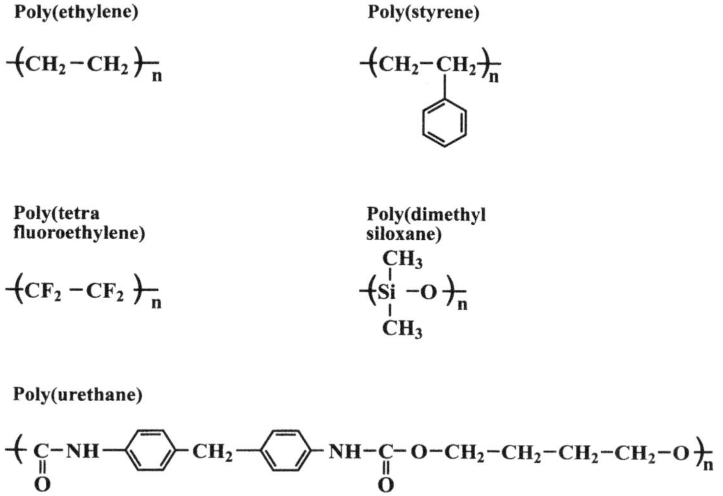

Poly(ethylene)

$-\!\!\left(CH_2-CH_2 \right)\!\!-_n$

Poly(styrene)

$-\!\!\left(CH_2-CH(C_6H_5) \right)\!\!-_n$

Poly(tetra fluoroethylene)

$-\!\!\left(CF_2-CF_2 \right)\!\!-_n$

Poly(dimethyl siloxane)

$-\!\!\left(Si(CH_3)_2-O \right)\!\!-_n$

Poly(urethane)

$-\!\!\left(C(=O)-NH-C_6H_4-CH_2-C_6H_4-NH-C(=O)-O-CH_2-CH_2-CH_2-CH_2-O \right)\!\!-_n$

Fig. 23.1. Chemical structures of some common polymers.

and the long-range order are responsible for many properties of polymers and their surfaces and can be evident in spectral data. While nearest-neighbour interactions are most prominent, often it is essential to also take into account longer-distance effects. In XPS, secondary shifts can be quite marked and may lead to erroneous interpretations of functional group signals particularly in the C 1*s* spectral region.

Polymeric surfaces tend to be more difficult to analyze than metals and ceramics because polymer surfaces generally are mobile and unpredictable. The atoms that constitute polymers can change their spatial positions with time over considerable distances [5,6,1]. Motions that alter the spatial positions of constituent atoms in polymers are of two types, viz. rotational and translational (reptation). Polymer chains have freedom of rotation around the central polymeric backbone (the long chain of covalently connected atoms that form the principal cohesive linkage) unless such rotation is hindered by conjugation (π-bonds) or steric crowding. There also are concerted motions of polymeric chain segments which, like rotational movements, are driven by random Brownian motion. The relatively low packing density of polymers (compared with inorganic materials) provides void spaces of atomic dimensions which enable such motions to occur. Rotational and segmental motions are, however, short distance; longer distances are not easily traversed by the "sideways" motions typical of segmental flexibility. Long-distance diffusion is achieved by reptation, a worm-like motion in which a polymer chain end diffuses into a temporary gap in-between other chains and drags its entire chain along. The two motions, rotational and translational, are illustrated in Fig. 23.2.

(a) Rotational motion

$$-CH_2-\underset{\underset{\underset{\underset{CH_2-CH_2-OH}{|}}{O}}{|}}{\overset{\overset{CH_3}{|}}{C}}-CH_2- \quad \longleftrightarrow \quad -CH_2-\underset{\underset{CH_3}{|}}{\overset{\overset{\overset{\overset{CH_2-CH_2-OH}{|}}{O}}{|}}{C}}-CH_2-$$

(with a $C=O$ group between the backbone carbon and the ester oxygen in each structure)

(b) Translational motion (reptation)

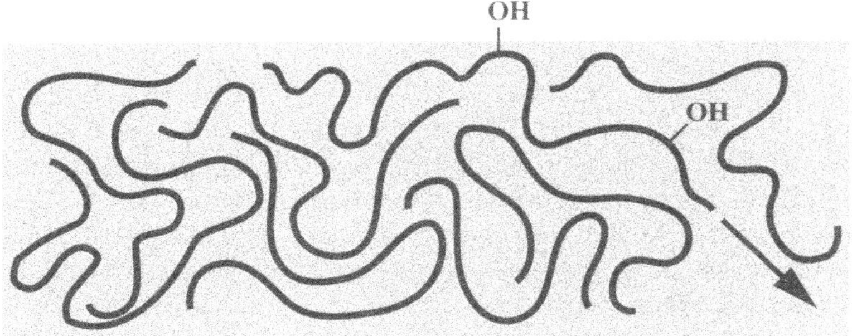

Fig. 23.2. Two possible motions of polymer chains (see text for details).

These motions can be neglected when analyzing compositionally uniform polymers such as polyolefins and polytetrafluoroethylene; all the polymer chain "worms" are of identical composition and it does not matter that they may exchange places between surface layers and sub-surface regions during or in-between surface analyses. For many polymers of interest, however, not all chains are identical, or the chains may be identical overall (apart from a polydispersity of length) but be comprised of non-identical segments. The first case is typically found with surface-modified polymers, and the second case applies to segmented block copolymers. In both cases, complications arise because the intrinsic rotational and translational mobility of the polymer chains enables them to respond to interfacial forces.

The mobility of polymer chains allows interfacial energy minimization to occur: the polymer surface layers undergo rotational and diffusional motions to adopt a chemical composition that minimizes the interfacial energy. In a nonpolar environment such as air, polymers will minimize the density of polar groups at the surface, whereas in an aqueous environment polymer surfaces become enriched in polar groups and reduce the density of nonpolar (dispersive) groups [7]. Rotational motions around the backbone and diffusional motions both are thought to contribute to this interfacial energy minimization.

One result is that the surface layers of a polymer may differ compositionally from the "bulk" composition; surface analysis data may thus differ substantially from the theoretical elemental ratios expected on the basis of the polymer's overall elemental composition. The differences may relate to fabrication and storage conditions. Another result is that polymer surface compositions may reflect the history of the sample. A third consequence is that polymer surfaces can show time-dependence.

This issue is relevant because most surface analytical techniques require a high vacuum environment, which is very different from the aqueous environment applicable to for instance biomedical applications. As a result, the surface compositions determined by surface analytical techniques may bear little relationship to the effective surface compositions that determine the interfacial interactions when a polymer contacts an aqueous medium. Hence, one might draw erroneous interpretations about relationships between surface compositions and performance if the polymer surface mobility/adaptability was neglected. For this reason, Ratner and colleagues have pioneered the method of cold-stage (freeze-hydration) XPS which enables assessment of the polymer surface composition as it exists in contact with (frozen) water. The method of cold-stage XPS will be described below (section 23.4.4).

The simplest case of a polymeric surface responding to environmental changes is that of a rotational motion around the polymer backbone. Thus, for poly(hydroxyethyl methacrylate) (pHEMA) Holly and Refojo [7] inferred from contact angle measurements that the methyl and hydroxyethyl groups rotated around the backbone C-C links to expose a maximum of the former groups in contact with air and a maximum of the latter in contact with water (Fig. 23.2a). For polyether-polyurethanes, the surface layers were found to be enriched in the polyether component when the material is in contact with air and enriched with the polyurethane component when in contact with water [8].

The surface modification of polymers has become a very attractive route towards expanding the use of existing polymers into applications for which they possess suitable bulk properties but inadequate surface properties [1]. Examples are the corona discharge treatment of various hydrophobic polymers to convert the surface such that it can be printed or wetted, and the provision of polar groups on polyolefins to enable adhesive bonding. However, such modified surface layers represent non-equilibrium structures; even when exposed to a constant environment, surface-modified polymers typically show a time-dependence of their surface properties. The best known example is the gradual loss of the hydrophilic character conferred to hydrophobic polymers

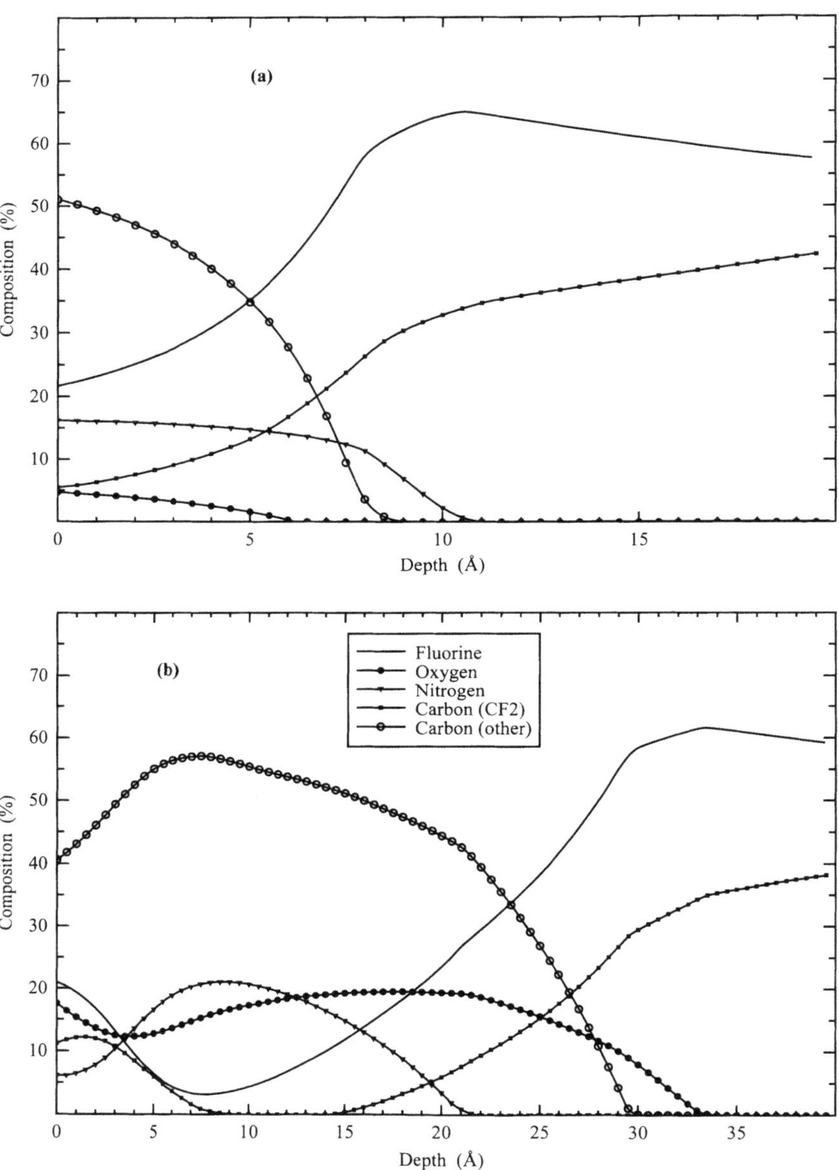

Fig. 23.3. Compositional depth profile of an NH_3 plasma treated fluorinated ethylene propylene copolymer (FEP) determined by angle-resolved XPS **(a)** immediately after the surface treatment and **(b)** after storage in air for several months. In the case of the aged sample, nitrogen-carrying polymer segments, formed at the surface by the plasma treatment, have partially moved below the surface and have been replaced by fluorocarbon segments. Oxygen which is being incorporated due to oxidative processes during aging also moves toward the bulk.

by corona or plasma treatments [1]. This loss is a consequence of the diffusion, by reptation, into the polymer material of polymeric chains that were at the surface during treatment and therefore carry a number of new polar groups along their backbone. Concurrently, untreated polymer chains emerge from the "inside" of the polymer to the surface. This loss can occur even when the treated polymer is stored in water [9], even though one might expect polar groups to be energetically favoured at the interface when in contact with a polar medium. The reason for the (partial) disappearance of polar groups from the surface even in contact with water has been postulated to be the translational entropy gain associated with reptation of treated chains [9]: the presence of a high density of treated chains within a narrow surface region is equivalent to a chemical potential gradient; as in solutions, such concentration gradients are subject to reduction by diffusional motions. In other words, the redistribution of some of the treated chains into the polymer and the concurrent emergence of untreated chains to the surface layers dilutes the concentrations of two chemically dissimilar species of polymer chains. This effect can cause marked time dependences in surface compositions [9] and the analyst needs to be aware of the effect, its rate constant, and the time lag between surface modification and analysis (see Fig. 23.3).

In summary, the chemical composition of a polymer surface is generally not identical to and predictable from the composition of the bulk polymer. Furthermore, polymer surfaces can respond to changes in their environment or to surface modification processes by undergoing rotational and diffusional segmental and chain motions. As a result, polymer surface compositions can be unpredictable as well as varying with time. To provide meaningful feedback in applied studies on optimization of polymer surfaces intended for industrial usage, analysis methods need to take into account the history of the sample and the intended environment of the polymer surface in question.

23.2 Surface Contamination and Additives

In order to obtain meaningful surface analytical results, it is vital that surfaces are free from contamination. An overview of general sample handling procedures as applicable to most samples has been detailed in Chapter 2. However, due to the intrinsic mobility and organic nature of polymers, contaminants, and additives, there are a number of additional specific requirements and limitations with regard to sample preparation of polymers for surface analysis. Not only do environmental contaminants, such as volatile hydrocarbons, adsorb on polymer surfaces, but also polymers contain various extraneous low molecular weight compounds ("additives") that are deliberately added during synthesis or processing. Considerable care must also be taken in the interpretation of analytical results obtained on polymers, particularly when the possibility of inadvertent introduction of surface contamination exists, since contaminant signals can at times be hard to distinguish from

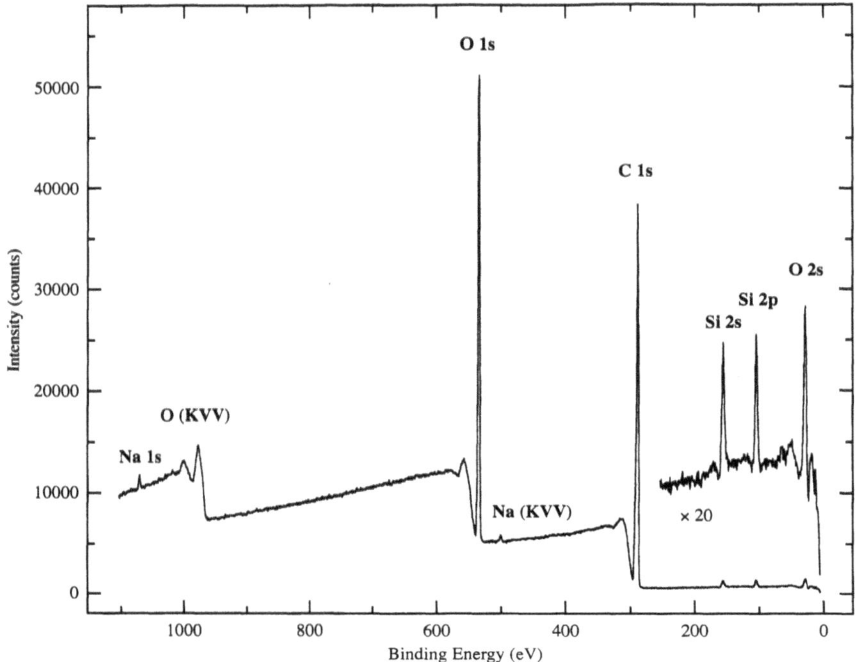

Fig. 23.4. XPS survey spectrum of a commercial contact lens (poly hydroxyethylmethacrylate). Photoelectron signals due to silicon are clear evidence for contamination. The Si $2p$ binding energy of about 102 eV indicates the presence of an organosilicon compound. Also detected is residual sodium from the saline solution which is used to store the contact lens.

signals due to the polymer itself and can lead to erroneous interpretations of surface segregation and mobility effects, for instance.

Silicones are the most common contaminants observed on polymer samples, particularly those from an industrial/commercial source. Fortunately, the presence of Si signals in XPS provides a clear indicator, unless the polymer itself contains this element; and the high positive ion yield of silicones in SSIMS allows unambiguous identification. The pervasiveness of this class of contaminants is linked to its common usage in a variety of applications such as lubricants and mold release agents, its low surface tension, and its ability to migrate across a surface [10]. XPS data obtained from a contaminated contact lens are shown in Fig. 23.4. Other commonly observed contaminants are additives, including plasticizers, catalysts, antioxidants, light stabilizers, slip extrusion agents, and others which may be included at a very low concentration as a bulk additive, but preferentially migrate to the outermost surface of the polymer. Some of these additives, particularly slip extrusion agents which act as lubricants, can be very surface active and are either applied to the surface or accumulate at the surface by diffusion. Thus, they interfere with surface analyses and can cause erroneous interpretations of the results of surface modification processes [11], since additives, being low molecular weight organic molecules, often do not possess a unique elemental signal. The propensity of these substances to surface segregate results in increased distortion of results for techniques most sensitive to the outermost atomic layer, such as static SIMS and contact angle measurements. Furthermore, the high ion yield of PDMS fragments in static SIMS relative to other organic polymers means that even for trace contamination (e.g. 0.2 % Si as measured by XPS) the PDMS spectrum swamps the signals of interest.

As outlined in Chapter 2, techniques such as solvent cleaning or ion beam etching are often used to remove contamination from samples prior to analysis. These approaches are not suitable for polymers. Solvent cleaning can lead to swelling of the polymer and enhance the mobility and rearrangement of the surface region, or solubilise low molecular weight components. Thus the cleaned surface analysed does not reflect the 'true' surface [12]. Ion beam etching is also not suitable for cleaning polymer surfaces. Chemical functionalities of organic materials are easily degraded leading to preferential sputtering and subsequent chemical reactions. The type of chemical reactions induced depend on the initial composition of the material. Argon ion irradiation may result in dehydrogenation, loss of heteroatoms and loss of aromaticity. Therefore neither atomic concentration data nor molecular species information reliably reflect the composition of the original material after ion beam etching [13,14].

The surface of as-supplied polymers often is far from what one may expect it to be. In fact, experience shows that very few polymers possess clean surfaces; most must be rigorously cleaned from contaminants and additives before reliable analyses are possible. A good clean reference polymer is Teflon FEP, which can be used for calibration of XPS elemental F/C ratios. Presumably its very low surface energy prevents adsorption of the typical adventitious contaminants (see Fig. 23.5). Teflon PTFE, on the other hand, is not as suitable as it requires addition of a hydrocarbon compound for processing.

Therefore, surface analysis of polymer samples requires rigorous adherence to the protocols of clean sample handling as outlined in Chapter 2. In addition, prior to commencement of any surface modification experimentation it is worthwhile to remove any potential source of surface contaminants from the bulk material itself: This can be efficiently achieved by sequential Soxhlet extraction with a polar solvent followed by a non-polar solvent. In all subsequent stages of sample treatment it is wise to adopt the philosophy that tools, beakers, etc, are a potential source of contamination unless proven otherwise. The time between sample treatment and analysis should be kept as brief as possible to minimise the extent of surface reorientation or out-diffusion.

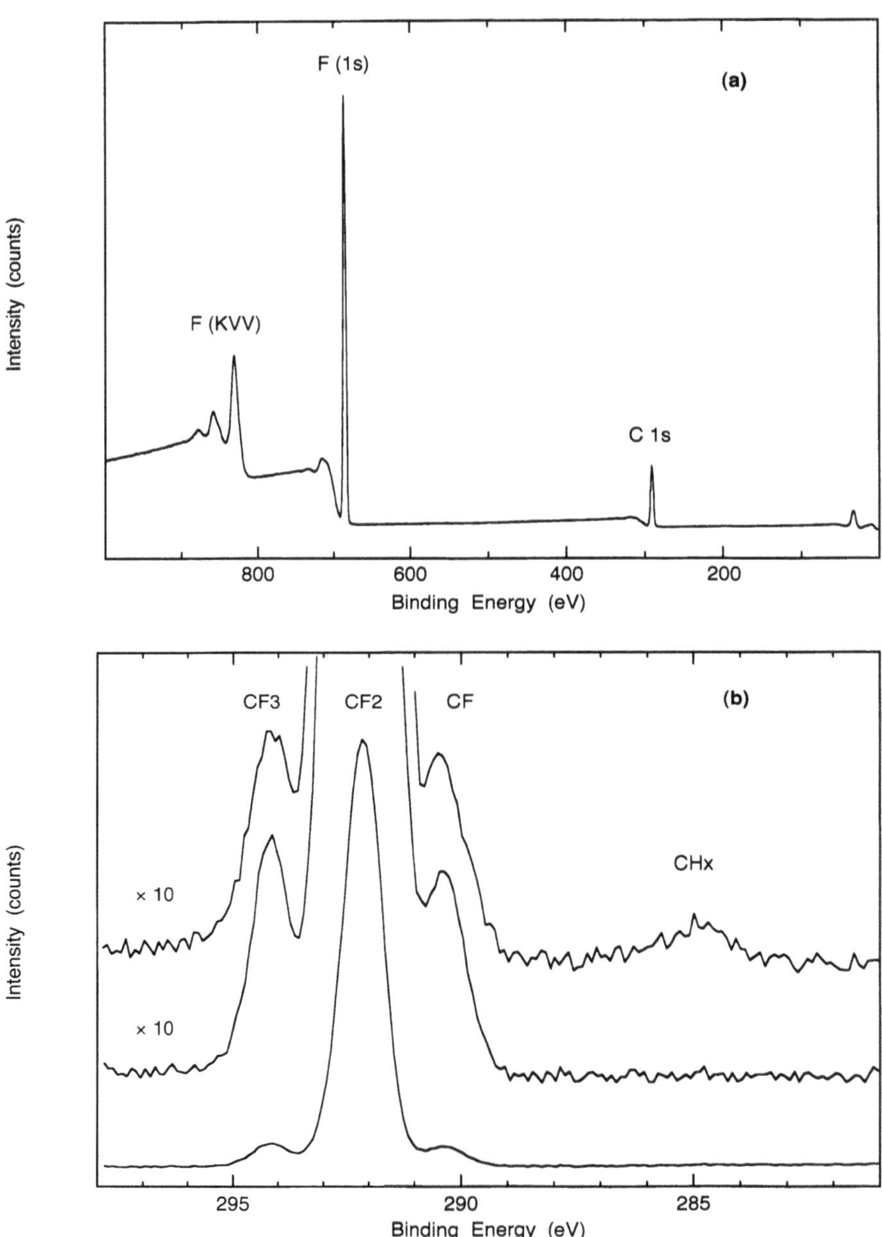

Fig. 23.5. XPS spectra of Teflon FEP: **(a)** survey spectrum and **(b)** high resolution carbon 1*s* spectra. In the latter case data from two different specimens are plotted, the first obtained from a section of clean FEP, the second from a slightly contaminated sample. The spectrum of the latter clearly shows a hydrocarbon signal at around 285.0 eV (top spectrum).

Surface analysis, particularly by XPS, has been of tremendous value in identifying surface contamination in a myriad of cases and has allowed identification of the reasons for numerous industrial problems related to interfacial interactions. Some contaminants, however, can be difficult to detect by XPS when their elemental composition is similar to that of the polymer. An illustrative example is a study which showed by TOF-SSIMS that the surface of commercial, biomedical grade Pellethane polyurethane was covered with a bis-ethylene stearamide lubricant [15], and that the good biocompatibility of Pellethane may well be due to this additive rather than the base polymer material! Surface contaminants can also interfere in adhesive bonding and in surface treatments [11].

23.3 Contact Angle Measurements

This is the simplest, least expensive and most rapid method for assessing the surface of polymers. It is, however, chemically least informative, basically providing only information about the surface energy of the material and thus about the presence of hydrophobic (nonpolar) and hydrophilic (polar) groups on the surface. From the difference between advancing and receding contact angles, one may estimate the fraction of polar and nonpolar surface segments. Contact angles are also affected by the surface topography of the material, and hence, for chemically well characterized surfaces, one can also infer a roughness factor as a measure of the topography [1,6].

Contact angle determinations are most valuable for rapidly detecting the presence – but not identity – of surface contaminants, and for studying the extent and stability of surface modification. Often contaminants which adsorb from the vapour phase are less polar, and thus an air/water contact angle higher than expected is an indication of contamination. Contact angles also enable the rapid verification of the efficacy of surface modification procedures which, by the insertion of new groups, usually alter the air/water contact angles considerably.

23.4 X-Ray Photoelectron Spectroscopy (XPS)

23.4.1 General Aspects

Most polymer surfaces of interest are multifunctional. The chemical shifts that are caused by covalent bonds with elements with different electronegativity are indicators of the chemical functional groups present in a polymer. However, a limitation of XPS is the relatively poor chemical specificity of its information due to the small dynamic range of core level chemical shifts when analyzing polymer samples. In principle, chemical functional groups can be identified on the basis of tabulated chemical shifts [16]. For instance, in the N 1s region, a signal at $\sim$399.1 eV can reasonably be assigned to amine groups while a signal at ~400.0 eV can be indicative of amide groups. However, one needs to be cautious since secondary shifts may cause significant deviation from "typical" binding energy values for particular chemical groups.

The most useful information for polymer surface analysis is in the C 1*s* region. For example, in the case of C bound to H and/or O one can differentiate carbon atoms of the following major classes: "neutral" C (bound to C and H neighbours only), C in C–O bonds (hydroxyl, ether, etc.), C in C=O (carbonyl, aldehyde), C in COO (acid, ester) and C in OC(O)O (carbonate). Using a monochromatic XPS instrument these classes can usually be discerned as separate peaks (Fig. 23.6); for instance, the difference between the signals from "neutral" C atoms and C–O carbon atoms is 1.5 eV while the typical linewidth of modern monochromatic spectrometers is 0.8–1.0 eV when analysing polymeric materials. However, there can be substantial overlap of the peaks, particularly when secondary shifts apply or the polymer is characterized by an inhomogeneous distribution of local micro-environments for the chemical groups in question. Examples of the latter are plasma polymer coatings; they possess a random, amorphous structure which leads to relatively large XPS peak widths compared with conventional polymers of more ordered structures.

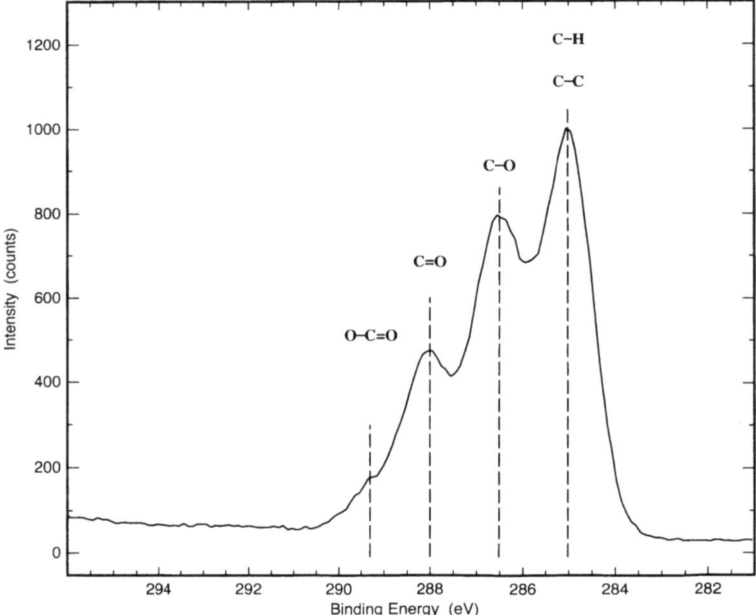

Fig. 23.6. XPS carbon 1*s* spectrum of a polysaccharide coating on a contact lens. Labelled are the four categories of carbon-oxygen structures which are at least partially resolved by XPS (see text for details).

Curve fitting protocols can be used to quantify the contributions of different components to one particular photoelectron peak. However, it is not possible to distinguish between C atoms that possess analogous bonding environments, for instance C in hydroxyl groups and C in ether groups. If such distinction and quantification is desired, derivatization reactions must be utilized (see section 23.4.5).

23.4.2 Angle-Resolved XPS

Angle-resolved (or angle-dependent) XPS (ARXPS) enables the surface scientist to probe the compositional depth profile of any solid surface. Various methods have been proposed to evaluate ARXPS data; these range from simple models to estimate the thickness of an overlayer on an infinitely thick substrate (e.g. relative ratio and patchy overlayer algorithms) to more complex algorithms based on Laplace transforms or regularization methods which attempt to generate concentration profiles. A comparative review has been published by Tielsch and Fulghum [17]. All algorithms are invariably based on simplifying and idealising assumptions regarding the sample and its interaction with photoelectrons. It is therefore important to establish whether theoretical models and the associated assumptions are a realistic approximation of the particular system under investigation. This is particularly true for the case of polymeric materials. Polymers almost always have characteristics which complicate an angle-resolved analysis: the surface is not smooth on a molecular scale, surface restructuring due to mobility of molecular segments results in additional compositional gradients at the surface, and contaminants are usually present on the surface. In most cases interpretation of ARXPS data is limited to qualitative statements, comparing, for example, data obtained at normal and grazing emission. However, more sophisticated approaches have been successfully developed and applied to polymer analysis. Tyler et al. [18] have developed an algorithm for generating depth profiles from ARXPS data using a regularization method. This method was shown to yield realistic depth profiles from polymeric samples and proved to be stable with respect to both noise and the number of data points used. The application of the algorithm is, however, restricted to samples with a microscopically flat surface; few commercial polymers are sufficiently smooth. One example is given in (Fig. 23.3); it shows the depth distribution of the elements present in ammonia-plasma-modified Teflon FEP [9]. One approach to overcoming the limitations imposed by a non-ideal surface topography employs atomic force microscopy (AFM) to characterise and quantify roughness by slope histograms. This information is fed into an overlayer algorithm in order to correct for the effects of surface roughness [19]. Quantitative interpretation of ARXPS data also requires knowledge of the inelastic mean free path of photoelectrons travelling through solid matter.

23.4.3 Inelastic Mean Free Path in Polymers

As discussed in Chapter 1, the distance that a photoelectron travels through the sample material is an important parameter in quantitation by XPS. An important application of XPS in polymer science is the measurement of the thickness of polymer coatings, and this requires knowledge of the value of the IMFP for the material studied. Most direct experimental measurements of IMFP have been obtained for inorganics and metals, but it is clear that the IMFP is considerably different for organic materials [20]. Due to the paucity of experimental data providing direct measurement of IMFP in polymeric materials, predictive equations provide a valuable alternative estimation of IMFP. A number of predictive equations have been developed and refined over the past 20 years, however debate continues as to the most appropriate algorithm, hence a variety of different algorithms are employed by different XPS users. For this reason, the main algorithms employed for polymeric materials will be reviewed here.

It is useful at the outset to clarify the difference between the terms IMFP and attenuation length, as these are often used interchangeably but have distinctly different meanings [21]. As the electron travels though the solid it is undergoing both elastic and inelastic scattering events. The inelastic mean free path is the distance that an electron travels through a solid before undergoing an inelastic scattering event. This term is usually derived from theory using models of electron scattering and physical constants such as refractive indices etc. In practice, the related value, the effective attenuation length (EAL) is used for most experimental measurements and calculations. This is an experimentally determined value which ignores the effects of elastic scattering, hence it is less than the IMFP. Monte Carlo simulations can be used to calculate the effective attenuation length from a given IMFP.

One of the first algorithms for prediction of EAL in organic compounds was developed by Seah and Dench by measuring the signal attenuation in overlayers of a given thickness. From this data they developed a fitting equation for organic compounds: [20].

$$\lambda_{\mathrm{EAL}} = 49E^{-2} + 0.11E^{0.5} \tag{23.1}$$

where λ_{EAL} is the effective attenuation length in mg/m^2, and E is the energy of photoelectron (in eV).

The relationship between IMFP and material properties was given a more fundamental grounding by Ashley, who related the IMFP to valence-electron excitations and hence the valence electron density. This relationship is appealing in its simplicity, being separated into a material-dependent term and an energy-dependent term, enabling the IMFP for linear polymers to be calculated from a knowledge of the structure of the repeat unit and the density [22].

$$\lambda = \frac{M}{\rho n} \cdot \frac{E_K}{14.1 \ln E_K - 25.5 - 1500/E_K} \tag{23.2}$$

where M is the molecular weight of the repeat unit, ρ the density, n the no. of valence electrons in the repeat unit, λ the IMFP in Angstroms, and E_K denotes the kinetic energy of the photoelectron. Tabulations of the Ashley parameter $(M/\rho n)$ are available for common organic polymers, making the calculation of λ for a given energy straightforward.

To be scientifically rigorous, the best predicitive equation currently available is that developed by Tanuma, Powell and Penn, known as TPP-2M. This is the refined version of their earlier equations, and is appropriate for organic materials. It is based on a modified form of the Bethe equation for elastic scattering [23].

$$\lambda = \frac{E}{E_p^2 \left[\beta \ln(\gamma E) - (C/E) + (D/E^2)) \right]} \qquad (23.3)$$

where λ is the IMFP in Å, and γ, β, C, D and E_p are material related parameters that take into account atomic weight, density, number of valence electrons and the bandgap energy (see reference for full definition of variables)

As this algorithm is somewhat unwieldy and the bandgap energy may not be known, Gries has recently proposed a simplified predictive algorithm (G1) based on an atomistic model of inelastic electron scattering [24]. It is claimed to be more widely applicable than the TPP-2M equation. Though Tanuma and coworkers have criticised this model for being inappropriate for metals, semiconductors and inorganic compounds, they acknowledge it may be relevant for molecular solids where the interactions between constituent molecules are weak. As for the Ashley equation, it also predicts that the IMFP will be inversely proportional to density.

$$\lambda = k_1 (V_a/Z^*) E/(\lg E - k_2) \qquad (23.4)$$

where V_a is the atomic volume in $\mathrm{cm^3\,mol^{-1}}$ (i.e. M/ρ), $k_1 = 0.0018$, $k_2 = 1.00$ for organic compounds. Z^* is regarded as the 'nominal effective' number of interaction-prone electrons per atom, and can be approximated by $\sqrt{Z}$.

23.4.4 Cold Stage XPS

As discussed previously, polymer surfaces are dynamic and able to restructure in order to minimise their interfacial energy. The surface composition in air or vacuum may therefore be very different to that in water. For example, a pHEMA/PS diblock copolymer will rearrange to preferentially expose the polar pHEMA groups in an aqueous environment, but the non-polar polystyrene component will rearrange back to the surface upon drying [25]. The time scale for rearrangements can range from seconds to months [26].

Knowledge of the surface composition of the hydrated polymer surface is of particular importance for polymers designed for use in an aqueous environment, such as biomaterials or anti-fouling coatings for marine applications.

If the surface rearranges rapidly on drying, then results obtained by traditional analytical techniques performed in air or vacuum, such as FTIR, XPS or SIMS, may bear little relation to the hydrated surface of interest. In these cases, cryogenic sample handling techniques may be employed to 'freeze-in' the hydrated structure throughout handling and analysis [25–31].

The technique of 'frozen (hydrated)' XPS developed by Ratner and coworkers employs a spectrometer with cold-stage facilities in both the entry chamber and analysis chamber [25,27]. The wet specimen is placed in the entry chamber under a dry nitrogen purge, and frozen rapidly by liquid nitrogen cooling of the sample holder. Once frozen, the sample can then be exposed to UHV to "etch" the ice off the surface by sublimation. During this stage the temperature can be raised to facilitate sublimation, but should remain below the glass transition temperature of the sample to prevent rearrangement. The sample is held at $-12°$ C during XPS analysis to prevent rearrangement. Caution is required during this process so that extensive contamination is not adsorbed onto the cold surface [25].

To date this approach has been used primarily for characterizing polymers for biomaterials applications, including radiation-grafted hydrogels [25,27], methacrylate copolymers [32], amphiphilic networks [29] and polyurethanes [25,30,31]. Extensive rearrangement was observed for a poly(hydroxyethyl methacrylate) radiation grafted onto a silicone rubber [25–27]. When analysed in the frozen (hydrated) form the surface composition consisted of 95at% HEMA. When warmed in-situ, the surface quickly rearranged to predominantly silicone. When rehydrated and refrozen the pHEMA spectrum was regained, thus demonstrating the reversibility of the surface structure. The extent and rate of reorientation of polyurethanes is very dependent on the composition of the hard and soft segments, and is generally slower [25,26,31]. The cold stage approach has also been used to characterize proteins adsorbed onto polymer surfaces [33].

23.4.5 Derivatization of Chemical Groups

A primary interest in the analysis of polymer surfaces is often the identification and quantification of the functional groups present as this will determine the surface reactivity. A number of functional groups have very similar binding energies, such as hydroxyl, amine and ether groups, and their overlapping bands make the results of curve fitting ambiguous. In complex systems such as plasma treated surfaces, the wide range of functional groups not only results in a large number of overlapping component peaks, but also broadening of each component due to secondary shift effects. In these cases curve fitting is prone to high levels of error. An alternative approach to enable the quantitation of the different functional groups is chemical derivatization, which differentiates functional groups on the basis of their different chemical reactivities. In this approach a particular functional group is 'tagged' quantitatively with a unique atom which has a high photoionization cross

Table 23.1. Common Derivatization Reagents

Functional Group	Reagent	Product
Hydroxyl $-CH-OH$	Trifluoroacetic anhydride $(CF_3CO)_2O$	$-OCOCF_3$
	Acyl chloride CH_3OCl	$-OCOCH_3$
Carbonyl $-CH{=}O$	Hydrazine NH_2NH_2	$-CH{=}NNH_2$
	Pentafluorophenylhydrazine $NH_2NHC_6F_5$	$-CH{=}NNHC_6F_5$
Carboxylic acid $-COOH$	Trifluoroethanol CF_3CH_2OH	$-CO_2CH_2CF_3$
	$AgNO_3$	$-COO^-Ag^+$
Unsaturation $-CH{=}CH-$	Bromine Br_2	$-CHBr-CHBr-$
Amine $-C-NH_2$	Pentafluorobenzaldehyde C_6H_5CHO	$-C-N{=}CHC_6F_5$

section and is stable under the conditions of analysis. This approach is analogous to the derivatization methods employed in other disciplines such as gas chromatography, or nitroxide derivatives used as spin labels in ESR.

A number of potential reagents have been proposed for derivatization, and these have been summarised in reviews by Batich [34], Andrade [2] and Briggs and Seah [3]. However, many of these have been proposed on the basis of solution organic chemistry, and the analagous reactions may not proceed as expected at a polymer surface. Evaluation of the reactions on model polymer surfaces is required. The key criteria for surface analytical derivatization reagents are summarised below, with examples drawn from common derivatization reactions. These selected derivatization schemes are listed in Table 23.1.

There are several key criteria in the design of a derivatization scheme:

1. Ease of identification. Usually a derivatization reagent introduces a new element with a high photoionization cross section, such as fluorine, giving the added bonus of improving the sensitivity of the technique as well as providing a unique tag element. Multiple substitution on the reagent enhances this even further, as in the labelling of amine groups by pentafluorobenzaldehyde [34].

2. Stoichiometry. Ideally the reaction would proceed to 100 % completion. At the very least, the yield of the reaction must be determined on an analogous polymer surface, noting that it is not only the functionality but also the degree of substitution which will affect the ultimate yield. For example, Alexander's study of the derivatization of carboxylic acid groups by vapour phase trifluoroethanol showed that acid groups of polyacrylic acid were completely labelled after 10 hours, while the conversion for polymethacrylic acid was only around 60 % even after 75 hours [35].

3. Selectivity. Ideally the reagent should react only with the functional group of interest. At the very least, cross-reactions with other functional groups must be known so that their contribution can be considered. This is particularly important for complex multifunctional surfaces, but unfortunately very few reagents have been compared systematically on a range of model polymers. One instructive example is the investigation by Chilkoti and Ratner on the reactivity of trifluoracetic anhydride [36]. Previously it was thought that TFAA could be used to selectively label hydroxyl groups on highly oxidised surfaces (on nitrogen-free samples, as TFAA is known to also react with amines, amides and ureas). However, in their study encompassing 15 different polymer surfaces it was shown that TFAA also reacted rapidly and quantitatively with epoxide groups, as well as to a lesser extent with ketone and carboxylic acid groups. Non-specific adsorption can also result in significant background signal in some derivatization schemes.

4. Rate. Ideally, the reaction should be rapid at room temperature. The time to reach stoichiometric completion must be known, and this will depend on both the rate of diffusion of the reagent into the polymer surface layers, as well as the rate of reaction with the functional group. For example, on model polymer surfaces of electropolymerized substituted phenols, Zeggane and Delamar have shown that the reaction of TFAA with hydroxyl groups was rapid, while the reaction of PFPH with carbonyls was slow, even though the penetration of both reagents was rapid [37].

5. Distribution. In most studies it is desirable for the derivatization reaction to label groups uniformly at least to the depth of XPS analysis. This can be checked by angular resolved studies on a derivatized model polymer surface to confirm that the depth profile of the tag element is uniform. For example, the derivatization of hydroxyl groups in poly(acrylic acid) by TFE shows a sampling depth-dependent conversion, with incomplete labelling at greater depths [36]. Alternatively, the derivatization scheme may be designed to label only a specific region. For example, by using a large reagent in a non-swelling solvent it may be possible to distinguish between groups in the outermost polymer layer and subsurface regions. Batich and Wendt [38] observed that in the $AgNO_3$ labelling of carboxylic acid groups in a model ethylene/methacrylic acid copolymer, the majority of the silver signal was detected near the surface. The derivatization scheme may also be designed to determine the number of functional groups which will be available for a particular interfacial covalent coupling reaction, by replicating the conditions of attachment and using a probe of similar size and reactivity [39].

6. Stability. Ideally, the product of the derivatization reaction should be stable to the conditions of analysis, i.e. X-rays, vacuum and heat. Though fluorinated compounds are popular choices for derivatization reagents because of their high cross-section, the C-F bond is unfortunately labile under X-ray irradiation and so the time of analysis should be kept as brief as possible, or a correction applied for the loss of fluorine during analysis [16]. The stability of the product during storage prior to analysis must also be considered. For example, the hydrazone tag produced by labelling carbonyl groups with hydrazine is subject to partial hydrolysis when stored under ambient conditions [36].

7. Rearrangement. Ideally the derivatization reaction should result in no change in the surface composition other than the intended labelling reaction. In their study of nitrogen plasma modified polyethylene, Everhart and Reilly [12] demonstrated that the mobility of surface amine groups, and hence the results of derivatization, were strongly dependent on the solvent used in the reaction. Extraction of low molecular weight oligomeric species is also a concern in solvent-based derivatization reactions. To avoid these problems of rearrangement and extraction, gas-phase derivatization schemes are employed wherever possible.

23.5 Secondary Ion Mass Spectrometry (SIMS)

During a SIMS experiment the sample is irradiated with primary ions (Ar^+ being the most widely used), and the secondary ions emitted from the sample surface are collected and mass analyzed. For polymer surface characterization it is important to operate the spectrometer in the "static" mode, i.e. the accumulated ion dose during spectral acquisition must be kept sufficiently low for the surface to be essentially unperturbed for the duration of the experiment. Under typical conditions (2 keV $<$ ion energy $<$ 4 keV) the primary ion penetrates to a depth of $\sim$3 nanometers below the surface. Transfer of kinetic energy within the material, as well as electronic interactions, lead to primary events such as bond cleavage reactions and the emission of small fragments, followed by indirect processes such as the desorption of large molecular fragments some distance from the primary ion impact site. About 5 % of the material sputtered from the surface consists of positively and negatively charged ions (atoms or molecular fragments), and these are mass analyzed to produce positive and negative, respectively, SIMS spectra. Quadrupole mass spectrometers are most commonly used at present but their mass range is limited; they are not ideally suited to the surface analysis of polymers. State-of-the-art instruments employ time-of-flight (TOF)

Table 23.2. Assignment of major peaks observed in the positive ion spectrum of a pHEMA contact lens (see Fig. 23.7).

m/z	Hydrocarbon	Oxygen-containing fragments
15	CH_3^+	
27	$C_2H_3^+$	
29	$C_2H_5^+$	CHO^+
39	$C_3H_3^+$	
41	$C_3H_5^+$	
43	$C_3H_7^+$	$C_2H_3O^+$
45		$C_2H_5^+$ (pendant hydroxyethyl group)
53	$C_4H_5^+$	C_3HO^+
55	$C_4H_7^+$	
69		$C_4H_5O^+$ (methacryloyl ion)
71		$C_4H_7O^+$
89		$C_4H_9O_2^+$
91	$C_7H_7^+$	
113		$[M-OH]^+$ (M=backbone subunit of pHEMA)

mass analysers which offer superior collection and transmission properties (i.e. much higher detection efficiency) and, in principle, an unlimited mass range. The high mass resolution of some instruments allows the separation of fragments differing by much less than 1 amu and thus, by isotopic identification, better assignments.

As in conventional mass spectrometry, spectra are complex, and the information content is high but interpretation is not straightforward. The analysis of polymer surfaces by SIMS is still in its infancy, and only recently have reference spectra been collected on reliable, clean materials for aiding in interpretation. Quantitation is still a problem; so far, the compositional accuracy of ESCA cannot be matched by SIMS but, on the other hand, SIMS excels by providing detailed information about molecular structures from the fragmentation patterns. The most successful application of static SIMS so far has been in the area of fingerprinting: spectra from pure polymers usually contain peaks assignable to multiple repeat units (main chain scission), intact side chains and fragments of these. Careful spectral interpretation combined with existing knowledge about fragmentation pathways (e.g. based on pyrolysis / electron impact mass spectrometry) therefore allows different classes of polymers to be distinguished as well as individual members of one class to be identified. As an example, Fig. 23.7 shows the positive TOF-SIMS spectrum of a poly(hydroxyethyl methacrylate) (pHEMA) contact lens with assignments of major peaks listed in Table 23.2 [40].

The interpretation of SIMS spectra becomes more uncertain in the study of modified polymer surfaces which may contain several different functional

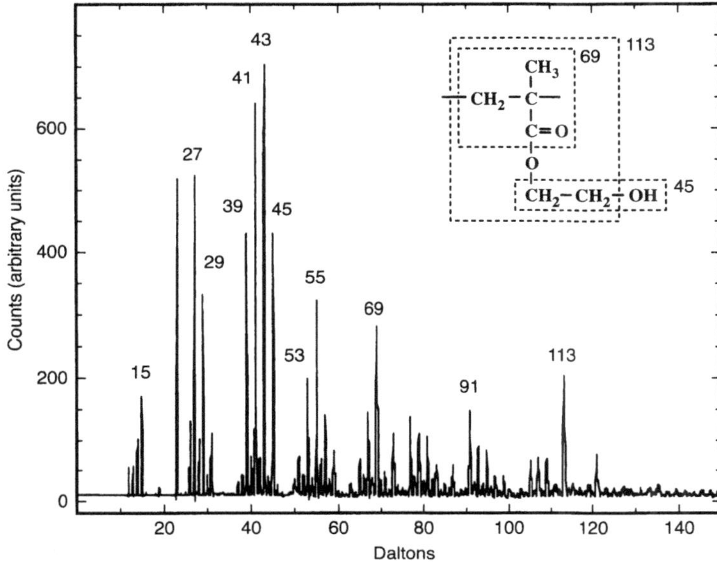

Fig. 23.7. Positive TOF-SIMS spectrum of a poly(hydroxyethyl methacrylate) contact lens. The insert shows the structure of the monomer unit; several characteristic molecular fragments are indicated, corresponding to major peaks as labelled.

groups and for which no similar, well characterized reference material is available.

The information depth of static SIMS is assumed to be of the order of 1 nm, i.e. the very surface layers of a polymer are probed. This high surface sensitivity is of particular interest where one needs to characterize the surface chemistry of a polymer on a depth scale of less than 1 nm, which is typically the range of interfacial forces. The low information depth, high absolute sensitivity, high molecular specificity, and a spatial resolution of the order of several μm which allows imaging to be done, make SIMS a relatively recent but very valuable addition to existing polymer surface analysis techniques. The complementary combination of XPS and static SIMS has already proven to be powerful in the analysis of modified polymer surfaces.

Static SIMS is also a powerful method for detecting and identifying low molecular weight organic contaminants on polymeric surfaces [15]; the mass of the molecular parent ion, if observed, and characteristic SIMS fragmentation patterns of classes of organic compounds can provide information on the nature of the contaminant. Figure 23.8 shows an example [40] of a lipid adsorbed onto a commercial hydrogel contact lens material. The peak at 284 m.u. is assigned to the molecular ion. The characteristic fragmentation pattern of a series of peaks separated by 14 m.u. indicates an extended linear

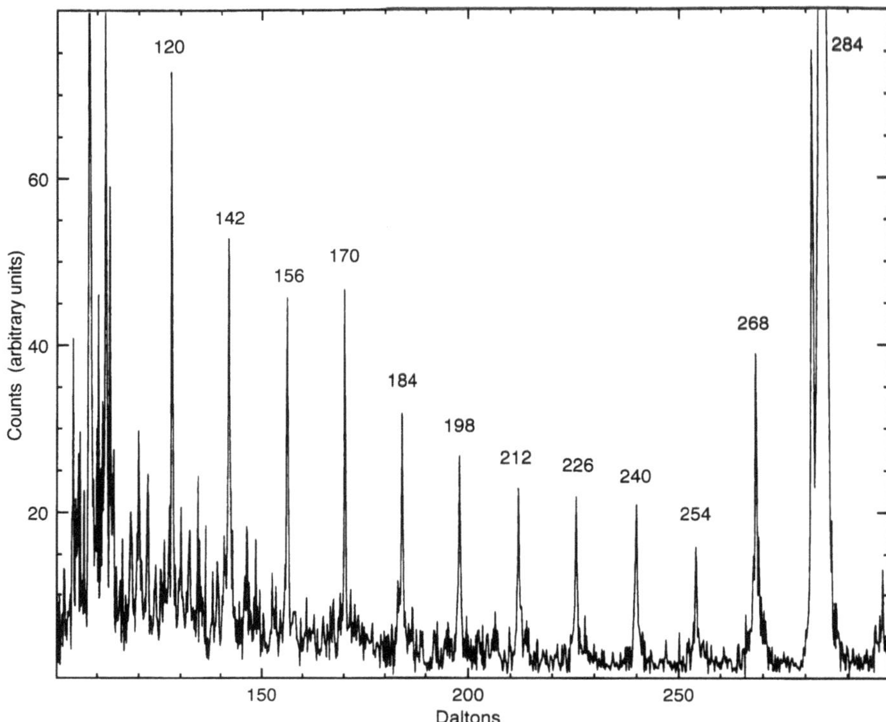

Fig. 23.8. SIMS spectrum of a lipid adsorbed onto a contact lens (see text for details).

hydrocarbon structure. The fragmentation pattern indicates that it is the hexadecyltrimethylammonium cation.

The main limitation of SIMS, particularly in the case of organic surface analysis, is the difficulty in quantitation of SIMS intensity data because of the strong matrix effects on secondary ion yields. There can also be challenges in the interpretation of complex spectra, although the conventional mass spectrometry literature provides a large amount of relevant data that can be used to interpret SSIMS fragmentation patterns and thereby identify parent molecules.

23.6 Scanning Probe Microscopy Methods: Scanning Tunneling Microscopy (STM) and Atomic Force Microscopy (AFM)

Scanning probe microscopes, of which the STM and AFM are examples, provide excellent tools for the study of polymer surfaces. Both of these techniques employ an ultrasharp tip, which is raster scanned across the surface of interest. In STM, changes in the height of surface features are measured by monitoring the tunneling current resulting from an applied voltage between the tip and sample. AFM, on the other hand, extracts the height of surface features by measuring the deflection of a tip in intimate contact with the sample during scanning. This is usually achieved using a laser beam reflected from the back of the tip on to a position sensitive photodiode. A feedback loop to a piezoelectric displacement actuator allows the tip to be held at a constant height above the sample or at constant applied force to the sample. In both types of microscope, the height (Z) and X-Y scan position data, are then used to reconstruct a three dimensional topological representation of the surface of interest at sub-nanometer resolution [41].

The principal limitation of STM is that it requires a conducting surface in order for tunneling between it and the tip to be possible. For most polymer surfaces, this requires a thin conducting layer to be deposited on the sample, which may mask features of interest. The AFM has the advantage of not being restricted to conducting surfaces, since it does not require an electrical circuit to be established between tip and sample. For this reason, AFM has seen wider application in the study of polymer interfaces.

One problem associated with the 'contact mode' of AFM operation outlined above, however, is that of sample damage. This will be discussed in more detail below (section 23.7: Sample Damage). One of the techniques developed in order to minimise damage is the intermittent contact (tapping)

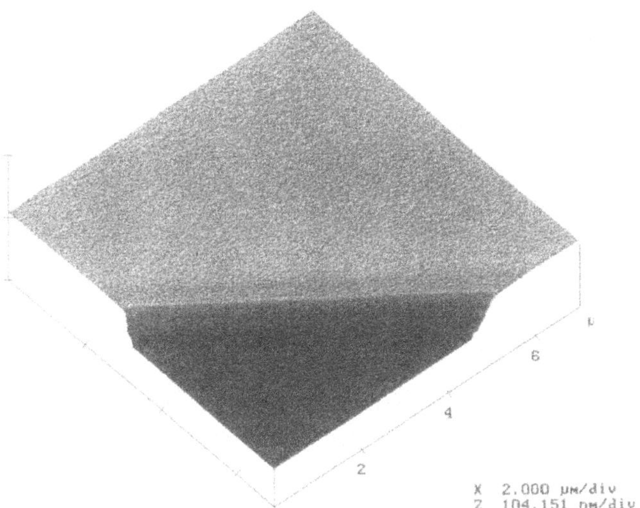

Fig. 23.9. AFM Tapping Mode Image of a Plasma Polymer film deposited on a freshly cleaved mica surface. The step was created by masking one side of the sample during the deposition, then removing the mask prior to imaging.

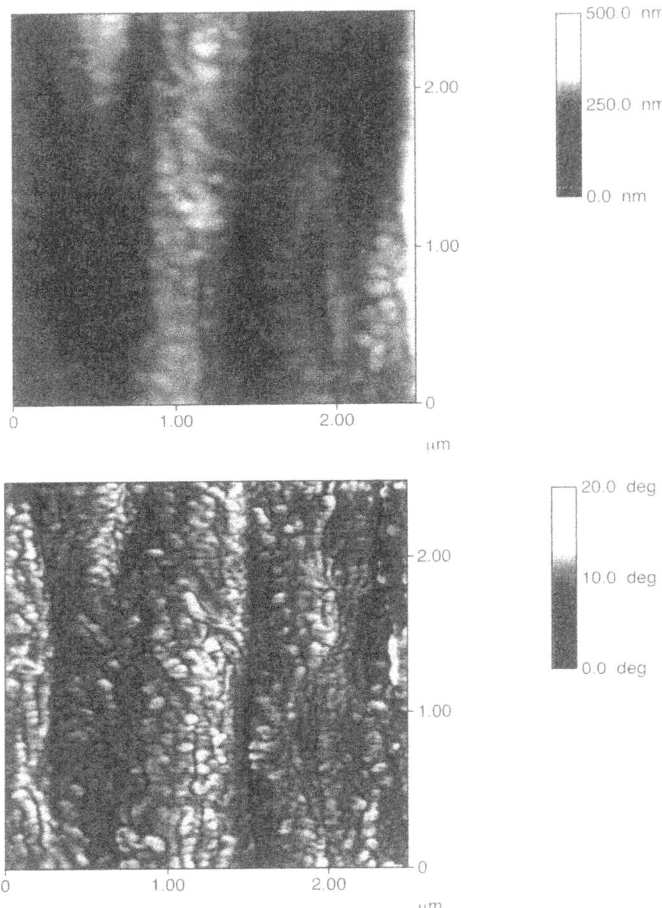

Fig. 23.10. $2.5 \times 2.5\,\mu m$ Tapping mode *(top)* and Phase mode *(bottom)* image of a stretched polyethylene film, highlighting the increased two dimensional resolution attainable using phase imaging.

mode of operation, in which the tip oscillates normal to the sample surface, thereby reducing contact time (see Fig. 23.9). In this case, variations in both the amplitude and phase of the tip oscillation contain useful information. Indeed, so-called 'Phase Imaging' has shown great promise in the analysis of polymer surfaces due to its exquisite sensitivity to variations in composition, adhesion, friction and viscoelastic properties (see Fig. 23.10) [42].

The control of tip-sample interactions has also been used to gain insights into interfacial chemical and mechanical properties which are frequently not accessible from the topology information. In force modulation imaging, the tip is oscillated whilst in contact with the sample. By monitoring the compliance of the surface through the bending of the AFM tip, two-dimensional images of the local elasticity of the sample surface can then be constructed. Such imaging is particularly useful in detecting differences in crystallinity across the surface of a polymer sample [43].

The torsional forces experienced by the AFM tip as it is scanned across the sample also provide a measure of the frictional force between it and the surface. This Lateral Force Microscopy (LFM) allows chemically distinct regions on surfaces to be differentiated through their affinity for the tip. By modifying the surface chemistry of AFM tips, it has also been possible to compare the frictional forces resulting from the interactions operating between surfaces bearing specific functionalities, and to construct two dimensional images of chemically patterned surfaces despite minimal differences in topology [44].

Measurements of the long range forces operating between AFM tips of well defined geometry and surfaces of interest also yields valuable information regarding the interactions of polymer surfaces, particularly in solvent environments. In this mode of operation, native AFM tips or AFM tips modified with micrometre sized colloidal spheres are brought into close proximity with the surface, and force versus separation curves are constructed from the deflection of the tip at different positions normal to the surface [45]. Such force versus separation curves reveal information regarding the extension of polymer chains from a surface, as well as providing insights into other properties of the polymer surface, such as charge density and hydrophobicity.

23.7 Specimen Damage

Most polymeric materials are rather susceptible to degradation caused by ionising radiation. Since the latter is used in some form or other in all spectroscopic surface analytical techniques the analyst needs to be aware of the potential presence and magnitude of artefacts due to sample decomposition during the experiment. Indeed, some techniques such as Auger Electron Spectroscopy or Scanning Electron Microscopy are unsuitable for polymer analysis: these techniques use electron beams with beam energies of several keV or more as the excitation radiation, usually leading to a rapid degradation of the polymer surface. For the two most common spectroscopic techniques, XPS and SSIMS, sample degradation and its dependence on variables such as flux, dose or energy of the primary (as well as secondary) radiation has been characterised extensively.

In the case of XPS, studies have revealed extensive sample damage in certain cases. This phenomenon usually manifests itself in a decrease of the X/Carbon atomic ratio (X = heteroatom) as shown in Fig. 23.11. Some authors have observed extremely rapid decomposition, affecting up to 90 % of the polymer being analysed within one hour or less of analysis time [46–49]. For most polymeric materials, however, degradation rates are substantially lower [16]. The type of damage suffered by a particular material depends

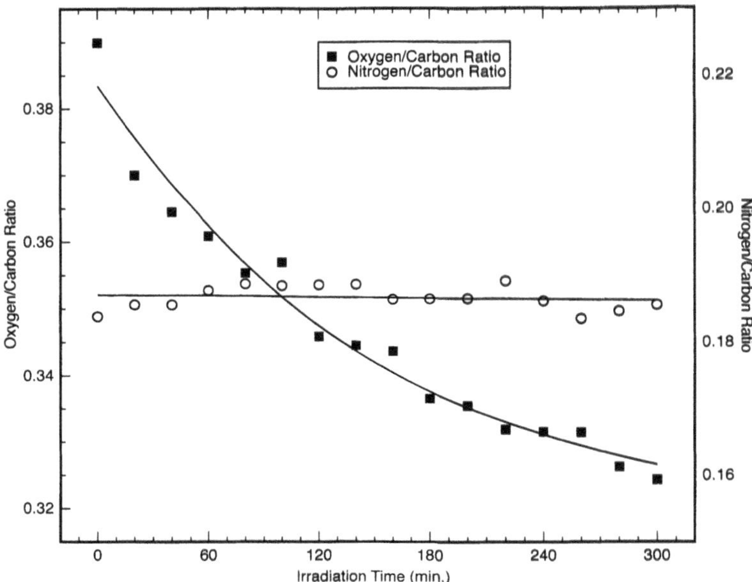

Fig. 23.11. Decomposition of a plasma polymer coating during XPS analysis. The coating was deposited from a N,N dimethylacetamide radio-frequency glow-discharge and consists of a crosslinked hydrocarbon network with a range of oxygen-, nitrogen- and oxygen/nitrogen-based functional groups. Note that whereas oxygen is lost rather rapidly, the nitrogen content appears to be not affected by radiation at all.

strongly on its chemical structure. Many different mechanisms have been identified. Polyvinylchloride degrades mainly via dehydrochlorination. Fluorine abstraction and chain scission in fluoropolymers lead to the generation of free radicals and subsequent cross linking. Polymethylmethacrylate degrades via main chain scission, resulting in cross linking, volatile product formation and cyclization. In perfluoropolyethers cleavage of the C–O bond generates mainly CO_2 and volatile fragments such as OCF_2 leaving an oxygen-depleted fluorocarbonaceous material.

The following contributions to specimen damage have been identified and reported in the literature (see for example reviews by Koenig and Grant [50], Briggs [3] and by Frydman et al. [49]: X-radiation, consisting of principal and satellite lines of the characteristic radiation as well as Bremsstrahlung, heat due to the close proximity of the hot X-ray source casing, secondary electrons from surrounding parts of the spectrometer, and photoelectrons emitted in the sample itself. Whereas with non-monochromatized X-ray sources all radiative processes contribute to sample decomposition, with monochromatized sources, used in modern spectrometers, only the characteristic Ka1,2 X-ray lines affect the sample. Figure 23.12 compares the degradation of a fluorinated

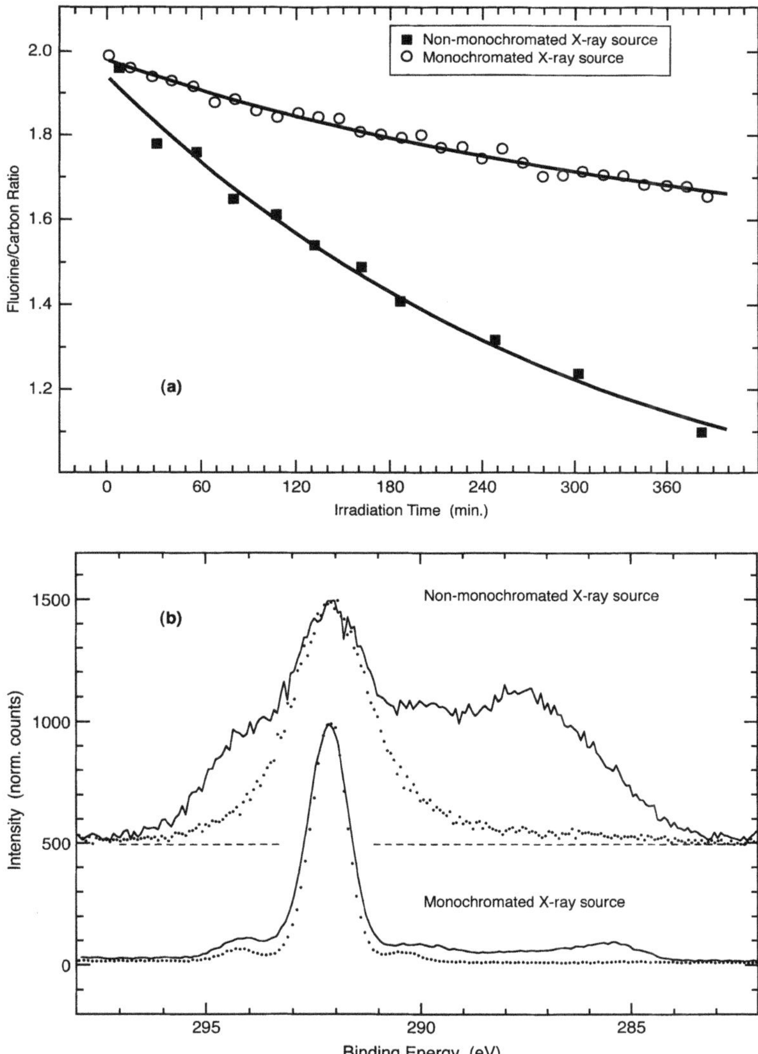

Fig. 23.12. Irradiation induced degradation of a fluorinated ethylene propylene copolymer (FEP Teflon) in two different XPS instruments under the respective standard analysis conditions. One spectrometer employs a non-monochromatized, the other a monochromatized X-ray source. **(a)** Fluorine/Carbon atom number ratio as a function of irradiation time. **(b)** Evolution of the C 1*s* spectrum with evidence for fluorine abstraction, followed by various rearrangement, branching and crosslinking reactions, particularly in the case of the non-monochromatized X-ray source. Shown are the spectra recorded at the beginning of the experiment (dotted curves) and after approximately 6 hours of continuous irradiation (continuous curves).

ethylene propylene copolymer (FEP Teflon) in two different spectrometers under the respective standard analysis conditions. Clearly, decomposition in the modern instrument is substantially reduced. In addition to X-ray source related processes other factors may influence the sample surface during analysis. Exposure of the sample to ultrahigh vacuum may result in desorption of volatile species from its surface. Similarly, dehydration of samples may also result in rearrangement of polymeric segments. Finally, there is the danger of reaction with or contamination from residual gases in the spectrometer.

With due care (eg. by reducing analysis time) and using modern equipment, most of these potential contributions to sample damage can either be eliminated or at least minimised enabling a reliable and artefact-free analysis of most organic materials. One should, however, always be aware that degradation mechanisms and rates can differ markedly between polymers; their respective susceptibility needs to be determined on a case-by-case basis.

In the case of static SIMS (SSIMS) the situation is slightly different since the nature of the analysis process intrinsically involves damaging the sample surface. In addition, SSIMS is more sensitive than XPS with respect to structural changes, i.e. more subtle effects of radiation damage are detected quite readily. Initially it was claimed that a primary ion current density of $< 10^{14}$ ions/cm^2 would ensure an analysis free of artefacts [51]. However, this threshold value was based on only one type of degradation, viz. actual removal of material (sputtering). Later studies revealed more structural degradation at even lower ion densities due to other processes (reviewed in [52]). Damage mechanisms identified include sputtering by the primary ion beam and molecular changes, including the creation of free radicals, caused by the collision cascade following the impact of the primary ion. In addition, the electron beam which is used to control the surface potential during a SSIMS analysis can give rise to degradation since these electrons may have energies of up to several hundred eV.

Suggestions on how to minimise specimen damage include using a primary *atom* beam instead of ions since to some extent damage is caused by the neutralisation of the primary ion in the material being analysed [52]. Due to the much higher sensitivity of Time-Of-Flight (TOF) mass analysers, spectra may be acquired with these instruments at much reduced primary ion currents (10^8 ions/cm^2) thereby almost eliminating the problem in the spectroscopic mode.

Scanning Probe Microscopy techniques rely on the interaction between the microscope tip or stylus and the sample surface in order to probe the upper atomic or molecular layer of the specimen. Depending on the type of interaction this can result in damage to the surface layer and, hence, to experimental artefacts. Organic materials, being rather soft and possessing some structural mobility, are particularly susceptible in that respect. Problems are frequently exacerbated by the presence of adhesive forces between tip and sample. The most extensive damage has been observed with atomic force microscopes used in the contact mode; in this mode the tip being scanned across the sample is in actual contact with the surface atoms. It may therefore penetrate the surface of soft organic specimens and even partially sweep away material during a scan. Since a minimum load force is necessary for the tip to stay in contact with the surface (up to 100 nN) a certain degree of damage is inevitable. Several methods have been developed to reduce sample damage (see for example [53,54]):

(1) In the non-contact mode of operation the tip 'floats' above the surface, sampling the long-range surface forces operating between tip and sample.
(2) Tapping Mode AFM: in this mode the tip oscillates with a drive amplitude which is sufficiently high to ensure periodical contact with the surface, thus providing the necessary feedback signal for the AFM. Elimination of the friction force results in a reduction of the overall force by at least an order of magnitude.
(3) Scanning under liquid removes capillary forces which are present in air due to a thin hydrated layer on the polymer surface. The load force may be reduced by up to two orders of magnitude.

23.8 Charge Correction

Surface analysis of (non-conducting) organic materials involving charged particles invariably leads to a charge build-up on the sample surface. The causes and consequences of this charged layer depend on the technique used.

In XPS, the build-up of charge is due to the lack of sufficient delocalized, conduction-band electrons which could neutralise the positive holes created by the emission of photo- and Auger electrons during analysis. The residual positive charge on the sample surface forms an energy barrier which the photoelectrons have to overcome in addition to the binding energy. As a consequence peak positions appear to be shifted to higher binding energy. In spectrometers equipped with non-monochromatized X-ray sources the observed shifts are usually uniform across the surface and invariant with time; a stable charge balance is almost instantaneously established on the specimen between emission of electrons on the one hand and supply of low-energy electrons from the surrounding spectrometer parts (e.g. the X-ray source window) on the other. In addition to the analysis geometry, the magnitude of this shift is influenced by the surface composition of the analysed material and is of the order of several eV to tens of eV. Since the radiation produced by monochromated X-ray sources is usually focussed on a small spot on the sample, no secondary electrons are generated; this not only gives rise to extensive build-up of a positive charge (tens to hundreds of eV) but charging may be non-uniform across the surface as well as time-dependent. Modern instruments therefore employ charge neutralisation facilities to achieve a stable and uniform charge balance; these range from simple electron flood guns supplying low kinetic energy electrons to the sample surface to flood guns either combined with a metal mesh above the specimen or with a magnetic immersion lens which focusses the electrons on the surface.

As for all insulating materials, the use of a reliable energy reference is crucial in determining binding energies. The most reliable method of spectral referencing is the use of an internal standard; this is particularly well suited to the analysis of most polymeric materials because of the prominence of a signal with known binding energy originating from the polymer backbone. For example, the binding energy of a 1*s* photoelectron emitted by a carbon atom in aliphatic structures bonded to carbon and/or hydrogen (285.00 eV) is now well established [16]. Various other methods of binding energy calibration have been investigated, eg. the use of adventitious carbon contamination which is detected on almost every surface or the deliberate deposition of a reference element onto the surface prior to analysis (evaporation of gold, implantation of Ar ions). However, they all suffer from specific problems such as ill-defined binding energy (adventitious carbon) or relaxation effects [55,56]. Deposition of a reference element on the surface may even partially degrade organic materials.

Charge neutralisation or, more appropriately, the control of surface potential and its effects on the information content of spectra, is considerably more difficult in SIMS than in XPS. Here the main problem is not a shift of peak positions but a shift of the energy distribution of secondary ions; as a consequence, relative peak intensities change or a complete loss of signal might occur [52,57,58]. The objective is to adjust the surface potential in order to match the maximum of the secondary ion distribution to the acceptance "window" of the mass analyzer. This is usually achieved by a combination of an electron flood gun (electron energies of up to several hundred eV) and the application of a bias potential to the specimen. The use of electron sources employing rather high electron energies and current densities for charge neutralisation, however, can also give rise to potentially serious problems such as electron stimulated ion emission and sample degradation. In some cases using a primary atom beam instead of ions can facilitate surface potential control. Generally, controlling the surface potential is more difficult in the negative ion mode compared to the positive ion mode and more difficult in imaging mode compared to spectroscopic mode. Due to the substantially lower primary ion beam currents and its pulsed nature in TOF instruments, charging is much less of a problem but the surface potential still needs to be controlled. The most successful method is based on combining the pulsed primary ion beam with a pulsed secondary ion extraction field and a pulsed electron flood source in a precisely controlled sequence.

23.9 Grazing Angle Infrared Spectroscopy

While not a true surface analytical technique, grazing angle IR is of interest because its probe depth is intermediate between those of ESCA and conventional ATR-IR. It is of merit for the analysis of ultrathin coatings, such as plasma polymers, which often give poor, weak ATR spectra but can readily be analyzed by grazing angle IR [59]. A drawback of this method is the difficulty of quantification; polarization effects and the absence of reliable standards at present allow only qualitative work. This method can, however, be very useful in complementing ESCA data by probing for groups which cannot be differentiated by ESCA, while relying on ESCA for quantitation.

References

1. F. Garbassi, M. Morra, E. Occhiello: *Polymer Surfaces - From Physics to Technology*, (John Wiley & Sons, Chichester, 1994)
2. J.D. Andrade in: *Surface and Interfacial Aspects of Biomedical Polymers, Volume 1, Surface Chemistry and Physics*, J.D. Andrade (Ed) (Plenum Press, New York, 1985)
3. D. Briggs in: *Practical Surface Analysis, Volume 1: Auger and X-ray Photoelectron Spectroscopy*, D. Briggs and M.P. Seah (Eds), 2nd Edn. (John Wiley & Sons, Chichester, 1990),437–484
4. J.C. Vickerman (Ed), *Surface Analysis - The Principal Techniques*, (John Wiley & Sons, Chichester, 1997)
5. J.D. Andrade (Ed.), *Polymer Surface Dynamics*, (Plenum Press, New York, 1988)
6. M. Morra, E. Occhiello, F. Garbassi: *Adv. Coll. Interface Sci.*, **32**, 79 (1990)
7. F.J. Holly, M. Refojo: *J. Biomed. Mater. Res.*, **9**, 315 (1975)
8. J.H. Chen, E. Ruckenstein: *J. Colloid Interface Sci.*, **135**, 496 (1990)
9. T.R. Gengenbach, X. Xie, R.C. Chatelier, H.J. Griesser: *J. Adhesion Sci. Tech.*, **8**, 305 (1994)
10. D.T. Clark, in: *Photon, Electron and Ion Probes of Polymer Structure and Properties*, D.W. Dwight, T.J. Fabish and H.R. Thomas (Eds.), *ACS Symposium Series,* **162**, 225 (1981)
11. M.A. Golub, T. Wydeven, R.D. Cormia: *Langmuir*, **7**, 1026 (1991)
12. D.S. Everhart, C.N. Reilley: *Surf. Interface Anal.*, **3**, 126 (1981)
13. G. Marletta, S.M. Catalano, S. Pignataro: *Surf. Interface Anal.*, **16**, 407 (1990)
14. G. Marletta: *Nucl. Instrum. Methods Phys. Res.*, **B46**, 295 (1990)
15. D. Briggs, "Characterization of Surfaces", in: *Comprehensive Polymer Science, Vol. 1: Polymer Characterization*, C. Booth and C. Price (Eds.), (Pergamon Press, 1989)
16. G. Beamson, D. Briggs: *High Resolution XPS of Organic Polymers. The Scienta ESCA300 Database*, 1st edition (John Wiley & Sons Ltd, (1992).
17. B.J. Tielsch, J.E. Fulghum: *Surf. Interface Anal.*, **21**, 621 (1994)
18. B.J. Tyler, D.G. Castner, B.D. Ratner: *Surf. Interface Anal.*, **14**, 443 (1989)
19. R.C. Chatelier, H.A.W. StJohn, T.R. Gengenbach, P. Kingshott, H.J. Griesser: *Surf. Interface Anal.*, **25**, 741 (1997)
20. M.P. Seah, W.A. Dench: *Surf. Interface Anal.*, **1**, 2 (1979)
21. A. Jablonski, C.J. Powell: *Surf. Interface Anal.*, **20**, 771 (1993)
22. J.C. Ashley: *J. Electron Spectrosc. Relat. Phenom.*, **28**, 177 (1982)
23. S. Tanuma, C.J. Powell, D.R. Penn: *Surf. Interface Anal.*, **21**, 165 1993)

24. W.H. Gries: *Surf. Interface Anal.*, **24**, 38 (1996)
25. K. Lewis, B.D. Ratner: *J. Colloid Interface Sci.*, **159**, 77 (1993)
26. B.D. Ratner, D.G. Castner: *Colloids and Surfaces B: Biointerfaces*, **2**, 333 (1994)
27. B.D. Ratner, P.K. Weathersby, A.S. Hoffman, M.A. Kelly, L.H. Scharpen, *J. Appl. Polym. Sci.*, **22**, 643 (1978)
28. B.D. Ratner, D.G. Castner, T.A. Horbett, T.J. Lenk, K.B. Lewis, R.J. Rapoza: *J. Vac. Sci. Technol. A*, **8**, 2306 (1990)
29. D. Park, B. Keszler, V. Galiatsatos, J.P. Kennedy, B.D. Ratner: *Macromolecules*, **28**, 2595 (1995)
30. A. Takahara, N.-J. Jo, K. Takamori, T. Kajiyama, in: *Progress in Biomedical Polymers*, C.G. Gebelein and R.L. Dunn (Eds) (Plenum Press, New York, 1990)
31. J.H. Silver, K.B. Lewis, B.D. Ratner, S.L. Cooper: *J. Biomed. Mater. Res.*, **27**, 735 (1993)
32. J. Lukas, R.N.S. Sodhi, M.V. Sefton: *J. Colloid Interface Sci.*, **174**, 421 (1995)
33. A. Baty, P.M. Leavitt, C.A. Siedlecki, B.J. Tyler, P.A. Suci, R.E. Marchant, G.G. Geesey: *Langmuir*, **13**, 5702 (1997)
34. C.D. Batich: *Appl. Surface Sci.*, **32**, 57 (1988)
35. M. Alexander: *Photoemissions*, **8**, 8 (1996)
36. A. Chilkoti, B.D. Ratner: *Surf. Interface Anal.*, **17**, 567 (1991)
37. S. Zeggane, M. Delamar: *Appl. Surface Sci.*, **31**, 151 (1988)
38. C.D. Batich, R.C. Wendt: *ACS Symp. Ser.*, **162**, 221 (1981)
39. R.C. Chatelier, T.R. Gengenbach, Z.R. Vasic, H.J. Griesser: *J. Biomat. Sci. Polymer Edn.*, **7**, 601 (1995)
40. P. Kingshott: PhD Thesis, University of New South Wales, Sydney, Australia, 1998.
41. S.N. Magonov, M.-H. Whangbo: *Surface Analysis with STM and AFM*, (Verlagsgesellschaft GmbH, Weinheim, Federal Republic of Germany, 1996)
42. K.L. Babcock, C.B. Prater: *Phase Imaging: Beyond Topography*, (Digital Instruments Application Note, 1995)
43. P. Maivald, H.J. Butt, S.A.C. Gould, C.B. Prater, B. Drake, J.A. Gurley, V.B. Elings, P.K. Hansma: *Nanotechnology*, **2**, 103 (1991)
44. C.D. Frisbie, L.F. Rozsnyai, A. Noy, M.S. Wrighton, C.M. Lieber: *Science*, **265** (5181), 2071 (1994)
45. P.G. Hartley, "Measurement of Colloidal Interactions Using the Atomic Force Microscope", in *Colloid Polymer Interactions*, P.L. Dubin and R. Farinato (Eds.) (Wiley & Sons, New York, 1999)
46. T. Takahagi, Y. Nakayama, F. Soeda, A. Ishitani: *J. Appl. Polym. Sci.*, **41**, 1451 (1990)
47. F.-M. Pan, Y.-L. Lin, S.-R. Horng: *Appl. Surf. Sci.*, **47**, 9 (1991)
48. J. Lacoste, Y. Deslandes, P. Black, D.J. Carlsson: *Polym. Degrad. Stab.*, **49**, 21 (1995)
49. E. Frydman, H. Cohen, R. Maoz, J. Sagiv: *Langmuir*, **13**, 5089 (1997)
50. M.F. Koenig, J.T. Grant: *Appl. Surf. Sci.*, **25**, 455 (1986)
51. A. Benninghoven: *Phys. Status Solidi*, **34**, K169 (1969)
52. D. Briggs in: D. Briggs and M.P. Seah: *Practical Surface Analysis. Vol. 2: Ion and Neutral Spectroscopy*. 2nd edition (John Wiley & Sons, Chichester, England, 1992) Chapter 7

53. F. Sommer, T. Minh Duc, R. Pirri, G. Meunier, C. Quet: *Langmuir*, **11**, 440 (1995)
54. J.P. Spatz, S. Sheiko, M. Möller: *Langmuir*, **13**, 4699 (1997)
55. M.P. Seah in: D. Briggs and M.P. Seah: *Practical Surface Analysis. Vol. 1: Auger and X-Ray Photoelectron Spectroscopy*. 2nd edition (John Wiley & Sons, Chichester, England, 1990) Appendix 2
56. A.E. Hughes, B.A. Sexton: *J. Electron Spectrosc. Relat. Phenom.*, **50**, C15 (1990)
57. F. Garbassi, M. Morra, E. Occhiello: *Polymer Surfaces. From Physics to Technology*. 1st edition (John Wiley & Sons, Chichester, England, 1994)
58. N.M. Reed, J.C. Vickerman in: L. Sabbatini and P.G. Zambonin: *Surface Characterization of Advanced Polymers*. 1st edition (VCH Verlagsgesellschaft GmbH, Weinheim, Federal Republic of Germany, 1993) chapter 3
59. T.R. Gengenbach, H.J. Griesser: *J. Polym. Sci.: Part A: Polym. Chem.*, **36**, 985 (1998)

24 Glow Discharge Optical Emission Spectrometry

T. Nelis and R. Payling

Glow Discharge Optical Emission Spectrometry (GD–OES) is a well established technique for surface and interface characterisation [1]. It was first introduced by Grimm in Germany in 1967 [2]. In the beginning it was only used for bulk analysis of solid conducting samples, but quickly the field of application was widened to include qualitative depth profiles of surfaces and coatings. French, German and Japanese teams of industrial researchers were very active in this field in the 1970s and 1980s, leading to the first methods for quantitative depth profiling [3]. The next big step forward was the development of the radio frequency (rf) version of GD–OES by Richard Passetemps at Renault in France and Ken Marcus at Clemson University in USA, allowing the analysis of insulating as well as conductive solid materials [4–6]. In recent years the technique has also found application in the semiconductor manufacturing process.

The sample is mounted on an o-ring on the outside of the source, making sample placement fast and simple. There is no ultra high vacuum system. Once the sample is placed, the source is pumped continuously with a mechanical pump and in a few seconds reaches a residual pressure of about 0.2 Pa (2 mTorr). Then argon is introduced into the source and when the pressure reaches about 500 Pa (6 Torr) the source is ready for analysis. A voltage is applied to the sample and sputtering begins and a glow discharge plasma is created.

Argon is flowing constantly through the source to the vacuum pump. Because of the sweeping action of the argon, the purity of the vacuum is essentially that of the introduced argon, ie a few ppm of contaminants. Unlike other surface techniques, once the voltage is applied to the sample, sputtering in GD–OES is so rapid that surface adsorption is negligible. In common with SNMS, sputtering and excitation in GD–OES occur in different parts of the source, allowing GD–OES to be fully quantitative.

24.1 Instrument

A typical instrument consists of a glow discharge (GD) source connected to one or more optical spectrometers. See Fig. 24.1.

A hollow anode glow discharge chamber serves as the light source for the spectrometer. The two electrodes of the discharge cavity comprise: a grounded copper tube (anode) and the sample to be analysed (cathode),

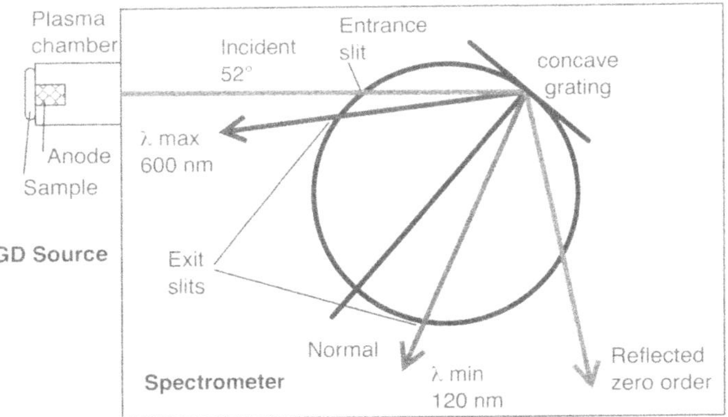

Fig. 24.1. Schematic diagram of a GD–OES instrument

e.g. a zinc coated steel sheet or a silicon wafer. The sample mounting places some restrictions on the types of samples that can be analysed. In particular they must be capable of withstanding evacuation, i.e. they cannot be loose powders, and ideally they should have a nearly flat face at least as large as the outside diameter of the anode, this eliminates some complex shapes. In addition, they must be capable of withstanding some heating of their surface by the plasma, perhaps $200°C$, this eliminates some polymers and low temperature metals.

An argon flow is maintained in the discharge cavity to stabilise the pressure between 200–1400 Pa (2–12 Torr). An rf power of 10–80 W, typically 30–50 W, is supplied to the sample and a plasma is created within the tubular anode. A dc voltage can also be used for conductive samples. The anode is in a fixed position close to the sample surface, forming a gap of 0.1–0.2 mm, so that no plasma is possible directly between the annular surface of the anode and the sample surface. The inner diameter of the anode – usually 4 mm, but 2–8 mm are also used – thereby defines the analysis spot size, as the plasma is restricted to this area. In rf operation, a negative potential, the auto bias potential or self-bias, is established at the sample surface (on both conductive and non-conductive samples) due to the difference in size between the two electrodes.

Argon ions, created in the negative glow of the plasma, are accelerated towards the sample where they bombard the surface. This bombardment with argon ions induces an erosion process called sputtering. Atoms from the sample surface diffuse through the cathode dark space of the plasma into the negative glow. Here they are excited through collisions with high energy electrons, metastable argon atoms and ions. Each element emits a characteristic electromagnetic spectrum. This spectrum is recorded by an optical spectrometer and the concentration distribution of the different elements is deduced from the intensities of the characteristic spectral lines.

The discharge cavity is designed to create a flat-bottomed erosion crater. For very flat homogeneous samples, typically used in the semiconductor industry, a depth resolution of 10 nm at the end of a 500 nm deep crater can be obtained. For other samples such as galvanized steel the roughness of the surface and the grain structure of the coating reduce the ultimate depth resolution typically to about 15% of depth.

Spectral interferences are generally not a severe problem in GD–OES as there are many optical emission lines to choose from. Background signals are much lower than for AES or XPS but higher than SIMS, hence the sensitivity is much better than AES or XPS and approaches that of SIMS.

The most common spectrometer used in a GD–OES instrument is a Paschen-Runge mounted polychromator based on a Rowland circle of fixed focal length. The entrance slit, the concave grating and the exit slits are all positioned on a circle of the same diameter as the focal length of the grating. Each exit slit selects a spectral line of interest. Behind each exit slit a photomultiplier (PM) tube is mounted to record the light intensity for the given wavelength. The signals from all PM tubes are recorded simultaneously.

Alternatively a tuneable Czerny–Turner mounted monochromator can be used. The Czerny–Turner mount is characterised by its high light throughput and high resolving power. The full spectrum can be recorded in two minutes or the monochromator can be used with the polychromator as an additional acquisition channel to measure elements or lines not available on the polychromator.

24.2 Theory

The recorded intensity is a measure of the number of atoms of a particular element in the plasma. This number is given by the product of the concentration of the element in the sample c_i and the sputtering rate q. Once equilibrium sputtering is reached (i.e. after some nanometres) the sputtering rate is the same for all elements. The relationship between $c_i \cdot q$ and intensity is determined beforehand by calibration. The measured intensities in a qualitative depth profile (intensity vs time) can then be converted immediately into $c_i \cdot q$ values from the calibration. The $c_i \cdot q$ values are then summed at each time interval to give the instantaneous value of q at each point (since $\sum_i c_i = 100\,\%$) and the $c_i \cdot q$ values then divided by q to give c_i. The depth is determined by integrating the q values over time and adjusting for density. This simple method appears to be unique to GD–OES. It produces quantitative depth profiles (concentration vs depth) whose concentrations and depths are accurate to 5–10% for major and minor elements.

24.3 Applications

24.3.1 Near Surface

A typical sputtering rate in GD–OES is 10–150 nm s^{-1}, i.e. 30–500 atomic layers per second, but since photomultiplier counting rates can be very fast, and integration times are typically 1–10 ms at high speed, fast GD–OES instruments are routinely capable of making 1–10 measurements per element per atomic layer. Hence GD–OES, in principle, is capable of surface analysis. Unfortunately argon sputtering is not a layer by layer process so GD–OES suffers similar sputter-induced effects as SIMS and XPS/AES when combined with Ar sputtering. GD–OES has the added complication that the plasma may take several atomic layers to establish, so the depth profile of the first few nanometres can be greatly distorted.

Several publications have compared near surface results from GD–OES with other surface techniques. Alexandre *et al.* compared GD–OES with SIMS and AES depth profiles of the first few nanometres of passive films on iron and found the results were remarkably similar [7]; as were results by Janssen on steel and aluminium surfaces [8], Justo and Ferreira on zinc [9], and Suzuki and Suzuki on steel [10]. Others have found significant qualitative differences [11]. The most recent comparison between GD–OES and SIMS is by Shimizu *et al.* on anodic alumina thin films relevant to the fabrication of electrolytic capacitors and liquid crystal displays [12]. One sample was produced by anodising pure aluminium in a Cr_2O_3 doped sodium chromate solution. In both GD–OES and SIMS, the chromium signal peaked at 19 nm with a half-width of 7 nm. Transmission electron microscopy gave a depth of 19 nm and width of 2–3 nm.

The general outcome of these comparisons is: (1) GD–OES is typically 100 times faster than the other techniques, (2) GD–OES sensitivity is much higher than XPS/AES and more comparable to SIMS, (3) GD–OES can detect hydrogen which is difficult or impossible for the other techniques, (4) the others can provide additional information on chemical bonds which GD–OES cannot, (5) the others can be focussed to small (micron or less) areas which GD–OES cannot, and (6) because of their different strengths and weaknesses, the techniques continue to be complementary.

24.3.2 Coatings

GD–OES is the only technique capable of bulk analysis and of rapid, quantitative depth profiling of thin films and commercial coatings [13]. All of the common industrial zinc-based coatings have now been successfully quantified: two examples are shown in Fig. 24.2. Note the enrichment of Al and Pb at the surface and interface of the galvanized steel sample and the uniformity of the ZnNi coating.

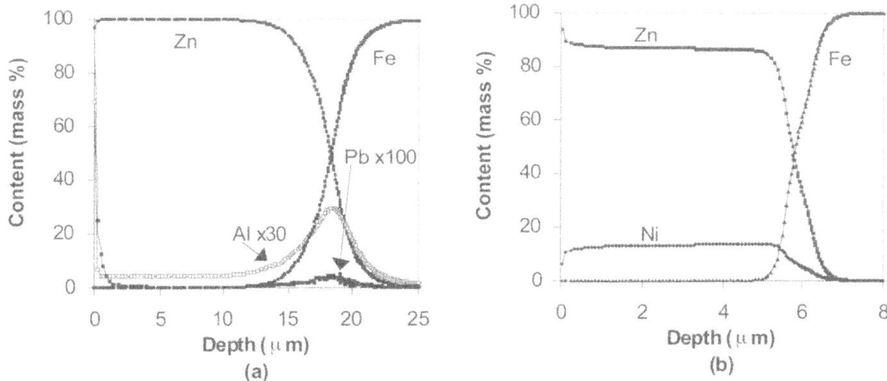

Fig. 24.2. Quantitative GD–OES depth profiles of **(a)** galvanized steel and **(b)** ZnNi coated steel.

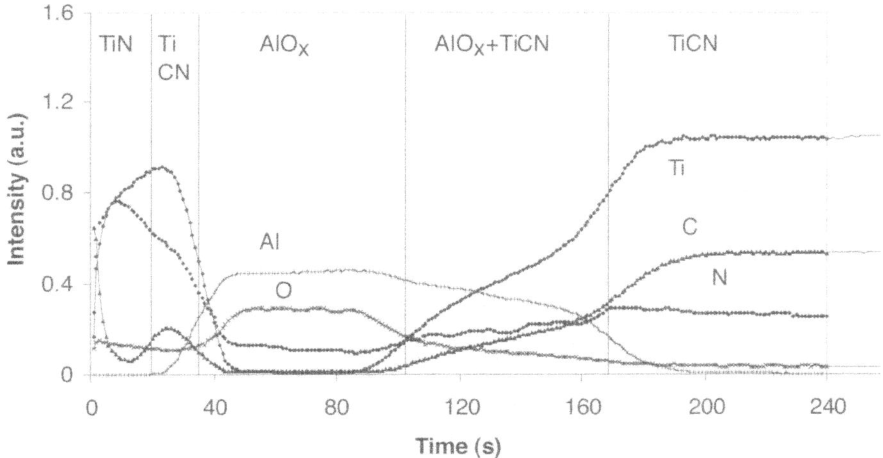

Fig. 24.3. GD–OES depth profile of a PVD multilayer sample.

An area of growing interest for GD–OES is PVD/CVD coatings, especially hard coatings, including tungsten and titanium carbide and carbonitride layers on tool steels. An example of a complex PVD coating is given in Fig. 24.3, where for simplicity only the first five layers and some elements are shown.

24.3.3 Semiconductor Processing

GD–OES can be used for the measurement of the elemental concentration profiles of thin films used in the manufacture of integrated circuits (ICs) [14]. The detailed information of element concentration profiles throughout the layer and at the interface often gives valuable information on the process

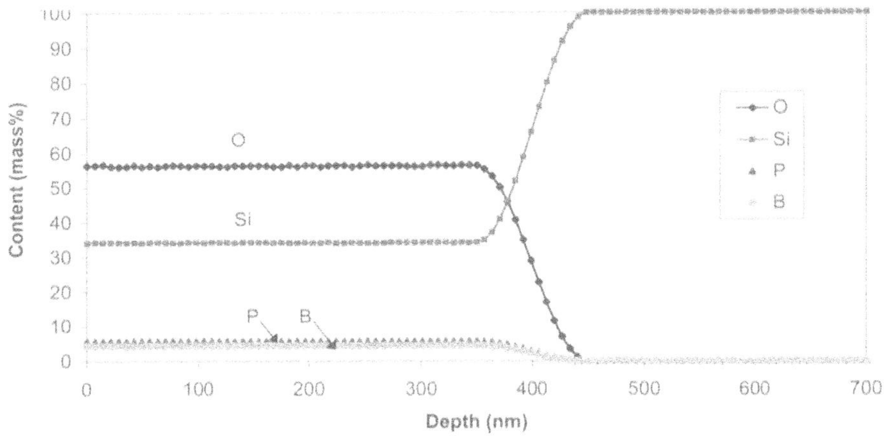

Fig. 24.4. Quantitative GD–OES depth profile of BPSG on silicon.

quality and stability. The homogeneity of the deposition process across the wafer can be checked by analysing different spots on the wafer. The instrument can be equipped with an automatic wafer handling system for large wafers.

One typical application of rf GD–OES in the 'memory' production process is the characterisation of BPSG (boron phosphorus silica glass) films, as these films cannot easily be analysed by any other analytical technique already used in IC manufacturing.

The boron and phosphorous concentrations can be determined precisely as a function of film thickness (see Fig. 24.4). Figure 24.4 shows a simulated depth profile because the original data is confidential. As only atomic species are measured, the analysis is independent of the molecular structure of the BPSG film and in particular of residual water. In addition to the control of the concentration profiles of the major dopants, diffusion processes between the silicon core and the BPSG film can be investigated. Effects at the layer surface and particularly at the interface can be studied. One example of the capabilities of GD–OES, though of only minor interest to IC manufacturing, is the control of alkali elements, such as sodium, at the interface.

To optimise deposition processes and other treatments, elements such as hydrogen, carbon, chlorine, fluorine, etc. can be measured. A few minutes are usually enough to obtain an analytical result, and the optimal process parameters can be quickly established.

A different application of rf GD–OES in semiconductors is the control of ion implantation concentration profiles. The detection limits for most elements are in the ppm range or 10^{17} atoms cm^{-3}. In this case the advantage of GD–OES compared to SIMS is the speed and cost of analysis rather than the detection limit of some elements.

References

1. R. Payling, D.G. Jones, A. Bengtson (Eds.) *Glow Discharge Optical Emission Spectrometry* (John Wiley & Sons, Chichester, 1997)
2. W. Grimm: Naturwiss. **54**, 586 (1967)
3. J. Pons-Corbeau: Surf. Interf. Anal. **32**, 365 (1985)
4. M. Chevrier, R. Passetemps: European Patent No. 0 296 920 A1 (1988)
5. D.C. Duckworth, R.K. Marcus: Anal. Chem. **61**, 1879 (1989)
6. R.K. Marcus, US Patent No. 5,006,706 (1991)
7. B. Alexandre, R. Berneron, J.C. Charbonnier, R. Namdar-Irani, L. Nevot: Mémoires Etudes Sci. Rev. Metall., Sept., 483 (1981)
8. E. Janssen: Mater. Sci. Engin. **42**, 309 (1980)
9. M.J. Justo, M.G.S. Ferreira: Corros. Sci. **34**, 533 (1993)
10. S. Suzuki, K. Suzuki: Surf. Interf. Anal. **17**, 551 (1991)
11. A. Benhamou, F. Bourelier, K. Vu Quang: Métaux **679**, 96 (1982)
12. K. Shimizu, G.M. Brown, H. Habazaki, K. Kobayashi, P. Skeldon, G.E. Thompson, G.C. Wood: Surf. Interf. Anal. **27**, 24 (1999)
13. J. Angeli, in R. Payling, D.G. Jones, A. Bengtson (Eds.), *Glow Discharge Optical Emission Spectrometry* (John Wiley & Sons, Chichester, 1997), p 565
14. I.M. Dharmadasa, M. Ives, J.S. Brooks, C. Breen: in R. Payling, D.G. Jones, A. Bengtson (Eds.), *Glow Discharge Optical Emission Spectrometry* (John Wiley & Sons, Chichester, 1997), p 668

Part IV

Appendix

Acronyms Used in Surface and Thin Film Analysis

A selection of acronyms encountered in surface and thin film analysis. The majority apply to specific experimental techniques or related aspects. A given technique may have more than one acronym in common use.

ABS	Atomic Beam Scattering
ACAR	Angular Correlation of Annihilation Radiation
AEAPS	Auger Electron Appearance Potential Spectroscopy
AEM	Analytical Electron Microscopy (Chap. 3)
AEM	Auger Electron Microscopy (Chap. 6)
AES	Auger Electron Spectroscopy (Chap. 6)
AEXAFS	Auger Emission X-ray Absorption Fine Structure
AFM	Atomic Force Microscopy
APS	Appearance Potential Spectroscopy
AR	Angle Resolved (technique prefix)
ASW	Acoustic Surface Wave measurements
ATR	Attenuated Total Reflectance (Chap. 8)
BIS	Bremsstrahlung Isochromat Spectroscopy
CAE	Constant Analyser Energy
CHA	Concentric Hemispherical Analyser (Chaps. 6, 7)
CIS	Characteristic Isochromat Spectroscopy
CL	Cathode Luminescence (Chap. 3)
CMA	Cylindrical Mirror Analyser (Chap. 6)
CPD	Contact Potential Difference (synonymous with work function) (Chap. 1)
CRR	Constant Retard Ratio
CVD	Chemical Vapour Deposition
DAPS	Disappearance Potential Spectroscopy
ΔH_{ads}	Heat of adsorption measurements
$\Delta\emptyset$	Change in work function (Chap. 1)
DIET	Desorption Induced by Electronic Transitions
DRIFT	Diffuse Reflectance Infrared Fourier Transform spectroscopy (Chap. 8)
ECM	Electrical Conductivity Microscope/y
EDAX	Energy Dispersive Analysis of X-rays (Chap. 3)
EDC	Electron Energy Distribution Curve

EDS	Energy Dispersive Spectroscopy (equivalent to EDAX)
EDX	Energy Dispersive X-ray (spectroscopy) (equivalent to EDAX)
EELS	Electron Energy Loss Spectroscopy (Chap. 3)
EID	Electron Impact Desorption (equivalent to ESD)
EIL	Electron Induced Luminescence
EL	ElectroLuminescence
ELEED	Elastic Low Energy Diffraction (Chap. 13)
ELL	Ellipsometry (Chap. 1)
ELS	Energy Loss Spectroscopy (usually low resolution EELS) (Chap. 3)
EM	Electron Microscopy (Chap. 3)
EMP	Electron MicroProbe analysis (equivalent to EDAX) (Chap. 3)
EPMA	Electron Probe MicroAnalysis (see EMP)
ES	Emission Spectroscopy (technique suffix)
ESCA	Electron Spectroscopy for Chemical Analysis (synonymous with XPS) (Chap. 7)
ESD	Electron Stimulated Desorption (may be of ions, neutrals or excited states)
ESDI	Electron Stimulated Desorption of Ions
ESDIAD	Electron Stimulated Desorption Ion Angular Distribution
ESDN	Electron Stimulated Desorption of Neutrals
ESR	Electron Spin Resonance
EXAFS	Extended X-ray Absorption Fine Structure (Chap. 1)
EXFAS	Extended Fine Auger Structures (Chap. 1)
FABMS	Fast Atom Bombardment Mass Spectrometry (Chap. 1)
FABS	Fast Atom Bombardment Spectrometry (see FABMS)
FD	Field Desorption
FD	Flash Desorption (equivalent to TDS/TPD at high ΔT rate)
F–d	Force versus Distance analysis
FDM	Field Desorption Microscopy
FDS	Field Desorption Spectrometry
FDS	Flash Desorption Spectrometry (equivalent to TDS/TPD at high ΔT rate)
FEED	Field Emission Energy Distribution
FEES	Field Electron Energy Spectroscopy
FEM	Field Emission Microscopy (Chap. 1)
FERP	Field Emission Retarding Potential
FeSEM	Field-emission Scanning Electron Microscope/y
FFM	Lateral/Friction Force Microscope/y
FIM	Field Ion Microscopy (Chap. 1)
FIM-APS	Field Ion Microscope-Atom Probe Spectroscopy
FT	Fourier Transform
FIS	Field Ion Spectroscopy (Chap. 1)
FTIR	Fourier Transform InfraRed (Chap. 8)

FT-RAIRS	Fourier Transform Reflection Absorption Infrared Spectroscopy
FTRS	Fourier Transform Raman Spectroscopy
GDMS	Glow Discharge Mass Spectrometry
GDOS	Glow Discharge Optical Emission Spectroscopy
GIXD	Grazing Incidence X-ray Diffraction
HEED	High Energy Electron Diffraction (Chap. 12)
HEIS	High Energy Ion Scattering (Chap. 9)
HPT	High Precision Translator
HREELS	High Resolution Electron Energy Loss Spectroscopy (equivalent to EELS)
IAES	Ion (excited) Auger Electron Spectroscopy
IBSCA	Ion Beam SpectroChemical Analysis (equivalent to SCANIIR)
ICISS	Impact Collision Ion Scattering Spectroscopy
IETS	Inelastic Electron Tunnelling Spectroscopy
IES	Fourier Transform Infrared Emission Spectroscopy
IID	Ion Impact Desorption
IIRS	Ion Impact Radiation Spectroscopy
IIXS	Ion Induced X-ray Spectroscopy
ILEED	Inelastic Low Energy Electron Diffraction (Chap. 13)
ILS	Ionization Loss Spectroscopy
IMFP	Inelastic Mean Free Path (Chap. 1)
IMMA	Ion Microprobe Mass Analysis (Chap. 5)
IMXA	Ion Microprobe X-ray Analysis
INA	Ion Neutral Analysis (equivalent to SMNS)
INS	Ion Neutralization Spectroscopy
IPS	Inverse Photoemission Spectroscopy
IR	InfraRed (technique prefix)
IRAS	Infrared Reflection Absorption Spectroscopy (Chap. 8)
IRRAS	InfraRed Reflection Absorption Spectroscopy (see IRAS)
IRS	Internal Reflectance Spectroscopy (Chap. 8)
IS	Ionization Spectroscopy
ISD	Ion Stimulated Desorption
ISS	Ion Scattering Spectroscopy (Chap. 11)
ITS	Inelastic Tunnelling Spectroscopy
L	Langmuir (10^{-6} Torr sec)
LAMMA	Laser Microprobe Mass Analysis
LEED	Low Energy Electron Diffraction (Chap. 13)
LEELS	Low Energy Electron Loss Spectroscopy (equivalent to EELS)
LEERM	Low Energy Electron Reflection Microscopy
LEIS	Low Energy Ion Scattering (Chap. 11)
LFM	Lateral/Friction Force Microscope/y
LID	Laser Induced Desorption
LIMA	Laser Induced Mass Analysis (equivalent to LAMMA)
LITD	Laser Induced Thermal Desorption

LMP	Laser MicroProbe
LS	Light Scattering
MBE	Molecular Beam Epitaxy
MBRS	Molecular Beam Reactive Scattering
MBSS	Molecular Beam Surface Scattering
MEED	Medium Energy Electron Diffraction
MEIS	Medium Energy Ion Scattering
MFM	Magnetic Force Microscope/y
MOMBE	Metal Organic Molecular Beam Epitaxy
MOSS	Mössbauer Spectroscopy
MS	Magnetic Saturation
NBS	Nuclear Backscattering Spectroscopy (Chap. 9)
NEXAFS	Near Edge X-ray Absorption Fine Structure
NIRS	Neutral Impact Radiation Spectroscopy
NMR	Nuclear Magnetic Resonance
NRA	Nuclear Reaction Analysis (Chap. 9)
PAES	Proton (excited) Auger Electron Spectroscopy
PAS	PhotoAcoustic Spectroscopy
PAX	Photoemission of Adsorbed Xenon
PD	PhotoDesorption
PD	Photoelectron Diffraction
PE	PhotoEmission
PEM	PhotoElectron Microscopy
PES	PhotoElectron Spectroscopy (synonymous with UPS, XPS and ESCA)
PESM	PhotoElectron SpectroMicroscopy
PIXE	Particle (Proton) Induced X-ray Emission
PSD	Photon Stimulated Desorption
PSEE	PhotoStimulated Exoelectron Emission
PSID	Photon Stimulated Ion Desorption
PVD	Physical Vapour Deposition
RAIRS	Reflection Absorption InfraRed Spectroscopy (Chap. 8)
RBS	Rutherford Backscattering Spectroscopy (Chap. 9)
REELS	Reflection Electron Energy Loss Spectroscopy
REM	Reflection Electron Microscopy
RFA	Retarding Field Analyser
RGA	Residual Gas Analyser
RHEED	Reflection High Energy Electron Diffraction (Chap. 12)
RVS	Raman Vibrational Spectroscopy
SAED	Selected Area Electron Diffraction (Chap. 3)
SALI	Secondary Atom Laser Ionization
SAM	Scanning Auger Microscopy (Chap. 6)
SAXPS	Selected Area X-ray Photoelectron Spectroscopy (Chap. 7)

SC	Surface Capacitance
SCM/EFM	Scanning Capacitive Microscope/y
SDMM	Scanning Desorption Molecular Microscope
SE	SpectroEllipsometry
SEE	Secondary Electron Emission
SEM	Scanning Electron Microscopy (Chap. 3)
SEMPA	Scanning Electron Microscopy with Polarization Analysis
SERS	Surface Enhanced Raman Spectroscopy
SEXAFS	Surface Extended X-ray Absorption Fine Structure (Chap. 1)
SFM	Scanning Force Microscope/y
SHG	Second Harmonic Generation (optical)
SI	Surface Ionization
SIIMS	Secondary Ion Imaging Mass Spectrometry (Chap. 5)
SIMS	Secondary Ion Mass Spectrometry (Chap. 5)
SKP	Scanning Kelvin Probe
SLEEP	Scanning Low Energy Electron Probe
SNIFTIRS	Subtractively Normalized Interfacial FTIR Spectroscopy
SNMS	Secondary Neutrals Mass Spectrometry (equivalent to INA)
SNOM	Scanning Near-field Optical Microscope/y
SP	Spin Polarized (technique prefix)
SP	Surface Potential
SPIES	Surface Penning Ionization Electron Spectroscopy
SPIPE	Spin Polarized Inverse PhotoEmission
SPM	Scanned Probe Microscope/y
SPPD	Spin Polarized Photoelectron Diffraction
SPV	Surface Photovoltage
SRPS	Synchrotron Radiation Photoelectron Spectroscopy
SRS	Surface Reflectance Spectroscopy
SSA	Spherical Sector Analyser
SSIMS	Static Secondary Ion Mass Spectrometry (Chap. 5)
STEM	Scanning Transmission Electron Microscopy (Chap. 3)
SThM	Scanning Thermal Microscope/y
STM	Scanning Tunnelling Microscopy (Chap. 10)
STS	Scanning Tunnelling Spectroscopy (Chap. 10)
STS	Surface Tunnelling Spectroscopy
SXAPS	Soft X-ray Appearance Potential Spectroscopy
SXES	Soft X-ray Emission Spectroscopy
TD	Thermal Desorption
TDMS	Thermal Desorption Mass Spectrometry
TDS	Thermal Desorption Spectrometry (equivalent to TPD)
TE	Thermionic Emission
TEAM	Thermal Energy Atomic and Molecular beam scattering
TEAS	Thermal Energy Atom Scattering
TED	Transmission Electron Diffraction

TELS	Transmission Energy Loss Spectroscopy
TEM	Transmission Electron Microscopy (Chap. 3)
TL	ThermoLuminescence
TPD	Temperature Programmed Desorption (equivalent to TDS)
TPRS	Temperature Programmed Reaction Spectrometry
UHV	Ultrahigh Vacuum (Chap. 2)
UPS	Ultraviolet Photoelectron Spectroscopy (Chap. 14)
XAES	X-ray (excited) Auger Electron Spectroscopy (Chap. 7)
XAFS	X-ray Absorption Fine Structure
XANES	X-ray Absorption Near Edge Structure (equivalent to NEXAFS)
XEM	eXoElectron Microscopy
XES	eXoElectron Spectroscopy
XES	X-ray Emission Spectroscopy
XPD	X-ray Photoelectron Diffraction
XPS	X-ray Photoelectron Spectroscopy (synonymous with ESCA) (Chap. 7)
XRD	X-ray Diffraction
XRF	X-ray Fluorescence
XTEM	Cross-sectional Transmission Electron Microscopy
WF	Work Function (Chap. 1)

Surface Science Bibliography

Auciello, O., R. Kelly (eds.): *Ion Bombardment Modification of Surfaces* (Elsevier, Amsterdam 1984)

Bamford, C.H., C.F.H. Tiper, R.G. Compton (eds.): Simple Processes at the Gas-Solid Interface, in *Comprehensive Chemical Kinetics* (Elsevier, Amsterdam 1984)

Barr, T.L., L.E. Davis (eds.): *Applied Surface Analysis* (American Society for Testing and Materials, 1978)

Bauer, E.: LEED and Auger Methods, in *Interaction on Metal Surfaces*, ed. by R. Gomer, Topics Appl. Phys. **4** (Springer, Berlin, Heidelberg 1975) pp. 225–274

Baun, W.C.: Ion Scattering Spectrometry: a Versatile Technique for a Variety of Materials, Surf. Interface Anal. **3**, 243 (1981)

Behm, R.J., W. Höseler: Scanning Tunneling Microscopy – a Review, in *Chemistry and Physics of Solid Surfaces VI*, ed. by R. Vanselow, R. Howe, Springer Ser. Surf. Sci. Vol. 5 (Springer, Berlin, Heidelberg 1986) pp. 361–412

Behrisch, R. (ed.): *Sputtering by Particle Bombardment*, Vols. I and II, Topics Appl. Phys., Vols. 47, 52 (Springer, Berlin, Heidelberg 1981, 1983)

Benninghoven, A.L.: Developments in Secondary Ion Mass Spectrometry and Applications to Surface Studies, Surf. Sci. **53**, 596 (1975)

Benninghoven, A.L., F.G. Rudenauer, H.W. Werner: *Secondary Ion Mass Spectrometry*, Vol. 86 of Chemical Analysis, ed. by P.J. Elving, J.D. Winefordner, I.M. Kothoff (Wiley, New York 1987)

Berkowitz, J.: *Photoabsorption, Photoionization and Photoelectron Spectroscopy* (Academic, New York 1979)

Bianconi, A., L. Inoccia, S. Stipchich (eds.): *EXAFS and Near Edge Structure*, Springer Ser. Chem. Phys., Vol. 27 (Springer, Berlin, Heidelberg 1983)

Blakely, J.M.: *Introduction to the Properties of Crystal Surfaces* (Pergamon, Oxford 1973)

Blakely, J.M. (ed.): *Surface Physics of Materials*, Vols. I and II (Academic, New York 1975)

Blakely, J.M. (ed.): *Surface Physics of Crystal Solids* (Academic, New York 1975)

Boehm, H.P., H. Knözinger: Nature and Estimation of Functional Groups on Solid Surfaces, in *Catalysis – Science & Technology*, Vol. 4 (Springer, Berlin, Heidelberg 1983)

de Boer, J.H.: *The Dynamical Character of Adsorption* (2nd edn.) (Oxford University Press, London 1968)

Briggs, D., M.P. Seah (eds.): *Practical Surface Analysis by Auger and X-ray Photoelectron Spectroscopy* (Wiley, Chichester 1983)

Briggs, D. (ed.): *Handbook of X-ray and Ultraviolet Photoelectron Spectroscopy* (Heyden, London 1977)

Brown, A., J.C. Vickerman: Static SIMS for Applied Surface Analysis, Surf. Interface Anal. **6**, 1 (1984)

Brümmer, D., O. Heydenreich, K.H. Krebs, H.G. Schneider (eds.): *Handbuch Festkörperanalyse mit Elektronen, Ionen und Röntgenstrahlen* (Vieweg, Braunschweig 1980)

Brundle, C.R., A.D. Baker (eds.): *Electron Spectroscopy: Theory, Techniques and Applications*, Vols. 1–4 (Academic, London 1977, 1978, 1979, 1981)

Brundle, C.R.: Elucidation of Surface Structure and Bonding by Photoelectron Spectroscopy, Surf. Sci. **48**, 99 (1975)

Carbonara, R.S., J.R. Cuthill (eds.): *Surface Analysis Techniques for Metallurgical Applications* (American Society for Testing and Materials, 1975)

Cardona, M., L. Ley (eds.): *Photoemission in Solids I (General Principles)*; *Photoemission in Solids II (Case Studies)*, Topics Appl. Phys. **26** and **27** (Springer, Berlin, Heidelberg 1978, 1979)

Carlson, T.A.: *Photoelectron and Auger Spectroscopy* (Plenum, New York 1975)

Chu, K.W., J.W. Mayer, M.A. Nicolet: *Backscattering Spectrometry* (Academic, New York 1978)

Clarke, L.J.: *Surface Crystallography – An Introduction to LEED* (Wiley, New York 1975)

Czanderna, A.W. (ed.): *Methods of Surface Analysis*, Methods and Phenomena Vol. 1 (Elsevier, Amsterdam 1975)

Davis, L.A., N.C. MacDonald, P.W. Palmberg, G.E. Riach, R.E. Weber: *Handbook of Auger Electron Spectroscopy*, 2nd edn. (Perkin-Elmer Corp, Eden Prairie 1979)

Delannay, F. (ed.): *Characterization of Heterogoeneous Catalysts*, Chem. Ind. Ser. **15** (Dekker, New York 1984)

Deviney, M.L., J.L. Grand (eds.): Catalyst Characterization Science, ACS Symp. Ser. **288**, Washington, D.C. (1985)

Dobrzynski, L.: *Handbook of Interfaces and Surfaces*, Vols. 1 and 2 (Garland, New York 1978)

Engel, T., K.H. Rieder: Structural Studies of Surfaces with Atomic and Molecular Beam Diffraction, in *Structural Studies of Surfaces*, G. Höhler (ed), Springer Tracts Mod. Phys. **91** (Springer, Berlin, Heidelberg 1982)

Ertl, G., J. Küppers: *Low Energy Electrons and Surface Chemistry*, Monographs in Modern Chemistry Vol. 4 (Verlag Chemie, Weinheim 1974)

Fadley, C.S.: Angle Resolved Photoelectron Spectroscopy, Progr. Surf. Sci. **16**, 275 (1984)

Fiermans, L., R. Hoogewijs, J. Vennik: Electron Spectroscopy of Transition Metal Oxides, Surf. Sci. **47**, 1 (1975)

Fiermans, L., J. Venniuk, W. Dekeyser (eds.): *Electron and Ion Spectroscopy in Solids* (Plenum, New York 1978)

Gibson, W.M.: Determination by Ion Scattering of Atom Positions at Surfaces and Interfaces, in *Chemistry and Physics of Solid Surfaces V*, ed. by R. Vanselow, R. Howe, Springer Ser. Chem. Phys. Vol. 35 (Springer, Berlin, Heidelberg 1984) pp. 427–454

Gomer, R. (ed.): *Interactions on Metal Surfaces*, Topics Appl. Phys. Vol. 4 (Springer, Berlin, Heidelberg 1975)

Gosh, P.K.: *Introduction to Photoelectron Spectroscopy* (Wiley, New York 1983)

Hannay, N.B.: *Treatise on Solid State Chemistry*, Vol. 6A, Surfaces I, Vol. 6B, Surfaces II (Plenum, New York 1976)

Heiland, W., E. Taglauer: The Backscattering of Low Energy Ions and Surface Structure, Surf. Sci. **68**, 96 (1977)

Heinz, K., K. Müller: LEED-Intensities – Experimental Progress, and New Possibilities of Surface Structure Determination, in *Structural Studies of Surfaces*, G. Höhler (ed), Springer Tracts Mod. Phys. **91** (Springer, Berlin, Heidelberg 1982) pp. 1–54

Hodgson, K.O., B. Hedman, J.E. Penner-Hahn (eds.): *EXAFS and Near Edge Structure III*, Springer Proc. Phys. Vol. 2 (Springer, Berlin, Heidelberg 1984)

Hofmann, S.: Practical Surface Analysis: State of the Art and Recent Developments in AES, XPS, ISS and SIMS, Surf. Interface Anal. **9**, 3 (1986)

Ibach, H. (ed.): *Electron Spectroscopy for Surface Analysis*, Topics Curr. Phys. Vol. 4 (Springer, Berlin, Heidelberg 1977)

Jona, F.: LEED Crystallography, J. Phys. C. **11**, 4271 (1978)

Joyner, R.W., G.A. Somorjai: Recent Trends in the Application of LEED, Surf. Def. Prop. Sol. **2**, 1 (1973)

Kane, P.F., G.B. Larrabee (eds.): *Characterization of Solid Surfaces* (Plenum, New York 1974)

Karchaudhari, S.N., K.L. Cheng: Recent Study of Solid Surfaces by Photoelectron Spectroscopy, Appl. Spectrosc. Rev. **16**, 187 (1980)

Kay, E., P. Bagus (eds.): Topics in Surface Chemistry (Plenum, New York 1978)

Kelley, M.J.: Chemtech **17**, 30 (1987); ibid **17**, 98 (1987); ibid **17**, 107 (1987); ibid **17**, 232 (1987); ibid **17**, 294 (1987); ibid **17**, 490 (1987)

Kemeny, G. (ed.): *Surface Analysis of High Temperature Materials: Chemistry and Topography* (Elsevier, London 1984)

Kimura, K., S. Katsumota, Y. Achiba, T. Yanazaki, S. Iwata: *Handbook of HeI Photoelectron Spectra of Fundamental Organic Molecules* (Japan Sci. Soc. Press, Halsted Press, New York 1981)

King, D.A., N.V. Richardson, S. Holloway: *Vibrations at Surfaces 1985*, Studies in Surface Science and Catalysis (Elsevier, Amsterdam 1986)

King, D.A., D.P. Woodruff: The Chemical Physics of Solid Surfaces and Heterogeneous Catalysis (Elsevier, Amsterdam) Vol. 1 Clean Solid Surfaces (1981); Vol. 2 Adsorption at Solid Surfaces (1983); Vol. 3 Chemisorption Systems (in press); Vol. 4 Fundamental Studies of Heterogeneous Catalysis (1982)

MacIntyre, N.S. (ed.): *Quantitative Surface Analysis of Materials* (American Society for Testing and Materials 1978)

Maradudin, A.A., R.F. Wallis, L. Dobrzynski: *Handbook of Interfaces and Surfaces*, Vol. 3 (Garland, New York 1980)

Marcus P.M., F. Jona (eds.): *Determination of Surface Structure by LEED* (Plenum, New York 1984)

Mayer, J.W., E. Rimini: *Ion Beam Handbook for Material Analysis* (Academic, New York 1977)

McCracken, G.M.: The Behaviour of Surfaces under Ion Bombardment, Rept. Progr. Phys. **38**, 24 (1975)

McGuire, G.E.: *Auger Electron Spectroscopy Reference Manual* (Plenum, New York 1979)

Nizzoli, F., K.H. Rieder, F.F. Willis (eds.): *Dynamical Phenomena at Surfaces, Interfaces and Superlattices*, Springer Ser. Surf. Sci. Vol. 3 (Springer, Berlin, Heidelberg 1985)

Oechsner, H. (ed.): *Thin Film and Depth Profile Analysis*, Topics Curr. Phys. Vol. 37 (Springer, Berlin, Heidelberg 1984)

Pendry, J.B.: *LEED – The Theory and its Application to Determination of Surface Structures* (Academic, London 1974)

Prutton, M.: *Surface Physics* (Clarendon, Oxford 1983)

Rhodin, T.N., G. Ertl (eds.): *The Nature of the Surface Chemical Bond* (North-Holland, Amsterdam 1979)

Roberts, M.W., C.S. McKee: *Chemistry of Metal-Gas Interface* (Clarendon, Oxford 1978)

Roberts, M.W.: Photoelectron Spectroscopy and Surface Chemistry, Adv. Catal. **29**, 55 (1980)

Sickafus, E.N., H.P. Bonzel: *Surface Analysis by Low Energy Electron Diffraction and Auger Electron Spectroscopy* (Academic, New York 1971)

Somorjai, G.A. (ed.): *The Structure and Chemistry of Solid Surfaces* (Wiley, New York 1969)

Somorjai, G.A.: *Principles of Surface Chemistry* (Prentice-Hall, New Jersey 1972)

Somorjai, G.A.: *Chemistry in Two Dimensions: Surfaces* (Cornell University Press, Ithaca 1981)

Spicer, W.E., K.Y. Yu, I. Lindau, P. Pianetta, D.M. Collins: Ultraviolet Photoemission Spectroscopy of Surfaces and Surface Sorption, Surf. Def. Prop. Solids **5**, 103 (1976)

Steele, W.A.: *The Interaction of Gases with Solid Surfaces* (Pergamon, Oxford 1974)

Stöhr, J.: *Surface Crystallography by SEXAFS and NEXAFS*, in R. Vanselow, R. Howe (eds.): *Chemistry and Physics of Solid Surfaces V*, Springer Ser. Chem. Phys. Vol. 35 (Springer, Berlin, Heidelberg 1984)

Teo, B.K., D.C. Joy (eds.): *EXAFS Spectroscopy – Techniques and Applications* (Plenum, New York 1981)

Teo, B.K.: *EXAFS – Basic Principles and Data Analysis* (Springer, Berlin, Heidelberg 1986)

Thomas, J.M., R.M. Lambert (eds.): *Characterization of Catalysts* (Wiley, New York 1980)

Thompson, M., M.D. Baker, A. Christie, T. Tyson: *Auger Electron Spectroscopy* (Wiley, New York 1985)

Tompkins, F.C.: *Chemisorption of Gases on Metals* (Academic, London 1978)

Turner, D.W., C. Baker, A.D. Baker, C.R. Brundle: *Molecular Photoelectron Spectroscopy* (Wiley, New York 1970)

Turner, N.H., R.J. Colton: Surface Analysis: X-ray Photoelectron Spectroscopy and Secondary Ion Mass Spectroscopy, Anal. Chem. **54**, 293R (1982)

Urch, D.S. and M.S.: ESCA-Auger Table, Queen Mary College, Chem. Dept., University of London (1981/2)

van der Veen, J.: Ion Beam Crystallography of Surfaces and Interfaces, Surf. Sci. Rep. **5**(5/6), 199 (1985)

Van Hove, M.A., S.Y. Tong: *Surface Crystallography by LEED – Theory, Computation and Structural Results*, Springer Ser. Chem. Phys. Vol. 2 (Springer, Berlin, Heidelberg 1979)

Van Hove, M.A., S.Y. Tong (eds.): *The Structure of Surfaces*, Springer Ser. on Surf. Sci. Vol. 2 (Springer, Berlin, Heidelberg 1985)

Vanselow, R., S.Y. Tong (eds.): *Chemistry and Physics of Solid Surfaces* (CRC, Cleveland 1977)

Vanselow, R. (ed.): *Chemistry and Physics of Solid Surfaces II* (CRC, Boca Raton 1979)

Vanselow, R., W. England (eds.): *Chemistry and Physics of Solid Surfaces III* (CRC, Boca Raton 1982)

Vanselow, R., R. Howe (eds.): *Chemistry and Physics of Solid Surfaces IV*, Springer Ser. Chem. Phys. Vol. 20 (Springer, Berlin, Heidelberg 1982)

Vanselow, R., R. Howe (eds.): *Chemistry and Physics of Solid Surfaces V*, Springer Ser. Chem. Phys. Vol. 35 (Springer, Berlin, Heidelberg 1984)

Vanselow, R., R. Howe (eds.): *Chemistry and Physics of Solid Surfaces VI, VII, VIII*, Springer Ser. Surf. Sci. Vols. 5, 10, 22 (Springer, Berlin, Heidelberg 1986, 1988, 1990)

Wagner, C.D., W.M. Riggs, L.E. Davis, J.F. Moulder, G.E. Muilenberg (eds.): *Handbook of X-ray Photoelectron Spectroscopy* (Perkin-Elmer Corp., Eden Prairie 1978)

Walls, J.M. (ed.): *Methods of Surface Analysis* (Cambridge University Press, Cambridge 1989)

Wandelt, K.: Photoemission Studies of Adsorbed Oxygen and Oxide Layers, Surf. Sci. Rep. **2**(1), 1 (1982)

Weissmann, R., K. Kümmer: Auger Electron Spectroscopy – a Local Probe for Solid Surfaces, Surf. Sci. Rep. **1**(5), 251 (1981)

Werner, H.W.: The Use of Secondary Ion Mass Spectrometry in Surface Analysis, Surf. Sci. **47**, 301 (1975)

Werner, H.W.: Quantitative Secondary Ion Mass Spectrometry: A Review, Surf. Interface Anal. **2**, 56 (1980)

Werner, H.W., R.P.H. Garten: Comprehensive Study of Methods for Thin-Film and Surface Analysis, Rep. Progr. Phys. **47**, 221 (1984)

Wittmaack, K.: Secondary Ion Mass Spectrometry as a Means of Surface Analysis, Surf. Sci. **89**, 668 (1979)

Williams, R.H., G.P. Srivastava, I.T. McGovern: Photoelectron Spectroscopy of Solids and their Surfaces, Rep. Progr. Phys. **43**, 1357 (1980)

Winograd, N.F., B.J. Garrison: Surface Structure Determination with Ion Beams, Acc. Chem. Res. **13**, 406 (1980)

Woodruff, D.P., T.A. Delchar: *Modern Techniques of Surface Science* (Cambridge University Press, Cambridge 1986)

Index

127° cylindrical analyser 180

II-VI semiconductors 490

Absorption coefficient 241
Activated carbons 511
Adhesion 5, 483
Adsorbate-surface bonding 507
Adsorbed methoxy on copper and platinum 514
Adsorption 9, 377
- from solution, 511
- isotherms, 50
Alloying 478
Alloys 55
Aluminium 11, 189, 309
- Al–Zn Coating, 482
- Alumina, Al_2O_3, 11, 26, 221, 296, 408–409, 426–429, 456, 556
- β-alumina, 409
- $Al_xGa_{1-x}As$, 312
- Al_2O_3, 26
- tungstate, 410
Amorphous materials 4
Analysis by XPS 408, 417
Angle resolved X-ray photoelectron spectroscopy (ARXPS) 35, 53
- of polymers, 531
Aqueous
- solutions, 206
- systems, 206, 207
Arsenic 309
ASTM standards 78
Atomic beam scattering (ABS) 20, 50, 56
Atomic force microscopy (AFM) 247, 440
- of biological materials, 273
-- fibroblasts 275
-- tobacco mosaic virus (TMV) 273
- of polymers, 540
- probe
-- aspect ratio 266
-- spring constants 265
-- tip artefacts 267
-- tip parameters 266
- thin film methodology, 269
Atomically clean surfaces 9
Atop sites 430
Auger 293
- effect, 155
- electron, 175
- electron energy, 156
- electron spectroscopy (AES), 155
- lines, 191
- microscopy, 164
- parameter, 186, 393
Auger electron spectrometry (AES) 295
Auger electron spectroscopy (AES) 28, 29, 32, 33, 39, 41, 46, 53, 405, 441, 442, 445, 446, 505
Aurocyanide 511
- adsorption on carbon, 512
Automotive emissions by catalysts 506

Backscattered electron images 13
Baking 74
Ball cratering 35
Band structure 337–343
- engineering, 338
Beer-Lambert expression 25
Bethe equation 168
Binding energy 156, 175
Bioceramics 392

Biological
– applications, 207
– materials, 273
Blended polymers 55
Boron phosphorus silica glass (BPSG) 558
Bragg peaks in LEED 325, 326
Bremsstrahlung radiation 178
Bridging sites 430
Brønsted acid sites 420, 422, 428
Bromine 498

Cadmium
– sulphate, 494
– sulphide, CdS, 491
– telluride, CdTe, 491
Calcite 307
Calcium
– $CaTiO_3$ perovskite, 60
Carbon 500
Carbonaceous
– deposits, 74
– overlayers, 72
Carousel 76
Catalysts 5, 55, 296, 299, 405–432
Catalytic reactions 208
Cathodes 296
Ceramics 6, 53, 55, 57, 377
Channel electron multiplier 182
Channelling effect 238
Characterisation of catalysts 405
Charge 407
– correction, 547
– exchange factor, 291
Charging 64, 412
– effects, 410
Chemical
– effects, 168
– shifts, 185, 409
Chemisorption 506, 508
Chromium
– Chromate, 12
– Chromia, Cr_2O_3, 456
Cladding 12
Clays 225, 386
Cleanliness 78
Cobalt 463
– on Kieselguhr Catalysts, 414
– Co_3O_4, 415
– CoO, 416
Coherence length 20, 311
Cold storage X-ray photoelectron spectroscopy 533
Colloid and surface chemistry 13
Colour centres 64
Composites 55, 57
Composition 291
Concentric hemispherical analyser (CHA) 159, 180
Conduction band 47
Conductive mask 80
Contact angle 399
Contacts 435
Contamination 295, 436
– time, 72
Contrast 88, 93, 101
Conversion coatings 11
Copper 342
– Cu_3Au, 345
– (100) surface, 514
– Cu_2O, 188
– CuO, 188
Corrosion 5, 11, 481
– of iron and steels, 5
Coulomb barrier 240
Coupling agents 398
Cross section 291, 295
Crystalline materials 4
Cylindrical mirror analyser (CMA) 159, 180

de Broglie wavelength 313
Decomposition 54
Defects 9, 42, 436
Degradation 437
de Haas van Alphen 344
Depth profiles 51, 53, 86, 191, 211, 217, 459
Depth resolution
– sputter profiling, 107
Detectors 323
Device technology 435
Diamond-like 7
Differential charging 408
Diffraction 101
– pattern, 309
– spots, 307
Dipole matrix element 340

Display type analysers 340
Dissociation revealed by Auger spectroscopy 506
Dissolution 377, 390
Doping profiles 495
Dose 72
Double-sided tapes 77
Dynamic secondary ion mass spectrometry 62
Dynode multiplier 182

$E(\mathbf{k})$ band structure 340
Effective attenuation length (EAL) 168
Electrodeposition 491
Electron
- diffraction, 101
- flood gun, 410
- flux, 63, 64
- gun, 182
- lenses, 89
- microscopy, 85
- multiplier, 182
- radiation damage, 63
- sources, 89, 158
- spectrometer, 407
Electron beam 308
Electron beam induced current (EBIC) microscopy 439
Electron energy
- analyser, 158, 180
- detector, 180
Electron energy loss spectroscopy (EELS) 36, 46, 51, 55, 158, 505, 514
Electron spectroscopy for chemical analysis (ESCA) 38
Electron spin resonance (ESR) 49
Electron stimulated desorption
- (ESD), 62
- ion angular distribution (ESDIAD), 62
Electronegative 289
Electronic
- angular momentum, 184
- stopping, 232
- structure, 46
Electrostatic charging 64
Electrostatic energy analyzer 303
Elemental composition 229
Elemental sensitivities 28–30
Ellipsometry 54
Enamel 477
Energy
- and angular distribution of secondary ions, 137
- distribution, 156
- referencing, 65
Energy dispersive spectroscopy (EDS) 55
Energy dispersive X-ray (EDX) spectrometry 98, 99, 443, 450, 451
Enhanced diffusion 54
Environmental scanning electron microscope (ESEM) 97, 98
Errors 41
Escape depth 178
Ewald sphere 309
Exposure 72
Extended X-ray absorption fine structure (EXAFS) 466–471

Failure 435
- analysis, 436
Fast atom bombardment mass spectrometry (FABMS) 40
Fatty acids 38
Fermi
- edge, 340
- energy, 342
- level, 46, 156, 445
- surface, 344
Fibre specimens 81
Fibroblasts 275
Field emission
- microscopy (FEM), 18, 44, 48
- spectroscopy (FES), 48
Field ion microscopy (FIM) 18, 44, 48
Field ionisation spectroscopy (FIS) 48
Fingerprint 39
Fischer-Tropsch synthesis 416, 420
Flotation 206
Fluence 72
Fluorescence 221, 224
Force–distance analysis (F–d) 258
Forward recoil 240
Fourier transform 184

Fourier transform infrared spectroscopy (FTIR) 36, 40, 203–205
Fourier transform reflection absorption infrared spectroscopy (FT–RAIRS) 407
Fracture 55
– faces, 57
Fresnel 33
Friction 5
Friction force microscopy (FFM) *see* Lateral force microscopy (LFM)

Gallium 309
– arsenide GaAs, 309
– GaAs, 28
– GaAs (100) surface, 314
Galvanized steel 480, 556
Gamma ray 241
Ghost lines 188
Gibbsite 221
Glass fibres 398
Glasses 4, 6, 53, 297, 377
Glow Discharge Optical Emission Spectrometry (GD–OES) 553
Gold 491, 500, 511–513
– Aurocyanide, 511
– Aurocyanide adsorption on carbon, 512
Grain boundaries 55, 377, 474
Grazing angle infrared spectroscopy 548

Helium 31
Herz-Knudsen equation 71
High energy ion scattering (HEIS) 20, 50, 51, 54
Hollow cathode discharge lamp 337
Hollow sites 430
Hydrogen 31
– impurity, 55
– profile, 243
Hydrophobicity 387
Hydroxyapatite 395
Hydroxyl species 497

Ideal gas equation 71
Identification of adsorbed species 504
Imaging 87, 102
Imaging XPS 183
Impact collision ion surface scattering (ICISS) 300
Implantation 436
Impurities 211
Indiffusion 448
Indium 309
– foil, 77
– InGaAs, 343
– tin oxide (ITO), 491
Inelastic mean free path (IMFP) 24–26, 33, 35, 53, 72, 158, 178
– in polymers, 532
Infrared spectroscopy 203
Insulating sample 407
Integrated circuits 435
Interdiffusion 452
Interfacial 436, 452
Intergranular films 377
International Standards Organisation (ISO) 171
Ion
– beam, 33, 53, 231
– exchange, 390
– fraction, 292
– milling, 459
Ion neutralisation spectroscopy (INS) 48
Ion scattering spectroscopy (ISS) 41, 405, 411
Ionization potential 292
Iron 506
– catalysts, 416
– Fe_2O_3, 417
– Oxides, 380

Junctions 436

k–space resolution 344
Kaolinite 222, 225
Kelvin probe 46
Kikuchi lines 317
Kinematic factor 234
Kink sites 42

Laboratory standards 41
Langmuir 50
Langmuir (L) 73
Lapped face 35
Lateral force microscopy (LFM) 260

– carbide-like ceramic, 280
– diamond-like carbon films, 278
– multi-asperity regime, 261
– single asperity regime, 261
Leaching 51, 377, 390
Lewis acid sites 426
Line defects 428
Longitudinal optical vibrations 426
Low energy electron diffraction (LEED) 8, 19–21, 44–46, 50, 53, 63, 103, 156, 159, 311, 319–335
– apparatus, 322
– applications, 329–333
– data collection, 323
– experiment, 320, 321
– fine structure, 328, 332, 333
– intensity analysis, 326, 330, 333
– pattern, 506
– sample preparation, 323
Low energy electron reflection microscopy (LEERM) 19
Low energy ion scattering (LEIS) 20, 24, 50, 53, 56, 114, 119

Magnesium
– MgO, 9, 17, 53
Magnetic
– lens, 184
– shielding, 339
Magnetic force microscopy (MFM) 252
Mass resolution 289
Materials science 5
Matrix correction factor 160
Mean free path 72, 311
Medium energy electron diffraction (MEED) 20
Medium energy ion scattering (MEIS) 20, 50
Mercury
– HgCdTe film, 178
Metal–support interaction 412
Metal-ceramic bonding 393
Metallic coatings 478
Metals 6, 55
Methanol 498
Methoxy species (CH_3O) 514
Migration energy 312
Miller index 44
Millibar (mbar) 73
Minerals 6, 53, 55, 219, 377
– processing, 381
– surface, 225
– surfaces, 206
Molecular and dissociated forms 507
Molecular beam epitaxy (MBE) 309
Molecular flux 71
Momentum 340
Moore's Law 436
Mounting of powder or granular samples 407
Multilayer formation 49
Multiphase materials 56
Multiple scattering 316

Near edge X-ray absorption fine structure spectroscopy (NEXAFS) 348, 361
Near-surface region 51
Nickel 48, 345, 506
– NiO, 21, 42, 43, 49
Nitric acid 498
Nitric oxide 505–511
– adsorption on metals, 505
– on (110) iron by UPS, 509
– on nickel, 506
Non-resonant reaction 240
Nuclear reaction analysis (NRA) 31, 51, 54, 229
Nuclear stopping 232
Nucleation 60

Optical absorption 340
Optoelectronic 443
Ores 55
Orthophosphoric acid 498
Oxidation 377, 456
– of methanol, 514
Oxide 53, 386, 479
– domain boundaries, 428
– point defects, 428
Oxyhydroxides 12

Palladium
– Ag/Pd alloy, 339, 340
– on magnesia, 412
Particle induced x-ray emission (PIXE) 443

Pascals (Pa) 73
Passivation 51, 483
Periodic potential 342
Perovskite 60
Phase transitions 345
Phosphor 308
– Phosphate, 12
Photo-degradation 221
Photoacoustic spectroscopy (PAS) 207, 216
Photoelectron
– microscopy (PEM), 33
– spectromicroscope, 183
– spectroscopy, 175
Photoemission process 175
Photoionization cross-section 27
Photoluminescence 48
Photovoltaic 490
Pitting corrosion 11
Plasma 387
Plasmon energy loss 189
Platinum 496
– (111) surface, 516
Polarisation 54
Polycrystalline nickel 507
Polyethylene 57
Polymers 4, 6, 9, 53, 55, 57
– matrices, 389
– surfaces, 484, 519
Polystyrene 57
Pore structures 57
Pores 17
Position sensitive detector 182, 342
Potassium
– dichromate, 498
– $KAu(CN)_2$, 64
Powder 93
Practical detection limits 31
Praseodymium 417
Precipitates 17, 457
Preferential sputtering 35, 54, 501
Promotors such as MgO and ThO_2 414
Protein 211
Purity 436
PVD/CVD coatings 557

Quadrupole mass spectrometer 74
Quality control 203
Quantitative 555
Quantitative, standardless analysis 229
Quartz-halogen incandescent lamp 7

Radiation damage 61–64
Raman 221
Rare earth promoter 417
Reaction of nitric oxide with carbon monoxide 506
Reciprocal lattice 309
Recoils 289
Reconstruction 8, 20, 54
Recrystallisation 60
Recrystallised Surfaces 395
Reflection absorption infrared spectroscopy (RAIRS) 407
Reflection electron microscopy (REM) 103
Reflection high energy electron diffraction (RHEED) 20, 21, 54, 307
Refractory 48
– metals, 44
Relaxation 8, 20
Reliability of data 41
Resolution 32, 93, 101, 158, 437
Resolving power 88
Resonance 293
Resonant scattering 238
Retarding field analyser (RFA) 180
Rocks 57
Rutherford backscattering 464
Rutherford backscattering spectrometry (RBS) 229, 295, 442, 443, 446, 448
Rutherford backscattering spectroscopy (RBS) 20, 31, 50, 51, 54
Rutherford scattering cross section 234

Sample treatment chamber 406
Scanned probe microscopy (SPM) 247
– figures of merit, 250
– instrument calibration, 262
Scanners
– calibration, 263
– spatial characteristics, 263

Scanning Auger microprobe (SAM) 441
Scanning Auger microscopy (SAM) 15, 18, 28, 33, 34, 37, 50, 55–57, 60, 458
Scanning electron microscopy (SEM) 13, 15, 50, 55, 60, 87, 91
Scanning transmission electron microscopy (STEM) 15
Scanning tunneling electron microscopy (STEM) 88
Scanning tunnelling microscopy (STM) 18, 19, 32, 48, 50, 55, 56, 63, 247
- of polymers, 540
- principles of operation, 253
Scanning tunnelling spectroscopy (STS) 48, 252
Scatter diagrams 37
Schottky 444, 445
Schrödinger equation 343
Secondary electrons 159
- yield, 64
Secondary ion imaging mass spectrometry (SIIMS) 34
Secondary ion mass spectrometry (SIMS) 28–33, 39, 41, 50, 53, 56, 61, 62, 108, 121, 295
- dynamic, 62
- of polymers, 537
- static, 62
Secondary ion mass spectroscopy (SIMS) 127–154, 442, 448, 449, 491
- Advantages and disadvantages, 130
- Construction of analyser, 138
- Dynamic SIMS, 131
- Imaging SIMS, 132
- Static SIMS, 128
Secondary ions 147
- Yield, 131
Segregation 51, 53, 54, 211, 474
Selected area X-ray photoelectron spectroscopy (SAXPS) 33, 34
Selvedge 23
- depth, 4
- layer, 3
Semiconductor 6, 435, 557
- contacts, 443
- II-VI, 490
Sensitivity 155
- factor, 160, 297
- limit, 28
Shadow cone 301–302
Shake-up lines 187
Sickafus 159
Signal enhancement 145
Signal to noise ratio 30, 205, 221
Silica 296, 429
Silicon 309
- Silane, 387
- Silica, SiO_2, 456
- Silicones, 526
- Siloxane, 387
Silver
- Ag/Pd alloy, 339, 340
Single crystal 340, 506
Single junction solar cells 491
Skeletal isomerization of alkenes 409
Sodium
- as impurity, 409
Solar energy 490
Sources of contamination 79
Space-charge
- effects, 48
- layer, 46
Spatial limits 32
Spatial resolution 33, 155
Specimen
- contamination, 79
- handling, 76
- history, 78
- mounting, 80
- rod, 76
- storage and transfer, 79
- stubs, 76
Spectral information 184
Spectrometer 175, 555
- transmission, 41
Spin-orbit splitting 184
Spot patterns 321, 323–325
Sputter
- damage, 54
- depth profiling, 107
- profiling, 35, 80, 491
- yield, 35, 108, 132
Standards 297

Static secondary ion mass spectrometry 519
Steel sheet 473
Stopping power 232
Straggling effect 233
Substitutionally disordered alloys 345
Sulfide 381
Superalloys 456
Supported tungsten catalysts 410
Surface
- adsorption, 145
- carbon, 476–477, 481
- charging, 49
- compositional maps, 163
- crystallography, 50
- dipoles, 47
- layers, 381
- modification, 7, 398
- morphology, 164
- oxide, 455
- photovoltage, 49
- plasmon, 189
- reconstructions, 312
- roughness, 312
- sensitivity, 158
- states, 343, 445
- treatments, 50
Surface barrier analysis with LEED 332–333
Surface enhanced X-ray absorption fine structure (SEXAFS) 22, 50, 366
Surface extended energy loss fine structure (SEELFS) 22, 50
Surface vibrational spectroscopy 420
Synchrotron
- beamlines, 352
- radiation, 183, 350
-- facility 343
Synroc 56, 58, 61

Tantalum oxide standard 167
Target factor analysis (TFA) 169
Tellurium 187
Thermal desorption spectrometry (TDS) 31
Thin film 269, 489
- magnetic tapes, 270
- photovoltaics, 490
- wool fibre, 270
Titanium
- TiO_2, 44
Tobacco mosaic virus (TMV) 273
Topology of the Fermi surface 344
Toroidal electron energy analyser 340
Torr 73
Transmission electron microscopy (TEM) 15, 42, 44, 50, 55–57, 61, 87, 99, 441, 450, 451
Transverse optical vibrations 426
Tungsten
- oxide catalysts, 409
- oxide on alumina, 410
- WTe_2, 267
Two-dimensional growths 312

Ultrahigh vacuum (UHV) 47, 71
Ultraviolet photoelectron spectroscopy (UPS) 38, 39, 44, 46, 337–345, 507
Undulators 351
Universal curve 158, 168
Universal inelastic mean free path curve 24
UV microscopy 437

Vacancy sites 42, 43
Vacuum technology 76
Valence states 38
Versailles project on Advanced Materials and Standards (VAMAS) 171
Very large scale integrated circuits (VLSI) 436
Very low energy electron diffraction (VLEED) 333
Vibrational level structure 38
Vibrational modes 514
Virtual leaks 74
Voltage contrast (VC) microscopy 439

Wave vector 337, 340
Weathering 484
Wigglers 351
Work function 46, 175

X-ray
- attenuation, 27
- line satellites, 187

– photoelectron spectroscopy (XPS), 175
– source, 179
X-ray absorption fine structure spectroscopy (XAFS) 347–371
– Near edge (NEXAFS), 361
X-ray absorption near edge structure (XANES) 22, 347, 348, 361
X-ray diffraction (XRD) 450, 451
X-ray fluorescence (XRF) 155
X-ray photoelectron spectroscopy (XPS) 24, 26, 28–39, 41, 44, 46, 49, 53, 56, 60, 63, 64, 168, 169, 405, 412–414, 442, 505, 511
– of polymers, 529

Young's modulus 272, 276

Zeolites 420–425
Zinc
– containing glass, 395
– ZnO, 44, 50, 53
ZSM-5 421–425

SPRINGER SERIES IN SURFACE SCIENCES

Editors: G. Ertl, H.Lüth and D.L. Mills

Founding Editor: H.K.V. Lotsch

1 **Physisorption Kinetics**
By H. J. Kreuzer, Z. W. Gortel

2 **The Structure of Surfaces**
Editors: M. A. Van Hove, S. Y. Tong

3 **Dynamical Phenomena at Surfaces, Interfaces and Superlattices**
Editors: F. Nizzoli, K.-H. Rieder, R. F. Willis

4 **Desorption Induced by Electronic Transitions, DIET II**
Editors: W. Brenig, D. Menzel

5 **Chemistry and Physics of Solid Surfaces VI**
Editors: R. Vanselow, R. Howe

6 **Low-Energy Electron Diffraction**
Experiment, Theory and Surface Structure Determination
By M. A. Van Hove, W. H. Weinberg, C.-M. Chan

7 **Electronic Phenomena in Adsorption and Catalysis**
By V. F. Kiselev, O. V. Krylov

8 **Kinetics of Interface Reactions**
Editors: M. Grunze, H. J. Kreuzer

9 **Adsorption and Catalysis on Transition Metals and Their Oxides**
By V. F. Kiselev, O. V. Krylov

10 **Chemistry and Physics of Solid Surfaces VII**
Editors: R. Vanselow, R. Howe

11 **The Structure of Surfaces II**
Editors: J. F. van der Veen, M. A. Van Hove

12 **Diffusion at Interfaces: Microscopic Concepts**
Editors: M. Grunze, H. J. Kreuzer, J. J. Weimer

13 **Desorption Induced by Electronic Transitions, DIET III**
Editors: R. H. Stulen, M. L. Knotek

14 **Solvay Conference on Surface Science**
Editor: F. W. de Wette

15 **Surfaces and Interfaces of Solids**
By H. Lüth*)

16 **Atomic and Electronic Structure of Surfaces**
Theoretical Foundations
By M. Lannoo, P. Friedel

17 **Adhesion and Friction**
Editors: M. Grunze, H. J. Kreuzer

18 **Auger Spectroscopy and Electronic Structure**
Editors: G. Cubiotti, G. Mondio, K. Wandelt

19 **Desorption Induced by Electronic Transitions, DIET IV**
Editors: G. Betz, P. Varga

20 **Scanning Tunneling Microscopy I**
General Principles and Applications to Clean and Adsorbate-Covered Surfaces
Editors: H.-J. Güntherodt, R. Wiesendanger
2nd Edition

21 **Surface Phonons**
Editors: W. Kress, F. W. de Wette

22 **Chemistry and Physics of Solid Surfaces VIII**
Editors: R. Vanselow, R. Howe

23 **Surface Analysis Methods in Materials Science**
Editors: D. J. O'Connor, B. A. Sexton, R. St. C. Smart
2nd Edition

24 **The Structure of Surfaces III**
Editors: S. Y. Tong, M. A. Van Hove, K. Takayanagi, X. D. Xie

25 **NEXAFS Spectroscopy**
By J. Stöhr

26 **Semiconductor Surfaces and Interfaces**
By W. Mönch
3rd Edition

27 **Helium Atom Scattering from Surfaces**
Editor: E. Hulpke

28 **Scanning Tunneling Microscopy II**
Further Applications and Related Scanning Techniques
Editors: R. Wiesendanger, H.-J. Güntherodt
2nd Edition

29 **Scanning Tunneling Microscopy III**
Theory of STM and Related Scanning Probe Methods
Editors: R. Wiesendanger, H.-J. Güntherodt
2nd Edition

30 **Concepts in Surface Physics**
By M. C. Desjonquères, D. Spanjaard*)

31 **Desorption Induced by Electronic Transitions, DIET V**
Editors: A. R. Burns, E. B. Stechel, D. R. Jennison

32 **Scanning Tunneling Microscopy and Its Applications**
By C. Bai
2nd Edition

33 **Adsorption on Ordered Surfaces of Ionic Solids and Thin Films**
Editors: H.-J. Freund, E. Umbach

34 **Surface Reactions**
Editor: R. J. Madix

35 **Applications of Synchrotron Radiation**
High-Resolution Studies of Molecules and Molecular Adsorbates on Surfaces
Editor: W. Eberhardt

36 **Kinetics of Metal-Gas Interactions at Low Temperatures: Hydriding, Oxidation, Poisoning**
By E. Fromm

37 **Magnetic Multilayers and Giant Magnetoresistance**
Fundamentals and Industrial Applications
Editor: U. Hartmann*)

*) Available as a textbook

www.ingramcontent.com/pod-product-compliance
Ingram Content Group UK Ltd.
Pitfield, Milton Keynes, MK11 3LW, UK
UKHW021902190726
13853UKWH00003B/1383

* 9 7 8 3 6 6 2 0 5 2 2 8 0 *